Lecture Notes in Computer Science 12062

More information about this series at http://www.springer.com/series/7407

Uli Fahrenberg · Peter Jipsen ·
Michael Winter (Eds.)

Relational and Algebraic Methods in Computer Science

18th International Conference, RAMiCS 2020
Palaiseau, France, October 26–29, 2020
Proceedings

Springer

Editors
Uli Fahrenberg
Laboratoire d'informatique (LIX)
École Polytechnique
Palaiseau, France

Peter Jipsen
Faculty of Mathematics
Chapman University
Orange, CA, USA

Michael Winter
Department of Computer Science
Brock University
St. Catharines, ON, Canada

ISSN 0302-9743 ISSN 1611-3349 (electronic)
Lecture Notes in Computer Science
ISBN 978-3-030-43519-6 ISBN 978-3-030-43520-2 (eBook)
https://doi.org/10.1007/978-3-030-43520-2

LNCS Sublibrary: SL1 – Theoretical Computer Science and General Issues

This Springer imprint is published by the registered company Springer Nature Switzerland AG
The registered company address is: Gewerbestrasse 11, 6330 Cham, Switzerland

Preface

This volume contains the proceedings of the 18th International Conference on Relational and Algebraic Methods in Computer Science (RAMiCS 2020), which was to be held at École polytechnique in Palaiseau, close to Paris, during April 8–11, 2020. The conference was postponed to October due to the COVID-19 pandemic and was ultimately held virtually during October 26–29, 2020.

The RAMiCS conferences aim to bring a community of researchers together to advance the development and dissemination of relation algebras, Kleene algebras, and similar algebraic formalisms. Topics covered range from mathematical foundations to applications such as conceptual and methodological tools in computer science and beyond. More than 25 years after its formation in 1991 in Warsaw, Poland initially as "Relational Methods in Computer Science" RAMiCS remains a main venue in this field. The series merged with the workshops on Applications of Kleene Algebra in 2003 and adopted its current name in 2009. Previous events were organized in Dagstuhl, Germany (1994), Paraty, Brazil (1995), Hammamet, Tunisia (1997), Warsaw, Poland (1998), Québec, Canada (2000), Oisterwijk, The Netherlands (2001), Malente, Germany (2003), St. Catharines, Canada (2005), Manchester, UK (2006), Frauenwörth, Germany (2008), Doha, Qatar (2009), Rotterdam, The Netherlands (2011), Cambridge, UK (2012), Marienstatt, Germany (2014), Braga, Portugal (2015), Lyon, France (2017), and Groningen, The Netherlands (2018).

RAMiCS 2020 attracted 29 submissions, of which 20 were selected for presentation by the Program Committee. Each submission was evaluated according to high academic standards by at least three independent reviewers and scrutinized further during two weeks of intense electronic discussion. The organizers are very grateful to all Program Committee members for this hard work, including the lively and constructive debates, as well as to the external reviewers for their generous help and expert judgments. Without this dedication we could not have assembled such a high-quality program; we hope that all authors have benefitted from these efforts.

Apart from the submitted articles, this volume features the abstracts of the presentations of the three invited speakers. The three abstracts are on "Probabilistic bisimulation with silent moves" by Christel Baier, TU Dresden, Germany; "Weighted automata and quantitative logics" by Manfred Droste, University of Leipzig, Germany; and "Combining probabilistic and non-deterministic choice via weak distributive laws" by Daniela Petrisan (joint work with Alexandre Goy), Université Paris Diderot, France. We are delighted that all three invited speakers accepted our invitation to present their work at the conference.

Last, but not least, we would like to thank the members of the RAMiCS Steering Committee for their support and advice. We gratefully acknowledge financial and administrative support by the Chaire Ingénerie des Systémes Complexes and by the Department of Computer Science of École polytechnique.

We also appreciate the excellent facilities offered by the EasyChair conference administration system and Alfred Hofmann's help in publishing this volume with Springer. Finally, we are indebted to all authors and participants for supporting this conference.

February 2020

Michael Winter
Uli Fahrenberg
Peter Jipsen

Organization

Program Committee

Roland Backhouse	University of Nottingham, UK
Rudolf Berghammer	Kiel University, Germany
Manuel Bodirsky	TU Dresden, Germany
Jules Desharnais	Laval University, Canada
Amina Doumane	University of Lyon, EnsL, UCBL, CNRS, LIP, France
Uli Fahrenberg	École polytechnique, France
Hitoshi Furusawa	Kagoshima University, Japan
Mai Gehrke	Université Côte d'Azur, France
Walter Guttmann	University of Canterbury, New Zealand
Robin Hirsch	University College London, UK
Peter Höfner	Data61 CSIRO, Australia
Marcel Jackson	La Trobe University, Australia
Jean-Baptiste Jeannin	University of Michigan, USA
Peter Jipsen	Chapman University, USA
Stef Joosten	Open University of the Netherlands, The Netherlands
Wolfram Kahl	McMaster University, Canada
Dexter Kozen	Cornell University, USA
Tadeusz Litak	FAU Erlangen-Nürnberg, Germany
Wendy MacCaull	St. Francis Xavier University, Canada
Roger Maddux	Iowa State University, USA
Annabelle McIver	Macquarie University, Australia
Szabolcs Mikulas	University of London, UK
Ali Mili	NJIT, USA
Jose Oliveira	University of Minho, Portugal
Alessandra Palmigiano	Vrije Universiteit Amsterdam, The Netherlands
Damien Pous	CNRS, ENS Lyon, France
Mehrnoosh Sadrzadeh	University of London, UK
Luigi Santocanale	Aix-Marseille Université, France
John Stell	University of Leeds, UK
Georg Struth	University of Sheffield, UK
Michael Winter	Brock University, Canada

Additional Reviewers

Doberkat, Ernst-Erich
Doe, John
Frittella, Sabine
Fussner, Wesley
Glück, Roland
Greco, Giuseppe
Greiner, Johannes
Hodkinson, Ian
Jakl, Tomáš
Joosten, Sebastiaan
Knäuer, Simon
Kurz, Alexander
McLean, Brett
Nishizawa, Koki
Paoli, Francesco
Protin, Clarence
Quinn-Gregson, Thomas
Reggio, Luca
St. John, Gavin
Tsumagari, Norihiro

Abstracts of Invited Speakers

Probabilistic Bisimulation with Silent Moves

Christel Baier

Institute for Theoretical Computer Science, Technische Universität Dresden
christel.baier@tu-dresden.de

Abstract. Formal notions of bisimulation relations are a key concept of concurrency theory to equate, respectively distinguish processes according to the behaviour they exhibit when interacting with other processes, taking the stepwise behaviour of processes as a reference. Speaking roughly, bisimilar processes can mutually simulate each other by mimicking their visible stepwise behaviour. While notions of strong bisimilarity rely on the assumption that all transitions are visible, a variety of weaker notions of bisimilarity have been proposed that abstract away from internal computations ("silent moves" in the jargon of van Glabbeek). Analogous notions have been introduced for probabilistic automata and related models that share the idea that probabilistically bisimilar processes induce the same distributions over visible steps, and thus have the same observable quantitative behaviour.

The talk will first give a brief summary of classical results of concurrency theory on bisimulation relations with silent moves and their logical characterizations. The second part will report on analogous results in the probabilistic setting and recent results on the probabilistic bisimulation spectrum with silent moves.

Weighted Automata and Quantitative Logics

Manfred Droste

Institut für Informatik, Universität Leipzig, Germany
droste@informatik.uni-leipzig.de

Quantitative models and quantitative analysis in Computer Science are receiving increased attention. The goal of this talk is to investigate quantitative automata and quantitative logics. Weighted automata on finite words have already been investigated in seminal work of Schützenberger (1961) [15]. They consist of classical finite automata in which the transitions carry weights. These weights may model, e.g., the cost, the consumption of resources, or the reliability or probability of the successful execution of the transitions. This concept soon developed a flourishing theory, as is exemplified and presented in the books [1, 7, 9, 11, 13, 14,].

We investigate weighted automata and their relationship to weighted logics. For this, we present syntax and semantics of a quantitative logic; the semantics counts 'how often' a formula is true in a given word. Our main result [5], extending the classical result of Büchi [2] and Elgot [10], shows that if the weights are taken from an arbitrary semiring, then weighted automata and a syntactically defined fragment of our weighted logic are expressively equivalent. A corresponding result holds for infinite words. Moreover, this extends to quantitative automata investigated by Chatterjee, Doyen and Henzinger [3, 4] with (non-semiring) average-type behaviors, or with discounting or limit average objectives for infinite words [8]. Finally, recall that by fundamental results of Schützenberger [16] and McNaughton and Papert [12] from the 1970s, the classes of first-order definable and aperiodic languages coincide. Very recently, this equivalence could be extended to weighted automata [6].

References

1. Berstel, J., Reutenauer, C.: Noncommutative Rational Series with Applications. Encyclopedia of Mathematics and Its Applications, vol. 137. Cambridge University Press (2010). 10.1017/CBO9780511760860
2. Büchi, J.R.: Weak second-order arithmetic and finite automata. Math. Log. Q. **6**, 66–92 (1960). 10.1002/malq.19600060105
3. Chatterjee, K., Doyen, L., Henzinger, T.A.: Quantitative languages. In: Kaminski, M., Martini, S. (eds.) CSL 2008. LNCS, vol. 5213, pp. 385–400. Springer, Heidelberg (2008). 10.1007/978-3-540-87531-4_28
4. Chatterjee, K., Doyen, L., Henzinger, T.A.: Expressiveness and closure properties for quantitative languages. In: Logic In Computer Science (LICS 2009), pp. 199–208 (2009). 10.1109/LICS.2009.16
5. Droste, M., Gastin, P.: Weighted automata and weighted logics. Theor. Comp. Sci. (special issue of ICALP'05) **380**, 69–86 (2007). 10.1016/j.tcs.2007.02.055

6. Droste, M., Gastin, P.: Aperiodic weighted automata and weighted first-order logic. In: Mathematical Foundations of Computer Science (MFCS 2019), LIPIcs, vol. 138, pp. 76:1–76:15. Schloss Dagstuhl–Leibniz-Zentrum für Informatik (2019). 10.4230/LIPIcs.MFCS.2019.76
7. Droste, M., Kuich, W., Vogler, H. (eds.): Handbook of Weighted Automata. EATCS Monographs in Theoretical Computer Science. Springer, Heidelberg (2009). 10.1007/978-3-642-01492-5
8. Droste, M., Meinecke, I.: Weighted automata and weighted MSO logics for average and long-time behaviors. Inf. Comput. **220**, 44–59 (2012). 10.1016/j.ic.2012.10.001
9. Eilenberg, S.: Automata, Languages, and Machines, Pure and Applied Mathematics, vol. 59, Part A. Academic Press (1974). 10.1016/S0079-8169(08)60875-2
10. Elgot, C.C.: Decision problems of finite automata design and related arithmetics. Trans. Am. Math. Soc. **98**, 21–51 (1961). 10.2307/2270940
11. Kuich, W., Salomaa, A.: Semirings, Automata, Languages. EATCS Monographs on Theoretical Computer Science, vol. 5. Springer, Heidelberg (1986). 10.1007/978-3-642-69959-7
12. McNaughton, R., Papert, S.A.: Counter-Free Automata. The MIT Press (1971). ISBN 0262130769
13. Sakarovitch, J.: Elements of Automata Theory. Cambridge University Press (2009). 10.1017/CBO9781139195218
14. Salomaa, A., Soittola, M.: Automata-Theoretic Aspects of Formal Power Series. Texts and Monographs in Computer Science. Springer, New York (1978). 10.1007/978-1-4612-6264-0
15. Schützenberger, M.P.: On the definition of a family of automata. Inf. Control **4**(2), 245–270 (1961). 10.1016/S0019-9958(61)80020-X
16. Schützenberger, M.P.: On finite monoids having only trivial subgroups. Inf. Control **8**(2), 190–194 (1965). 10.1016/S0019-9958(65)90108-7

Combining Probabilistic and Non-deterministic Choice Via Weak Distributive Laws

Daniela Petrisan

joint work with Alexandre Goy
University Paris Diderot
petrisanirif.fr

Combining probabilistic choice and non-determinism is a long standing and challenging problem in denotational semantics. At a category theoretic level computational effects can be modeled using monads. In this particular instance one can use the powerset monad for non-determinism and the finite distribution monad for probabilistic choice. The problem with combing the two effects stems from the fact that the corresponding monads do not compose well. One way to compose monads is via distributive laws, but as shown in Varacca's PhD thesis [8], there is no distributive law of the powerset monad over the distribution monad. This entails in particular that the powerset monad does not lift to a monad on the category of convex (or barycentric) algebras – the Eilenberg-Moore algebras for the finite distribution monad. Nevertheless, various workarounds have been proposed [1, 2, 5–9,]. On the category of convex algebras one can define a monad that maps a convex algebra to the set of its convex subsets. This induces the composite monad of convex sets of distributions. In domain theory [5] this corresponds to the power Kegelspitzen construction.

In this talk we show the existence of a *weak* distributive law of the powerset monad $\mathcal{P}$ over the finite distribution monad $\mathcal{D}$. Since the finite distribution functor preserves weak pullback it has a canonical extension to the category of sets and relations. This induces a distributive law of the powerset monad $\mathcal{P}$ over the finite distribution *functor* $\mathcal{D}$. Using results from mathematical optimization, we show that this natural transformation interacts well with the multiplication of $\mathcal{D}$. We therefore obtain a canonical weak distributive law in the sense of [3, 4]. As a consequence, we retrieve the well-known convex powerset monad as a weak lifting of the powerset monad to the category of convex algebras.

We provide applications to the study of trace semantics and behavioral equivalences of systems with an interplay between probability and non-determinism.

References

1. Bonchi, F., Silva, A., Sokolova, A.: The power of convex algebras. In: 28th International Conference on Concurrency Theory, CONCUR 2017, 5–8 September 2017, Berlin, Germany, pp. 23:1–23:18 (2017)

2. Bonchi, F., Sokolova, A., Vignudelli, V.: The theory of traces for systems with nondeterminism and probability. In: 34th Annual ACM/IEEE Symposium on Logic in Computer Science, LICS 2019, Vancouver, BC, Canada, 24–27 June 2019, pp. 1–14 (2019)
3. Böhm, G.: The weak theory of monads. Adv. Math. **225**(1), 1–32 (2010)
4. Garner, R.: The Vietoris monad and weak distributive laws. Appl. Categor. Struct. (2019)
5. Keimel, K., Plotkin, G.D.: Mixed powerdomains for probability and nondeterminism. Log. Methods Comput. Sci. **13**(1) (2017)
6. Mislove, M.: Nondeterminism and probabilistic choice: obeying the laws. In: Palamidessi, C., (eds.) CONCUR 2000—Concurrency Theory, pp. 350–365. Springer, Heidelberg (2000)
7. Tix, R., Keimel, K., Plotkin, G.: Semantic domains for combining probability and non-determinism. Electron. Notes Theoret. Comput. Sci. **222,** 3–99 (2009)
8. Varacca, D.: Probability, nondeterminism and concurrency: two denotational models for probabilistic computation. Technical report, PhD thesis, Univ. Aarhus, 2003. BRICS Dissertation Series (2003)
9. Varacca, D., Winskel, G.: Distributing probability over non-determinism. Math. Struct. Comput. Sci. **16**(1), 87–113 (2006)

Contents

Commutative Doubly-Idempotent Semirings Determined by Chains and by Preorder Forests

Natanael Alpay and Peter Jipsen(✉)

Chapman University, Orange, CA, USA
jipsen@chapman.edu

Abstract. A commutative doubly-idempotent semiring (cdi-semiring) $(S, \vee, \cdot, 0, 1)$ is a semilattice $(S, \vee, 0)$ with $x \vee 0 = x$ and a semilattices $(S, \cdot, 1)$ with identity 1 such that $x0 = 0$, and $x(y \vee z) = xy \vee xz$ holds for all $x, y, z \in S$. Bounded distributive lattices are cdi-semirings that satisfy $xy = x \wedge y$, and the variety of cdi-semirings covers the variety of bounded distributive lattices. Chajda and Länger showed in 2017 that the variety of all cdi-semirings is generated by a 3-element cdi-semiring. We show that there are seven cdi-semirings with a $\vee$-semilattice of height less than or equal to 2. We construct all cdi-semirings for which their multiplicative semilattice is a chain with $n + 1$ elements, and we show that up to isomorphism the number of such algebras is the nth Catalan number $C_n = \frac{1}{n+1}\binom{2n}{n}$. We also show that cdi-semirings with a complete atomic Boolean $\vee$-semilattice on the set of atoms A are determined by singleton-rooted preorder forests on the set A. From these results we obtain efficient algorithms to construct all multiplicatively linear cdi-semirings of size n and all Boolean cdi-semirings of size 2^n.

Keywords: Idempotent semirings · Distributive lattices · Preorder forests

1 Introduction

The structure of distributive lattices is well understood since every distributive lattice is a subalgebra of a product of the 2 element lattice, i.e., a subalgebra of a Boolean lattice. The situation is more complicated for idempotent semirings $(A, \vee, \cdot, 0, 1)$, defined by the identities

$$(x \vee y) \vee z = x \vee (y \vee z) \quad x \vee y = y \vee x \quad x \vee 0 = x \quad x \vee x = x \quad x0 = 0 = 0x$$
$$(xy)z = x(yz) \quad x1 = x = 1x \quad (x \vee y)z = xz \vee yz \quad x(y \vee z) = xy \vee xz.$$

Note that xy stands for $x \cdot y$, $x^0 = 1$ and $x^{n+1} = x^n x$. The subclass of *commutative doubly idempotent semirings*, or *cdi-semirings* for short, is obtained by adding the identities $xy = yx$ and $x^2 = x$. Even for this much smaller class of cdi-semirings there is no general structure theory. The classes of idempotent

U. Fahrenberg et al. (Eds.): RAMiCS 2020, LNCS 12062, pp. 1–14, 2020.
https://doi.org/10.1007/978-3-030-43520-2_1

semirings and cdi-semirings are defined by a list of identities, hence they are varieties, i.e., closed under products, subalgebras and homomorphic images.

Since we are also assuming $\cdot$ is commutative and idempotent, there are two underlying semilattice orders $x \leq y \iff x \vee y = y$ and $x \sqsubseteq y \iff xy = x$. A cdi-semiring is a bounded distributive lattice if and only if the two orders coincide, or equivalently if the absorption laws $x \vee xy = x$ and $x(x \vee y) = x$ hold. While the variety of cdi-semirings is quite special, it includes all distributive lattices and is small enough that there is hope for a general description of its finite members.

The aim of this paper is to give structural descriptions for some subclasses of cdi-semirings. In particular, we show in Sect. 2 that there are, up to isomorphism, only seven cdi-semirings of height 2. In Sect. 3 we give a complete description of the finite cdi-semirings for which the monoidal semilattice order $\sqsubseteq$ is a chain (i.e., linearly-ordered). Finally, in Sect. 4 we describe all finite Boolean cdi-semirings by certain preorder forests on the set of atoms.

Recall that *Kleene algebras* are idempotent semirings with a unary operation x^* such that (i) $1 \vee x \vee x^*x^* = x^*$, (ii) $xy \leq y \implies x^*y = y$ and (iii) $yx \leq y \implies yx^* = y$ hold. It is well known that the class KA of all Kleene algebras is not closed under homomorphic images, hence (ii), (iii) cannot be replaced by identities and the class KA of Kleene algebras is only a quasivariety. Our first observation is that the results in this paper also apply to a special class of Kleene algebras.

Lemma 1. *Let $\mathcal{V}$ be the variety of idempotent semirings that satisfy $x^2 \leq 1 \vee x$, and define a unary * on members of $\mathcal{V}$ by the term $x^* = 1 \vee x$. Then $\mathcal{V} \subseteq \mathsf{KA}$, and cdi-semirings are precisely the members of $\mathcal{V}$ that satisfy the identities $xy = yx$ and $x^2 = x$.*

Proof. We first prove that $\mathcal{V} \subseteq \mathsf{KA}$ by showing that $x^2 \leq 1 \vee x$ and $x^* = 1 \vee x$ imply (i)–(iii) in the definition of Kleene algebras. Let $\mathbf{A} \in \mathcal{V}$ and $x, y \in A$. Then

$$1 \vee x \vee x^*x^* = 1 \vee x \vee (1 \vee x)(1 \vee x) = 1 \vee x \vee x^2 = 1 \vee x = x^*.$$

Assuming $xy \leq y$, we have $y \vee xy = y$ and $x^*y = (1 \vee x)y = y \vee xy = y$. Similarly $yx \leq y \Rightarrow yx^* = y$.

For the last part, observe that all cdi-semirings are members of $\mathcal{V}$ since $x^2 = x$ implies $x^2 \leq 1 \vee x$. □

There are two 3-element cdi-semirings, and in [1] it is proved that the variety CDI of cdi-semirings is generated by one of them, denoted by S_3, (the other one is the 3-element distributive lattice). In the literature of semirings there are several definitions depending on whether the algebra contains an identity and/or a zero element. Polin [10] studied minimal varieties of semirings without 0, 1 as constant operations. A variety is minimal if it has no proper subvarieties other than the variety of one-element algebras. Polin showed there are 8 minimal varieties of semirings (without 0, 1) generated by 2-element semirings and 2 countable sequences of minimal varieties of rings generated by finite prime fields and by finite prime additive cyclic groups with constantly zero multiplication. If the constants are included, then there are still the two countable sequences and only one more minimal variety: the variety of bounded distributive lattices.

McKenzie and Romanovska [6] proved that the variety of doubly idempotent semirings without 0, 1 has exactly 4 proper subvarieties: the trivial variety, the variety of distributive lattices (without constants for top, bottom), the variety of semilattices (defined by $xy = x \vee y$), and the join of the previous two varieties, called *distributive bisemilattices* and defined as commutative doubly idempotent semirings (without constants) where $x \vee yz = (x \vee y)(x \vee z)$. When 0 is in the signature of semirings with $0 \vee x = x$ and $x0 = 0$, then the distributivity of $\vee$ over $\cdot$ implies the absorption laws since

$$x \vee xy = (x \vee x)(x \vee y) = x(x \vee y) = (x \vee 0)(x \vee y) = x \vee 0y = x \vee 0 = x.$$

Hence the variety of distributive bisemilattices with 0 coincides with the variety of distributive lattices with 0. Likewise the identity $xy = x \vee y$ implies $0 = x0 = x \vee 0 = x$ hence the variety of semilattices coincides with the trivial variety. So with constants, the variety CDI has only two subvarieties, namely the variety of bounded distributive lattices, generated by the 2-element lattice **2** and the variety of one-element algebras.

2 Cdi-Semirings of Height Two

Recall that in an idempotent semiring **S**, the join-semilattice order is denoted by $x \le y$. If $(S, \le)$ is a linear order (or *chain* for short) then the *height* of **S** is $|S| - 1$. In general the height of an idempotent semiring is the maximal height over all subchains of $(S, \le)$. The top element in the $\le$-order is denoted by $\top$.

It follows from a result of Stanovsky [11] about idempotent residuated lattices that there are only a small number of cdi-semirings of height 2. The proof below is self-contained and constructs all nonisomorphic cdi-semirings of height ≤ 2.

Recall that an *atom* of a poset with bottom element 0 is an element $a \neq 0$ such that $x < a$ implies $x = 0$.

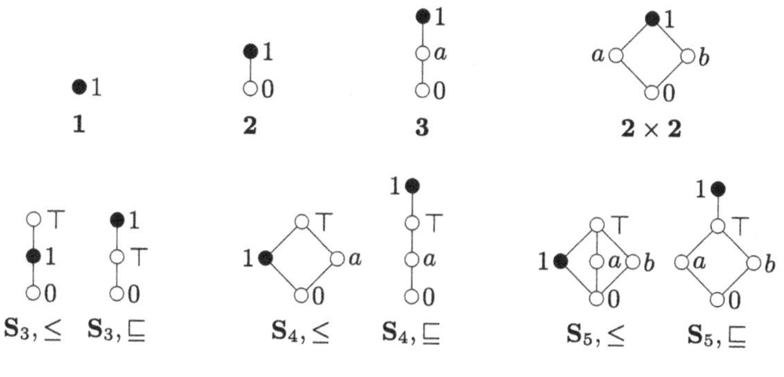

Fig. 1. All cdi-semirings of height 2 or less, ordered by $\le$ and $\sqsubseteq$, with 1 marked by $\bullet$. The top row are bounded distributive lattices, hence $\le$ and $\sqsubseteq$ coincide.

Theorem 2. *There are, up to isomorphism, seven cdi-semirings of height two or less (Fig. 1).*

Proof. Let $\mathbf{S}$ be a cdi-semirings of height ≤ 2. For any elements $x, y \in S$ such that $x \in \{0,1\}$ or $y \in \{0,1\}$ the multiplication xy is fixed by the semiring axioms and $xx = x$, hence the structure of $\mathbf{S}$ is determined by the join-semilattice order and the products of distinct elements $x, y \in S \setminus \{0,1\}$. If $\mathbf{S}$ has height 0, it is the one-element semiring $(0 = 1)$, and if $\mathbf{S}$ has height 1, it is the 2-element lattice with $0 \neq 1$. In the remaining cases, $\mathbf{S}$ has height 2, so let A be the set of atoms of $\mathbf{S}$.

If $|A| = 1$, then $\mathbf{S}$ has three elements and either $A = \{1\}$ or $A = \{a\}$ for some $a \neq 1$. Therefore $\mathbf{S}$ is $\mathbf{S}_3$ or $\mathbf{3}$.

If $|A| = 2$, then $A = \{1, a\}$ for some $a \neq 1$ or $A = \{a, b\}$ for $a \neq 1$ and $b \neq 1$. In the first case $a\top = a(1 \vee a) = a \vee a = a$, and in the second case $a, b \leq \top = 1$, hence $ab \leq a, b$ and it follows that $ab = 0$. Therefore $\mathbf{S}$ is S_4 or $\mathbf{2} \times \mathbf{2}$.

If $|A| \geq 3$, then we have distinct elements $a, b, c \in A$. If $\top = 1$ then as in the previous case $ab = 0$ and similarly $ac, bc = 0$. We also have $b \vee c = 1$ since $\mathbf{S}$ has height 2. But now $0 = ab \vee ac = a(b \vee c) = a1 = a$ contradicts the assumption that a is an atom, hence we conclude that $\top \neq 1$ and therefore 1 is an atom. Since $\mathbf{S}$ has height 2, we have $a \vee 1 = \top = b \vee 1$ and

$$ab \vee b = ab \vee 1b = (a \vee 1)b = (b \vee 1)b = b \vee b = b.$$

It follows that $ab \leq b$, and similarly $ab \leq a$, hence $ab = 0$. In the case when $A = \{1, a, b\}$ we again have $a\top = a$ as well as $b\top = b$, therefore $\mathbf{S}$ is $\mathbf{S}_5$.

In all other cases $|A| > 3$, hence we have distinct $1, a, b, c \in A$ and $a \vee 1 = \top$. The same argument as above shows that $ab = 0$ and $ac = 0$, so

$$0 = ab \vee ac = a(b \vee c) = a\top = a(a \vee 1) = a \vee a = a$$

which again contradicts the assumption that a is an atom, so no further cdi-semirings of height 2 exist. □

3 Catalan Semirings

As mentioned in the introduction, cdi-semirings have a multiplicative semilattice order defined by $x \sqsubseteq y$ if and only if $x \cdot y = x$. A cdi-semiring is called a *Catalan semiring* if this multiplicative order is a chain. A search with Prover9/Mace4 [5] shows there are $1, 1, 2, 5, 14, 42$ such cdi-semirings of size up to 6. This sequence coincides with the sequence of Catalan numbers $C_n = \frac{1}{n+1}\binom{2n}{n}$ [8] and our next result shows that this coincidence continues for all n. Using a result of [2] we construct all finite Catalan semirings by defining a *Catalan sum* Ⓒ on this class. To distinguish the operations and constants in several semirings, we superscript them with the name of the semiring.

Let $\mathbf{A}$ and $\mathbf{B}$ be two Catalan semirings and define $\mathbf{C} = \mathbf{A}$ Ⓒ $\mathbf{B}$ to be the structure over the disjoint union of A and B given in the following way. Then $0^{\mathbf{C}} = 0^{\mathbf{A}}$, $1^{\mathbf{C}} = 1^{\mathbf{A}}$ and the operations are given by

$$x \vee^{\mathbf{C}} y = \begin{cases} x \vee^{\mathbf{A}} y & \text{if } x, y \in A\backslash\{0\} \\ x \vee^{\mathbf{B}} y & \text{if } x, y \in B \\ 1^{\mathbf{B}} \vee^{\mathbf{B}} y & \text{if } x \in A\backslash\{0\}, y \in B \\ 1^{\mathbf{B}} \vee^{\mathbf{B}} x & \text{if } x \in B, y \in A\backslash\{0\} \\ y & \text{if } x = 0^{\mathbf{A}} \\ x & \text{if } y = 0^{\mathbf{A}} \end{cases} \qquad x \cdot^{\mathbf{C}} y = \begin{cases} x \cdot^{\mathbf{A}} y & \text{if } x, y \in A\backslash\{0\} \\ x \cdot^{\mathbf{B}} y & \text{if } x, y \in B \\ y & \text{if } x \in A\backslash\{0\}, y \in B \\ x & \text{if } x \in B, y \in A\backslash\{0\} \\ 0^{\mathbf{A}} & \text{if } x = 0^{\mathbf{A}} \text{ or } y = 0^{\mathbf{A}} \end{cases}$$

Recall that for two partially ordered sets P_1, P_2 the *ordinal sum* $P_1 \oplus P_2$ is given by the disjoint union of P_1, P_2 with every element of P_1 below every element of P_2. Using this construction, the multiplicative semilattice of $\mathbf{C}$ is simply the ordinal sum $\{0^{\mathbf{A}}\} \oplus (B, \sqsubseteq) \oplus (A \setminus \{0^{\mathbf{A}}\}, \sqsubseteq)$, and the join-semilattice of $\mathbf{C}$ is described by Fig. 2. Note that if $\mathbf{A}$ or $\mathbf{B}$ is a one-element algebra, the underlying lattice of $\mathbf{A} © \mathbf{B}$ is the ordinal sum of the lattices of $\mathbf{A}$ and $\mathbf{B}$.

The next lemma is proved in [2] for finite commutative Catalan idempotent residuated lattices. Every finite idempotent semiring uniquely expands to a finite residuated lattice, hence we can state the result in the following way.

Lemma 3. (i) *If* $\mathbf{A}, \mathbf{B}$ *are finite Catalan semirings then* $\mathbf{A} © \mathbf{B}$ *is a Catalan semiring of size* $|A| + |B|$.
(ii) *Suppose* $\mathbf{C}$ *is a finite Catalan semiring of cardinality* $n \geq 2$. *Then* $\mathbf{C} = \mathbf{A} © \mathbf{B}$ *for a unique pair* $\mathbf{A}, \mathbf{B}$ *of smaller Catalan semirings.*

Proof. (i) Assume that $\mathbf{A}, \mathbf{B}$ are finite Catalan semirings and let $\mathbf{C} = \mathbf{A} © \mathbf{B}$. Then by construction, $\mathbf{C}$ has a linear monoidal order $\sqsubseteq^{\mathbf{C}}$, and $\leq^{\mathbf{C}}$ is a join-semilattice order (Fig. 2). Hence $\cdot^{\mathbf{C}}$ and $\vee^{\mathbf{C}}$ are associative, commutative and idempotent. The least element of the lattice order $(C, \leq^{\mathbf{C}})$ is the least element of the monoidal order $(C, \sqsubseteq^{\mathbf{C}})$. Thus all we need to prove in order to show that $\mathbf{C}$ is a Catalan semiring is distributivity, i.e. $x(y \vee z) = xy \vee xz$. In principle there are eight cases to check, but when x, y, z are all in either A or B then distributivity holds. By commutativity of $\vee$ there are four cases left to check:

1. Let $x \in A \setminus \{0^{\mathbf{A}}\}$, and $y, z \in B$. Then $x(y \vee z) = x(y \vee^{\mathbf{B}} z) = y \vee^{\mathbf{B}} z$ and $xy \vee xz = y \vee^{\mathbf{B}} z$ since $y, z \sqsubseteq x$.
2. Let $y \in A \setminus \{0^{\mathbf{A}}\}$ and $x, z \in B$. Then $x(y \vee z) = x(1^{\mathbf{B}} \vee^{\mathbf{B}} z)$ and $xy \vee xz = x \vee^{\mathbf{B}} xz = x(1^{\mathbf{B}} \vee^{\mathbf{B}} z)$.
3. Let $x, y \in A \setminus \{0^{\mathbf{A}}\}$, and $z \in B$. Then $x(y \vee z) = x(1^{\mathbf{B}} \vee^{\mathbf{B}} z) = 1^{\mathbf{B}} \vee^{\mathbf{B}} z$ and $xy \vee xz = xy \vee^{\mathbf{B}} z = 1^{\mathbf{B}} \vee^{\mathbf{B}} z$.
4. Let $y, z \in A \setminus \{0^{\mathbf{A}}\}$, and $x \in B$. Then $x(y \vee z) = x(y \vee^{\mathbf{A}} z) = x$ and $xy \vee xz = x \vee^{\mathbf{B}} x = x$.

Finally, when one of x, y, z is $0^{\mathbf{A}}$ then the distributivity also holds.

(ii) Assume $\mathbf{C}$ is a finite nontrivial Catalan semiring, hence the $\sqsubseteq$-semilattice order is a chain. Let $b \in C$ be the unique atom in this chain, and define the sets $B = \{x \in C : b \leq x\}$ and $A = C \setminus B$. The operations $\cdot, \vee$ are defined on A and B by restriction from $\mathbf{C}$. To show that these operations are well defined, it suffices to show that A, B are closed under $\cdot^{\mathbf{C}}, \vee^{\mathbf{C}}$. This is true for $\cdot^{\mathbf{C}}$ since $x \cdot^{\mathbf{C}} y \in \{x, y\}$. Moreover, B is closed under $\vee^{\mathbf{C}}$ since it is upward closed.

Suppose that $b \le x \vee y$ and $x \ne 0^{\mathbf{C}} \ne y$. Since b is an atom of $(C, \sqsubseteq)$, we have $b \sqsubseteq x, y$. If $xy = x$, then $b = xb \le x(x \vee y) = x^2 \vee xy = x \vee x = x$ by distributivity and idempotency. Similarly, if $xy = y$ then $b \le y$. Also, if $x = 0^C$ then $b \le x \vee y = y$, and if $y = 0^C$ then $b \le x \vee y = x$. Hence if $x \vee y \in B$, then $x \in B$ or $y \in B$, thus A is closed under $\vee^{\mathbf{C}}$. Let $\mathbf{A}$ and $\mathbf{B}$ be the Catalan semirings with the operations $\cdot, \vee$ induced by restriction of $\cdot^{\mathbf{C}}, \vee^{\mathbf{C}}$. Note that $0^{\mathbf{A}} = 0^{\mathbf{C}}$ and $0^{\mathbf{B}} = b$. The identity elements $\mathbf{A}, \mathbf{B}$ are defined below.

We now want to show that $\mathbf{B}$ is an interval of $(C, \sqsubseteq^{\mathbf{C}})$. If $b' \in B$ then $b \le b'$, and $b \sqsubseteq x \sqsubseteq b'$ implies that $b = xb \le xb' = x$, hence $x \in B$. Since $\mathbf{C}$ is finite, it follows that for some $c \in C$ we have $B = \{x : b \sqsubseteq x \sqsubseteq c\}$. Hence for every $x \in B$ we have $xc = x$, i.e., $1^{\mathbf{B}} = c$ is the identity of $\mathbf{B}$. If $1^{\mathbf{C}} \in B$, then $c = 1^{\mathbf{C}}$ and $A = \{0^{\mathbf{C}}\}$ and otherwise $1^{\mathbf{C}}$ is the identity of $\mathbf{A}$. The elements of $A \setminus \{0^{\mathbf{C}}\}$ are linearly ordered by $\sqsubseteq^{\mathbf{C}}$ and they are above the interval of $(B, \sqsubseteq)$.

Let $x \in A \setminus \{0^{\mathbf{C}}\}$, then $cx = c$ and $0^{\mathbf{C}} \le c$. For $y \in B$ if $x \le y$, then $c = cx \le cy = y$, hence c is above every element of $\mathbf{A}$ and any element of B that is above some element of $A \setminus \{0^{\mathbf{C}}\}$ is above c. Moreover, for $y \in B$ and $x \in A \setminus \{0^{\mathbf{C}}\}$, we have $x \le x \vee y \in B$. Thus $c \le x \vee y$ and therefore $c \vee y \le x \vee y$. Since $x \le c$, we have $x \vee y \le c \vee y$, hence $x \vee y = c \vee y$. It follows that $\mathbf{C} = \mathbf{A} \circledcirc \mathbf{B}$. □

For $n, i > 0$, the Catalan semiring $\mathbf{C}_i^n$ is defined to be the ith Catalan semiring with n elements, starting with the one-element Catalan semiring $\mathbf{C}_1^1$. The next Catalan semiring would be $\mathbf{C}_1^2 = \mathbf{C}_1^1 \circledcirc \mathbf{C}_1^1$, the two-element distributive lattice. The two 3-element cdi-semirings are $\mathbf{C}_1^3 = \mathbf{C}_1^1 \circledcirc \mathbf{C}_1^2$ and $\mathbf{C}_2^3 = \mathbf{C}_1^2 \circledcirc \mathbf{C}_1^1$. In general, the Catalan semirings $\mathbf{C}_i^n$ of size n are built by constructing all Catalan sums of algebras $\mathbf{A}$ and $\mathbf{B}$ of size $n - k$ and k respectively, as k ranges from 1 to $n - 1$ (see Fig. 2). This yields the following result.

Theorem 4. *The number of Catalan semirings with $n + 1$ elements, up to isomorphism, is the nth Catalan number $C_n = \frac{1}{n+1}\binom{2n}{n}$.*

Proof. Let $CS(n)$ denote the number of Catalan semirings of cardinality n. The result is proved by induction. The sequence $\langle C_i : i \ge 0\rangle$ of Catalan numbers is determined recursively by $C_0 = 1$ and $C_{n+1} = \sum_{i=0}^{n} C_i C_{n-i}$. Obviously, $CS(1) = 1 = C_0$. Suppose now that $n \ge 1$ and $CS(n) = C_{n-1}$. Using the preceding lemma and the induction hypothesis, we have that

$$CS(n+1) = \sum_{k=1}^{n} CS(k) \cdot CS(n+1-k) = \sum_{k=1}^{n} C_{k-1} C_{n-k} = \sum_{i=0}^{n-1} C_i C_{n-1-i} = C_n.$$

□

The number of algebras for each size (up to isomorphism), along the number of cdi-semirings and distributive lattices, tell us how many cdi-semirings are described using the result. As one can see from the table below, this result helps us to understand a big portion of the cdi-semirings for small number of elements (Table 1).

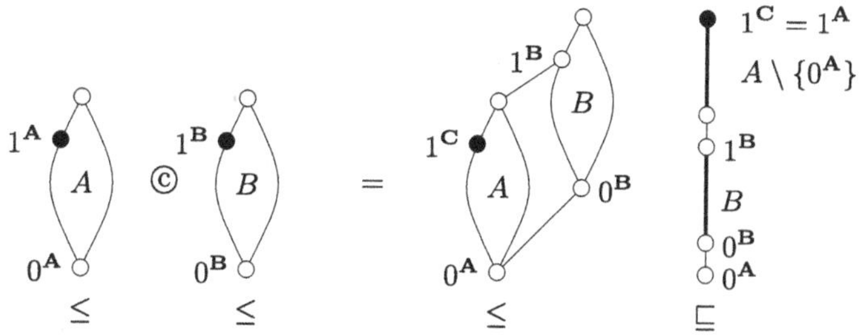

Fig. 2. The Catalan sum $\mathbf{C} = \mathbf{A} \copyright \mathbf{B}$

Table 1. Number of algebras up to isomorphism with n elements

# of elements $n =$	1	2	3	4	5	6	7	8
# of distr. lattices	1	1	1	2	3	5	8	15
# of Catalan semirings	1	1	2	5	14	42	132	429
# of cdi-semirings	1	1	2	6	20	77	333	1589

The construction of finite Catalan semirings is very efficient and can be implemented, for example, with the following short Python program that computes all Catalan semirings of size $\leq n$. The output (after conversion to TikZ) is shown in Fig. 3. The black dot marks the identity element and the elements are numbered in increasing order of the multiplicative semilattice. Note that these algebras are rigid (i.e., have trivial automorphism group) and are all pairwise nonisomorphic.

```
def catalan_sum(A,B):
# A,B are tuples with A[0] a list of upper covers, topologically sorted
# A[1]=[s[0],...,s[n-1]] a permutation of range(n) s.t. x*y=x iff s[x]<=s[y]
# A[2]=[p[0],...,p[n-1]] a list of coordinates p[i]=(x,y) for display
    m = len(A[0])
    n = len(B[0])
    id_B = B[1].index(n-1)
    uc = [A[0][0]+([m] if n!=1 or m==1 else [])] + A[0][1:-1]\
        + ([A[0][-1]+[id_B+m]] if m!=1 else [])\
        + [[x+m for x in u] for u in B[0]]
    s = [0] + [x+n for x in A[1][1:]] + [x+1 for x in B[1]]
    x = A[2][-1][0] if m==1 or n==1 else max([p[0] for p in A[2]]) + 1
    y = (A[2][-1][1] + 1) if m==1 or n==1 else \
        max(1, A[2][-1][1] - B[2][id_B][1] + 1)
    pos = A[2] + [(B[2][0][0]+x, 1 if m!=1 and n!=1 else B[2][0][1]+y)] + \
        [(p[0]+x,p[1]+y) for p in B[2][1:]]
    return (uc,s,pos)

def catalan_semirings(n):
# calculate all Catalan semirings of size 1 to n
    if n==0: return [[([[]],[0],[(0,0)])]]
    CL = catalan_semirings(n-1)
    return CL + [[catalan_sum(A,B) for i in range(len(CL))
        for A in CL[i] for B in CL[n-1-i]]]
```

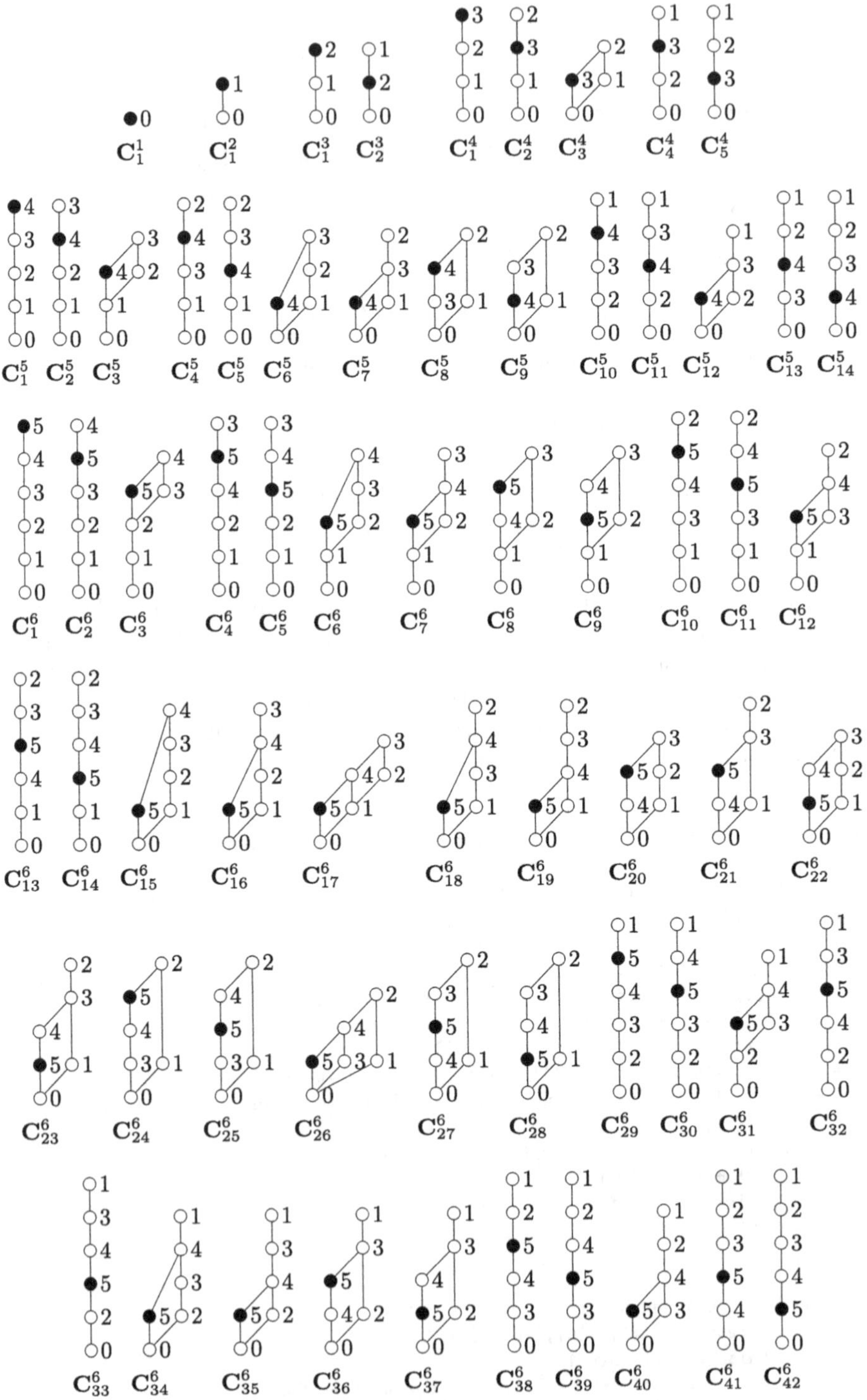

Fig. 3. $\leq$-order of Catalan semirings of size ≤ 6

4 Boolean Cdi-Semirings and Directed Graphs

An idempotent semiring is *Boolean* if its join-semilattice is the reduct of a Boolean algebra. In this section we analyze the structure of finite Boolean cdi-semirings. We use ideas from the theory of Boolean algebras with operators and relation algebras [3,4] to recover the semiring operations from a ternary relation on the atoms of the Boolean algebra. Lemma 5 below is a standard result that states this works in general for nonassociative nonunital complete atomic Boolean idempotent semirings. These algebras are also known as nonassociative atomic Boolean quantales. A *nonassociative quantale* $\mathbf{B} = (B, \bigvee, \cdot)$ is a complete join-semilattice $(B, \bigvee)$ with a binary operation $\cdot$ such that $x(\bigvee Y) = \bigvee_{y\in Y} xy$ and $(\bigvee Y)x = \bigvee_{y\in Y} yx$ for all $x \in B$ and $Y \subseteq B$. A *quantale* in addition satisfies the identity $(xy)z = x(yz)$. By completeness, every quantale has a least and a greatest element, denoted by 0 and $\top$ respectively. The complete distributivity of $\cdot$ over $\bigvee$ implies $x0 = 0 = 0x$. If it also has a *left identity* $1x = x$ and/or *right identity* $x1 = x$ then it is a *left/right unital* quantale. Hence a join-complete idempotent semiring is the same as a unital quantale. As for semirings, a quantale is *Boolean* if its join-semilattice order is that of a complete Boolean algebra, and *atomic* if every nonzero element has an atom below it. The set of atoms of $\mathbf{B}$ is denoted by $At(\mathbf{B})$.

Lemma 5. *1. Let* $\mathbf{B}$ *be a nonassociative atomic Boolean quantale with* $A = At(\mathbf{B})$ *and define a ternary relation* $R \subseteq A^3$ *by* $R(x, y, z) \iff x \leq yz$. *Then for all* $b, c \in B$,

$$bc = \bigvee\{x : \exists y \leq b\, \exists z \leq c\ R(x, y, z)\}.$$

2. Suppose $R \subseteq A^3$ *is a ternary relation on a set* A, *and define* $\mathbf{B} = (\mathcal{P}(A), \bigcup, \cdot)$ *where for* $Y, Z \in P(A)$

$$Y \cdot Z = \{x : \exists y \in Y\, \exists z \in Z\ R(x, y, z)\}.$$

Then $\mathbf{B}$ *is a nonassociative atomic Boolean quantale.*

As in the theory of Boolean algebras with operators or modal logic, the relational structure $\mathbf{A} = (A, R)$ from the preceding lemma is called the *atom structure* or *Kripke frame* of the Boolean quantale $\mathbf{B}$. Correspondence theory from modal logic also applies to Boolean quantales. For example, $\mathbf{B}$ is commutative if and only if $R(x, y, z) \Leftrightarrow R(x, z, y)$ for all $x, y \in At(\mathbf{B})$. It is convenient to split associativity into two inequalities $(ab)c \leq a(bc)$, called *subassociativity*, and $(ab)c \geq a(bc)$, called *supassociativity*, where $a, b, c \in B$.

Theorem 6. *Let* $\mathbf{B}$ *be a nonassociative atomic Boolean quantale with* R *defined on* $A = At(\mathbf{B})$ *as in the preceding lemma. Then for* $x, y, z \in A$, $\mathbf{B}$ *is*

(i) *mult. idempotent* $\Leftrightarrow$ $R(x,x,x) \mathbin{\&} (R(x,y,z) \Rightarrow x = y \text{ or } x = z)$
(ii) *subassociative* $\Leftrightarrow$ $(R(u,x,y) \mathbin{\&} R(w,u,z) \Rightarrow \exists v(R(v,y,z) \mathbin{\&} R(w,x,v)))$
(iii) *left unital* $\Leftrightarrow$ $\exists I \subseteq A(x = z \Leftrightarrow \exists y \in I\, R(x,y,z))$
(iv) *right unital* $\Leftrightarrow$ $\exists I \subseteq A(x = y \Leftrightarrow \exists z \in I\, R(x,y,z))$

Proof. (i) Assume $\mathbf{B}$ in multiplicatively idempotent, let $x, y, z \in A = At(\mathbf{B})$ be atoms and assume $x \le yz$. Then $y \vee z = (y \vee z)^2 = y^2 \vee yz \vee z^2 = y \vee z \vee yz$. Therefore $yz \le y \vee z$. Since $x \le yz$ we have $x \le y \vee z$, and we assumed x, y, z are atoms, hence it follows that $x = y$ or $x = z$.

Now suppose $R(x,x,x)$ and $(R(x,y,z) \Rightarrow x = y$ or $x = z)$ holds for all atoms $x, y, z \in A$. Then by Lemma 5.1, for any $c \in B$ we have $c \le cc$ since $R(x,x,x)$ holds for all atoms $x \le c$. Now let x be an atom such that $x \le c \cdot c$. Again by Lemma 5.1, $x \le y \cdot z$ for some atoms $y, z \le c$, therefore $R(x,y,z)$ holds and by assumption $x = y$ or $x = z$. Hence $x \le c$ and it follows that $cc = c$.

(ii) Since all variables in subassociativity are distinct, this property holds for all elements of $\mathbf{B}$ if and only if it holds for all atoms. Now let $x, y, z \in A$. Then $(xy)z \le x(yz)$ is equivalent to $w \le (xy)z \Rightarrow w \le x(yz)$ for all $w \in A$. This in turn is equivalent to

$$\exists u \in A\,(u \le xy \mathbin{\&} w \le uz) \Rightarrow \exists v \in A\,(v \le yz \mathbin{\&} w \le xv).$$

The first existential quantifier can move out of the premise to the front of the formula and switches to a universal quantifier, hence the formula translates to the given condition for R.

(iii) If $\mathbf{B}$ is left unital then it has a 1 such that $1b = b$ for all $b \in B$ and we can define $I = \{z \in A : z \le 1\}$. For atoms $x, z \in A$ if $x = z$ then $x = 1z \le 1z$, so by Lemma 5.1 there exists an atom $y \in I$ such that $x \le yz$, which shows $R(x,y,z)$. Conversely, assume $x \le yz$ where $y \in I$. Then $yz \le 1z = z$ implies $x \le z$ and since both are atoms, $x = z$. This proves the forward direction of (iii).

Now assume a set $I \subseteq A$ with the given property exists and define $1 = \bigvee I$. It suffices to show that $z = 1z$ for all atoms $z \in A$ since this equality lifts to all of $\mathbf{B}$. Let $x \le 1z$, then by Lemma 5.1 $x \le yz$ for some $y \in I$. Hence $x = z$, which shows that z is the only atom below $1z$. It follows that $z = 1z$.

(iv) This proof is similar to (iii). □

From now on a ternary relation R is called commutative, (multiplicatively) idempotent, subassociative or (left/right) unital if its corresponding Boolean quantale has the same property.

We now observe that if multiplication is idempotent then the ternary relation can be replaced by two reflexive binary relations P and Q. In the commutative case they coincide, so the structure of nonassociative Boolean cdi-semirings is determined by a single reflexive relation Q. The proof follows directly from the formula $R(x,y,z) \Rightarrow x = y$ or $x = z$.

Lemma 7. *An idempotent ternary relation $R \subseteq A^3$ is definitionally equivalent to a pair of reflexive binary relations $P, Q \subset A^2$ via the definitions*

$$(\mathsf{Pdef}) \quad P(x,y) \Leftrightarrow R(x,y,x) \qquad (\mathsf{Qdef}) \quad Q(x,y) \Leftrightarrow R(x,x,y)$$
$$(\mathsf{Rdef}) \quad R(x,y,z) \Leftrightarrow (x=y \ \&\ Q(y,z)) \text{ or } (x=z \ \&\ P(z,y)).$$

Moreover, the relation R is commutative if and only if $P = Q$.

The existentially quantified subassociative property for ternary relation is not easy to work with, hence it is noteworthy that, in the presence of idempotence, subassociativity can be replaced by the following three universal formulas for P and Q.

Theorem 8. *An idempotent ternary relation $R \subseteq A^3$ is subassociative if and only if the corresponding reflexive relations P, Q satisfy*

$$(\mathsf{P}_1) \quad P(x,y) \ \&\ P(y,z) \Rightarrow P(x,z) \quad \text{i.e. } P\text{-transitivity}$$
$$(\mathsf{P}_2) \quad Q(x,y) \ \&\ Q(x,z) \Rightarrow Q(y,z) \text{ or } P(z,y)$$
$$(\mathsf{P}_3) \quad P(x,y) \ \&\ Q(y,z) \ \&\ x \neq y \Rightarrow P(x,z)$$

To characterize supassociativity of R, it suffices to interchange P, Q in these conditions to obtain (P'_1), (P'_2), (P'_3). Hence R is associative if and only if P, Q satisfy all six conditions.

Proof. Suppose (P_1)–(P_3) hold and recall that subassociativity of R is given by

$$R(u,x,y) \,\&\, R(w,u,z) \Rightarrow \exists v(R(v,y,z) \,\&\, R(w,x,v)).$$

Assume $R(u,x,y)$ and $R(w,u,z)$ holds. From (Rdef) we get

$$[u = x \ \&\ Q(x,y) \ \text{ or } \ u = y \ \&\ P(y,x)] \ \text{ and}$$
$$[w = u \ \&\ Q(u,z) \ \text{ or } \ w = z \ \&\ P(z,u)].$$

We consider 4 cases, with the aim of showing that in each case there exists a v that satisfies the conclusion of subassociativity, i.e.,

$$[(\mathsf{A}) \ v = y \ \&\ Q(y,z) \text{ or } (\mathsf{B}) \ v = z \ \&\ P(z,y)] \ \text{ and}$$
$$[(\mathsf{C}) \ w = x \ \&\ Q(x,v) \text{ or } (\mathsf{D}) \ w = v \ \&\ P(v,x)].$$

Case 1: Suppose $u = x$, $Q(x,y)$, $w = u$ and $Q(u,z)$. Then we have $u = x = w$, $Q(x,y)$ and $Q(x,z)$. From (P_2) we deduce $Q(y,z)$ or $P(z,y)$, and we want to find v such that $[(\mathsf{A})$ or $(\mathsf{B})]$ and $[(\mathsf{C})$ or $(\mathsf{D})]$. If $Q(y,z)$ holds, we choose $v = y$, then (A) and (C) hold, and if $P(z,y)$, we choose $v = z$, then (B) and (C) hold.

Case 2: Suppose $u = y$, $P(y,x)$, $w = u$ and $Q(u,z)$. Then $u = w = y$ and $P(y,x)$ and $Q(y,z)$ holds. Taking $v = y$ we get $v = y$ and $Q(y,z)$ and $w = v$ and $P(v,x)$. Hence (A) and (D) are true.

Case 3: Suppose $u = x$, $Q(x, y)$, $w = z$ and $P(z, u)$, hence $P(z, x)$. First, assuming $z \neq x$, we have $P(z, x)$, $Q(x, y)$ so by (P_3) it follows that $P(z, y)$. Now choosing $v = z$ shows (B) and (D) hold.

If remains to handle the case when $z = x$. Since $Q(x, y)$ and $Q(x, x)$ hold, (P_2) implies $Q(y, x)$ or $P(x, y)$. In case $Q(y, x)$ holds we choose $v = y$ to get (C) and (A) (since $z = x$). In the other case $P(x, y)$ holds, and then we choose $v = x$ to get (B) and (D).

Case 4: Suppose $u = y$, $P(y, x)$, $w = z$ and $P(z, u)$, hence $P(z, y)$. From (P_1) (transitivity) we deduce $P(z, x)$. Now taking $v = z$ we see that (B) and (D) are true.

Hence in all four cases we have proved subassociativity.

Conversely, assume that subassociativity holds for R:

$$R(u, x, y) \mathbin{\&} R(w, u, z) \Rightarrow \exists v(R(v, y, z) \mathbin{\&} R(w, x, v)).$$

We show that (P_1)–(P_3) hold.

For (P_1) assume $P(x, y)$ and $P(y, z)$. Then we have $R(x, y, x)$ and $R(y, z, y)$ by definition of P. Matching $R(y, z, y)$ & $R(x, y, x)$ to the premise of subassociativity with $u := y$, $x := z$, $w := x$ and $z := x$, there exists v such that $R(v, y, x)$ and $R(x, z, v)$ holds. By idempotence of R and Theorem 6(i) it follows that $x = z$ or $x = v$ hold and hence we get $P(x, z)$ (from $x = z$ or from (Pdef) and $R(x, z, x)$).

For (P_2) assume $Q(x, y)$ and $Q(x, z)$. By definition of Q we get $R(x, x, y)$ and $R(x, x, z)$. Let $u := x$ and $w := x$, then by subassociativity there exists v such that $R(v, y, z)$ and $R(x, x, v)$ holds. By idempotence there are two options for v: if $v = y$ we have $Q(y, z)$ and if $v = z$ we have $P(z, y)$. Hence (P_2) holds.

For (P_3) assume $Q(y, z)$ and $P(x, y)$ and $x \neq y$ hold. From the definition of Q and P we get $R(y, y, z)$ and $R(x, y, x)$. Let $u := y$, $x := y$, $y := z$, $w := x$ and $z := x$, then by subassociativity there exists v such that $R(v, z, x)$ and $R(x, y, v)$ hold. Since $x \neq y$, it follows from $R(x, y, v)$ and by *mult. idempotent* that $v = x$, so $P(x, z)$ follows from the first conjunct. Hence (P_3) is true. □

Corollary 9. *An atomic Boolean idempotent quantale is determined by two reflexive binary relations P, Q on its set of atoms such that the condition* (P_1), (P_2), (P_3), (P_1'), (P_2'), (P_3') *from the previous theorem hold.*

However the conditions (P_1), (P_2), (P_3) are nonintuitive, and it is fortunate that in the commutative case they reduce to a much simpler pair of axioms. Recall that a *preorder* is a reflexive transitive binary relation and a *partial order* is a preorder that is antisymmetric: $P(x, y) \mathbin{\&} P(y, x) \Rightarrow x = y$. A *forest* is a partial order such that

$$(*) \qquad P(x, y) \mathbin{\&} P(x, z) \Rightarrow P(y, z) \text{ or } P(z, y)$$

i.e., all the elements above a given element are linearly ordered. A forest can have many connected components, each of which is a *tree*. If each tree has a top element (called the root) then forest is said to be *rooted*. Finite forests are always rooted and they are easy to enumerate up to isomorphism. In fact they are in one-one correspondence with finite trees since one can add a new root to convert any forest into trees with one more element. The number of finite trees with n unlabeled elements (i.e., up to isomorphism) is the sequence A00081 [7].

A preorder $P \subset A^2$ is determined by the equivalence relation $\equiv = P \cap P^{-1}$ and the induced partial order on the set of equivalence classes $P/\equiv$.

A *preorder forest* is a preorder that satisfies property $(*)$ and it is *rooted* if each component has a largest equivalence class. Finite preorder forests are always rooted, and the number of finite preorder forests with n unlabeled elements (i.e., up to isomorphism) is also easy to count, given by the sequence A052855 [9]. Finally, a preorder forest is said to have *singleton roots* if it is rooted and all largest equivalence classes contain only one element.

Theorem 10. *Atomic Boolean commutative idempotent unital quantales are definitionally equivalent to preorder forests with singleton roots.*

In the finite case these algebras are Boolean cdi-semirings, hence all finite Boolean cdi-semirings can be constructed by enumerating preorder forests with singleton roots.

Proof. Let P, Q be the reflexive binary relations on the atoms that exist by idempotence. From commutativity it follows that $P = Q$ hence (P_2) reduces to $(*)$ and (P_1) implies (P_3). This means the relation P is a preorder forest. For any atom z below 1, $P(z, x)$ implies $R(z, x, z)$, and it follows from unitality that $x = z$. Hence z is a unique maximal element of the preorder.

Conversely, from a preorder forest with singleton roots we define I to be the set of all roots of the forest to get a unit element for the quantale. □

Figure 4 shows the preorder forests with singleton roots up to cardinality 4. They correspond to Boolean semirings of size $2, 4, 8$ and 16. It is interesting to note that there are $1, 2, 5, 14, \ldots$ such semirings of each size, but this is *not* related to the Catalan numbers since the sequence continues with 41 followed by 127 (while the Catalan numbers are $42, 132$).

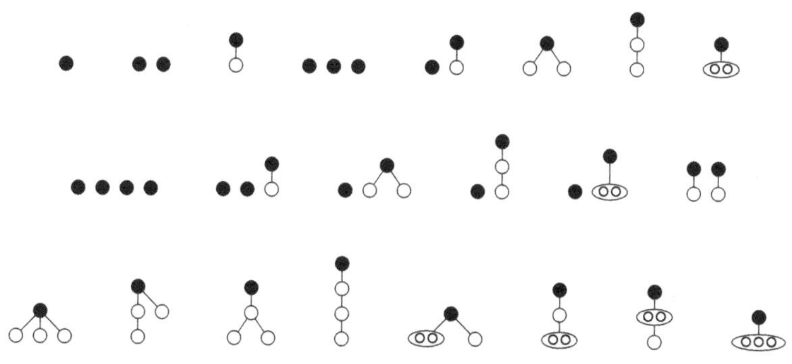

Fig. 4. Preorder forests with singleton roots represented by black dots

Note that every finite forest is a preorder forest with singleton roots, and it is interesting to investigate the multiplicative semilattices obtained from specific finite forests. As a simple example, the forests where each component is a singleton poset correspond to cdi-semirings that are Boolean lattices.

5 Conclusion

In the theory of rings and other algebras, multiplicatively idempotent elements often play a central role in controlling some structural aspects of the algebra. The structure of idempotent semirings in general is quite challenging, but with suitable restrictions some nice characterizations can be found. Here we considered commutative doubly idempotent semirings of height ≤ 2, or with a multiplicative linear order or with a Boolean join-semilattice. In each case it was possible to give detailed descriptions of the finite members that allow them to be enumerated easily up to isomorphism. It is likely that some of the techniques explored here can be applied to larger classes of idempotent semirings by, for example, weakening the assumption of commutativity or allowing distributive join-semilattices.

References

1. Chajda, I., Länger, H.: The variety of commutative additively and multiplicatively idempotent semirings. Semigroup Forum **96**(2), 409–415 (2017). https://doi.org/10.1007/s00233-017-9905-2
2. Jipsen, P., Gil-Férez, J., Metcalfe, G.: Structures theorems for idempotent residuated lattices, preprint. http://math.chapman.edu/~jipsen/preprints/GJMsubmitted.pdf
3. Jónsson, B., Tarski, A.: Boolean algebras with operators. II. Amer. J. Math. **74**, 127–162 (1952)
4. Maddux, R.: Some varieties containing relation algebras. Trans. Amer. Math. Soc. **272**(2), 501–526 (1982)
5. McCune, W.: Prover9 and Mace4 2005–2010. http://www.cs.unm.edu/~mccune/Prover9
6. McKenzie, R., Romanowska, A.: Varieties of -distributive bisemilattices. Contrib. Gen. Algebra **1**, 213–218 (1979). Proceedings of the Klagenfurt Conference Klagenfurt. Heyn, Klagenfurt (1978)
7. OEIS Foundation Inc.: Number of unlabeled rooted trees with n nodes. In: The On-Line Encyclopedia of Integer Sequences (2019). https://oeis.org/A000081
8. OEIS Foundation Inc.: Catalan numbers. In: The On-Line Encyclopedia of Integer Sequences (2019). https://oeis.org/A000108
9. OEIS Foundation Inc.: Number of forests of rooted trees of nonempty sets with n points. In: The On-Line Encyclopedia of Integer Sequences (2019). https://oeis.org/A052855
10. Polin, S.V.: Minimal varieties of semirings. Math. Notes **27**(3–4), 259–264 (1980). https://doi.org/10.1007/BF01140525. Mat. Zametki. 27(4), pp. 524–537 (1980) (in Russian)
11. Stanovský, D.: Commutative idempotent residuated lattices. Czech. Math. J. **57**(132), 191–200 (2007). no. 1

A Relation-Algebraic Treatment of the Dedekind Recursion Theorem

Rudolf Berghammer(✉)

Institut für Informatik, Christian-Albrechts-Universität Kiel,
Olshausenstraße 40, 24098 Kiel, Germany
rub@informatik.uni-kiel.de

Abstract. The recursion theorem of Richard Dedekind is fundamental for the recursive definition of mappings on natural numbers since it guarantees that the mapping in mind exists and is uniquely determined. Usual set-theoretic proofs are partly intricate and become lengthy when carried out in full detail. We present a simple new proof that is based on a relation-algebraic specification of the notions in question and combines relation-algebraic laws and equational reasoning with Scott induction. It is very formal and most parts of it consist of relation-algebraic calculations. This opens up the possibility for mechanised verification. As an application we prove a relation-algebraic version of the Dedekind isomorphism theorem. Finally, we consider two variants of the recursion theorem to deal with situations which frequently appear in practice but where the original recursion theorem is not applicable.

1 Introduction

The so-called recursion theorem of Richard Dedekind, first formulated and proved in [6], pertains to the method of recursively defining mappings $f : \mathbb{N} \to A$ on the set of natural numbers $\mathbb{N}$ by first defining the value of $f(0)$ (in [6] $f(1)$, since there the natural numbers start with 1) and then defining the value of $f(n+1)$ (in [6] $f(n')$, with n' as the successor of n) subject to the value of $f(n)$, for an arbitrary natural number $n \in \mathbb{N}$. It states that there exists precisely one such mapping and this guarantees the correctness of the method. Besides the Peano axioms, Dedekind's original proof (see [6], Satz 126) decisively depends on the linear ordering of the natural numbers which, in contrast with modern approaches, is specified before addition is introduced. About fifty years later proofs have been published which do not use the order but are based only on the zero/one element and the successor mapping, that is, on the vocabulary of the Peano axioms. Two of them can be found in [9,11]. The reader interested in the history of the Dedekind recursion theorem is referred to [7,8], for example.

Nowadays the Dedekind recursion theorem is frequently presented using *Peano structures*. These are algebraic structures (N, z, s) with a non-empty carrier set N, an element $z \in N$ (the *zero element*) and a mapping $s : N \to N$ (the *successor mapping*) such that the following three axioms hold:

U. Fahrenberg et al. (Eds.): RAMiCS 2020, LNCS 12062, pp. 15–30, 2020.
https://doi.org/10.1007/978-3-030-43520-2_2

$$\left.\begin{array}{l}\forall x, y \in N : s(x) = s(y) \Rightarrow x = y \\ \neg \exists x \in N : s(x) = z \\ \forall A \in 2^N : z \in A \wedge (\forall x \in A : s(x) \in A) \Rightarrow A = N\end{array}\right\} \quad (1)$$

Then the recursion theorem states that, given a Peano structure (N, z, s), a non-empty set A, an element $c \in A$ and a mapping $F : A \to A$, there exists precisely one mapping $f : N \to A$ with the following two properties:

$$f(z) = c \qquad \forall x \in N : f(s(x)) = F(f(x)) \quad (2)$$

Modern proofs of the recursion theorem define the mapping f of (2) as a relation, viz. as the intersection of all relations R with source N and target A such that $z\,R\,c$ and for all $x \in N$ and $y \in A$ from $x\,R\,y$ it follows that $s(x)\,R\,F(y)$. For example, in [3], pages 346–348, the partly intricate proof that this intersection in fact is a univalent and total relation (that is, a mapping) and satisfies the two formulae of (2) is carried out in great detail.

Specifying the notions in question in the language of relation algebra and combining relation-algebraic calculations with Scott induction, in Sect. 3 of this paper we present a new proof of the Dedekind recursion theorem that is simpler than the purely set-theoretic proof of [3] or similar proofs. A further advantage of the new proof is that it is very formal and most parts of it consist of equational reasoning. This opens up the possibility for its mechanised verification by means of a theorem-proving tool. As an application of our relation-algebraic version of the recursion theorem we present in Sect. 4 a relation-algebraic version of the Dedekind isomorphism theorem, i.e., prove that all (relational) Peano structures are isomorphic. Finally, in Sect. 5 we consider two cases of recursive definitions of mappings which frequently appear in practice but where the original Dedekind recursion theorem is not applicable since either the mapping f to be defined is not unary or the result of $f(s(x))$ depends not only on the value of $f(x)$ but also on x. For each case we give an example and prove a corresponding variant of the relation-algebraic recursion theorem.

2 Mathematical Preliminaries

We assume the reader to be familiar with the basic concepts of partially ordered sets and complete lattices, including monotone mappings on them, basic fixpoint theory (fixpoint calculus) and the construction of direct products. Otherwise we refer to standard textbooks on ordered sets and lattices, e.g., [4,5], and to [13].

Given a partially ordered set $(A, \leq)$ that is a complete lattice, we denote the least element of A by the symbol $\bot$, the least upper bound of the subset B of A by $\bigsqcup B$ and the greatest lower bound of B by $\bigsqcap B$. Alfred Tarski's well-known fixpoint theorem (see [16]) states that each monotone mapping $f : A \to A$ has a least fixpoint, denoted as $\mu(f)$, and $\mu(f) = \bigsqcap\{x \in A \mid f(x) \leq x\}$ holds. For proving properties of $\mu(f)$ we will apply the principle of *Scott induction*, sometimes also called computational induction or fixpoint induction. Usually the principle is formulated for complete partial orders (CPOs), that is, for partially ordered sets with a least element and the property that each chain possesses a least upper bound. See [10], for example. Scott induction also works in the case of complete lattices, since complete lattices are CPOs.

Assume $(A, \leq)$ to be a complete lattice. Then a predicate P on its carrier set A is called *admissible* (for Scott induction) if for every chain C in $(A, \leq)$ the following implication is true: if for all $x \in C$ it holds $P(x)$, then $P(\bigsqcup C)$ holds, too. Now, Scott induction states that for each monotone mapping $f : A \to A$ and each admissible predicate P on A from the two conditions

$$P(\bot) \qquad\qquad \forall x \in A : P(x) \Rightarrow P(f(x)) \tag{3}$$

it follows that $P(\mu(f))$. The left condition of (3) is called the *induction base* and the right one the *induction step* with induction hypothesis $P(x)$. Besides the above version we will also apply a version which in [10] is called *simultaneous*. We consider the case of two complete lattices $(A, \leq_1)$ and $(B, \leq_2)$ with least elements $\bot_1 \in A$ and $\bot_2 \in B$ and two monotone mappings $f_1 : A \to A$ and $f_2 : B \to B$ only. Then if P is an admissible predicate on the direct product $A \times B$, which is ordered by the product order (that is, by $(x_1, x_2) \leq (y_1, y_2)$ iff $x_1 \leq_1 y_1$ and $x_2 \leq_2 y_2$, for all $x_1, y_1 \in A$ and $x_2, y_2 \in B$), then from the two conditions

$$P(\bot_1, \bot_2) \qquad\qquad \forall x \in A, y \in B : P(x, y) \Rightarrow P(f_1(x), f_2(y)) \tag{4}$$

it follows that $P(\mu(f_1), \mu(f_2))$. This principle is obtained from the original one by taking in (3) the least element $(\bot_1, \bot_2)$ of the product lattice $(A \times B, \leq)$ as $\bot$ and the product of the two mappings $f_1 : A \to A$ and $f_2 : B \to B$, defined by

$$f_1 \otimes f_2 : A \times B \to A \times B \qquad\qquad (f_1 \otimes f_2)(x, y) = (f_1(x), f_2(y)),$$

as mapping f. Namely, from the monotonicity of f_1 with respect to $\leq_1$ and of f_2 with respect to $\leq_2$ and the definition of the product order $\leq$ it follows that $f_1 \otimes f_2$ is monotone with respect to $\leq$ and $\mu(f_1 \otimes f_2) = (\mu(f_1), \mu(f_2))$.

Given complete lattices $(A, \leq_1)$ and $(B, \leq_2)$, a predicate P on the carrier set $A \times B$ of the product lattice $(A \times B, \leq)$ is admissible (for the simultaneous Scott induction described by (4)) if there exist $\bigsqcup$-distributive mappings $\alpha : A \to C$ and $\beta : B \to C$ into a complete lattice $(C, \leq_3)$ such that $P(x, y)$ iff $\alpha(x) \leq_3 \beta(y)$, for all $x \in A$ and $y \in B$, or $P(x, y)$ iff $\alpha(x) = \beta(y)$, for all $x \in A$ and $y \in B$. See e.g., [10] for a proof of this property.

We assume the reader also to be familiar with the basic concepts of (axiomatic) relation algebra as introduced in [15] by Alfred Tarski. Otherwise we refer again to standard textbooks, e.g., to [12,14].

As in [14] we work with typed relations. For given sets (or objects in case of axiomatic relation algebra) A and B we denote the set of all relations with source A and target B by $[A \leftrightarrow B]$ and write $R : A \leftrightarrow B$ instead of $R \in [A \leftrightarrow B]$. As operations and predicates on relations we use transposition R^T, complementation $\overline{R}$, union $R \cup S$, intersection $R \cap S$, composition $R ; S$, inclusion $R \subseteq S$ and equality $R = S$, and as special relations we use the empty relation O, the universal relation L and the identity relation I. As usual, in the latter cases we overload the symbols, i.e., avoid the binding of types to them, since all types can be derived from the context by means of the typing rules of the operations. All basic relation-algebraic laws we will apply in the remainder of the paper are well known for set-theoretic relations; their proofs from the axioms of an (axiomatic) relation algebra can be found in [14], for example.

Many important properties of relations can be specified in a quantifier-free manner using (conjunctions of) inclusions and equations between relation-algebraic expressions only. In this paper we will use that a relation $R : A \leftrightarrow B$ is univalent iff $R^{\mathsf{T}} ; R \subseteq \mathsf{I}$, total iff $R ; \mathsf{L} = \mathsf{L}$ or, equivalently, iff $\mathsf{I} \subseteq R ; R^{\mathsf{T}}$, injective iff $R ; R^{\mathsf{T}} \subseteq \mathsf{I}$ and surjective iff $R^{\mathsf{T}} ; \mathsf{L} = \mathsf{L}$ or, equivalently, iff $\mathsf{I} \subseteq R^{\mathsf{T}} ; R$. For all R and S the following implication is shown in [14] as Proposition 4.2.2.iv:

$$R \subseteq S \wedge S \text{ univalent} \wedge R \text{ total} \implies R = S. \tag{5}$$

Other results of [14] we will apply are Proposition 4.2.2.iii, stating that

$$Q \text{ univalent} \implies R ; Q \cap S = (R \cap S ; Q^{\mathsf{T}}) ; Q, \tag{6}$$

for all Q, R and S, and Proposition 2.4.2.i, stating that

$$(Q \cap R ; \mathsf{L}) ; S = Q ; S \cap R ; \mathsf{L}, \tag{7}$$

for all Q, R and S.

We also need *relational vectors*, which are relations $v : A \leftrightarrow B$ with $v = v ; \mathsf{L}$, and *relational points*, which are injective and surjective relational vectors. In case of set-theoretic relations a little reflection shows that $v : A \leftrightarrow B$ is a relational vector iff there exists a subset V of the set A such that $v = V \times B$, and it is a relational point iff additionally V is a singleton set. Hence, a set-theoretic relational vector models a subset of its source and a set-theoretic relational point models an element of its source. Therefore, the targets are irrelevant and in most applications, also of (axiomatic) relation algebra, relational vectors and points are from a set $[A \leftrightarrow \mathbf{1}]$, where $\mathbf{1}$ is a singleton set (a specific object, respectively). In this case the demand $v = v ; \mathsf{L}$ can be dropped, since it holds because the identity relation and the universal relation from $[\mathbf{1} \leftrightarrow \mathbf{1}]$ coincide.

To treat mappings with more than one argument relation-algebraically, we will use constructions related to direct products, viz. projection relations, products and pairings. Their formal introduction is postponed to Sect. 5.

3 Relation-Algebraic Version of the Recursion Theorem

In this section we formulate the recursion theorem of Dedekind in the language of relation algebra and present a proof that combines relation-algebraic calculations and Scott induction. We start with the following definition of a relational Peano structure. In a similar form its axioms can be found already in [2]. Since the Dedekind recursion theorem is a theorem on sets, in Definition 3.1 and all results we will prove in the remainder of the paper we consider relations as set-theoretic ones. But we will use only the operations of (axiomatic) relation algebra and its laws. As a consequence, our results remain true in this more general setting.

Definition 3.1. *A triple* (N, z, S) *is called a* relational Peano structure *if* N *is a non-empty set,* $z : N \leftrightarrow \mathbf{1}$ *is a relational point,* $S : N \leftrightarrow N$ *is a univalent, total and injective relation,* $S\,;z = \mathsf{O}$ *and for all relational vectors* $v : N \leftrightarrow \mathbf{1}$ *from* $z \cup S^{\mathsf{T}};v \subseteq v$ *it follows that* $v = \mathsf{L}$.

Compared with the notion of a Peano structure formulated in the introduction we see that the relational point $z : N \leftrightarrow \mathbf{1}$ models the zero element and the univalent, total and injective relation $S : N \leftrightarrow N$ equals the injective successor mapping. The equation $S\,;z = \mathsf{O}$ is the relation-algebraic version of the second formula of (1) and that for all relational vectors $v : N \leftrightarrow \mathbf{1}$ from $z \cup S^{\mathsf{T}};v \subseteq v$ it follows $v = \mathsf{L}$ is the relation-algebraic version of the third formula of (1). To be able to prove totality of relations by means of Scott induction, in the next lemma (following [2]) we specify the last axiom of a relational Peano structure as a least fixpoint equation. Notice, that in the remainder of the paper monotonicity of a mapping on relations always supposes inclusion as order.

Lemma 3.1. *Assume* $z : N \leftrightarrow \mathbf{1}$ *to be a relational vector,* $S : N \leftrightarrow N$ *to be a relation and the mapping* g *to be defined as follows:*

$$g : [N \leftrightarrow \mathbf{1}] \to [N \leftrightarrow \mathbf{1}] \qquad\qquad g(v) = z \cup S^{\mathsf{T}};v \tag{8}$$

Then g *is monotone. Furthermore, we have* $\mu(g) = \mathsf{L}$ *iff for all relational vectors* $v : N \leftrightarrow \mathbf{1}$ *from* $z \cup S^{\mathsf{T}};v \subseteq v$ *it follows that* $v = \mathsf{L}$.

Proof. The monotonicity of the mapping g follows from the monotonicity of union and composition. To show the second claim, we calculate as follows:

$$\begin{aligned}
\mu(g) = \mathsf{L} &\iff \bigcap\{v \in [N \leftrightarrow \mathbf{1}] \mid g(v) \subseteq v\} = \mathsf{L} && \text{fixpoint theorem}\\
&\iff \bigcap\{v \in [N \leftrightarrow \mathbf{1}] \mid z \cup S^{\mathsf{T}};v \subseteq v\} = \mathsf{L} && \text{by (8)}\\
&\iff \forall\, v \in [N \leftrightarrow \mathbf{1}] : z \cup S^{\mathsf{T}};v \subseteq v \Rightarrow v = \mathsf{L} && \Box
\end{aligned}$$

Having specified Peano structures in the language of relation algebra, we now consider the two formulae of the recursive definition of the mapping $f : N \to A$ via (2). If we model the element $z \in N$ by the relational point $z : N \leftrightarrow \mathbf{1}$ of a relational Peano structure (N, z, S), use the univalent, total and injective relation $S : N \leftrightarrow N$ instead of the injective successor mapping $s : N \to N$, model the element $c \in A$ by the relational point $c : A \leftrightarrow \mathbf{1}$, take the mapping $F : A \to A$ as univalent and total relation from $[A \leftrightarrow A]$ and take the mapping $f : N \to A$ as univalent and total relation from $[N \leftrightarrow A]$, then the two formulae of (2) are relation-algebraically specified as follows:

$$z\,;c^{\mathsf{T}} \subseteq f \qquad\qquad S\,;f = f\,;F \tag{9}$$

As next result we show how the two formulae of (9) can be specified by a single fixpoint equation.

Lemma 3.2. *Assume* (N, z, S) *to be a relational Peano structure,* $c : A \leftrightarrow \mathbf{1}$ *to be a relational point,* $F : A \leftrightarrow A$ *to be univalent and total and the mapping* h *to be defined as follows:*

$$h : [N \leftrightarrow A] \to [N \leftrightarrow A] \qquad\qquad h(X) = z\,;c^{\mathsf{T}} \cup S^{\mathsf{T}}\,;X\,;F \tag{10}$$

Then h is monotone and $\mu(h) : N \leftrightarrow A$ is total. Furthermore, for all univalent and total relations $f : N \leftrightarrow A$ the two formulae of (9) hold iff $f = h(f)$.

Proof. The monotonicity of the mapping h follows again from the monotonicity of union and composition.

With regard to the totality of the relation $\mu(h)$ we prove $\mu(g) \subseteq \mu(h) ; \mathsf{L}$, with the mapping g defined by (8). We apply Scott induction (of the form (4)) with the predicate P on the direct product $[N \leftrightarrow \mathbf{1}] \times [N \leftrightarrow A]$ defined by $P(v, X)$ iff $v \subseteq X ; \mathsf{L}$, for all relational vectors $v : N \leftrightarrow \mathbf{1}$ and relations $X : N \leftrightarrow A$. Since the two equations $\alpha(v) = v$ and $\beta(X) = X ; \mathsf{L}$ define two $\bigcup$-distributive mappings $\alpha : [N \leftrightarrow \mathbf{1}] \to [N \leftrightarrow \mathbf{1}]$ and $\beta : [N \leftrightarrow A] \to [N \leftrightarrow \mathbf{1}]$, respectively, the predicate P is admissible due to the criterion mentioned in Sect. 2.

A proof of the induction base $P(\mathsf{O}, \mathsf{O})$ is trivial. For a proof of the induction step, assume an arbitrary relational vector $v : N \leftrightarrow \mathbf{1}$ and an arbitrary relation $X : N \leftrightarrow A$ such that $P(v, X)$ holds. Then we get $P(g(v), h(X))$ by the following calculation:

$$
\begin{aligned}
g(v) &= z \cup S^{\mathsf{T}} ; v && \text{by (8)} \\
&\subseteq z \cup S^{\mathsf{T}} ; X ; \mathsf{L} && \text{as } P(v, X) \\
&= z \cup S^{\mathsf{T}} ; X ; F ; \mathsf{L} && F \text{ total} \\
&= z ; \mathsf{L} \cup S^{\mathsf{T}} ; X ; F ; \mathsf{L} && z \text{ relational point (i.e., vector)} \\
&= z ; c^{\mathsf{T}} ; \mathsf{L} \cup S^{\mathsf{T}} ; X ; F ; \mathsf{L} && c \text{ relational point (i.e., surjective)} \\
&= (z ; c^{\mathsf{T}} \cup S^{\mathsf{T}} ; X ; F) ; \mathsf{L} && \\
&= h(X) ; \mathsf{L} && \text{by (10)}
\end{aligned}
$$

Therefore, we have $P(\mu(g), \mu(h))$, i.e., $\mu(g) \subseteq \mu(h) ; \mathsf{L}$. Now, $\mathsf{L} = \mu(h) ; \mathsf{L}$ follows from the last axiom of a relational Peano structure and Lemma 3.1.

For a proof of the remaining claim, assume an arbitrary univalent and total relation $f : N \leftrightarrow A$ to be given. To show implication "$\Longrightarrow$", suppose the two formulae of (9) to be true. We start with the following calculation:

$$
\begin{aligned}
h(f) &= z ; c^{\mathsf{T}} \cup S^{\mathsf{T}} ; f ; F && \text{by (10)} \\
&= z ; c^{\mathsf{T}} \cup S^{\mathsf{T}} ; S ; f && \text{second formula of (9)} \\
&\subseteq z ; c^{\mathsf{T}} \cup f && S \text{ univalent} \\
&= f && \text{first formula of (9)}
\end{aligned}
$$

In combination with Tarski's fixpoint theorem from $h(f) \subseteq f$ we get $\mu(h) \subseteq f$. Now, the desired equation $f = h(f)$ follows from the univalence of f, the totality of $\mu(h)$, inclusion $\mu(h) \subseteq f$ and implication (5). With regard to implication "$\Longleftarrow$", assume $f = h(f)$. The following proof of the first formula of (9) uses definition (10) of the mapping h and $f = h(f)$:

$$
z ; c^{\mathsf{T}} \subseteq z ; c^{\mathsf{T}} \cup S^{\mathsf{T}} ; f ; F = h(f) = f
$$

The second formula of (9) is shown by the following calculation:

$$\begin{aligned}
S;f &= S;h(f) && \text{as } f = h(f)\\
&= S;(z;c^{\mathsf{T}} \cup S^{\mathsf{T}};f;F) && \text{by (10)}\\
&= S;z;c^{\mathsf{T}} \cup S;S^{\mathsf{T}};f;F\\
&= S;S^{\mathsf{T}};f;F && \text{axiom } S;z = \mathsf{O}\\
&= f;F && S \text{ total and injective}
\end{aligned}$$

□

Notice, that in this proof only the univalence of the relation f is used. But from $\mu(h) \subseteq f$ and the totality of $\mu(h)$ the totality of f follows. For F only totality is applied. Now, we are able to prove the following relation-algebraic version of the recursion theorem of Dedekind. Here univalence of F is used, too.

Theorem 3.1. *Let (N, z, S) be a relational Peano structure, $c : A \leftrightarrow \mathbf{1}$ be a relational point and $F : A \leftrightarrow A$ be univalent and total. Then there exists precisely one univalent and total relation $f : N \leftrightarrow A$ that satisfies the two formulae of (9), viz. the least fixpoint $\mu(h)$ of the mapping h of (10).*

Proof. From Lemma 3.2 we already know that $\mu(h)$ is total. To prove that $\mu(h)$ is also univalent, we use Scott induction (of the form (3)) with the predicate P on the set $[N \leftrightarrow A]$ defined by $P(X)$ iff $X^{\mathsf{T}};X \subseteq \mathsf{I}$, for all relations $X : N \leftrightarrow A$. To verify that P is admissible, assume the subset $\mathcal{C}$ of $[N \leftrightarrow A]$ to be a chain of univalent relations. Then the following calculation shows that also the union (i.e., least upper bound) $\bigcup\mathcal{C}$ is a univalent relation:

$$\begin{aligned}
(\textstyle\bigcup\mathcal{C})^{\mathsf{T}};(\bigcup\mathcal{C}) &= (\textstyle\bigcup\{R^{\mathsf{T}} \mid R \in \mathcal{C}\});(\bigcup\mathcal{C})\\
&= \textstyle\bigcup\{R^{\mathsf{T}};(\bigcup\mathcal{C}) \mid R \in \mathcal{C}\}\\
&= \textstyle\bigcup\{\bigcup\{R^{\mathsf{T}};S \mid S \in \mathcal{C}\} \mid R \in \mathcal{C}\}\\
&\subseteq \mathsf{I} && \text{see below}
\end{aligned}$$

The last step uses $\bigcup\{R^{\mathsf{T}};S \mid S \in \mathcal{C}\} \subseteq \mathsf{I}$, for all relations $R \in \mathcal{C}$. This inclusion holds as, given any $R \in \mathcal{C}$, it holds that $R^{\mathsf{T}};S \subseteq \mathsf{I}$, for all relations $S \in \mathcal{C}$. The latter, in turn, follows from the chain property of $\mathcal{C}$ and since all relations of $\mathcal{C}$ are univalent. Namely, given any $S \in \mathcal{C}$, in case $R \subseteq S$ we get $R^{\mathsf{T}};S \subseteq S^{\mathsf{T}};S \subseteq \mathsf{I}$ and in case $S \subseteq R$ we get $R^{\mathsf{T}};S \subseteq R^{\mathsf{T}};R \subseteq \mathsf{I}$.

A proof of the induction base $P(\mathsf{O})$ is obvious. To show the induction step, assume an arbitrary relation $X : N \leftrightarrow A$ with $P(X)$. Then $P(h(X))$ holds because of the following calculation:

$$\begin{aligned}
h(X)^{\mathsf{T}};h(X) &= (z;c^{\mathsf{T}} \cup S^{\mathsf{T}};X;F)^{\mathsf{T}};(z;c^{\mathsf{T}} \cup S^{\mathsf{T}};X;F) && \text{by (10)}\\
&= (c;z^{\mathsf{T}} \cup F^{\mathsf{T}};X^{\mathsf{T}};S);(z;c^{\mathsf{T}} \cup S^{\mathsf{T}};X;F)\\
&= c;z^{\mathsf{T}};z;c^{\mathsf{T}} \cup c;z^{\mathsf{T}};S^{\mathsf{T}};X;F \cup{}\\
&\quad\; F^{\mathsf{T}};X^{\mathsf{T}};S;z;c^{\mathsf{T}} \cup F^{\mathsf{T}};X^{\mathsf{T}};S;S^{\mathsf{T}};X;F\\
&\subseteq \mathsf{I} && \text{see below.}
\end{aligned}$$

Concerning the last step, $c;z^{\mathsf{T}};z;c^{\mathsf{T}} \subseteq c;\mathsf{L};c^{\mathsf{T}} = c;c^{\mathsf{T}} \subseteq \mathsf{I}$ uses that c is a relational point (i.e., an injective relational vector). Equation $c;z^{\mathsf{T}};S^{\mathsf{T}};X;F = \mathsf{O}$

follows from $z^\mathsf{T}; S^\mathsf{T} = (S\,;z)^\mathsf{T} = \mathsf{O}$, where the axiom $S\,;z = \mathsf{O}$ of a relational Peano structure is applied. Also $F^\mathsf{T}; X^\mathsf{T}; S\,; z\,; c^\mathsf{T} = \mathsf{O}$ follows from this axiom. Finally, for $F^\mathsf{T}; X^\mathsf{T}; S\,; S^\mathsf{T}; X\,; F \subseteq F^\mathsf{T}; X^\mathsf{T}; X\,; F \subseteq F^\mathsf{T}; F \subseteq \mathsf{I}$ we use that S is injective, X is univalent (due to the induction hypothesis $P(X)$) and F is univalent.

Because of $\mu(h) = h(\mu(h))$ and since $\mu(h)$ is univalent and total, from implication "$\Longleftarrow$" of Lemma 3.2 we get that the two formulae of (9) hold for the univalent and total relation $\mu(h)$, that is, we have $z\,; c^\mathsf{T} \subseteq \mu(h)$ and $S\,; \mu(h) = \mu(h)\,; F$.

To show that $\mu(h)$ is the only univalent and total relation from $[N \leftrightarrow A]$ that satisfies the two formulae of (9), let an arbitrary univalent and total relation $f : N \leftrightarrow A$ be given such that $z\,; c^\mathsf{T} \subseteq f$ and $S\,; f = f\,; F$. Then implication "$\Longrightarrow$" of Lemma 3.2 shows $f = h(f)$, from which $\mu(h) \subseteq f$ follows. This inclusion, the univalence of f, the totality of $\mu(h)$ and implication (5) yield $\mu(h) = f$. □

The proofs of Lemma 3.2 and Theorem 3.1 contain the decisive ideas which also will be used in Sect. 5 for proving the variants of Theorem 3.1 we have mentioned in the introduction.

4 An Application: The Isomorphism Theorem

Besides the recursion theorem a second important result of [6] is the nowadays called Dedekind isomorphism theorem (see [6], Satz 132). In modern terminology it says that for each pair of Peano structures (N, z, s) and (N_1, z_1, s_1) there exists a bijective mapping $\Phi : N \to N_1$ with the following two properties:

$$\Phi(z) = z_1 \qquad\qquad \forall x \in N : \Phi(s(x)) = s_1(\Phi(x)) \tag{11}$$

When translated into the language of relation algebra with relational Peano structures (N, z, S) and (N_1, z_1, S_1), the bijective mapping $\Phi : N \to N_1$ becomes a univalent, total, injective and surjective relation $\Phi : N \leftrightarrow N_1$ for which the following relation-algebraic versions of the two formulae of (11) hold:

$$z\,; z_1^\mathsf{T} \subseteq \Phi \qquad\qquad S\,; \Phi = \Phi\,; S_1 \tag{12}$$

To prove the existence of such a relation Φ, we consider the monotone mapping h of (10), where the set A is instantiated by N_1, the relational point c is instantiated by $z_1 : N_1 \leftrightarrow \mathbf{1}$ and the relation F is instantiated by $S_1 : N_1 \leftrightarrow N_1$. So, the mapping we consider is given as follows:

$$h_1 : [N \leftrightarrow N_1] \to [N \leftrightarrow N_1] \qquad\qquad h_1(X) = z\,; {z_1}^\mathsf{T} \cup S^\mathsf{T}; X\,; S_1 \tag{13}$$

Furthermore, we define Φ as least fixpoint of h_1, i.e. by $\Phi := \mu(h_1) : N \leftrightarrow N_1$. Then from Theorem 3.1 we get that Φ is the only univalent and total relation from $[N \leftrightarrow N_1]$ that satisfies the two formulae of (12). So, it remains to verify Φ as injective and surjective. To this end, we consider the following monotone mapping h_2 (that is again a specific instance of the mapping h of (10)):

$$h_2 : [N_1 \leftrightarrow N] \to [N_1 \leftrightarrow N] \qquad\qquad h_2(Y) = z_1\,; z^\mathsf{T} \cup S_1^\mathsf{T}; Y\,; S \tag{14}$$

It is easy to verify that the mapping $t : [N \leftrightarrow N_1] \rightarrow [N_1 \leftrightarrow N]$, defined by $t(X) = X^{\mathsf{T}}$ for all $X : N \leftrightarrow N_1$, is a lower adjoint of a Galois connection between the complete lattices $([N \leftrightarrow N_1], \subseteq)$ and $([N_1 \leftrightarrow N], \subseteq)$ and that $t \circ h_1 = h_2 \circ t$. Hence, the μ-fusion theorem of the fixpoint calculus (see [13]) yields

$$\Phi^{\mathsf{T}} = \mu(h_1)^{\mathsf{T}} = t(\mu(h_1)) = \mu(h_2).$$

This equation and the univalence and totality of $\mu(h_2)$ (a consequence of Theorem 3.1) yield the injectivity and surjectivity of Φ. Altogether, we have shown the following relation-algebraic version of the Dedekind isomorphism theorem.

Theorem 4.1. *Assume (N, z, S) and (N_1, z_1, S_1) to be relational Peano structures. Then there exists precisely one univalent, total, injective and surjective relation $\Phi : N \leftrightarrow N_1$ that satisfies the two formulae of (12), viz. the least fixpoint $\mu(h_1)$ of the mapping h_1 of (13).*

5 Variants of the Relation-Algebraic Recursion Theorem

When defining a mapping on natural numbers (or on a Peano structure) recursively, it frequently possesses, besides the argument that controls the recursion, additional arguments. An example is the following recursive definition of the addition-mapping $add : N \times N \rightarrow N$ on a Peano structure (N, z, s), where the first argument of add controls the recursion:

$$\forall y \in N : add(z, y) = y \qquad \forall x \in N, y \in N : add(s(x), y) = s(add(x, y)) \tag{15}$$

Since the original Dedekind recursion theorem only treats the recursive definition of unary mappings, it cannot immediately be applied to show that there exists precisely one mapping $add : N \times N \rightarrow N$ for which the two formulae of (15) hold. Therefore, in the following we present a corresponding variant – in terms of sets as well as in terms of relation algebra. To simplify the presentation, we consider mappings of the kind $f : N \times B \rightarrow A$ only. Taking B as a direct product $\prod_{i=1}^{n} B_i$, this also covers the case of mappings with more than two arguments.

The set-theoretic variant of the Dedekind recursion theorem we have in mind is as follows: Let (N, z, s) be a Peano structure, A and B be non-empty sets and mappings $d : B \rightarrow A$ and $G : A \rightarrow A$ be given. Then there exists precisely one mapping $f : N \times B \rightarrow A$ that satisfies the following two formulae:

$$\forall y \in B : f(z, y) = d(y) \qquad \forall x \in N, y \in B : f(s(x), y) = G(f(x, y)) \tag{16}$$

If this statement is translated into the language of relation algebra, with a relational Peano structure (N, z, S) and the mappings d and G as univalent and total relations, then we obtain the following variant of Theorem 3.1.

Theorem 5.1. *Assume (N, z, S) to be a relational Peano structure and $d : B \leftrightarrow A$ and $G : A \leftrightarrow A$ to be univalent and total. Then there exists precisely one univalent and total relation $f : N \times B \leftrightarrow A$ that satisfies the following two formulae:*

$$[\![z \,;\mathsf{L}, d]\!] \subseteq f \qquad (S \otimes \mathsf{I}) \,; f = f \,; G \tag{17}$$

The construction $[\![z\,;\mathsf{L},d]\!]$ of the first formula of (17) is known as the *left pairing* or *strict join* of the point $z\,;\mathsf{L} : N \leftrightarrow A$ and the relation $d : B \leftrightarrow A$. Using point-wise notation, it relates $(x_1,x_2) \in N \times B$ with $y \in A$ iff $x_1\,(z\,;\mathsf{L})\,y$ and $x_2\,d\,y$. In other words, it relates (x_1,x_2) with y iff x_1 is the zero element and d maps x_2 to y. The construction $S \otimes \mathsf{I}$ of the second formula of (17) is called the *product* or *parallel composition* of the relations $S : N \leftrightarrow N$ and $\mathsf{I} : B \leftrightarrow B$. In a point-wise notation it relates $(x_1,x_2) \in N \times B$ with $(y_1,y_2) \in N \times B$ iff $x_1\,S\,y_1$ and $x_2\,\mathsf{I}\,y_2$. Hence, the relation $S \otimes \mathsf{I} : N \times B \leftrightarrow N \times B$ is the relational counterpart of the product $s \otimes \mathsf{I} : N \times B \rightarrow N \times B$ of the successor mapping $s : N \rightarrow N$ with the identity relation / mapping on the set B in the sense of Sect. 2.

Using relation-algebraic specifications of the two projection relations, left pairings and products and following the lines of the proof of Theorem 3.1, also Theorem 5.1 can be proved with purely relation-algebraic means. To do so, we start with the relation-algebraic definitions $[\![z\,;\mathsf{L},d]\!] := \pi\,;z\,;\mathsf{L} \cap \rho\,;d : N \times B \leftrightarrow A$ of the left pairing and $S \otimes \mathsf{I} := \pi\,;S\,;\pi^{\mathsf{T}} \cap \rho\,;\mathsf{I}\,;\rho^{\mathsf{T}} : N \times B \leftrightarrow N \times B$ of the product, where $\pi : N \times B \leftrightarrow N$ and $\rho : N \times B \leftrightarrow B$ are the projection relations of the direct product $N \times B$. Up to isomorphism, the latter are specified relation-algebraically by the following four axioms (see also [2,14]):

$$\pi^{\mathsf{T}}\,;\pi = \mathsf{I} \qquad \rho^{\mathsf{T}}\,;\rho = \mathsf{I} \qquad \pi\,;\pi^{\mathsf{T}} \cap \rho\,;\rho^{\mathsf{T}} = \mathsf{I} \qquad \pi^{\mathsf{T}}\,;\rho = \mathsf{L} \tag{18}$$

From the first three formulae of (18) we get that the projection relations π and ρ are univalent, total and surjective. The definition of the left pairing $[\![z\,;\mathsf{L},d]\!]$ and the univalence of ρ and d imply

$$[\![z\,;\mathsf{L},d]\!]^{\mathsf{T}}\,;[\![z\,;\mathsf{L},d]\!] \subseteq (\rho\,;d)^{\mathsf{T}}\,;\rho\,;d = d^{\mathsf{T}}\,;\rho^{\mathsf{T}}\,;\rho\,;d \subseteq \mathsf{I},$$

such that $[\![z\,;\mathsf{L},d]\!]$ is univalent. Also the product $S \otimes \mathsf{I}$ is univalent, since its definition and the univalence of π and S imply

$$(S \otimes \mathsf{I})^{\mathsf{T}}\,;(S \otimes \mathsf{I}) \subseteq (\pi\,;S\,;\pi^{\mathsf{T}})^{\mathsf{T}}\,;\pi\,;S\,;\pi^{\mathsf{T}} = \pi\,;S^{\mathsf{T}}\,;\pi^{\mathsf{T}}\,;\pi\,;S\,;\pi^{\mathsf{T}} \subseteq \pi\,;\pi^{\mathsf{T}}$$

and its definition and the univalence of ρ imply

$$(S \otimes \mathsf{I})^{\mathsf{T}}\,;(S \otimes \mathsf{I}) \subseteq (\rho\,;\rho^{\mathsf{T}})^{\mathsf{T}}\,;\rho\,;\rho^{\mathsf{T}} = \rho\,;\rho^{\mathsf{T}}\,;\rho\,;\rho^{\mathsf{T}} \subseteq \rho\,;\rho^{\mathsf{T}}$$

such that the third formula of (18) yields $(S \otimes \mathsf{I})^{\mathsf{T}}\,;(S \otimes \mathsf{I}) \subseteq \pi\,;\pi^{\mathsf{T}} \cap \rho\,;\rho^{\mathsf{T}} = \mathsf{I}$. Similar calculations show that $S \otimes \mathsf{I}$ is total and injective.

After these preparations we are able to prove Theorem 5.1 with relation-algebraic means. The idea is the same as in case of Theorem 3.1. We define an appropriate monotone mapping on the set $[N \times B \leftrightarrow A]$ and verify that its least fixpoint satisfies the desired properties. Concretely, we consider the least fixpoint $\mu(h_3) : N \times B \leftrightarrow A$ of the following monotone mapping:

$$h_3 : [N \times B \leftrightarrow A] \rightarrow [N \times B \leftrightarrow A] \qquad h_3(X) = [\![z\,;\mathsf{L},d]\!] \cup (S \otimes \mathsf{I})^{\mathsf{T}}\,;X\,;G \tag{19}$$

The proof that $\mu(h_3)$ is the only univalent and total relation from $[N \times B \leftrightarrow A]$ that satisfies the two formulae of (17) is given by the following four lemmas.

Lemma 5.1. *The relation $\mu(h_3)$ is total.*

Proof. Besides the mapping h_3 of (19) we additionally consider the mapping g of (8) and show $\pi ; \mu(g) \subseteq \mu(h_3) ; \mathsf{L}$ using Scott induction (of the form (4)). Then the totality of the projection relation $\pi : N \times B \leftrightarrow N$ and the last axiom of a Peano structure in combination with Lemma 3.1 yield $\mathsf{L} = \pi ; \mathsf{L} = \pi ; \mu(g) \subseteq \mu(h_3) ; \mathsf{L}$.

For the Scott induction we use the admissible predicate P on the direct product $[N \leftrightarrow \mathbf{1}] \times [N \times B \leftrightarrow A]$ defined by $P(v, X)$ iff $\pi ; v \subseteq X ; \mathsf{L}$, for all relational vectors $v : N \leftrightarrow \mathbf{1}$ and relations $X : N \times B \leftrightarrow A$. The induction base $P(\mathsf{O}, \mathsf{O})$ is obvious. To show the induction step, assume an arbitrary relational vector $v : N \leftrightarrow \mathbf{1}$ and an arbitrary relation $X : N \times B \leftrightarrow A$ with $P(v, X)$. Then the following calculation shows $P(g(v), h_3(X))$:

$$\begin{aligned}
\pi ; g(v) &= \pi ; (z \cup S^{\mathsf{T}} ; v) && \text{by (8)}\\
&= \pi ; z \cup \pi ; S^{\mathsf{T}} ; v\\
&= (\pi ; z ; \mathsf{L} \cap \rho ; d ; \mathsf{L}) \cup (\pi ; S^{\mathsf{T}} \cap \rho ; \mathsf{L}) ; v && z \text{ vector and } \rho, d \text{ total}\\
&= (\pi ; z ; \mathsf{L} \cap \rho ; d) ; \mathsf{L} \cup (\pi ; S^{\mathsf{T}} \cap \rho ; \mathsf{L}) ; v && \text{by (7)}\\
&= [\![z ; \mathsf{L}, d]\!] ; \mathsf{L} \cup (\pi ; S^{\mathsf{T}} \cap \rho ; \mathsf{L}) ; v && \text{definition left pairing}\\
&= [\![z ; \mathsf{L}, d]\!] ; \mathsf{L} \cup (\pi ; S^{\mathsf{T}} \cap \rho ; \rho^{\mathsf{T}} ; \pi) ; v && \text{last formula of (18)}\\
&= [\![z ; \mathsf{L}, d]\!] ; \mathsf{L} \cup (\pi ; S^{\mathsf{T}} ; \pi^{\mathsf{T}} \cap \rho ; \rho^{\mathsf{T}}) ; \pi ; v && \pi \text{ univalent and (6)}\\
&= [\![z ; \mathsf{L}, d]\!] ; \mathsf{L} \cup (\pi ; S ; \pi^{\mathsf{T}} \cap \rho ; \rho^{\mathsf{T}})^{\mathsf{T}} ; \pi ; v\\
&= [\![z ; \mathsf{L}, d]\!] ; \mathsf{L} \cup (S \otimes \mathsf{I})^{\mathsf{T}} ; \pi ; v && \text{definition product}\\
&\subseteq [\![z ; \mathsf{L}, d]\!] ; \mathsf{L} \cup (S \otimes \mathsf{I})^{\mathsf{T}} ; X ; \mathsf{L} && \text{by } P(v, X)\\
&= [\![z ; \mathsf{L}, d]\!] ; \mathsf{L} \cup (S \otimes \mathsf{I})^{\mathsf{T}} ; X ; G ; \mathsf{L} && G \text{ total}\\
&= ([\![z ; \mathsf{L}, d]\!] \cup (S \otimes \mathsf{I})^{\mathsf{T}} ; X ; G) ; \mathsf{L}\\
&= h_3(X) ; \mathsf{L} && \text{by (19)} \quad \square
\end{aligned}$$

Lemma 5.2. *The relation $\mu(h_3)$ is univalent.*

Proof. We use Scott induction (of the form (3)) with the admissible predicate P on the set $[N \times B \leftrightarrow A]$ defined by $P(X)$ iff $X^{\mathsf{T}} ; X \subseteq \mathsf{I}$, for all relations $X : N \times B \leftrightarrow A$. The induction base $P(\mathsf{O})$ is obvious. To verify the induction step, let an arbitrary relation $X : N \times B \leftrightarrow A$ be given such that $P(X)$ is true. To get $P(h_3(X))$, we start with the calculation

$$\begin{aligned}
h_3(X)^{\mathsf{T}} ; h_3(X) &= ([\![z ; \mathsf{L}, d]\!] \cup (S \otimes \mathsf{I})^{\mathsf{T}} ; X ; G)^{\mathsf{T}} ; ([\![z ; \mathsf{L}, d]\!] \cup (S \otimes \mathsf{I})^{\mathsf{T}} ; X ; G)\\
&= [\![z ; \mathsf{L}, d]\!]^{\mathsf{T}} ; [\![z ; \mathsf{L}, d]\!] \cup [\![z ; \mathsf{L}, d]\!]^{\mathsf{T}} ; (S \otimes \mathsf{I})^{\mathsf{T}} ; X ; G \cup\\
&\quad\; G^{\mathsf{T}} ; X^{\mathsf{T}} ; (S \otimes \mathsf{I}) ; [\![z ; \mathsf{L}, d]\!] \cup G^{\mathsf{T}} ; X^{\mathsf{T}} ; (S \otimes \mathsf{I}) ; (S \otimes \mathsf{I})^{\mathsf{T}} ; X ; G\\
&\subseteq \mathsf{I} \cup (G^{\mathsf{T}} ; X^{\mathsf{T}} ; (S \otimes \mathsf{I}) ; [\![z ; \mathsf{L}, d]\!])^{\mathsf{T}} \cup G^{\mathsf{T}} ; X^{\mathsf{T}} ; (S \otimes \mathsf{I}) ; [\![z ; \mathsf{L}, d]\!]
\end{aligned}$$

using the definition (19) of the mapping h_3, some basic laws of relation algebra, that $[\![z ; \mathsf{L}, d]\!]$, G and X are univalent (X because of the induction hypothesis $P(X)$) and that $S \otimes \mathsf{I}$ is injective. Now, the definitions of $S \otimes \mathsf{I}$ and $[\![z ; \mathsf{L}, d]\!]$, the univalence of π and the axiom $S ; z = \mathsf{O}$ of a relational Peano structure imply

$$(S \otimes \mathsf{I}) ; [\![z ; \mathsf{L}, d]\!] \subseteq \pi ; S ; \pi^{\mathsf{T}} ; \pi ; z ; \mathsf{L} \subseteq \pi ; S ; z ; \mathsf{L} = \mathsf{O} \qquad (20)$$

and in combination with the above calculation we get $P(h_3(X))$. $\square$

Lemma 5.3. *The relation $\mu(h_3)$ satisfies the two formulae of (17).*

Proof. Using the definition of the mapping h_3 by (19) and that $\mu(h_3)$ is a fixpoint of h_3 we obtain

$$[\![z\,;\mathsf{L},d]\!] \subseteq [\![z\,;\mathsf{L},d]\!] \cup (S\otimes \mathsf{I})^{\mathsf{T}}\,;\mu(h_3)\,;G = h_3(\mu(h_3)) = \mu(h_3),$$

such that $\mu(h_3)$ satisfies the first formula of (17). The calculation

$$\begin{aligned}
(S\otimes \mathsf{I})\,;\mu(h_3) &= (S\otimes \mathsf{I})\,;h_3(\mu(h_3)) && \mu(h_3)\text{ fixpoint}\\
&= (S\otimes \mathsf{I})\,;([\![z\,;\mathsf{L},d]\!] \cup (S\otimes \mathsf{I})^{\mathsf{T}}\,;\mu(h_3)\,;G) && \text{by (19)}\\
&= (S\otimes \mathsf{I})\,;[\![z\,;\mathsf{L},d]\!] \cup (S\otimes \mathsf{I})\,;(S\otimes \mathsf{I})^{\mathsf{T}};\mu(h_3)\,;G && \\
&= \mathsf{O} \cup (S\otimes \mathsf{I})\,;(S\otimes \mathsf{I})^{\mathsf{T}};\mu(h_3)\,;G && \text{by (20)}\\
&= \mu(h_3)\,;G && S\otimes \mathsf{I}\text{ total, inj.}
\end{aligned}$$

shows that $\mu(h_3)$ satisfies the second formula of (17), too. □

Lemma 5.4. *Assume $f : N\times B \leftrightarrow A$ to be univalent and total. If it satisfies the two formulae of (17), then $f = \mu(h_3)$.*

Proof. We start with the calculation

$$\begin{aligned}
h_3(f) &= [\![z\,;\mathsf{L},d]\!] \cup (S\otimes \mathsf{I})^{\mathsf{T}}\,;f\,;G && \text{by (19)}\\
&\subseteq f \cup (S\otimes \mathsf{I})^{\mathsf{T}}\,;f\,;G && \text{first formula of (17)}\\
&= f \cup (S\otimes \mathsf{I})^{\mathsf{T}}\,;(S\otimes \mathsf{I})\,;f && \text{second formula of (17)}\\
&\subseteq f && S\otimes \mathsf{I}\text{ univalent}
\end{aligned}$$

and get $\mu(h_3) \subseteq f$ due to Tarski's fixpoint theorem. This, the univalence of f, the totality of $\mu(h_3)$ (i.e., Lemma 5.1) and implication (5) yield $\mu(h_3) = f$. □

A second situation in which the original Dedekind recursion theorem is not applicable is given when the result of the expression $f(s(x))$ not only depends on the value of $f(x)$ but also on x. The following recursive definition of a mapping $sum : N \to N$ that computes the sum $\sum_{i=z}^{n} i$ by means of the addition-mapping add of (15) is an example for this:

$$sum(z) = z \qquad \forall x \in N : sum(s(x)) = add(sum(x), s(x))$$

Such a situation also requires a generalisation of the original Dedekind recursion theorem. The mapping F has to be binary and of type $F : A\times N \to A$ and the recursive definition (2) of $f : N \to A$ changes to the following one:

$$f(z) = c \qquad \forall x \in N : f(s(x)) = F(f(x), x) \tag{21}$$

When translated into the language of relation algebra, the statement that there exists precisely one mapping $f : N \to A$ that satisfies the two formulae of (21), leads to the following second variant of Theorem 3.1.

Theorem 5.2. *Assume (N, z, S) to be a relational Peano structure, $c : A \leftrightarrow \mathbf{1}$ to be a relational point and $F : A \times N \leftrightarrow A$ to be univalent and total. Then there exists precisely one univalent and total relation $f : N \leftrightarrow A$ that satisfies the following two formulae:*

$$z ; c^{\mathsf{T}} \subseteq f \qquad\qquad S ; f = [f, \mathsf{I}]\!] ; F \tag{22}$$

Also Theorem 5.2 uses a relation-algebraic notion we have not introduced in Sect. 2. This is the *right pairing* or *fork* $[f, \mathsf{I}]\!]$ of the two relations $f : N \leftrightarrow A$ and $\mathsf{I} : N \leftrightarrow N$. Relation-algebraically it is defined by $[f, \mathsf{I}]\!] := f ; \pi^{\mathsf{T}} \cap \mathsf{I} ; \rho^{\mathsf{T}} = [\![f^{\mathsf{T}}, \mathsf{I}^{\mathsf{T}}]^{\mathsf{T}} : N \leftrightarrow A \times N$, where $\pi : A \times N \leftrightarrow A$ and $\rho : A \times N \leftrightarrow N$ are now the projection relations of the direct product $A \times N$; see again [2,14]. From the definition of right pairings (generalising that of $[f, \mathsf{I}]\!]$ to arbitrary relations with the same source) and the axioms (18) we get that right pairings of univalent relations are univalent and a composition with a univalent relation from the left distributes over right pairings. These are the only new relation-algebraic properties we will use in the following proof of Theorem 5.2. Concretely, we show that the least fixpoint $\mu(h_4) : N \leftrightarrow A$ of the monotone mapping

$$h_4 : [N \leftrightarrow A] \to [N \leftrightarrow A] \qquad\qquad h_4(X) = z ; c^{\mathsf{T}} \cup [S^{\mathsf{T}} ; X, S^{\mathsf{T}}]\!] ; F \tag{23}$$

is the only univalent and total relation from $[N \leftrightarrow A]$ that satisfies the two formulae of (22). As in case of Theorem 5.1 this is obtained by four lemmas.

Lemma 5.5. *The relation $\mu(h_4)$ is total.*

Proof. By means of the mapping g of (8) and Scott induction (of the form (4)) we show $\mu(g) \subseteq \mu(h_4) ; \mathsf{L}$, since then the totality of $\mu(g)$ yields $\mathsf{L} = \mu(h_4) ; \mathsf{L}$. We apply the admissible predicate P on the direct product $[N \leftrightarrow \mathbf{1}] \times [N \leftrightarrow A]$ defined by $P(v, X)$ iff $v \subseteq X ; \mathsf{L}$, for all relational vectors $v : N \leftrightarrow \mathbf{1}$ and relations $X : N \leftrightarrow A$. The induction base $P(\mathsf{O}, \mathsf{O})$ is obvious. To verify the induction step, let an arbitrary relational vector $v : N \leftrightarrow \mathbf{1}$ and an arbitrary relation $X : N \leftrightarrow A$ be given such that $P(v, X)$ holds. Then we have $P(g(v), h_4(X))$ due to the following calculation:

$$\begin{aligned}
g(v) &= z \cup S^{\mathsf{T}} ; v && \text{by (8)}\\
&\subseteq z \cup S^{\mathsf{T}} ; X ; \mathsf{L} && \text{by } P(v, X)\\
&= z \cup S^{\mathsf{T}} ; (X \cap \rho^{\mathsf{T}} ; \pi) ; \mathsf{L} && \text{last formula of (18)}\\
&= z \cup S^{\mathsf{T}} ; (X ; \pi^{\mathsf{T}} \cap \rho^{\mathsf{T}}) ; \pi ; \mathsf{L} && \pi \text{ univalent and (6)}\\
&= z \cup S^{\mathsf{T}} ; [X, \mathsf{I}]\!] ; \pi ; \mathsf{L} && \text{definition right pairing}\\
&= z \cup [S^{\mathsf{T}} ; X, S^{\mathsf{T}}]\!] ; \pi ; \mathsf{L} && \text{prop. right pairing } (S \text{ inj.})\\
&= z ; c^{\mathsf{T}} ; \mathsf{L} \cup [S^{\mathsf{T}} ; X, S^{\mathsf{T}}]\!] ; \pi ; \mathsf{L} && z \text{ and } c \text{ relational points}\\
&= z ; c^{\mathsf{T}} ; \mathsf{L} \cup [S^{\mathsf{T}} ; X, S^{\mathsf{T}}]\!] ; F ; \mathsf{L} && \pi \text{ and } F \text{ total}\\
&= (z ; c^{\mathsf{T}} \cup [S^{\mathsf{T}} ; X, S^{\mathsf{T}}]\!] ; F) ; \mathsf{L}\\
&= h_4(X) ; \mathsf{L} && \text{by (23)}
\end{aligned}$$

□

Lemma 5.6. *The relation $\mu(h_4)$ is univalent.*

Proof. We use Scott induction (of the form (3)) with the admissible predicate P on the set $[N \leftrightarrow A]$ defined by $P(X)$ iff $X^{\mathsf{T}} ; X \subseteq \mathsf{I}$, for all relations $X : N \leftrightarrow A$. The induction base $P(\mathsf{O})$ holds trivially. To show the induction step, let an arbitrary relation $X : N \leftrightarrow A$ with $P(X)$ be given. For $P(h_4(X))$ we then start with the following calculation that uses the definition of h_4 via (23):

$$\begin{aligned}
h_4(X)^{\mathsf{T}}; h_4(X) &= (z ; c^{\mathsf{T}} \cup [\![S^{\mathsf{T}}; X, S^{\mathsf{T}}]\!] ; F)^{\mathsf{T}}; (z ; c^{\mathsf{T}} \cup [\![S^{\mathsf{T}}; X, S^{\mathsf{T}}]\!] ; F) \\
&= c ; z^{\mathsf{T}}; z ; c^{\mathsf{T}} \cup c ; z^{\mathsf{T}} ; [\![S^{\mathsf{T}} ; X, S^{\mathsf{T}}]\!] ; F \cup \\
&\quad\; F^{\mathsf{T}}; [\![S^{\mathsf{T}} ; X, S^{\mathsf{T}}]\!]^{\mathsf{T}}; z ; c^{\mathsf{T}} \cup F^{\mathsf{T}}; [\![S^{\mathsf{T}}; X, S^{\mathsf{T}}]\!]^{\mathsf{T}}; [\![S^{\mathsf{T}} ; X, S^{\mathsf{T}}]\!] ; F \\
&= c ; z^{\mathsf{T}}; z ; c^{\mathsf{T}} \cup c ; z^{\mathsf{T}}; [\![S^{\mathsf{T}}; X, S^{\mathsf{T}}]\!] ; F \cup \\
&\quad\; (c ; z^{\mathsf{T}}; [\![S^{\mathsf{T}}; X, S^{\mathsf{T}}]\!]; F)^{\mathsf{T}} \cup F^{\mathsf{T}}; [\![S^{\mathsf{T}}; X, S^{\mathsf{T}}]\!]^{\mathsf{T}}; [\![S^{\mathsf{T}} ; X, S^{\mathsf{T}}]\!] ; F
\end{aligned}$$

From the proof of Theorem 3.1 we know already the inclusion $c ; z^{\mathsf{T}} ; z ; c^{\mathsf{T}} \subseteq \mathsf{I}$. That the second and third expression of the above union are empty follows from

$$z^{\mathsf{T}}; [\![S^{\mathsf{T}} ; X, S^{\mathsf{T}}]\!] = z^{\mathsf{T}}; (S^{\mathsf{T}} ; X ; \pi^{\mathsf{T}} \cap S^{\mathsf{T}}; \rho^{\mathsf{T}}) \subseteq z^{\mathsf{T}}; S^{\mathsf{T}} ; X ; \pi^{\mathsf{T}} = \mathsf{O},$$

where the definition of $[\![S^{\mathsf{T}} ; X, S^{\mathsf{T}}]\!]$ and the axiom $S ; z = \mathsf{O}$ of a relational Peano structure are applied. To conclude the proof of $h_4(X)^{\mathsf{T}}; h_4(X) \subseteq \mathsf{I}$ we calculate

$$F^{\mathsf{T}}; [\![S^{\mathsf{T}}; X, S^{\mathsf{T}}]\!]^{\mathsf{T}}; [\![S^{\mathsf{T}}; X, S^{\mathsf{T}}]\!] ; F \subseteq F^{\mathsf{T}}; F \subseteq \mathsf{I},$$

where the right pairing $[\![S^{\mathsf{T}}; X, S^{\mathsf{T}}]\!]$ is univalent as its components $S^{\mathsf{T}} ; X$ and S^{T} are univalent due to the injectivity of S and the induction hypothesis $P(X)$ and F is univalent by assumption. □

Lemma 5.7. *The relation $\mu(h_4)$ satisfies the two formulae of (22).*

Proof. The first formula of (22) holds due to

$$z ; c^{\mathsf{T}} \subseteq z ; c^{\mathsf{T}} \cup [\![S^{\mathsf{T}}; \mu(h_4), S^{\mathsf{T}}]\!] ; F = h_4(\mu(h_4)) = \mu(h_4),$$

where the definition (23) of the mapping h_4 and that $\mu(h_4)$ is a fixpoint of h_4 are applied. By means of the calculation

$$\begin{aligned}
S ; \mu(h_4) &= S ; h_4(\mu(h_4)) && \mu(h_4) \text{ fixpoint} \\
&= S ; (z ; c^{\mathsf{T}} \cup [\![S^{\mathsf{T}}; \mu(h_4), S^{\mathsf{T}}]\!] ; F) && \text{by (23)} \\
&= S ; z ; c^{\mathsf{T}} \cup S ; [\![S^{\mathsf{T}}; \mu(h_4), S^{\mathsf{T}}]\!] ; F && \\
&= S ; [\![S^{\mathsf{T}}; \mu(h_4), S^{\mathsf{T}}]\!] ; F && \text{as } S ; z = \mathsf{O} \\
&= S ; S^{\mathsf{T}}; [\![\mu(h_4), \mathsf{I}]\!] ; F && \text{prop. right pairing } (S \text{ inj.}) \\
&= [\![\mu(h_4), \mathsf{I}]\!] ; F && S \text{ total and injective}
\end{aligned}$$

the second formula of (22) is verified. □

Lemma 5.8. *Assume $f : N \leftrightarrow A$ to be univalent and total. If it satisfies the two formulae of (22), then $f = \mu(h_4)$.*

Proof. First, we calculate as follows:

$$\begin{aligned}
h_4(f) &= z ; c^{\mathsf{T}} \cup [S^{\mathsf{T}}; f, S^{\mathsf{T}}] ; F && \text{by (23)} \\
&\subseteq f \cup [S^{\mathsf{T}}; f, S^{\mathsf{T}}] ; F && \text{first formula of (22)} \\
&= f \cup S^{\mathsf{T}}; [f, \mathsf{I}] ; F && \text{property right pairing (S injective)} \\
&= f \cup S^{\mathsf{T}}; S ; f && \text{second formula of (22)} \\
&= f && \text{S univalent}
\end{aligned}$$

This yields $\mu(h_4) \subseteq f$ due to Tarski's fixpoint theorem. From this inclusion, the univalence of f, the totality of $\mu(h_4)$ (i.e., Lemma 5.5) and implication (5) we get $\mu(h_4) = f$. □

6 Concluding Remarks

In this paper we have presented a simple new proof of the Dedekind recursion theorem that is based on a relation-algebraic specification of the notions in question and combines relation-algebraic laws and equational reasoning with Scott induction. As a simple application and using the same means, we also have shown the Dedekind isomorphism theorem. Finally, we have treated two cases where the original Dedekind recursion theorem is not applicable and have presented two variants of the relation-algebraic version of the recursion theorem. Their proofs are variations of that of the latter theorem.

It is interesting to look at how Dedekind in [6] treats mappings with more than one argument. From his explanations to the definition of addition and multiplication (see [6], Erklärung 135 and Erklärung 147) it becomes clear that he implicitly uses currying and uncurrying. For example, in case of addition he does not define a binary operation. Instead of that he fixes a natural number m and then uses Satz 126 to define recursively a unary mapping that yields for each natural number n the sum $m + n$. In Erklärung 147 he explicitly speaks of an infinite set of new mappings on N found in such a way. Also in the proof of Satz 4 of [9], where again addition is recursively defined, implicitly currying and uncurrying are used. These approaches can be generalised as given below.

Consider the recursive definition

$$g(z) = d \qquad \forall x \in N : g(s(x)) = G \circ g(x) \tag{24}$$

of a mapping $g : N \to A^B$, where (N, z, s) is a Peano structure and the mappings $d : B \to A$ and $G : A \to A$ are given. Since g is unary, the original Dedekind recursion theorem shows that (24) has a unique solution. We have to instantiate in (2) the set A by the set of mappings A^B, the element c by the mapping d, the mapping F by the higher-order mapping $F : A^B \to A^B$ with $F(h) = G \circ h$, for all $h \in A^B$, and the mapping f by the mapping g. From the unique solution g of (24) we then obtain the unique solution f of (16) via uncurrying, i.e., by defining $f : N \times B \to A$ as $f(x, y) = g(x)(y)$, for all $x \in N$ and $y \in B$, or, shorter, by $f := \mathit{curry}^{-1}(g)$, where curry^{-1} is the inverse of the well-known bijective currying-mapping curry. The definition of f and curry^{-1} and the formulae of

(24) allow to show that f satisfies the two formulae of (16). That it is the only mapping with this property can be shown by means of the definition of f and *curry*, the formulae of (24) and $curry^{-1}(curry(h)) = h$, for all $h : N \times B \rightarrow A$.

All proofs of Sect. 3 to Sect. 5 are very formal and its decisive parts consist of equational reasoning using laws of relation algebra. These are ideal prerequisites for mechanised theorem proving. Concerning mathematical theorems, in the last years especially the proof assistant tools Coq and Isabelle/HOL have been used in this respect. A prominent example is the formal verification of Atle Selberg's elementary proof of the *Prime Number Theorem* in Isabelle/HOL; see [1]. For the future we also plan a mechanised verification of the proofs of this paper using Coq or Isabelle/HOL.

Acknowledgement. I thank the referees for carefully reading the paper and for their very valuable suggestions.

References

1. Avigad, J., Donnelly, K., Gray, D., Raff, P.: A formally verified proof of the prime number theorem. ACM Trans. Comput. Log. **9**(1:2), 1–23 (2007)
2. Berghammer, R., Zierer, H.: Relational algebraic semantics of deterministic and nondeterministic programs. Theor. Comput. Sci. **43**, 123–147 (1986)
3. Berghammer, R.: Mathematik für die Informatik, 3rd edn. Springer, Heidelberg (2019). https://doi.org/10.1007/978-3-658-16712-7
4. Birkhoff, G.: Lattice Theory, 3rd edn. American Mathematical Society Colloquium Publications, American Mathematical Society, New York (1967)
5. Davey, B.A., Priestley, H.A.: Introduction to Lattices and Order, 2nd edn. Cambridge University Press, Cambridge (2002)
6. Dedekind, R.: Was sind und was sollen die Zahlen? Vieweg, Braunschweig (1888)
7. Kolman, V.: Zahlen. Walter de Gruyter, Berlin (2016)
8. Lamm, C.: Karl Grandjot und der Dedekindsche Rekursionssatz. Mitt. DMV **24**(1), 37–45 (2016)
9. Landau, E.: Grundlagen der Analysis. Akademische Verlagsgesellschaft, Leipzig (1930)
10. Loeckx, J., Sieber, K.: The Foundations of Program Verification, 2nd edn. Wiley, Chichester (1987)
11. Lorenzen, P.: Die Definition durch vollständige Induktion. Monatsh. Math. Phys **47**(1), 356–358 (1939)
12. Maddux, R.D.: Relation Algebras. Elsevier, Amsterdam (2006)
13. Mathematics Program Construction Group: Fixed-point calculus. Inf. Process. Lett. **53**(3), 131–136 (1995)
14. Schmidt, G., Ströhlein, T.: Relations and Graphs. Monographs on Theoretical Computer Science EATCS. Springer, Heidelberg (1993). https://doi.org/10.1007/978-3-642-77968-8
15. Tarski, A.: On the calculus of relations. J. Symb. Log. **6**(3), 73–89 (1941)
16. Tarski, A.: A lattice-theoretical fixpoint theorem and its applications. Pac. J. Math. **5**(2), 285–309 (1955)

Hardness of Network Satisfaction for Relation Algebras with Normal Representations

Manuel Bodirsky and Simon Knäuer(✉)

Institut für Algebra, TU Dresden, 01062 Dresden, Germany
simon.knaeuer@tu-dresden.de

Abstract. We study the computational complexity of the general network satisfaction problem for a finite relation algebra A with a normal representation B. If B contains a non-trivial equivalence relation with a finite number of equivalence classes, then the network satisfaction problem for A is NP-hard. As a second result, we prove hardness if B has domain size at least three and contains no non-trivial equivalence relations but a symmetric atom a with a forbidden triple (a, a, a), that is, $a \not\leq a \circ a$. We illustrate how to apply our conditions on two small relation algebras.

1 Introduction

Many computational problems in temporal and spatial reasoning can be formulated as network satisfaction problems for a fixed finite relation algebra [Dün05,RN07,BJ17]. Famous examples of finite relation algebras that have been studied in this context are the Point Algebra, the Left Linear Point Algebra, Allen's Interval Algebra, RCC5, and RCC8, just to name a few; much more material about relation algebras can be found in [HH02]. Robin Hirsch [Hir96] asked in 1996 the *Really Big Complexity Problem (RBCP)*: can we classify the computational complexity of the network satisfaction problem for every finite relation algebra? For example, the network satisfaction problem for the Point Algebra and the Left Linear Point Algebra are polynomial-time tractable [VKvB89,BK07], while it is NP-complete for the other relation algebras mentioned above [All83,RN99]. A finite relation algebra with an undecidable network satisfaction problem has been found by Hirsch [Hir99].

An important notion in the theory of representability of finite relation algebras are *normal representations*, i.e., representations that are fully universal,

M. Bodirsky—The author has received funding from the European Research Council under the European Community's Seventh Framework Programme (FP7/2007-2013 Grant Agreement no. 681988, CSP-Infinity).
S. Knäuer—The author is supported by DFG Graduiertenkolleg 1763 (QuantLA).

U. Fahrenberg et al. (Eds.): RAMiCS 2020, LNCS 12062, pp. 31–46, 2020.
https://doi.org/10.1007/978-3-030-43520-2_3

square, and homogeneous [Hir96]. The network satisfaction problem for a relation algebra with a normal representation can be seen as the constraint satisfaction problem for an infinite structure $\mathfrak{B}$ that is homogeneous and finitely bounded (these concepts from model theory will be introduced in Sect. 3). The network satisfaction problem is in this case in NP and a complexity dichotomy has been conjectured [BPP14]. There is even a promising candidate condition for the boundary between NP-completeness and containment in P; the condition can be phrased in several equivalent ways [BKO+17,Bod18]. However, this conjecture has not yet been verified for the homogeneous finitely bounded structures that arise as the normal representation of a finite relation algebra.

We present some first steps towards a solution to the RBCP for relation algebras **A** with a normal representation $\mathfrak{B}$. Our approach is to study the automorphism group Aut($\mathfrak{B}$) of $\mathfrak{B}$ and to identify properties that imply hardness. Because of the homogeneity of $\mathfrak{B}$, one can translate back and forth between properties of **A** and properties of Aut($\mathfrak{B}$). For example, Aut($\mathfrak{B}$) is primitive if and only if **A** contains no equivalence relation which is different from the *trivial* equivalence relations Id and 1. Specifically, we show that the network satisfaction problem for **A** is NP-complete if

- Aut($\mathfrak{B}$) is primitive, $|B| > 2$ and **A** has a symmetric atom a with a forbidden triple (a, a, a), that is, $a \not\leq a \circ a$ (Sect. 5);
- Aut($\mathfrak{B}$) has a congruence with at least two but finitely many equivalence classes (Sect. 6).

In our proof we use the so-called *universal-algebraic approach* which has recently led to a full classification of the computational complexity of constraint satisfaction problems for $\mathfrak{B}$ if the domain of $\mathfrak{B}$ is *finite* [Bul17,Zhu17]. The central insight is that the complexity of the CSP is for finite $\mathfrak{B}$ fully determined by the *polymorphism clone* Pol($\mathfrak{B}$) of $\mathfrak{B}$. This result extends to homogeneous structures with finite relational signature (more generally, to ω-categorical structures [BN06]). Both of our hardness proofs come from the technique of factoring Pol($\mathfrak{B}$) with respect to a congruence with finitely many classes, and using known hardness conditions from corresponding finite-domain constraint satisfaction problems. The article is fully self-contained: we introduce the network satisfaction problem (Sect. 2), normal representations (Sect. 3), and the universal algebraic approach (Sect. 4).

2 The (General) Network Satisfaction Problem

Network satisfaction problems have been introduced in [LM94], capturing well-known computational problems, e.g., for Allen's Interval Algebra [All83]; see [Dün05] for a survey. An *algebra* in the sense of universal algebra is a set together with operations on this set, each equipped with an arity $n \in \mathbb{N}$. In this context, operations of arity zero are viewed as constants. The *type* of an algebra is a tuple that represents the arities of the operations. For the definitions concerning relation algebras, we basically follow [Mad06].

Definition 1. *Let D be a set and $E \subseteq D^2$ an equivalence relation. Let $(\mathcal{P}(E); \cup, \bar{\ }, 0, 1, \mathrm{Id}, \smile, \circ)$ be an algebra of type $(2,1,0,0,0,1,2)$ with the following operations:*

1. $A \cup B := \{(x,y) \mid (x,y) \in A \text{ or } (x,y) \in B\}$,
2. $\bar{A} := E \setminus A$,
3. $0 := \emptyset$,
4. $1 := E$,
5. $\mathrm{Id} := \{(x,x) \mid x \in D\}$,
6. $A^\smile := \{(x,y) \mid (y,x) \in A\}$,
7. $A \circ B := \{(x,z) \mid \exists y \in D : (x,y) \in A \text{ and } (y,z) \in B\}$.

A subalgebra of $(\mathcal{P}(E); \cup, \bar{\ }, 0, 1, \mathrm{Id}, \smile, \circ)$ is called a proper relation algebra.

A *representable relation algebra* is an algebra of type $(2,1,0,0,0,1,2)$ that is isomorphic (as an algebra) to a proper relation algebra. We denote algebras by bold letters, like $\mathbf{A}$; the underlying domain of an algebra $\mathbf{A}$ is denoted with the regular letter A. An algebra $\mathbf{A}$ is finite if A is finite. We do not need the more general definition of an *(abstract) relation algebra* (for a definition see for example [Mad06]) because the network satisfaction problem for relation algebras that are not representable is trivial. We use the language of model theory to define *representations* of relation algebras; the definition is essentially the same as the one given in [Mad06].

Definition 2. *A relational structure $\mathfrak{B}$ is called a* representation *of a relation algebra $\mathbf{A}$ if*

- *$\mathfrak{B}$ is an A-structure with domain B (i.e., each element $a \in A$ is used as a relation symbol denoting a binary relation $a^{\mathfrak{B}}$ on B);*
- *there exists an equivalence relation $E \subseteq B^2$ such that the set of relations of $\mathfrak{B}$ is the domain of a subalgebra of $(\mathcal{P}(E); \cup, \bar{\ }, 0, 1, \mathrm{Id}, \smile, \circ)$;*
- *the map that sends $a \in A$ to $a^{\mathfrak{B}}$ is an isomorphism between $\mathbf{A}$ and this subalgebra.*

Remark 3. For a relation algebra $\mathbf{A} = (A; \cup, \bar{\ }, 0, 1, \mathrm{Id}, \smile, \circ)$ the algebra $(A; \cup, \bar{\ }, 0, 1)$ is a Boolean algebra. With respect to this algebra there is a partial ordering on the elements of a relation algebra. We denote this with $\subseteq$ since in proper relation algebras this ordering is with respect to set inclusion. The minimal non-empty relations with respect to $\subseteq$ are called the *atomic relations* or *atoms*; we denote the set of atoms of $\mathbf{A}$ by A_0.

Definition 4. *Let $\mathbf{A}$ be a relation algebra. An $\mathbf{A}$-network $(V; f)$ is a finite set of nodes V together with a function $f \colon V \times V \to A$.*

Let $\mathfrak{B}$ be a representation of $\mathbf{A}$. An $\mathbf{A}$-network $(V; f)$ is satisfiable in $\mathfrak{B}$ *if there exists an assignment $s \colon V \to B$ such that for all $x, y \in V$*

$$(s(x), s(y)) \in f(x,y)^{\mathfrak{B}}.$$

An $\mathbf{A}$-network $(V; f)$ is satisfiable *if there exists some representation $\mathfrak{B}$ of $\mathbf{A}$ such that $(V; f)$ is satisfiable in $\mathfrak{B}$.*

Definition 5. *The* (general) network satisfaction problem *for a finite relation algebra* **A**, *denoted by* NSP(**A**), *is the problem of deciding whether a given* **A**-*network is satisfiable.*

3 Normal Representations and CSPs

We recall a connection between network satisfaction problems and constraint satisfaction problems that is presented in more detail in [BJ17, Bod18].

Definition 6 (from [Hir96]). *Let* **A** *be a relation algebra. An* **A**-*network* $(V; f)$ *is called* atomic *if the image of* f *only contains atoms and if*

$$f(a,c) \subseteq f(a,b) \circ f(b,c).$$

The last line ensures a "local consistency" of the atomic **A**-network with respect to the multiplication rules in the relation algebra **A**. This property is in the literature sometimes called "closedness" of an **A**-network [Hir97].

Definition 7 (from [Hir96]). *A representation* $\mathfrak{B}$ *of a relation algebra* **A** *is called*

- fully universal *if every atomic* **A**-*network is satisfiable in* $\mathfrak{B}$;
- square *if* $1^{\mathfrak{B}} = B^2$;
- homogeneous *if every isomorphism of finite substructures of* $\mathfrak{B}$ *can be extended to an automorphism;*
- normal *if it is fully universal, square and homogeneous.*

If a relation algebra **A** has a normal representation $\mathfrak{B}$ then the problem of deciding whether an **A**-network is satisfiable in *some* representation reduces to a question whether it is satisfiable in the concrete representation $\mathfrak{B}$. Such decision problems are known as constraint satisfaction problems, which are formally defined in the following.

Definition 8. *Let* $\mathfrak{B}$ *be a* τ-*structure for a finite relational signature* τ. *The* constraint satisfaction problem *of* $\mathfrak{B}$ *is the problem of deciding for a given finite* τ-*structure* $\mathfrak{C}$ *whether there exists a homomorphism from* $\mathfrak{C}$ *to* $\mathfrak{B}$.

To formulate the connection between NSPs and CSPs, we have to give a translation between networks and structures. On the one hand we may view an **A**-network $(V; f)$ as an A-structure $\mathfrak{C}$ with domain $C := V$ where $(a,b) \in f(a,b)^{\mathfrak{C}}$. On the other hand we can transform an A-structure $\mathfrak{C}$ into an **A**-network $(V; f)$ with $V = C$ and by defining the network function $f(x,y)$ for $x, y \in C$ as follows: let X be the set of all relations that hold on (x,y) in $\mathfrak{C}$. If X is non-empty we define $f(x,y) := \bigcup X$; otherwise $f(x,y) := 1$.

Proposition 9 (see [Bod18]). *Let* $\mathfrak{B}$ *be a normal representation of a finite relation algebra* **A**. *Then* NSP(**A**) *and* CSP($\mathfrak{B}$) *are the same problem (up to the translation showed above).*

The following is an important notion in model theory and the study of infinite-domain CSPs. Let $\mathcal{F}$ be a finite set of finite τ-structures. Then Forb($\mathcal{F}$) is the class of all finite τ-structures that embed no $\mathfrak{C} \in \mathcal{F}$. A class $\mathcal{C}$ of finite τ-structures is called *finitely bounded* if $\mathcal{C} = \text{Forb}(\mathcal{F})$ for a finite set $\mathcal{F}$. A structure $\mathfrak{B}$ is called *finitely bounded* if the class of finite structures that embed into $\mathfrak{B}$ is finitely bounded.

Proposition 10 (see [Bod18]). *Let* **A** *be a finite relation algebra with a normal representation* $\mathfrak{B}$. *Then* $\mathfrak{B}$ *is finitely bounded and* CSP($\mathfrak{B}$) *and* NSP(**A**) *are in NP.*

4 The Universal Algebraic Approach

This section gives a short overview of the important notions and concepts for the universal-algebraic approach to the computational complexity of CSPs.

4.1 Clones

We start with the definition of an operation clone.

Definition 11. *Let B be some set. Then $\mathcal{O}_B^{(n)}$ denotes the set of n-ary operations on B and $\mathcal{O}_B := \bigcup_{n\in\mathbb{N}} \mathcal{O}_B^{(n)}$. A set $\mathscr{C} \subseteq \mathcal{O}_B$ is called a* operation clone *(on B) if it contains all projections and is closed under composition, that is, for every $f \in \mathscr{C}$ and all $g_1, \ldots, g_k \in \mathscr{C}$ the n-ary operation $f(g_1, \ldots, g_k)$ with*

$$f(g_1, \ldots, g_k)(x_1, \ldots, x_n) := f(g_1(x_1, \ldots, x_n), \ldots, g_k(x_1, \ldots, x_n))$$

is also in $\mathscr{C}$. We denote the k-ary operations of $\mathscr{C}$ by $\mathscr{C}^{[k]}$.

Definition 12. *Let $\mathfrak{B}$ be a relational structure. Then f* preserves *a relation R of $\mathfrak{B}$ if the component-wise application of f on tuples $r_1, \ldots, r_k \in R$ results in a tuple of the relation. If f preserves all relations of $\mathfrak{B}$ then f is called a* polymorphism *of $\mathfrak{B}$. The set of all polymorphisms of arity $k \in \mathbb{N}$ is denoted by $\text{Pol}^{(k)}(\mathfrak{B})$ and $\text{Pol}(\mathfrak{B}) := \bigcup_{k\in\mathbb{N}} \text{Pol}^{(k)}(\mathfrak{B})$ is called the* polymorphism clone of $\mathfrak{B}$.

Polymorphisms are closed under the composition and a projection is always a polymorphism, therefore a polymorphism clone is indeed an operation clone.

Definition 13 *Let $\mathscr{C}$ and $\mathscr{D}$ be operation clones. A function $\mu\colon \mathscr{C} \to \mathscr{D}$ is called* minor-preserving *if it maps every operation to an operation of the same arity and satisfies for every $f \in \text{Pol}^k(\mathscr{C})$ and all projections $p_1, \ldots, p_k \in \text{Pol}^{(n)}(\mathfrak{B})$ the following identity:*

$$\mu(f(p_1, \ldots, p_k)) = \mu(f)(p_1, \ldots, p_k).$$

Operation clones $\mathscr{C}$ on countable sets B can be equipped with the following complete ultrametric d. Assume that $B = \mathbb{N}$. For two polymorphisms f and g of different arity we define $d(f, g) = 1$. If f and g are both of arity k we have

$$d(f,g) := 2^{-\min\{n \in \mathbb{N} \mid \exists s \in \{1,\dots,n\}^k : f(s) \neq g(s)\}}.$$

The following is a straightforward consequence of the definition.

Lemma 14. *Let $\mathscr{D}$ be an operation clone on B and $\mathscr{C}$ an operation clone on C and let $\nu\colon \mathscr{D} \to \mathscr{C}$ a map. Then ν is uniformly continuous (u.c.) if and only if*

$$\forall n \geq 1 \; \exists \text{ finite } F \subset D \forall f, g \in \mathscr{D}^{(n)} : f|_F = g|_F \Rightarrow \nu(f) = \nu(g).$$

In order to demonstrate the use of polymorphisms in the study of CSPs we have to define primitive positive formulas. Let τ be a relational signature. A first-order formula $\varphi(x_1, \dots, x_n)$ is called *primitive positive* if it has the form

$$\exists x_{n+1}, \dots, x_m (\varphi_1 \wedge \dots \wedge \varphi_s)$$

where $\varphi_1, \dots, \varphi_s$ are atomic formulas, i.e., formulas of the form $R(y_1, \dots, y_l)$ for $R \in \tau$ and $y_i \in \{x_1, \dots, x_m\}$, of the form $y = y'$ for $y, y' \in \{x_1, \dots x_m\}$, or of the form *false* and *true*. We have the following correspondence between polymorphisms and primitive positive formulas (or relations that are defined by them). Note that all of the statements in the following hold in a more general setting, but we only state them here for normal representations of finite relation algebras.

Theorem 15 (follows from [BN06]). *Let $\mathfrak{B}$ be a normal representation of a finite relation algebra* **A**. *Then the set of primitive positive definable relations in $\mathfrak{B}$ is exactly the set of relations that are preserved by* $\mathrm{Pol}(\mathfrak{B})$.

A special type of polymorphism plays an important role in our analysis.

Definition 16. *Let f be an n-ary operation on a countable set X. Then f is called cyclic if*

$$\forall x_1, \dots x_n \in X : f(x_1, \dots, x_n) = f(x_n, x_1 \dots, x_{n-1}).$$

We write Proj for the operation clone on a two-element set that consists of only the projections.

Theorem 17 (from [BK12,BOP18]). *Let $\mathscr{C}$ be an operation clone on a finite set C. If there exists no minor-preserving map $\mathscr{C} \to$* Proj *then $\mathscr{C}$ contains for every prime $p > |C|$ a p-ary cyclic operation.*

Note that every map between operation clones on finite domains is uniformly continuous.

Theorem 18 (from [BOP18]). *Let $\mathfrak{B}$ be normal representation of a finite relation algebra. If there is a uniformly continuous minor-preserving map* $\mathrm{Pol}(\mathfrak{B}) \to$ Proj, *then* $\mathrm{CSP}(\mathfrak{B})$ *is NP-complete.*

4.2 Canonical Functions

Let $\mathfrak{B}$ be a normal representation of a finite relation algebra $\mathbf{A}$.

Definition 19. *Let $a_1, \dots, a_k \in A$. Then $(a_1, \dots, a_k)^{\mathfrak{B}}$ denotes a binary relation on B^k such that for $x, y \in B^k$*

$$(a_1, \dots, a_k)^{\mathfrak{B}}(x, y) :\Leftrightarrow \bigwedge_{i \in \{1,\dots,k\}} a_i^{\mathfrak{B}}(x_i, y_i).$$

Recall that A_0 denotes the set of atoms of a representable relation algebra $\mathbf{A}$.

Definition 20. *Let $x, y \in B^k$. Since $\mathfrak{B}$ is square there are unique $a_1, \dots, a_k \in A_0$ such that $(a_1, \dots, a_k)^{\mathfrak{B}}(x, y)$. Then we call $(a_1, \dots, a_k)^{\mathfrak{B}}$ the* configuration *of (x, y). If $a_1, \dots, a_k \in X \subseteq A_0$ then $(a_1, \dots, a_k)$ is called an X-*configuration.

We specialise the concept of *canonical functions* (see, e.g., [BP16]) to our setting.

Definition 21. *Let f be a k-ary operation on B. Let $X \subseteq A_0$ and let T be the set of all X-configurations. Then f is called X-*canonical *if there exists a map $\overline{f}\colon T \to A_0$ such that for every $(a_1, \dots, a_k) \in T$ and $(x, y) \in (a_1, \dots, a_k)^{\mathfrak{B}}$ we have $(f(x), f(y)) \in \left(\overline{f}(a_1, \dots, a_k)\right)^{\mathfrak{B}}$. If $X = A_0$ then f is called* canonical.

An operation $f\colon B^n \to B$ is called *conservative* if for all $x_1, \dots, x_n \in B$

$$f(x_1, \dots, x_n) \in \{x_1, \dots, x_n\}.$$

If $\mathfrak{B}$ is a finite structure such that every polymorphism of $\mathfrak{B}$ is conservative, then CSP($\mathfrak{B}$) has been classified already before the proof of the Feder-Vardi conjecture, and there are several proofs [Bul03, Bul14, Bar11]. The polymorphisms of normal representations of finite relation algebras satisfy a strong property that resembles conservativity.

Proposition 22. *Let $\mathfrak{B}$ be a normal representation. Then every $f \in \mathrm{Pol}^{(n)}$ is* edge-conservative, *that is, for all $x, y \in B^n$ with configuration $(a_1, \dots, a_n)^{\mathfrak{B}}$ it holds that*

$$(f(x), f(y)) \in \left(\bigcup_{i \in \{1,\dots,n\}} a_i\right)^{\mathfrak{B}}.$$

Proof. By definition, $b := \bigcup_{i \in \{1,\dots,n\}} a_i$ is part of the signature of $\mathfrak{B}$. Moreover, for every $i \in \{1, \dots, n\}$ we have that $(x_i, y_i) \in b^{\mathfrak{B}}$ by the assumption on the configuration of x and y. Then $(f(x), f(y)) \in b^{\mathfrak{B}}$ because f preserves $b^{\mathfrak{B}}$. □

5 Finitely Many Equivalence Classes

In the following, **A** denotes a finite relation algebra with a normal representation $\mathfrak{B}$.

Theorem 23. *Suppose that $e \in A$ is such that $e^{\mathfrak{B}}$ is a non-trivial equivalence relation with finitely many classes. Then* CSP($\mathfrak{B}$) *is NP-complete.*

Proof. We use the notation $n := 1 \backslash e$. Let $\{c_1, \dots, c_m\}$ be a set of representatives of the equivalence classes of $e^{\mathfrak{B}}$. We denote the equivalence class of c_i by $\overline{c_i}$. A k-ary polymorphism $f \in \text{Pol}(\mathfrak{B})$ induces an operation $\overline{f}$ of arity k on $C = \{\overline{c_1}, \dots, \overline{c_m}\}$ in the following way:

$$\overline{f}(\overline{d_1}, \dots \overline{d_k}) := \overline{f(d_1, \dots d_k)}$$

for all $\overline{d_1}, \dots \overline{d_k} \in \{\overline{c_1}, \dots \overline{c_m}\}$. This definition is independent from the choice of the representatives since the polymorphisms preserve the relation $e^{\mathfrak{B}}$. We denote the set of all operations that are induced in this way by operations from Pol($\mathfrak{B}$) by $\mathscr{C}$. It is easy to see that $\mathscr{C}$ is an operation clone on a finite set. Moreover, the mapping μ: Pol($\mathfrak{B}$) $\to \mathscr{C}$ defined by $\mu(f) := \overline{f}$ is a minor-preserving map. To show that μ is uniformly continuous, we use Lemma 14; it suffices to observe that if two k-ary operations $f, g \in \text{Pol}(\mathfrak{B})$ are equal on $F := \{c_1, \dots, c_m\}$, then they induce the same operation on the equivalence classes.

Suppose for contradiction that $\mathscr{C}$ contains a p-ary cyclic operation for every prime $p > m$.

Case 1: $m = 2$. By assumption there exists a ternary cyclic operation $\overline{f} \in \mathscr{C}$. Since $e^{\mathfrak{B}}$ is non-trivial, one of the equivalence classes of $e^{\mathfrak{B}}$ must have size at least two. So we may without loss of generality assume that $\overline{c_1}$ contains at least two elements. Let $c_1' \in \overline{c_1}$ with $c_1 \neq c_1'$. We have that $\overline{f(c_1, c_1, c_2)} = \overline{f(c_2, c_1, c_1)}$ which means that

$$\big(f(c_1, c_1, c_2), f(c_2, c_1, c_1)\big) \in e^{\mathfrak{B}}. \tag{1}$$

On the other hand $(n, \text{Id}, n)^{\mathfrak{B}}\big((c_1, c_1, c_2), (c_2, c_1, c_1)\big)$. Since f is an edge conservative polymorphism we have that

$$\big(f(c_1, c_1, c_2), f(c_2, c_1, c_1)\big) \in (n \cup \text{Id})^{\mathfrak{B}}. \tag{2}$$

Combining (1) and (2) we obtain that

$$f(c_1, c_1, c_2) = f(c_2, c_1, c_1). \tag{3}$$

Similarly, $\overline{f(c_2, c_1, c_1)} = \overline{f(c_1, c_2, c_1)}$. Since f preserves the equivalence relation $e^{\mathfrak{B}}$ we also have $\big(f(c_1, c_2, c_1), f(c_1', c_2, c_1)\big) \in e^{\mathfrak{B}}$. But then $\big(f(c_2, c_1, c_1), f(c_1', c_2, c_1)\big) \in e^{\mathfrak{B}}$ holds. Also note that $(n, n, \text{Id})^{\mathfrak{B}}\big((c_2, c_1, c_1), (c_1', c_2, c_1)\big)$ implies that $\big(f(c_2, c_1, c_1), f(c_1', c_2, c_1)\big) \in (n \cup \text{Id})^{\mathfrak{B}}$. These two facts together imply $f(c_2, c_1, c_1) = f(c_1', c_2, c_1)$. By (3) and the transitivity of equality we get $f(c_1, c_1, c_2) = f(c_1', c_2, c_1)$. But this is impossible because $(e, n, n)^{\mathfrak{B}}((c_1, c_1, c_2), (c_1', c_2, c_1))$ implies that $f(c_1, c_1, c_2) \neq f(c_1', c_2, c_1)$.

Case 2: $m > 2$. Let f be a p-ary cyclic operation for some prime $p > m$. Consider the representatives c_1, c_2 and c_3. By the cyclicity of $\overline{f}$ we have

$$\overline{f(c_1, c_2, \dots, c_1, c_2, c_3)} = \overline{f(c_3, c_1, c_2 \dots, c_1, c_2)}$$

and therefore

$$\big(f(c_1, c_2, \dots, c_1, c_2, c_3), f(c_3, c_1, c_2 \dots, c_1, c_2)\big) \in e^{\mathfrak{B}}. \quad (4)$$

On the other hand,

$$(n, n, n, \dots, n, n)^{\mathfrak{B}}\big((c_1, c_2, \dots, c_1, c_2, c_3), (c_3, c_1, c_2 \dots, c_1, c_2)\big)$$

and since f preserves $n^{\mathfrak{B}}$ we get that

$$\big(f(c_1, c_2, \dots, c_1, c_2, c_3), f(c_3, c_1, c_2 \dots, c_1, c_2)\big) \in n^{\mathfrak{B}},$$

contradicting (4).

We showed that there exists a prime $p > m$ such that $\mathscr{C}$ does not contain a p-ary cyclic polymorphism and therefore Theorem 17 implies the existence of a (uniformly continuous) minor-preserving map $\nu \colon \mathscr{C} \to \mathrm{Proj}$. Since the composition of uniformly continuous minor-preserving maps is again uniformly continuous and minor-preserving, there exists a uniformly continuous minor-preserving map $\nu \circ \mu \colon \mathrm{Pol}(\mathfrak{B}) \to \mathrm{Proj}$. This map implies the NP-hardness of CSP($\mathfrak{B}$) by Theorem 18. □

6 No Non-trivial Equivalence Relations

In this section $\mathbf{A}$ denotes a finite relation algebra with a normal representation $\mathfrak{B}$ with $|B| > 2$.

Definition 24. *The automorphism group* $\mathrm{Aut}(\mathfrak{C})$ *of a relational structure* $\mathfrak{C}$ *is called* primitive *if* $\mathrm{Aut}(\mathfrak{C})$ *does not preserve a non-trivial equivalence relation, i.e., the only equivalence relations that are preserved by* $\mathrm{Aut}(\mathfrak{C})$ *are* Id *and* C^2.

Proposition 25. *Let* a *be an atom of* $\mathbf{A}$. *If* $\mathrm{Aut}(\mathfrak{B})$ *is primitive then* $a \subseteq \mathrm{Id}$ *implies* $a = \mathrm{Id}$.

Proof. If $a \subsetneq \mathrm{Id}$ then

$$c := \mathrm{Id} \cup (a \circ 1 \circ a)$$

would be such that $c^{\mathfrak{B}}$ is a non-trivial equivalence relation. □

Proposition 26. *Let* a *be a symmetric atom of* $\mathbf{A}$ *with* $a \cap \mathrm{Id} = 0$. *If* $\mathrm{Aut}(\mathfrak{B})$ *is primitive then* $a^{\mathfrak{B}} \circ a^{\mathfrak{B}} \neq \mathrm{Id}$.

Proof. Assume for contradiction $a^{\mathfrak{B}} \circ a^{\mathfrak{B}} = \mathrm{Id}^{\mathfrak{B}}$. This implies $(\mathrm{Id} \cup a)^{\mathfrak{B}} \circ (\mathrm{Id} \cup a)^{\mathfrak{B}} \subset (\mathrm{Id} \cup a)^{\mathfrak{B}}$ and therefore $(\mathrm{Id} \cup a)^{\mathfrak{B}}$ is an equivalence relation. Since $\mathfrak{B}$ is primitive $(\mathrm{Id} \cup a)^{\mathfrak{B}} = B^2$. By assumption B contains at least 3 elements. These elements are now all connected by the atomic relation $a^{\mathfrak{B}}$. This is a contradiction to our assumption $a^{\mathfrak{B}} \circ a^{\mathfrak{B}} = \mathrm{Id}^{\mathfrak{B}}$. □

Higman's lemma states that a permutation group G on a set B is primitive if and only if for every two distinct elements $x, y \in B$ the undirected graph with vertex set B and edge set $\{\{\alpha(x), \alpha(y)\} \mid \alpha \in G\}$ is connected (see, e.g., [Cam99]). We need the following variant of this result for $\mathrm{Aut}(\mathfrak{B})$; we also present its proof since we are unaware of any reference in the literature. If $a \in A$ then a sequence $(b_0, \ldots, b_n) \in B^{n+1}$ is called an *a-walk (of length n)* if $(b_i, b_{i+1}) \in a^{\mathfrak{B}}$ for every $i \in \{0, \ldots, n-1\}$ (we count the number of traversed edges rather than the number of vertices when defining the length).

Lemma 27. *Let $a \in A$ be a symmetric atom of* **A** *with $a \cap \mathrm{Id} = 0$ and suppose that $\mathrm{Aut}(\mathfrak{B})$ is primitive. Then there exists an $a^{\mathfrak{B}}$-walk of even length between any $x, y \in B$. Moreover, there exists $k \in \mathbb{N}$ such that for all $x, y \in B$ there exists an $a^{\mathfrak{B}}$-walk of length $2k$ between x and y.*

Proof. If R is a binary relation then $R^k = R \circ R \circ \cdots \circ R$ denotes the k-th relational power of R. The sequence of binary relations $L_n := \mathrm{Id}^{\mathfrak{B}} \cup \bigcup_{k=1}^{n} (a^{\mathfrak{B}})^{2k}$ is non-decreasing by definition and terminates because all binary relations are unions of at most finitely many atoms. Therefore, there exists $k \in \mathbb{N}$ such for all $n \geq k$ we have $L_n = L_k$. Note that L_k is an equivalence relation, namely the relation "there exists an $a^{\mathfrak{B}}$-walk of even length between x and y". Since $\mathfrak{B}$ is primitive L_k must be trivial. If $L_k = B^2$ then there exists an $a^{\mathfrak{B}}$-walk of length $2k$ between any two $x, y \in B$ and we are done. Otherwise,

$$L_k = \{(x, x) \mid x \in B\} = \mathrm{Id}^{\mathfrak{B}} .$$

Since a is symmetric $a^{\mathfrak{B}} \circ a^{\mathfrak{B}} \neq 0$ and $a^{\mathfrak{B}} \circ a^{\mathfrak{B}}$ contains therefore an atom. But then $a^{\mathfrak{B}} \circ a^{\mathfrak{B}} \subseteq L_k$ implies by Proposition 25 $a^{\mathfrak{B}} \circ a^{\mathfrak{B}} = L_k$. This is a contradiction to Proposition 26. □

Lemma 28. *Let $a \in A$ be a symmetric atom of* **A** *such that $\mathrm{Aut}(\mathfrak{B})$ is primitive and (a, a, a) is forbidden. Then all polymorphisms of $\mathfrak{B}$ are $\{\mathrm{Id}, a\}$-canonical.*

In the proof, we need the following notation. Let $a_1, \ldots, a_k \in A$ be such that $a_1 = \ldots = a_j$ and $a_{j+1} = \ldots = a_k$. Instead of writing $(a_1, \ldots, a_n)^{\mathfrak{B}}$ we use the shortcut $(a_1|_j a_{j+1})^{\mathfrak{B}}$.

Proof (of Lemma 28). The following ternary relation R on B is primitive positive definable in $\mathfrak{B}$.

$$R := \{(x_1, x_2, x_3) \in B^3 \mid (a \cup \mathrm{Id})^{\mathfrak{B}}(x_1, x_2) \wedge (a \cup \mathrm{Id})^{\mathfrak{B}}(x_2, x_3) \wedge a^{\mathfrak{B}}(x_1, x_3)\}$$

Observe that $c \in R$ if and only if $a^{\mathfrak{B}}(c_1, c_2) \wedge \mathrm{Id}^{\mathfrak{B}}(c_2, c_3)$ or $\mathrm{Id}^{\mathfrak{B}}(c_1, c_2) \wedge a^{\mathfrak{B}}(c_2, c_3)$.

Let f be a polymorphism of $\mathfrak{B}$ of arity n. Let $x, y, u, v \in B^n$ be arbitrary such that (x, y) and (u, v) have the same $\{\mathrm{Id}, a\}$-configuration. Without loss of generality we may assume that $(a|_j \mathrm{Id})^{\mathfrak{B}}(x, y)$ and $(a|_j \mathrm{Id})^{\mathfrak{B}}(u, v)$. Now consider $p, q \in B^n$ such that $(\mathrm{Id}\,|_j a)^{\mathfrak{B}}(p, q)$ holds.

Note that by the edge-conservativeness of f the following holds:

$$(f(x), f(y)) \in (a \cup \mathrm{Id})^{\mathfrak{B}}, (f(u), f(v)) \in (a \cup \mathrm{Id})^{\mathfrak{B}} \text{ and } (f(p), f(q)) \in (a \cup \mathrm{Id})^{\mathfrak{B}}.$$

By Lemma 27 there exists a $k \in \mathbb{N}$ such that for every $i \in \{1, \ldots, n\}$ there exists an $a^{\mathfrak{B}}$-walk $(s_i^0, \ldots, s_i^k)$ with $s_i^0 = y_i$ and $s_i^k = p_i$. Now consider the following walk in B^n:

$$\begin{aligned}
&(a|_j \mathrm{Id})^{\mathfrak{B}}(x, y)\\
&(\mathrm{Id}\,|_j a)^{\mathfrak{B}}(y, (s_1^0, \ldots s_j^0, s_{j+1}^1, \ldots s_n^1))\\
&(a|_j \mathrm{Id})^{\mathfrak{B}}((s_1^0, \ldots s_j^0, s_{j+1}^1, \ldots s_n^1), (s_1^1, \ldots s_j^1, s_{j+1}^1, \ldots s_n^1))\\
&\quad\vdots\\
&(a|_j \mathrm{Id})^{\mathfrak{B}}((s_1^i, \ldots s_j^i, s_{j+1}^{i+1}, \ldots s_n^{i+1}), (s_1^{i+1}, \ldots s_j^{i+1}, s_{j+1}^{i+1}, \ldots s_n^{i+1}))\\
&(\mathrm{Id}\,|_j a)^{\mathfrak{B}}((s_1^{i+1}, \ldots s_j^{i+1}, s_{j+1}^{i+1}, \ldots s_n^{i+1}), (s_1^{i+1}, \ldots s_j^{i+1}, s_{j+1}^{i+2}, \ldots s_n^{i+2}))\\
&\quad\vdots\\
&(a|_j \mathrm{Id})^{\mathfrak{B}}((s_1^{k-1}, \ldots, s_j^{k-1}, s_{j+1}^k, \ldots, s_n^k), p)\\
&(\mathrm{Id}\,|_j a)^{\mathfrak{B}}(p, q)
\end{aligned}$$

Every three consecutive elements on this walk are component wise in the relation R. Since R is primitive positive definable the polymorphism f preserves R by Theorem 15. This means that f maps this walk on a walk where the atomic relations are an alternating sequence of $a^{\mathfrak{B}}$ and $\mathrm{Id}^{\mathfrak{B}}$, which implies

$$(f(x), f(y)) \in a^{\mathfrak{B}} \Leftrightarrow (f(p), f(q)) \in \mathrm{Id}^{\mathfrak{B}}.$$

If we repeat the same argument with a walk from q to v we get:

$$(f(p), f(q)) \in a^{\mathfrak{B}} \Leftrightarrow (f(u), f(v)) \in \mathrm{Id}^{\mathfrak{B}}.$$

Combining these two equivalences gives us

$$(f(x), f(y)) \in a^{\mathfrak{B}} \Leftrightarrow (f(u), f(v)) \in a^{\mathfrak{B}}.$$

Since the tuples $x, y, u, v \in B^n$ were arbitrary this shows that f is $\{\mathrm{Id}, a\}$-canonical. □

Theorem 29. *Let* $\mathrm{Aut}(\mathfrak{B})$ *be primitive and let* a *be a symmetric atom of* $\mathbf{A}$ *such that* (a, a, a) *is forbidden. Then* $\mathrm{CSP}(\mathfrak{B})$ *is NP-hard.*

∘	Id	a	b
Id	Id	a	b
a	a	$\neg b$	b
b	b	b	$\neg b$

∘	Id	a	b
Id	Id	a	b
a	a	$\neg a$	$0'$
b	b	$0'$	1

Fig. 1. Multiplication tables of relation algebras #13 (left) and #17 (right).

Proof. By Lemma 28 we know that all polymorphisms of $\mathfrak{B}$ are $\{a, \text{Id}\}$-canonical. This means that every $f \in \text{Pol}(\mathfrak{B})$ induces an operation $\overline{f}$ of the same arity on the set $\{a, \text{Id}\}$. Let $\mathscr{C}_2$ be the set of induced operations. Note that $\mathscr{C}_2$ is an operation clone on a Boolean domain. The mapping $\mu\colon \text{Pol}(\mathfrak{B}) \to \mathscr{C}_2$ defined by $\mu(f) := \overline{f}$ is a uniformly continuous minor-preserving map.

Assume for contradiction that there exists a ternary cyclic polymorphism $\overline{s}$ in $\mathscr{C}_2$. Let $x, y, z \in B^3$ be such that

$$\begin{aligned} &(a, a, \text{Id})^{\mathfrak{B}}(x, y),\\ &(\text{Id}, a, a)^{\mathfrak{B}}(y, z),\\ \text{and} \quad &(a, \text{Id}, a)^{\mathfrak{B}}(x, z). \end{aligned}$$

By the cyclicity of the operation $\overline{s}$ and the edge-conservativeness of s we have that either

$$(s(x), s(y)) \in a^{\mathfrak{B}}, (s(y), s(z)) \in a^{\mathfrak{B}} \text{ and } (s(x), s(z)) \in a^{\mathfrak{B}}$$

or

$$(s(x), s(y)) \in \text{Id}^{\mathfrak{B}}, (s(y), s(z)) \in \text{Id}^{\mathfrak{B}} \text{ and } (s(x), s(z)) \in \text{Id}^{\mathfrak{B}}.$$

Since (a, a, a) is forbidden, the second case holds. Note that **A** must have an atom $b \neq \text{Id}$ such that the triple (a, a, b) is allowed, because otherwise a would be an equivalence relation. Now consider $u, v, w \in B^3$ such that

$$\begin{aligned} &(a, a, \text{Id})^{\mathfrak{B}}(u, v),\\ &(\text{Id}, a, a)^{\mathfrak{B}}(v, w),\\ \text{and} \quad &(a, b, a)^{\mathfrak{B}}(u, w). \end{aligned}$$

Since s is $\{a, \text{Id}\}$-canonical and with the observation from before we have

$$(s(u), s(v)) \in \text{Id}^{\mathfrak{B}} \text{ and } (s(v), s(w)) \in \text{Id}^{\mathfrak{B}}.$$

Now the transitivity of equality contradicts $(s(u), s(w)) \in (a \cup b)^{\mathfrak{B}}$.

We conclude that $\mathscr{C}_2$ does not contain a ternary cyclic operation. Since the domain of $\mathscr{C}_2$ has size two, Theorem 17 implies the existence of a u.c. minor-preserving map $\nu\colon \mathscr{C}_2 \to \text{Proj}$. The composition $\nu \circ \mu\colon \text{Pol}(\mathfrak{B}) \to \text{Proj}$ is also a u.c. minor-preserving map and therefore by Theorem 18 the CSP($\mathfrak{B}$) is NP-hard. □

7 Examples

Andréka and Maddux classified *small relation algebras*, i.e., finite relation algebras with at most 3 atoms [AM94]. We consider the complexity of the network satisfaction problem of two of them, namely the relation algebras #13 and #17 (we use the enumeration from [AM94]). Both relation algebras have normal representations (see below) and fall into the scope of our hardness criteria. Cristani and Hirsch [CH04] classified the complexities of the network satisfaction problems for small relation algebras, but due to a mistake the algebras #13 and #17 were left open.

Example 1 (Relation Algebra #13). The relation algebra #13 is given by the multiplication table in Fig. 1. This finite relation algebra has a normal representation $\mathfrak{B}$ defined as follows. Let V_1 and V_2 be countable, disjoint sets. We set $B := V_1 \cup V_2$ and define the following atomic relations:

$$\begin{aligned}
\mathrm{Id}^{\mathfrak{B}} &:= \{(x,x) \in B^2\},\\
a^{\mathfrak{B}} &:= \{(x,y) \in B^2 \setminus \mathrm{Id}^{\mathfrak{B}} \mid (x \in V_1 \wedge y \in V_1) \vee (x \in V_2 \wedge y \in V_2)\},\\
b^{\mathfrak{B}} &:= \{(x,y) \in B^2 \setminus \mathrm{Id}^{\mathfrak{B}} \mid (x \in V_1 \wedge y \in V_2) \vee (x \in V_2 \wedge y \in V_1)\}.
\end{aligned}$$

It is easy to check that this structure is a square representation for #13. Moreover, this structure is fully universal for #13 and homogeneous, and therefore a normal representation.

Note that the relation $(\mathrm{Id} \cup a)^{\mathfrak{B}}$ is an equivalence relation where V_1 and V_2 are the two equivalence classes. Therefore we get by Theorem 23 that the (general) network satisfaction problem for the relation algebra #13 is NP-hard. We mention that this result can also be deduced from the results in [BMPP19].

Example 2 (Relation Algebra #17). The relation algebra #17 is given by the multiplication table in Fig. 1. Let $\mathfrak{N} = (V; E^{\mathfrak{N}})$ be the countable, homogeneous, universal triangle-free, undirected graph (see [Hod97]), also called a *Henson graph.* We use this Henson graph to obtain a square representation $\mathfrak{B}$ with domain V for the relation algebra #17 as follows:

$$\begin{aligned}
\mathrm{Id}^{\mathfrak{B}} &:= \{(x,x) \in V^2\},\\
a^{\mathfrak{B}} &:= \{(x,y) \in V^2 \mid (x,y) \in E^{\mathfrak{N}}\},\\
b^{\mathfrak{B}} &:= \{(x,y) \in B^2 \setminus \mathrm{Id}^{\mathfrak{B}} \mid (x,y) \notin E^{\mathfrak{N}}\}.
\end{aligned}$$

This structure is homogeneous and fully universal since $\mathfrak{N}$ is homogeneous and embeds every triangle free graph. It is easy to see that there exists no non-trivial equivalence relation in this relation algebra. For the atom a the triangle (a, a, a) is forbidden, which means we can apply Theorem 29 and get NP-hardness for the (general) network satisfaction problem for the relation algebra #17. Also in this case, the hardness result can also be deduced from the results in [BMPP19].

8 Conclusion and Future Work

To the best of our knowledge the computational complexity of the (general) network satisfaction problem was previously only known for a small number of isolated finite relation algebras, for example the point algebra, Allens interval algebra, or the 18 small relation algebras from [AM94]. Both of our criteria, Theorems 23 and 29, show the NP-hardness for relatively large classes of finite relation algebras. In Sect. 7 we applied these results to settle the complexity status of two problems that were left open in [CH04].

To obtain our general hardness conditions we used the universal algebraic approach for studying the complexity of constraint satisfaction problems. This approach will hopefully lead to a solution of Hirsch's RBCP for all finite relation algebras **A** with a normal representation $\mathfrak{B}$. It is also relatively easy to prove that the network satisfaction problem for **A** is NP-complete if $\mathfrak{B}$ has an equivalence relation with an equivalence class of finite size larger than two. Hence, the next steps that have to be taken with this approach are the following.

- Classify the complexity of the network satisfaction problem for finite relation algebras **A** where the normal representation has a primitive automorphism group.
- Classify the complexity of the network satisfaction problem for relation algebras that have equivalence relations with infinitely many classes of size two.
- Classify the complexity of the network satisfaction problem for relation algebras that have equivalence relations with infinitely many infinite classes.

References

[All83] Allen, J.F.: Maintaining knowledge about temporal intervals. Commun. ACM **26**(11), 832–843 (1983)

[AM94] Andréka, H., Maddux, R.D.: Representations for small relation algebras. Notre Dame J. Formal Log. **35**(4), 550–562 (1994)

[Bar11] Barto, L.: The dichotomy for conservative constraint satisfaction problems revisited. In: Proceedings of the Symposium on Logic in Computer Science (LICS), Toronto, Canada (2011)

[BJ17] Bodirsky, M., Jonsson, P.: A model-theoretic view on qualitative constraint reasoning. J. Artif. Intell. Res. **58**, 339–385 (2017)

[BK07] Bodirsky, M., Kutz, M.: Determining the consistency of partial tree descriptions. Artif. Intell. **171**, 185–196 (2007)

[BK12] Barto, L., Kozik, M.: Absorbing subalgebras, cyclic terms and the constraint satisfaction problem. Log. Methods Comput. Sci. **8/1**(07), 1–26 (2012)

[BKO+17] Barto, L., Kompatscher, M., Olšák, M., Van Pham, T., Pinsker, M.: The equivalence of two dichotomy conjectures for infinite domain constraint satisfaction problems. In: Proceedings of the 32nd Annual ACM/IEEE Symposium on Logic in Computer Science - LICS 2017 (2017). Preprint arXiv:1612.07551

[BMPP19] Bodirsky, M., Martin, B., Pinsker, M., Pongrácz, A.: Constraint satisfaction problems for reducts of homogeneous graphs. SIAM J. Comput. **48**(4), 1224–1264 (2019). A conference version appeared in the Proceedings of the 43rd International Colloquium on Automata, Languages, and Programming, ICALP 2016, pp. 119:1–119:14

[BN06] Bodirsky, M., Nešetřil, J.: Constraint satisfaction with countable homogeneous templates. J. Log. Comput. **16**(3), 359–373 (2006)

[Bod18] Bodirsky, M.: Finite relation algebras with normal representations. In: Relational and Algebraic Methods in Computer Science - 17th International Conference, RAMiCS 2018, Groningen, The Netherlands, October 29 – November 1, 2018, Proceedings, pp. 3–17 (2018)

[BOP18] Barto, L., Opršal, J., Pinsker, M.: The wonderland of reflections. Isr. J. Math. **223**(1), 363–398 (2018)

[BP16] Bodirsky, M., Pinsker, M.: Canonical Functions: a Proof via Topological Dynamics (2016). Preprint available under http://arxiv.org/abs/1610.09660

[BPP14] Bodirsky, M., Pinsker, M., Pongrácz, A.: Projective clone homomorphisms. Accepted for publication in the Journal of Symbolic Logic (2014). Preprint arXiv:1409.4601

[Bul03] Bulatov, A.A.: Tractable conservative constraint satisfaction problems. In: Proceedings of the Symposium on Logic in Computer Science (LICS), Ottawa, Canada, pp. 321–330 (2003)

[Bul14] Bulatov, A.A.: Conservative constraint satisfaction revisited (2014). Manuscript, ArXiv:1408.3690v1

[Bul17] Bulatov, A.A.: A dichotomy theorem for nonuniform CSPs. In: 58th IEEE Annual Symposium on Foundations of Computer Science, FOCS 2017, Berkeley, CA, USA, 15–17 October 2017, pp. 319–330 (2017)

[Cam99] Cameron, P.J.: Permutation Groups. LMS Student Text 45. Cambridge University Press, Cambridge (1999)

[CH04] Cristiani, M., Hirsch, R.: The complexity of the constraint satisfaction problem for small relation algebras. Artif. Intell. J. **156**, 177–196 (2004)

[Dün05] Düntsch, I.: Relation algebras and their application in temporal and spatial reasoning. Artif. Intell. Rev. **23**, 315–357 (2005)

[HH02] Hirsch, R., Hodkinson, I.: Relation Algebras by Games. North Holland, Amsterdam (2002)

[Hir96] Hirsch, R.: Relation algebras of intervals. Artif. Intell. J. **83**, 1–29 (1996)

[Hir97] Hirsch, R.: Expressive power and complexity in algebraic logic. J. Log. Comput. **7**(3), 309–351 (1997)

[Hir99] Hirsch, R.: A finite relation algebra with undecidable network satisfaction problem. Log. J. IGPL **7**(4), 547–554 (1999)

[Hod97] Hodges, W.: A Shorter Model Theory. Cambridge University Press, Cambridge (1997)

[LM94] Ladkin, P.B., Maddux, R.D.: On binary constraint problems. J. Assoc. Comput. Mach. **41**(3), 435–469 (1994)

[Mad06] Maddux, R.D.: Relation Algebras. Elsevier, Amsterdam (2006)

[RN99] Renz, J., Nebel, B.: On the complexity of qualitative spatial reasoning: a maximal tractable fragment of the region connection calculus. Artif. Intell. **108**(1–2), 69–123 (1999)

[RN07] Renz, J., Nebel, B.: Qualitative spatial reasoning using constraint calculi. In: Aiello, M., Pratt-Hartmann, I., van Benthem, J. (eds.) Handbook of Spatial Logics, pp. 161–215. Springer, Berlin (2007). https://doi.org/10.1007/978-1-4020-5587-4_4

[VKvB89] Vilain, M., Kautz, H., van Beek, P.: Constraint propagation algorithms for temporal reasoning: a revised report. In: Reading in Qualitative Reasoning About Physical Systems, pp. 373–381 (1989)

[Zhu17] Zhuk, D.: A proof of CSP dichotomy conjecture. In: 58th IEEE Annual Symposium on Foundations of Computer Science, FOCS 2017, Berkeley, CA, USA, 15–17 October 2017, pp. 331–342 (2017)

The θ-Join as a Join with θ

Jules Desharnais[1] and Bernhard Möller[2(✉)]

[1] Université Laval, Québec, Canada
jules.desharnais@ift.ulaval.ca
[2] Universität Augsburg, Augsburg, Germany
bernhard.moeller@informatik.uni-augsburg.de

Abstract. We present an algebra for the classical database operators. Contrary to most approaches we use (inner) join and projection as the basic operators. Theta joins result by representing theta as a database table itself and defining theta-join as a join with that table. The same technique works for selection. With this, (point-free) proofs of the standard optimisation laws become very simple and uniform. The approach also applies to proving join/projection laws for preference queries. Extending the earlier approach of [16], we replace disjointness assumptions on the table types by suitable consistency conditions. Selected results have been machine-verified using the CalcCheck tool.

1 Introduction

The paper deals with an algebra for the classical operators of relational algebra as used in databases. While in most approaches the join operator is defined as a combination of direct product, selection and projection, we take a different approach, using (inner) join and projection as the basic operators. Theta joins are incorporated by simply representing (mathematically) theta as a database table itself and defining theta-join as a join with that table. The same can be done with selection by representing the corresponding condition as the table of all tuples that satisfy it. With this, (point-free) proofs of the standard laws become very simple and uniform. The approach is also suitable for proving join/projection laws for preference queries.

The paper builds upon [16]. While many of the laws there required disjointness of the types of the tables involved, we are here more general and replace disjointness of types by suitable consistency conditions. Technically, we extend the techniques there by deploying variants of the split and glue operators introduced in [3,4]. This allows point-free formulations of the new conditions and corresponding point-free proofs of the ensuing laws. Selected results have been machine-verified using the CalcCheck tool [9,10].

2 Preliminaries

Our approach is based on the algebra of binary relations, see e.g. [17]. A *binary relation* between sets M and N is a subset $R \subseteq M \times N$. We denote the empty

U. Fahrenberg et al. (Eds.): RAMiCS 2020, LNCS 12062, pp. 47–64, 2020.
https://doi.org/10.1007/978-3-030-43520-2_4

relation $\emptyset$ by 0 and the universal relation $M \times N$ by $\top_{M\times N}$, omitting the subscript when it is clear from the context. Domain, codomain and relational composition ; are defined as usual, the latter binding stronger than union and intersection. The *converse* of R is $R^{\smile} \subseteq N \times M$, given by $R^{\smile} = \{(y,x) \mid (x,y) \in R\}$.

If $M = N$ then R is called *homogeneous*. In this case there is the *identity relation* $1_M = \{(x,x) \mid x \in M\}$, which is neutral w.r.t. ; . If M is clear from the context we omit the subscript $_M$.

A *test* over M is a sub-identity $P \subseteq 1$ which encodes the subset $\{x \mid (x,x) \in P\}$. The *negation* $\neg P$ of test P is the complement of P relative to 1, i.e., $1 - P$, where $-$ is set difference. It encodes the complement of the set encoded by P. When convenient we do not distinguish between tests and the encoded sets.

Domain and codomain can be encoded as the tests

$$\ulcorner R = R \,;\, \top_{N\times M} \cap 1_M, \qquad R^{\urcorner} = \top_{M\times N} \,;\, R \cap 1_N. \tag{1}$$

We list a few properties of domain; symmetric ones hold for the codomain operator which, however, we do not use in this paper. For proofs see [6].

Lemma 2.1. *Consider relations R, S and test P.*
1. $\ulcorner(R \cup S) = \ulcorner R \cup \ulcorner S$. *Hence* $\ulcorner$ *is isotone, i.e., monotonically increasing, w.r.t.* $\subseteq$.
2. $\ulcorner R \,;\, R = R$ *and* $\neg\ulcorner R \,;\, R = 0$.
3. $\ulcorner P = P$. (stability)
4. $\ulcorner R = 0 \Leftrightarrow R = 0$. (full strictness)
5. $\ulcorner(P \,;\, R) = P \,;\, \ulcorner R$. (import/export)
6. $\ulcorner(R \,;\, S) = \ulcorner(R \,;\, \ulcorner S)$. (locality)
7. $R \,;\, P \cap S = (R \cap S) \,;\, P = R \cap S \,;\, P$. (restriction)

3 Typed Tuples

In this section we present the formal model of database objects as typed tuples. The types represent attributes, i.e., columns of a database relation. Conceptually and notationally, we largely base on [11].

Definition 3.1. Let $\mathcal{A}$ be a set of *attribute names* and $(D_A)_{A\in\mathcal{A}}$ be a family of nonempty sets, where for $A \in \mathcal{A}$ the set D_A is called the *domain* of A.
1. The set $\mathcal{U} =_{df} \bigcup_{A\in\mathcal{A}} D_A$ is called the *universe*.
2. A *type* T is a subset $T \subseteq \mathcal{A}$.
3. A T-*tuple* is a mapping $t : T \to \mathcal{U}$ where $\forall A \in T : t(A) \in D_A$. For $T = \emptyset$ the only T-tuple is the empty mapping $\emptyset$.
4. The domain D_T for a type T is the set of all T-tuples, i.e., the Cartesian product $D_T = \prod_{A\in T} D_A$.
5. For a T-tuple t and a sub-type $T' \subseteq T$ we define the projection $\pi_{T'}(t)$ to T' as the restriction of the mapping t to T'. By this $\pi_\emptyset(t) = \emptyset$. Projections π are not to be confused with the Cartesian product operator $\prod$.
6. A set of tuples of the same type is called a *table* and is relationally encoded as a test.

7. For a tuple t and a table P of T-tuples we introduce the abbreviations

$$t :: T \Leftrightarrow_{df} t \in D_T, \qquad P :: T \Leftrightarrow_{df} P \subseteq D_T.$$

Definition 3.2. Two tuples $t_i :: T_i$ $(i = 1, 2)$ are called *matching*, in signs $t_1 \# t_2$, iff $\pi_T(t_1) = \pi_T(t_2)$, where $T =_{df} T_1 \cap T_2$. In this case we define $t_1 \bowtie t_2 =_{df} t_1 \cup t_2$. The join of nonmatching tuples is undefined. If $T = \emptyset$, i.e., the types T_i are disjoint, then the t_i are trivially matching. The empty tuple $\emptyset$ matches every tuple and hence is the neutral element of $\bowtie$.

The *join* of two types T_1, T_2 is the union of their attributes, i.e., $T_1 \bowtie T_2 =_{df} T_1 \cup T_2$. For tables $P_i :: T_i$ $(i = 1, 2)$, the join $\bowtie$, binding stronger than union and intersection, is defined as the set of all matching combinations of P_i-tuples:

$$\begin{aligned} P_1 \bowtie P_2 &=_{df} \{t :: T_1 \bowtie T_2 \mid \pi_{T_i}(t) \in P_i \,(i = 1, 2)\} \\ &= \{t_1 \bowtie t_2 \mid t_i \in P_i \,(i = 1, 2), t_1 \# t_2\}. \end{aligned}$$

When we want to avoid numerical indices we use the convention that table P has type T_P, etc. The table $\{\emptyset\}$ is the neutral element of $\bowtie$ on tables.

Lemma 3.3. *$D_{T_1 \bowtie T_2} = D_{T_1} \bowtie D_{T_2}$. Hence $T_2 \subseteq T_1 \Rightarrow D_{T_1} \bowtie D_{T_2} = D_{T_1}$.*

Proof. Immediate from the definition of type join and Definition 3.1.4. □

Lemma 3.4. *Consider tables $P :: T_P, Q :: T_Q$ and an arbitrary type T'.*
1. *Every tuple is characterised by its projections: for $t \in D_{T_P \bowtie T_Q}$ we have $t = \pi_{T_P}(t) \bowtie \pi_{T_Q}(t)$. For $t, u \in D_{T_P \bowtie T_Q}$ this entails $t = u \Leftrightarrow \pi_{T_P}(t) = \pi_{T_P}(u) \wedge \pi_{T_Q}(t) = \pi_{T_Q}(u)$.*
2. *Projection sub-distributes over join: $\pi_{T'}(P \bowtie Q) \subseteq \pi_{T'}(P) \bowtie \pi_{T'}(Q)$.*
3. *If $T_P \cap T_Q = \emptyset$ then this strengthens to an equality.*

Proof.
1. Straightforward calculation.
2. By distributivity of restriction over union, for any two matching tuples t_1, t_2 (not necessarily from P, Q) we have $\pi_{T'}(t_1 \bowtie t_2) = \pi_{T'}(t_1) \bowtie \pi_{T'}(t_2)$. Hence if we take matching $t_1 \in P, t_2 \in Q$, then $t =_{df} \pi_{T'}(t_1 \bowtie t_2) \in \pi_{T'}(P \bowtie Q)$. Because $\pi_{T'}(t_1 \bowtie t_2) = \pi_{T'}(t_1) \bowtie \pi_{T'}(t_2)$, also $t \in \pi_{T'}(P) \bowtie \pi_{T'}(Q)$.
3.
$$\begin{aligned} & t \in \pi_{T'}(P) \bowtie \pi_{T'}(Q) \\ \Leftrightarrow\ & \exists\, u, v : u \in P \wedge v \in Q \wedge t = \pi_{T'}(u) \bowtie \pi_{T'}(v) && \{\!\!\{ \text{ definition of join } \}\!\!\} \\ \Rightarrow\ & \exists\, u, v : u \in P \wedge v \in Q \wedge t = \pi_{T'}(u \bowtie v) && \{\!\!\{ u \# v \text{ by } T_P \cap T_Q = \emptyset \}\!\!\} \\ \Rightarrow\ & t \in \pi_{T'}(P \bowtie Q) && \{\!\!\{ \text{ definitions } \}\!\!\} \end{aligned}$$
□

Lemma 3.5. *For $P_i :: T_i$ $(i = 1, 2)$ with disjoint T_i, i.e., with $T_1 \cap T_2 = \emptyset$, the join $P_1 \bowtie P_2$ is isomorphic to the Cartesian product of P_1 and P_2.*

Proof. For $t \in P_1 \bowtie P_2$, the conditions $\pi_{T_i}(t) \in P_i (i = 1, 2)$ are independent. Hence all elements of P_1 can be joined with all elements of P_2. Thus, by definition,

$$t \in P_1 \bowtie P_2 \Leftrightarrow \pi_{T_1}(t) \in P_1 \wedge \pi_{T_2}(t) \in P_2 \Leftrightarrow (\pi_{T_1}(t), \pi_{T_2}(t)) \in P_1 \times P_2. \quad \square$$

Lemma 3.6 [16]. *The following laws hold:*
1. $\bowtie$ *is associative, commutative and distributes over* $\cup$.
2. $\bowtie$ *is isotone in both arguments.*
3. *Assume* $P_i, Q_i :: T_i$ $(i = 1, 2)$. *Then the following interchange law holds:*

$$(P_1 \cap Q_1) \bowtie (P_2 \cap Q_2) = (P_1 \bowtie P_2) \cap (Q_1 \bowtie Q_2).$$

4. *For* $P, Q :: T$ *we have* $P \bowtie Q = P \cap Q$. *In particular,* $P \bowtie P = P$.

4 The θ-Join

For simplicity we restrict ourselves to θ-joins with binary relations θ. Assume tables $P :: T_P, Q :: T_Q$ with $T_P \cap T_Q = \emptyset$ as well as $A \in T_P, B \in T_Q$ and a binary relation $\theta \subseteq D_A \times D_B$. Note that the assumptions imply $A \neq B$[1]. We want to model an expression that in standard database theory would be written "$P \bowtie_{\theta(P.A, Q.B)} Q$". The corresponding table contains exactly those tuples t of table $P \bowtie Q$ in which the values $t(A) \in P.A$ and $t(B) \in Q.B$ (remember that t is a function from attribute names to values) are in relation θ.

The idea is to consider θ mathematically again as table of type $A \bowtie B$. Then the above expression can simply be represented as $P \bowtie \theta \bowtie Q$.

Example 4.1. Here is a simple database of persons and ages with $>$ as θ.

P:

Name1	Age1
A	50
B	55
C	60

$<$:

Age1	Age2
50	55
50	60
55	60

Q:

Age2	Name2
50	E
55	F
55	G

$P \bowtie Q$:

Name1	Age1	Age2	Name2
A	50	50	E
A	50	55	F
A	50	55	G
B	55	50	E
B	55	55	F
B	55	55	G
C	60	50	E
C	60	55	F
C	60	55	G

$P \bowtie <$:

Name1	Age1	Age2
A	50	55
A	50	60
B	55	60

$< \bowtie Q$:

Age1	Age2	Name2
50	55	F
50	55	G

$P \bowtie < \bowtie Q$:

Name1	Age1	Age2	Name2
A	50	55	F
A	50	55	G

□

[1] With this we follow the SQL standard. Note, however, that $P \bowtie \theta \bowtie Q$ is defined even if this disjointness condition does not hold. It is not even necessary to require $A \neq B$, although having $A = B$ is not interesting.

We use our view of the θ-join for algebraic proofs of two standard optimisation rules for projections applied to joins.

Theorem 4.2.
1. *If $Q :: L \subseteq T_P$ then $\pi_L(P \bowtie Q) = \pi_L(P) \bowtie Q$.*
2. *Assume $T_P \cap T_Q = \emptyset$ and $\theta :: L$ for some $L \subseteq T_P \cup T_Q$. This means that θ is to provide the "glue" between the type-disjoint P and Q. Set $L_P =_{df} T_P \cap L$ and $L_Q =_{df} T_Q \cap L$. Then we have the transformation rule*

$$\pi_L(P \bowtie \theta \bowtie Q) = \pi_{L_P}(P) \bowtie \theta \bowtie \pi_{L_Q}(Q) \qquad \textit{(push projection over join).}$$

Proof.
1. ($\subseteq$) Immediate from Lemma 3.4.2 and $\pi_L(Q) = Q$ by $Q :: L$.
($\supseteq$)

$$\begin{array}{ll}
& \pi_L(P) \\
= & \pi_L(P \bowtie D_P) & \{\!|\ \text{definition of } D_P\ |\!\} \\
= & \pi_L(P \bowtie D_P \bowtie D_L) & \{\!|\ \text{assumption } L \subseteq T_P, \text{ definition of } D_L\ |\!\} \\
= & \pi_L(P \bowtie D_L) & \{\!|\ \text{definition of } D_P\ |\!\} \\
= & \pi_L(P \bowtie (Q \cup \overline{Q})) & \{\!|\ \text{Boolean algebra, setting} \\
& & \quad \overline{X} =_{df} D_L - X \text{ for } X :: L\ |\!\} \\
= & \pi_L(P \bowtie Q) \cup \pi_L(P \bowtie \overline{Q})) & \{\!|\ \text{distributivity of join and projection}\ |\!\} \\
\subseteq & \pi_L(P \bowtie Q) \cup (\pi_L(P) \bowtie \pi_L(\overline{Q})) & \{\!|\ \text{Lemma 3.4.2}\ |\!\} \\
= & \pi_L(P \bowtie Q) \cup (\pi_L(P) \cap \pi_L(\overline{Q})) & \{\!|\ \text{Lemma 3.6.4}\ |\!\} \\
\subseteq & \pi_L(P \bowtie Q) \cup \pi_L(\overline{Q}) & \{\!|\ \text{Boolean algebra}\ |\!\} \\
= & \pi_L(P \bowtie Q) \cup \overline{Q} & \{\!|\ \pi_L(\overline{Q}) = \overline{Q} \text{ by } \overline{Q} :: L\ |\!\}
\end{array}$$

By Lemma 3.6.4 and shunting we obtain from this $\pi_L(P) \bowtie Q = \pi_L(P) \cap Q \subseteq \pi_L(P \bowtie Q)$.

2.

$$\begin{array}{ll}
& \pi_L(P \bowtie \theta \bowtie Q) \\
= & \pi_L(P \bowtie Q \bowtie \theta) & \{\!|\ \text{associativity and commutativity of } \bowtie\ |\!\} \\
= & \pi_L(P \bowtie Q) \bowtie \theta & \{\!|\ \text{Part 1}\ |\!\} \\
= & \pi_L(P) \bowtie \pi_L(Q) \bowtie \theta & \{\!|\ \text{assumption } T_P \cap T_Q = \emptyset \text{ with Lemma 3.4.3}\ |\!\} \\
= & \pi_L(P) \bowtie \theta \bowtie \pi_L(Q) & \{\!|\ \text{associativity and commutativity of } \bowtie\ |\!\} \\
= & \pi_{L_P}(P) \bowtie \theta \bowtie \pi_{L_Q}(Q) & \{\!|\ P :: T_P, Q :: T_Q, \text{ definition of projection}\ |\!\}
\end{array}$$ □

5 Selection as Join

Since the representation of the θ-join as a join with θ has proved useful, we will now treat selection $\sigma_C(P)$ for table P and condition C analogously. A *condition*, i.e., a predicate on tuples, is simply represented as a subset $C \subseteq D_L$ for some type L, which means $C :: L$. Conjunction and disjunction of $C, C' :: L$ are then represented by $C \bowtie C'$ and $C \cup C'$, resp. (see Lemma 3.6.4). For $P :: T$ and $L \subseteq T$, we can now just set $\sigma_C(P) =_{df} P \bowtie C$.

Lemma 5.1. *Assume again $P :: T$.*
1. *Selections commute, i.e., $\sigma_C(\sigma_{C'}(P)) = \sigma_{C'}(\sigma_C(P))$.*

2. *Selections can be combined, i.e.,* $\sigma_C(\sigma_{C'}(P)) = \sigma_{C \bowtie C'}(P)$.
3. *If* C *uses only attributes from* $L \subseteq T$*, i.e.,* $C \subseteq D_L$*, then* $\pi_L(\sigma_C(P)) = \sigma_C(\pi_L(P))$.

Proof.
1. Immediate from associativity/commutativity of $\bowtie$.
2. Ditto.
3. By definitions, Theorem 4.2.1 and definitions again:

$$\pi_L(\sigma_C(P)) = \pi_L(P \bowtie C) = \pi_L(P) \bowtie C = \sigma_C(\pi_L(P)). \qquad \square$$

6 Inverse Image and Maximal Elements

The tools developed in the preceding sections will now be applied to a subfield of database theory, namely to preference queries. They serve to remedy a well known problem for queries with *hard constraints*, by which the objects sought in the database are clearly and sharply characterised. If there are no exact matches the empty result set is returned, which is very frustrating for users.

Instead, over the last decade queries with *soft constraints* have been studied. These arise from a formalisation of the *user's preferences* in the form of partial strict orders [12,13]. Instead of returning an empty result set, one can then present the user with the maximal or "best" tuples w.r.t. her preference order.

We now show how to express the maximality operator algebraically and then prove a sample optimisation rule for it. The idea has already been described thoroughly in the predecessor paper [16]; hence we only give a brief presentation of it. After that we develop substantially new laws for it. The main ingredient is an inverse image operator on relations.

Definition 6.1. For a type T a T*-relation* is a homogeneous binary relation R on D_T; we abbreviate this by $R :: T^2$. In analogy to the notation in Sect. 2 we also write $\top_T$ instead of $D_T \times D_T$. For a relation $R :: T^2$ the *image* of a test $P :: T$ under R is obtained using the forward diamond operator as

$$|R\rangle P =_{df} \{(x,x) \mid \exists y \in P : x\, R\, y\} = \ulcorner(R\,;P).$$

Two immediate consequences of the definition and Lemma 2.1 are

$$|0\rangle P = 0, \qquad |\top\rangle P = \begin{cases} D_T & \text{if } P \neq 0, \\ 0 & \text{otherwise.} \end{cases} \tag{2}$$

The inverse image of a set P under a relation R consists of the elements that have an R-successor in P, i.e., are R-related to some object in P. Assume that R is a strict order (irreflexive and transitive), which is the case in our application domain of preferences. Then the inverse image of P consists of the tuples *dominated* by some tuple in P. This allows the following definition.

Definition 6.2. For a relation $R :: T^2$ and a set $P :: T$ the *R-maximal* objects of P form the relative complement of the set of R-dominated objects, viz.

$$R \rhd P =_{df} P \cap \neg|R\rangle P.$$

The mnemonic behind the $\rhd$ symbol is that in an order diagram for a preference relation R the maximal objects within P are the peaks in P; rotating the diagram clockwise by 90° puts the peaks to the right. Hence $R \rhd P$ might also be read as "R-peaks in P". From (2) we obtain

$$0 \rhd P = P, \qquad \mathsf{T} \rhd P = 0. \tag{3}$$

A central ingredient for the preference approach is a possibility for defining complex preference relations out of simpler ones. An example would be "I prefer cars that are green and, equally important, have low fuel consumption". The following sections deal with such construction mechanisms, notably with the join of relations.

7 The Join of Relations

Definition 7.1. The *join* $R_1 \bowtie R_2 :: (T_1 \bowtie T_2)^2$ of relations $R_i :: T_i^2$ $(i = 1, 2)$ is

$$t\,(R_1 \bowtie R_2)\,u \Leftrightarrow_{df} \pi_{T_1}(t)\,R_1\,\pi_{T_1}(u) \wedge \pi_{T_2}(t)\,R_2\,\pi_{T_2}(u).$$

Example 7.2. We model the above simple database of cars. Consider the set $\mathcal{A} = \{\mathrm{Col}, \mathrm{Fuel}\}$ of attribute names with $D_{\mathrm{Col}} = \{\mathsf{black}, \mathsf{blue}, \mathsf{green}, \mathsf{red}, \mathsf{white}\}$ and $D_{\mathrm{Fuel}} = \{4.0, 4.1, \ldots, 9.9, 10.0\}$. The comparison relation R_{Col} is given by the Hasse diagram

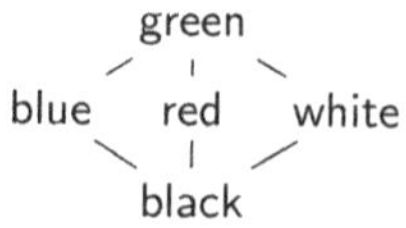

while as R_{Fuel} we choose $>$. A user uttering the preference R_{Col} does not like black at all, likes green best and otherwise is indifferent about blue, red, white. Hence $s\,(R_{\mathrm{Col}} \bowtie R_{\mathrm{Fuel}})\,t$ iff the colour of t is closer to green than that of s and the fuel value of t is less than that of s. □

Definition 7.3. Based on join we can define the two standard preference constructors $\otimes$ of *Pareto* and & of *prioritised* composition as

$$R \otimes S =_{df} (R \bowtie (1 \cup S)) \cup ((1 \cup R) \bowtie S),$$
$$R\&S =_{df} (R \bowtie \mathsf{T}) \cup (1 \bowtie S).$$

Pareto composition corresponds to the product order on pairs, with two variations: it does not consider pairs, but tuples from which parts are extracted by the projections involved in $\bowtie$; moreover, it is more liberal than the product of strict orders, since it also admits equality in one part of the tuples as long as there is a strict order relation between the other parts. Prioritised composition corresponds to the lexicographic order on pairs.

We seek a set of algebraic laws that allow proving optimisation rules similar to "push projection over join" from Theorem 4.2.2. As an example consider tables $P :: T_P, Q :: T_Q$ and a preference relation $R :: T_P^2$. Then we would like to show

$$(R \bowtie \mathsf{T}_Q) \rhd (P \bowtie Q) = (R \rhd P) \bowtie Q \tag{4}$$

under suitable side conditions on P, Q, R. The preference $R \bowtie \mathsf{T}_Q$, which also occurs as a part of the & constructor, expresses that the user does not care about the attributes in T_Q and is only interested in the T_P part. Therefore the preference query can be pushed to that part as shown on the right hand side. This may speed up the query evaluation considerably.

To achieve the mentioned algebraic laws we need to investigate the interaction between the $\bowtie$ and $\rhd$ operators involved. Of particular importance are so-called interchange laws: the above rule can, by (3), be written as

$$(R \bowtie \mathsf{T}_Q) \rhd (P \bowtie Q) = (R \rhd P) \bowtie (0 \rhd Q);$$

a maximum between joins is equal to a join between maxima[2].

8 Split, Glue and Pair Relations

To formulate and prove rules about the join of relations in an algebraic style we bring the pointwise definition into a more manageable point-free form. For this we deploy techniques from [3,4]. First we introduce relations for connecting tuples and pairs of tuples.

Definition 8.1. For types T_1, T_2 we define *split* $\prec$ and its converse *glue* $\succ$ with the functionalities

$$\begin{aligned} {}_{T_1 \bowtie T_2}\!\prec_{T_1 \times T_2} &\subseteq D_{T_1 \bowtie T_2} \times (D_{T_1} \times D_{T_2}), \\ {}_{T_1 \times T_2}\!\succ_{T_1 \bowtie T_2} &\subseteq (D_{T_1} \times D_{T_2}) \times D_{T_1 \bowtie T_2}. \end{aligned}$$

Again we suppress the type indices for readability. The behaviour is given by

$$t \prec (t_1, t_2) \;\Leftrightarrow_{df}\; (t_1, t_2) \succ t \;\Leftrightarrow_{df}\; t_1 = \pi_{T_1}(t) \wedge t_2 = \pi_{T_2}(t).$$

Hence $\prec$ relates every tuple to all its possible splits into matching pairs of subtuples. The definition is stronger than the corresponding one in [3,4], and this results in more useful laws which are detailed below: [3,4] allow arbitrary splittings on the left and right of $\succ\,;\prec$, whereas ours are "synchronised" by the projections so that the same splits are used on the left and right. By the difference in approach the forward interchange rule of Theorem 9.2 does not hold in their setting. For the purposes of database algebra, however, the stronger definition is quite adequate.

While split and glue tell us how to decompose or recompose tuples or tuple parts, we also want to relate corresponding parts "in parallel".

---

[2] We use this example only for motivation; strictly speaking an interchange law needs to have the same variables on both sides.

Definition 8.2. A *pair relation* over types T_1, T_2 is a subset of $(D_{T_1} \times D_{T_2}) \times (D_{T_1} \times D_{T_2})$. The *parallel product* $R_1 \times R_2$ of relations $R_i :: T_i^2$ is the pair relation

$$(t_1, t_2)\,(R_1 \times R_2)\,(u_1, u_2) \Leftrightarrow_{df} t_1\, R_1\, u_1 \,\wedge\, t_2\, R_2\, u_2.$$

By $1_{T_1 \times T_2}$ we denote the identity pair relation. When the T_i are clear from the context we omit the type index.

The parallel product is a standard construct in relation algebra; it occurs, for instance, in [2] and [7] and is also called a Kronecker product [8]. With its help we can express the lifting of join to relations in Definition 7.1 more compactly.

Lemma 8.3. *The join of relations $R_i :: T_i^2$ can be expressed point-free as*

$$R_1 \bowtie R_2 =_{df} \prec ; (R_1 \times R_2) ; \succ .$$

The proof is immediate from the definitions. From this relational representation it follows that join is strict w.r.t. 0 and distributes through union in both arguments. We note that for relational tests P, Q the lifting $P \bowtie Q$ is a test in the algebra of relations. Details are given in Lemma 10.5.

Next to this, we also use the concept of tests for pair relations. These are again sub-identities, i.e., subsets of $1_{T_1 \times T_2}$; as usual they are idempotent and commute under ; (e.g. [5]). The parallel product of tests is a test in the set of pair relations.

Definition 8.4. Another test in the set of pair relations is the lifted *matching check* ${}_{T_1}\circledast_{T_2}$: for tuples $t_i, u_i :: T_i$

$$(t_1, t_2)\ {}_{T_1}\circledast_{T_2}\ (u_1, u_2) \ \ \Leftrightarrow_{df}\ \ t_1 = u_1 \,\wedge\, t_2 = u_2 \,\wedge\, t_1 \# t_2.$$

To ease notation, we suppress the type indices.

We now present the essential laws for all these constructs.

Lemma 8.5.
1. $\succ\, = \,\prec^{\smile}$.
2. $\succ ; \prec\, = \circledast$ *and hence* $\succ ; \prec\, \subseteq 1$.
3. $\prec ; \succ\, = 1$.
4. $\circledast ; \succ\, = \,\succ$ *and symmetrically* $\prec ; \circledast = \,\prec$.
5. $\prec$ *and* $\succ$ *are deterministic and injective; in addition* $\prec$ *is total and* $\succ$ *is surjective.*
6. $\prec ; C ; \succ\, \subseteq R \ \Leftrightarrow\ \circledast ; C ; \circledast \,\subseteq\, \succ ; R ; \prec$. *In particular,* $\prec ; C ; \succ\, \subseteq\, \prec ; D ; \succ\ \Leftrightarrow\ \circledast ; C ; \circledast \,\subseteq\, \circledast ; D ; \circledast$.
7. $\prec ; \mathsf{T} ; \succ\ =\ \mathsf{T}$.
8. $\prec ; C ; \circledast ; \mathsf{T} ; \succ\ =\ \prec ; C ; \succ ; \mathsf{T}$.

Proof. The proofs of Parts 1—3 are straightforward pointwise calculations.
4. By 2 and 3, $\circledast ; \succ\, = \,\succ ; \prec ; \succ\, = \,\succ ; 1 = \,\succ$.
5. These are standard relation-algebraic consequences of Parts 1—3.

6. By isotony, Part 2, isotony and Parts 4 and 3,

$$\begin{aligned} &\prec ; C ; \succ \subseteq R \;\Rightarrow\; \succ ; \prec ; C ; \succ ; \prec \subseteq \succ ; R ; \prec \\ \Leftrightarrow\;& \textcircled{\#} ; C ; \textcircled{\#} \subseteq \succ ; R ; \prec \;\Rightarrow\; \prec ; \textcircled{\#} ; C ; \textcircled{\#} ; \succ \subseteq \prec ; \succ ; R ; \prec ; \succ \\ \Leftrightarrow\;& \prec ; C ; \succ \subseteq R. \end{aligned}$$

For $R = \prec ; D ; \succ$ the second claim results again by Part 2.

7. This is direct by totality of $\prec$ and surjectivity of $\succ$ (Part 5).
8. By Parts 2 and 7, $\prec ; C ; \textcircled{\#} ; \mathsf{T} ; \succ \;=\; \prec ; C ; \succ ; \prec ; \mathsf{T} ; \succ \;=\; \prec ; C ; \succ ; \mathsf{T}$. □

Lemma 8.6

1. $1_{T_1 \times T_2} = 1_{T_1} \times 1_{T_2}$.
2. $\mathsf{T}_{T_1} \times \mathsf{T}_{T_2}$ *is the universal pair relation.*
3. *The operators* $\times$ *and* $\cap$ *satisfy an equational interchange law:*

$$(R_1 \cap R_2) \times (S_1 \cap S_2) \;=\; (R_1 \times S_1) \cap (R_2 \times S_2).$$

4. *The operators* $\times$ *and* $;$ *satisfy an equational interchange law:*

$$(R_1 ; R_2) \times (S_1 ; S_2) \;=\; (R_1 \times S_1) ; (R_2 \times S_2).$$

Again, the proofs are straightforward calculations. In addition, we have the following result.

Lemma 8.7. *Identity and top behave nicely w.r.t.* $\bowtie$*, i.e.,* $1_{T_1} \bowtie 1_{T_2} = 1_{T_1 \bowtie T_2}$. *Similarly,* $\mathsf{T}_{T_1} \bowtie \mathsf{T}_{T_2} = \mathsf{T}_{T_1 \bowtie T_2}$*; equivalently,* $\prec ; \mathsf{T}_{T_1 \times T_2} ; \succ = \mathsf{T}_{T_1 \bowtie T_2}$.

Proof. For the first claim we calculate, using Lemmas 8.3, 8.6.1 and 8.5.3,

$$1_{T_1} \bowtie 1_{T_2} \;=\; \prec ; (1_{T_1} \times 1_{T_2}) ; \succ \;=\; \prec ; 1_{T_1 \times T_2} ; \succ \;=\; \prec ; \succ \;=\; 1_{T_1 \bowtie T_2}.$$

The second claim was shown in Lemma 8.5.7. □

9 Interchange Laws for Join

We have already seen some interchange laws. It turns out that join inherits many of them, sometimes as inclusions rather than equations.

Lemma 9.1. *Relations* $R_i, S_i :: T_i^2$ *satisfy the equational interchange law*

Proof.

$$(R_1 \bowtie R_2) \cap (S_1 \bowtie S_2) \;=\; (R_1 \cap S_1) \bowtie (R_2 \cap S_2).$$

$$\begin{aligned} & (R_1 \bowtie R_2) \cap (S_1 \bowtie S_2) \\ =\;& \prec ; (R_1 \times R_2) ; \succ \cap \prec ; (S_1 \times S_2) ; \succ && \{\!\!\{\ \text{Lemma 8.3}\ \}\!\!\} \\ =\;& \prec ; ((R_1 \times R_2) \cap (S_1 \times S_2)) ; \succ && \{\!\!\{\ \text{determinacy of } \prec \text{ and} \\ &&& \quad \text{injectivity of } \succ \text{ (Lemma 8.5.5)}\ \}\!\!\} \\ =\;& \prec ; ((R_1 \cap S_1) \times (R_2 \cap S_2)) ; \succ && \{\!\!\{\ \times\text{-}\cap\text{-interchange (Lemma 8.6)}\ \}\!\!\} \\ =\;& (R_1 \cap S_1) \bowtie (R_2 \cap S_2) && \{\!\!\{\ \text{Lemma 8.3}\ \}\!\!\} \end{aligned}$$

□

Theorem 9.2 (Forward Interchange). *Relations $R_i, S_i :: T_i^2$ satisfy the inclusional interchange law*

$$(R_1 \bowtie R_2) ; (S_1 \bowtie S_2) \subseteq (R_1 ; S_1) \bowtie (R_2 ; S_2).$$

Proof. We calculate as follows.

$$\begin{array}{ll}
(R_1 \bowtie R_2) ; (S_1 \bowtie S_2) & \\
= \prec ;(R_1 \times R_2) ; \succ ; \prec ; (S_1 \times S_2) ; \succ & \{\!\![\ \text{Lemma 8.3}\ ]\!\!\} \\
\subseteq \prec ;(R_1 \times R_2) ; 1 ; (S_1 \times S_2) ; \succ & \{\!\![\ \text{Lemma 8.5.2}\ ]\!\!\} \\
= \prec ;(R_1 \times R_2) ; (S_1 \times S_2) ; \succ & \{\!\![\ \text{neutrality of } 1\ ]\!\!\} \\
= \prec ;((R_1 ; S_1) \times (R_2 ; S_2)) ; \succ & \{\!\![\ ;\text{-}\times\text{-interchange (Lemma 8.6.3)}\ ]\!\!\} \\
= (R_1 ; S_1) \bowtie (R_2 ; S_2) & \{\!\![\ \text{Lemma 8.3}\ ]\!\!\}
\end{array}$$
□

Using this we can show a subdistribution law for domain over join.

Theorem 9.3. *For $R_i :: T_i^2$ $(i = 1, 2)$ the domain of their join satisfies*

$$\ulcorner(R_1 \bowtie R_2) \subseteq \ulcorner R_1 \bowtie \ulcorner R_2.$$

Proof. By (1) and Lemma 8.7, Theorem 9.2, $\bowtie$-$\cap$-interchange (Lemma 9.1) and (1):

$$\begin{array}{rl}
\ulcorner(R_1 \bowtie R_2) & = (R_1 \bowtie R_2) ; (\top_{T_1} \bowtie \top_{T_2}) \cap (1_{T_1} \bowtie 1_{T_2}) \\
& \subseteq ((R_1 ; \top_{T_1}) \bowtie (R_2 ; \top_{T_2})) \cap (1_{T_1} \bowtie 1_{T_2}) \\
& = (R_1 ; \top_{T_1} \cap 1_{T_1}) \bowtie (R_2 ; \top_{T_2} \cap 1_{T_2}) = \ulcorner R_1 \bowtie \ulcorner R_2
\end{array}$$
□

Next we present conditions under which these inclusions become equations.

10 Compatibility and Matching

Definition 10.1

1. We call R_1, R_2 *weakly matching* if for all $x_i \in \ulcorner R_i$ $(i = 1, 2)$ with $x_1 \# x_2$ there are $y_i :: T_i$ $(i = 1, 2)$ with $y_1 \# y_2$ and $x_i \, R_i \, y_i$. This means that starting from matching tuples one can always reach corresponding matching tuples via the R_i.
2. R_1, R_2 are *strongly matching* if for all $x_i \in \ulcorner R_i$ $(i = 1, 2)$ with $x_1 \# x_2$ all tuples $y_i :: T_i$ $(i = 1, 2)$ with $x_i \, R_i \, y_i$ satisfy $y_1 \# y_2$. This means that starting from matching tuples all corresponding tuples reachable via the R_i are matching again.

We want to find algebraic characterisations of these forms of matching.

Definition 10.2. Relations R_1, R_2 are *forward compatible* iff

$$\circledast ; (R_1 \times R_2) \subseteq (R_1 \times R_2) ; \circledast,$$

and *backward compatible* iff $(R_1 \times R_2) ; \circledast \subseteq \circledast ; (R_1 \times R_2)$. Finally, R_1 and R_2 are *compatible* iff they are forward and backward compatible.

We can now give point-free characterisations of matching.

Lemma 10.3

1. *All test relations are compatible with each other.*
2. *Two relations are strongly matching iff they are forward compatible.*
3. *R_1, R_2 are weakly matching iff $\textcircled{\#} \,;\, (\ulcorner R_1 \times \ulcorner R_2) \subseteq \ulcorner((R_1 \times R_2) \,;\, \textcircled{\#})$ iff $\ulcorner(\textcircled{\#} \,;\, (R_1 \times R_2)) \subseteq \ulcorner((R_1 \times R_2) \,;\, \textcircled{\#})$.*
4. *Strongly matching relations are also weakly matching.*

Proof.

1. For test relations P, Q the relation $P \times Q$ is a test in the algebra of pair relations. Since $\textcircled{\#}$ is a test there too, they commute, which means forward and backward compatibility of P and Q.
2. Straightforward predicate calculus with the definitions.
3. Ditto for the first inclusion. The second one results from the first by distributivity of domain over $\times$ and the import/export law of Lemma 2.1.5.
4. Immediate from the second inclusion of Part 3 and isotony of domain. □

Now we can show a reverse interchange law between $\bowtie$ and ;.

Theorem 10.4. (Backward Interchange). *Let $R_i, S_i :: T_i^2$. If R_1, R_2 are forward compatible or S_1, S_2 are backward compatible then*

$$(R_1 \,;\, S_1) \bowtie (R_2 \,;\, S_2) \;\subseteq\; (R_1 \bowtie R_2) \,;\, (S_1 \bowtie S_2).$$

In particular, if R_1, R_2 or S_1, S_2 are tests then the inclusion holds.

Proof. We assume R_1, R_2 to be forward compatible.

$$\begin{array}{ll}
(R_1 \,;\, S_1) \bowtie (R_2 \,;\, S_2) & \\
= \prec ;(R_1 \,;\, S_1) \times (R_2 \,;\, S_2) \,;\, \succ & \{\!\![\ \text{Lemma 8.3}\ ]\!\!\} \\
= \prec ;(R_1 \times R_2) \,;\, (S_1 \times S_2) \,;\, \succ & \{\!\![\ ;\text{-}\times\text{-interchange (Lemma 8.6.4)}\ ]\!\!\} \\
= \prec ;\, \textcircled{\#} \,;(R_1 \times R_2) \,;\, (S_1 \times S_2) \,;\, \succ & \{\!\![\ \text{Lemma 8.5.4}\ ]\!\!\} \\
\subseteq \prec ;(R_1 \times R_2) \,;\, \textcircled{\#} \,;\, (S_1 \times S_2) \,;\, \succ & \{\!\![\ \text{forward compatibility}\ ]\!\!\} \\
= \prec ;(R_1 \times R_2) \,;\, \succ \,;\, \prec ;(S_1 \times S_2) \,;\, \succ & \{\!\![\ \text{Lemma 8.5.2}\ ]\!\!\} \\
= (R_1 \bowtie R_2) \,;\, (S_1 \bowtie S_2) & \{\!\![\ \text{Lemma 8.3}\ ]\!\!\}
\end{array}$$

The proof under backward compatibility of S_1, S_2 is symmetric. □

Finally, we show the announced result on the join of tests.

Lemma 10.5. *If $P_i :: T_i$ $(i = 1, 2)$ are tests then $P_1 \bowtie P_2 :: T_1 \bowtie T_2$ is a test with $\neg(P_1 \bowtie P_2) = \neg P_1 \bowtie 1_{T_2} \cup 1_{T_1} \bowtie \neg P_2$, where $\neg P = 1 - P$.*

Proof. First,

$$\begin{array}{ll}
& (P_1 \bowtie P_2) \,;\, (\neg P_1 \bowtie 1_{T_2} \cup 1_{T_1} \bowtie \neg P_2) \\
= & \quad \{\!\![\ \text{distributivity}\ ]\!\!\} \\
& (P_1 \bowtie P_2) \,;\, (\neg P_1 \bowtie 1_{T_2}) \cup (P_1 \bowtie P_2) \,;\, (1_{T_1} \bowtie \neg P_2) \\
\subseteq & \quad \{\!\![\ \text{forward interchange (Theorem 9.2)}\ ]\!\!\}
\end{array}$$

$$
\begin{array}{ll}
& (P_1\,;\neg P_1) \bowtie (P_2\,;1_{T_2}) \cup (P_1\,;1_{T_1}) \bowtie (P_2\,;\neg P_2) \\
= & \quad \{\!|\ P_i \text{ tests and strictness of join } |\!\} \\
& 0_{T_1 \bowtie T_2}
\end{array}
$$

Second,

$$
\begin{array}{ll}
& P_1 \bowtie P_2 \cup \neg P_1 \bowtie 1_{T_2} \cup 1_{T_1} \bowtie \neg P_2 \\
= & \quad \{\!|\ \text{Boolean algebra and distributivity of join } |\!\} \\
& P_1 \bowtie P_2 \cup \neg P_1 \bowtie P_2 \cup \neg P_1 \bowtie \neg P_2 \cup P_1 \bowtie \neg P_2 \cup \neg P_1 \bowtie \neg P_2 \\
= & \quad \{\!|\ \text{distributivity of join and Boolean algebra } |\!\} \\
& 1_{T_1} \bowtie P_2 \cup 1_{T_1} \bowtie \neg P_2 \\
= & \quad \{\!|\ \text{distributivity of join and Boolean algebra } |\!\} \\
& 1_{T_1} \bowtie 1_{T_2} \\
= & \quad \{\!|\ \text{Lemma 8.7 } |\!\} \\
& 1_{T_1 \bowtie T_2}
\end{array}
$$

□

11 About Weak Matching

We have seen that strong matching turns $\bowtie$-;-interchange from inclusion to equation form (Lemma 10.3, Theorems 9.2 and 10.4). We now show that weak matching does the same for distributivity of domain over join.

Theorem 11.1. *Weakly matching $R_i :: T_i^2$ satisfy $\ulcorner R_1 \bowtie \ulcorner R_2 \subseteq \ulcorner(R_1 \bowtie R_2)$.*

Proof. By Lemmas 8.3, 8.5.4, weak matching with Lemma 10.3.3, domain representation (1), isotony, Lemmas 8.5(8,3) and 8.3 with domain representation (1):

$$
\begin{aligned}
\ulcorner R_1 \bowtie \ulcorner R_2 &= \prec;(\ulcorner R_1 \times \ulcorner R_2)\,;\succ\; = \;\prec;\textcircled{\#}\,;(\ulcorner R_1 \times \ulcorner R_2)\,;\succ \\
&\subseteq \prec;\ulcorner((R_1 \times R_2)\,;\textcircled{\#})\,;\succ\; = \;\prec;((R_1 \times R_2)\,;\textcircled{\#}\,;\mathsf{T} \cap 1)\,;\succ \\
&\subseteq \prec;(R_1 \times R_2)\,;\textcircled{\#}\,;\mathsf{T}\,;\succ \cap \prec;1\,;\succ\; = \;\prec;(R_1 \times R_2)\,;\succ;\mathsf{T} \cap 1 \\
&= \ulcorner(R_1 \bowtie R_2)
\end{aligned}
$$

□

Weak matching is even equivalent to distributivity of domain.

Theorem 11.2. *If $\ulcorner R_1 \bowtie \ulcorner R_2 \subseteq \ulcorner(R_1 \bowtie R_2)$ then R_1, R_2 are weakly matching.*

Proof. We first prove that an injective relation S and an arbitrary relation R satisfy $S;\ulcorner(S^{\smile};R) = \ulcorner R;S$. By domain representation (1), then by $R;(S;\mathsf{T}\cap 1) = R \cap \mathsf{T}\,;S^{\smile}$ and laws of $^{\smile}$, right distributivity due to injectivity of S, $P^{\smile} = P$ for any test P and domain representation (1):

$$
\begin{aligned}
S\,;\ulcorner(S^{\smile}\,;R) &= S\,;(S^{\smile}\,;R\,;\mathsf{T} \cap 1) = S \cap \mathsf{T}\,;R^{\smile}\,;S \\
&= (1 \cap \mathsf{T}\,;R^{\smile})\,;S = (1 \cap R\,;\mathsf{T})\,;S = \ulcorner R\,;S.
\end{aligned}
$$

To prove the theorem, we assume $\ulcorner R_1 \bowtie \ulcorner R_2 \subseteq \ulcorner(R_1 \bowtie R_2)$ and prove $\textcircled{\#}\,;(\ulcorner R_1 \times \ulcorner R_2) \subseteq \ulcorner((R_1 \times R_2)\,;\textcircled{\#})$ (see Lemma 10.3.3).

$$
\begin{array}{ll}
\quad ⊕ ; (\ulcorner R_1 \times \ulcorner R_2) & \\
= ⊕ ; (\ulcorner R_1 \times \ulcorner R_2) ; ⊕ & \{\!\!\{\ ⊕ \text{ and } \ulcorner R_1 \times \ulcorner R_2 \text{ are tests, idempotence} \\
 & \quad \text{and commutativity of tests } \}\!\!\} \\
= \succ ; \prec ; (\ulcorner R_1 \times \ulcorner R_2) ; \succ ; \prec & \{\!\!\{\ \text{Lemma 8.5.2 } \}\!\!\} \\
= \succ ; (\ulcorner R_1 \bowtie \ulcorner R_2) ; \prec & \{\!\!\{\ \text{Lemma 8.3 } \}\!\!\} \\
\subseteq \succ ; \ulcorner(R_1 \bowtie R_2) ; \prec & \{\!\!\{\ \text{assumption and isotony } \}\!\!\} \\
= \succ ; \ulcorner(\prec ; (R_1 \times R_2) ; \succ) ; \prec & \{\!\!\{\ \text{Lemma 8.3 } \}\!\!\} \\
= \ulcorner((R_1 \times R_2) ; \succ) \succ ; \prec & \{\!\!\{\ \text{Lemma 8.5(1,5) and preliminary result } \}\!\!\} \\
= \ulcorner((R_1 \times R_2) ; ⊕ ; \succ) ; ⊕ & \{\!\!\{\ \text{Lemma 8.5(2,4) } \}\!\!\} \\
\subseteq \ulcorner((R_1 \times R_2) ; ⊕) & \{\!\!\{\ \ulcorner(R ; S) \subseteq \ulcorner R,\ ⊕ \text{ is a test and isotony } \}\!\!\} \quad \square
\end{array}
$$

12 Join and Maximal Elements

We now study how join and the maximum operator interact. First we show an interchange law for join and diamond.

Lemma 12.1. *For $R_i :: T_i^2$ and $P_i :: T_i$ $(i = 1, 2)$,*

$$|R_1 \bowtie R_2\rangle(P_1 \bowtie P_2) \ \subseteq\ |R_1\rangle P_1 \bowtie |R_2\rangle P_2.$$

If the $R_i ; P_i$ are weakly matching then this strengthens to an equality.

Proof. By definition of inverse image, Theorem 9.2 with Lemma 10.3.1 and Theorem 10.4, Theorem 9.3 and definition of inverse image:

$$
\begin{aligned}
|R_1 \bowtie R_2\rangle(P_1 \bowtie P_2) &= \ulcorner((R_1 \bowtie R_2) ; (P_1 \bowtie P_2)) = \ulcorner((R_1 ; P_1) \bowtie (R_2 ; P_2)) \\
&\subseteq \ulcorner(R_1 ; P_1) \bowtie \ulcorner(R_2 ; P_2) = |R_1\rangle P_1 \bowtie |R_2\rangle P_2)
\end{aligned}
$$

The claim when the $R_i ; P_i$ are weakly matching follows by using Theorem 11.1 in the third step. □

This is used to derive an interaction law for join and maximum.

Lemma 12.2. *Consider tables $P :: T_P, Q :: T_Q$ and relations $R :: T_P^2$ and $S :: T_Q^2$ such that $R ; P$ and $S ; Q$ are weakly matching. Then*

$$(R \bowtie S) \rhd (P \bowtie Q) = (R \rhd P) \bowtie Q \cup P \bowtie (S \rhd Q).$$

Proof.

$$
\begin{array}{ll}
\quad (R \bowtie S) \rhd (P \bowtie Q) & \\
= (P \bowtie Q) - |R \bowtie S\rangle(P \bowtie Q) & \{\!\!\{\ \text{definition of } \rhd \ \}\!\!\} \\
= (P \bowtie Q) - (|R\rangle P \bowtie |S\rangle Q) & \{\!\!\{\ \text{Lemma 12.1 } \}\!\!\} \\
= (P \bowtie Q) ; \neg(|R\rangle P \bowtie |S\rangle Q) & \{\!\!\{\ \text{definition of } - \ \}\!\!\} \\
= (P \bowtie Q); & \{\!\!\{\ \text{complement of test (Lemma 10.5) } \}\!\!\} \\
\quad (\neg|R\rangle P \bowtie 1_Q \cup 1_P \bowtie \neg|S\rangle Q) & \\
= (P ; \neg|R\rangle P) \bowtie (Q ; 1_Q) & \{\!\!\{\ \text{distributivity and interchange laws of} \\
\quad \cup\ (P ; 1_P) \bowtie Q ; \neg|S\rangle Q) & \quad \text{Theorems 9.2 and 10.4, since } P, Q \text{ are tests } \}\!\!\} \\
= (R \rhd P) \bowtie Q \cup P \bowtie (S \rhd Q) & \{\!\!\{\ \text{neutrality of 1 and definition of } \rhd \ \}\!\!\} \quad \square
\end{array}
$$

Corollary 12.3. *Consider tables $P :: T_P, Q :: T_Q$ and a relation $R :: T_P^2$ such that $R\,;P$ and $\mathsf{T}_Q\,;Q$ are weakly matching. Then*

$$(R \bowtie \mathsf{T}_Q) \triangleright (P \bowtie Q) = (R \triangleright P) \bowtie Q.$$

Proof. Immediate from Lemma 12.2, (3), strictness of $\bowtie$ and neutrality of 0. □

This shows (4)—the only question is how to establish weak matching. For this we introduce a sufficient condition.

Definition 12.4. Assume tables $P :: T_P, Q :: T_Q$. We call P *joinable* with Q if $P \subseteq |\#\rangle Q$, where $\#$ is the matching relation between tuples. Pointwise, P is joinable with Q iff $\forall p \in P : \exists q \in Q : p \# q$. Informally this means that every tuple in P has a join partner in Q.

Lemma 12.5. *If P is joinable with Q then $R;P$ and $\mathsf{T}_Q;Q$ are weakly matching.*

Since the proof needs additional notions we defer it to the Appendix.
Now we can state an optimisation rule involving a θ-join.

Theorem 12.6. *Consider $P :: T_P, Q :: T_Q$, $R :: T_P^2$ as well as $\theta :: \{A\} \bowtie \{B\}$ with $A \in T_P, B \in T_Q$ with $T_P \cap T_Q = \emptyset$. If P is joinable with $\theta \bowtie Q$ then*

$$(R \bowtie \mathsf{T}_{\theta \bowtie Q}) \triangleright (P \bowtie \theta \bowtie Q) = (R \triangleright P) \bowtie \theta \bowtie Q.$$

This is immediate from Lemma 12.5 and Corollary 12.3.
Without the premise of joinability the theorem need not hold.

Example 12.7. Choose, for instance, θ as equality and $T_P = \{A\}$, $P = \{1, 2\}$, $T_Q = \{B\}$, $Q = \{1\}$ as well as $D_A = D_B = \{1, 2\}$. Here $\{2\}$ has no join partner in $\theta \bowtie Q$. Now for a preference R with $1\,R\,2$ we have the differing expressions

$$(R \bowtie \mathsf{T}_{\theta \bowtie Q}) \triangleright (P \bowtie \theta \bowtie Q) = (R \bowtie \mathsf{T}_{\theta \bowtie Q}) \triangleright \{(1, 1)\} = \{(1, 1)\},$$
$$(R \triangleright P) \bowtie \theta \bowtie Q = \{2\} \bowtie \theta \bowtie \{1\} = 0.$$

□

13 Conclusion and Outlook

We have presented a new and simple approach to an algebraic treatment of the theta join in databases. This is a piece that was missing in the predecessor paper [16], because there mostly only joins of tables with disjoint attribute sets were treated. However, overlapping types are mandatory for coping with theta joins. And so other important outcomes of the present paper are the more liberal notions of weak and strong matching of binary relations over database tuples.

With the help of the developed tools we have algebraically proved the correctness of two sample optimisation rules, namely "push projection over join" and "push preference over join".

Further work will be to treat the large catalogue of preference optimisation rules in [14] with these techniques. This also concerns the complex preference relation constructors of Pareto and prioritised composition. In fact, the relation $R \bowtie \mathsf{T}_U$ in Theorem 12.6 is equal to the prioritised preference $R \,\&\, 0$.

The present treatment was performed in the setting of concrete binary relations. While mostly point-free, some of the basic lemmas in Sect. 3 still were proved in a pointwise fashion. A next step to a more abstract view would be to axiomatise the projections and then reason point-free in terms of these. Another more abstract approach could be based on the concept of typed join algebras from the predecessor paper [16].

Acknowledgement. Helpful comments were provided by Patrick Roocks, Andreas Zelend and the anonymous referees.

14 Appendix

For types T_P, T_Q we use the notion of a *direct product* of D_P and D_Q (e.g. [17]). This is a pair (ρ_P, ρ_Q) of relations with $\rho_P \subseteq (D_P \times D_Q) \times D_P$ and $\rho_Q \subseteq (D_P \times D_Q) \times D_Q$ such that

$$\begin{array}{ll} \breve{\rho_P} ; \rho_P = 1, & \breve{\rho_Q} ; \rho_Q = 1, \\ \rho_P ; \breve{\rho_P} \cap \rho_Q ; \breve{\rho_Q} = 1, & \breve{\rho_P} ; \rho_Q = \mathsf{T}. \end{array}$$

Using this concept the parallel product can be represented as

$$P \times Q \;=\; \rho_P ; P ; \breve{\rho_P} \cap \rho_Q ; Q ; \breve{\rho_Q}. \tag{5}$$

The following properties of direct products are used in the main proof[3]:

$$\rho_P ; \mathsf{T} = \mathsf{T} = \rho_Q ; \mathsf{T}, \tag{6}$$

$$(R_1 ; \breve{\rho_P} \cap R_2 ; \breve{\rho_Q}) ; (\rho_P ; S_1 \cap \rho_Q ; S_2) \;=\; R_1 ; S_1 \,\cap\, R_2 ; S_2. \tag{7}$$

Proof of **Lemma** 12.5. The proof consists in showing $\circledast ; (\ulcorner(R;P) \times \ulcorner(\mathsf{T}_Q;Q)) \subseteq \ulcorner(((R;P) \times (\mathsf{T}_Q;Q));\circledast)$ (see Lemma 10.3.3). We do this by showing the stronger property $\ulcorner(R;P) \times \ulcorner(\mathsf{T}_Q;Q) \subseteq \ulcorner(((R;P) \times (\mathsf{T}_Q;Q));\circledast)$, from which the original claim follows by $\circledast \subseteq 1$ and isotony of ;.

Since "joinable" is defined with # and the formula to prove uses $\circledast$, we have to make a connection between the two:

$$\circledast \;=\; \ulcorner(\rho_P ; \# \cap \rho_Q). \tag{8}$$

---

[3] Equation (7) is valid for concrete relations. For abstract relations, only $\subseteq$ holds. This phenomenon is called *unsharpness* in the literature (an early mention is [18], a further elaboration [1]). The situation is similar with Lemma 8.6.4. The paper [15] constructs an RA that does not satisfy sharpness.

This is analogous to the conversion of a relation to a vector explained in [17], which would give $\circledast\,;\top = (\rho_P\,;\#\cap\rho_Q)\,;\top$. The inverse transformation is $\# = \rho_P\breve{}\,;(\circledast\,;\top\cap\rho_Q)$. Both equations are easily verified. Using restriction (Lemma 2.1.7) and Boolean algebra, the second one can be simplified to $\# = \rho_P\breve{};\circledast\,;\rho_Q$. Then by Definition 12.4 and the definition of diamond (Definition 6.1) P is joinable with Q iff

$$P \subseteq {}^{\ulcorner}(\rho_P\breve{}\,;\circledast\,;\rho_Q\,;Q). \tag{9}$$

Now we calculate as follows.

$$\begin{aligned}
& {}^{\ulcorner}(R\,;P) \times {}^{\ulcorner}(\top_Q\,;Q) \\
= \quad & \{\!\![\text{ distributivity of domain over } \times \;]\!\!\} \\
& {}^{\ulcorner}((R\,;P) \times (\top_Q\,;Q)) \\
= \quad & \{\!\![\text{ (5) }]\!\!\} \\
& {}^{\ulcorner}(\rho_P\,;R\,;P\,;\rho_P\breve{} \cap \rho_Q\,;\top_Q\,;Q\,;\rho_Q\breve{}) \\
\subseteq \quad & \{\!\![\text{ Boolean algebra and isotony of } {}^{\ulcorner} \;]\!\!\} \\
& {}^{\ulcorner}(\rho_P\,;R\,;P\,;\rho_P\breve{}) \\
= \quad & \{\!\![\text{ locality (Lemma 2.1.6) }]\!\!\} \\
& {}^{\ulcorner}(\rho_P\,;R\,;P\,;{}^{\ulcorner}(\rho_P\breve{})) \\
= \quad & \{\!\![\; \rho_P \text{ is surjective, hence } \rho_P\breve{} \text{ is total }]\!\!\} \\
& {}^{\ulcorner}(\rho_P\,;R\,;P\,;1) \\
= \quad & \{\!\![\text{ neutrality of 1 and (9) with Boolean algebra }]\!\!\} \\
& {}^{\ulcorner}(\rho_P\,;R\,;P\,;{}^{\ulcorner}(\rho_P\breve{}\,;\circledast\,;\rho_Q\,;Q)) \\
= \quad & \{\!\![\text{ locality (Lemma 2.1.6) twice }]\!\!\} \\
& {}^{\ulcorner}(\rho_P\,;R\,;P\,;\rho_P\breve{}\,;{}^{\ulcorner}(\circledast\,;\rho_Q\,;Q)) \\
= \quad & \{\!\![\text{ domain representation (1) }]\!\!\} \\
& {}^{\ulcorner}(\rho_P\,;R\,;P\,;\rho_P\breve{}\,;(\circledast\,;\rho_Q\,;Q\,;\top\cap 1)) \\
= \quad & \{\!\![\; \circledast \text{ is a test, restriction (Lemma 2.1.7) and neutrality of 1 }]\!\!\} \\
& {}^{\ulcorner}(\rho_P\,;R\,;P\,;\rho_P\breve{}\,;(\rho_Q\,;Q\,;\top\cap\circledast)) \\
= \quad & \{\!\![\; R_1\,;(R_2\,;\top\cap R_3) = (R_1\cap\top\,;R_2\breve{})\,;R_3 \text{ for all } R_1, R_2, R_3, \\
& \quad \text{laws of converse and } Q \text{ is a test }]\!\!\} \\
& {}^{\ulcorner}((\rho_P\,;R\,;P\,;\rho_P\breve{}\cap\top\,;Q\,;\rho_Q\breve{})\,;\circledast) \\
= \quad & \{\!\![\text{ (6) }]\!\!\} \\
& {}^{\ulcorner}((\rho_P\,;R\,;P\,;\rho_P\breve{}\cap\rho_Q\,;\top_Q\,;Q\,;\rho_Q\breve{})\,;\circledast) \\
= \quad & \{\!\![\text{ (5) }]\!\!\} \\
& {}^{\ulcorner}(((R\,;P)\times(\top_Q\,;Q))\,;\circledast)
\end{aligned}$$

□

References

1. Berghammer, R., Haeberer, A., Schmidt, G., Veloso, P.: Comparing two different approaches to products in abstract relation algebra. In: Nivat, M., Rattray, C., Rus, T., Scollo, G. (eds.) AMAST 1993, pp. 167–176. Springer, Heidelberg (1993). https://doi.org/10.1007/978-1-4471-3227-1_16

2. Berghammer, R., von Karger, B.: Relational semantics of functional programs. In: Brink, C., Kahl, W., Schmidt, G. (eds.) Relational Methods in Computer Science. Advances in Computing Science, pp. 115–130. Springer, Heidelberg (1997). https://doi.org/10.1007/978-3-7091-6510-2_8
3. Dang, H.-H., Höfner, P., Möller, B.: Algebraic separation logic. J. Logic Algebraic Program. **80**(6), 221–247 (2011)
4. Dang, H.-H., Möller, B.: Reverse exchange for concurrency and local reasoning. In: Gibbons, J., Nogueira, P. (eds.) MPC 2012. LNCS, vol. 7342, pp. 177–197. Springer, Heidelberg (2012). https://doi.org/10.1007/978-3-642-31113-0_10
5. Desharnais, J., Möller, B., Struth, G.: Modal Kleene algebra and applications – a survey. J. Relational Methods Comput. Sci. **1**, 93–131 (2004)
6. Desharnais, J., Möller, B., Struth, G.: Kleene algebra with domain. ACM Trans. Comput. Logic **7**, 798–833 (2006)
7. Haeberer, A., Frias, M., Baum, G., Veloso, P.: Fork algebras. In: Brink, C., Kahl, W., Schmidt, G. (eds.) Relational Methods in Computer Science. Advances in Computing Science, pp. 54–69. Springer, Heidelberg (1997)
8. Horn, R., Johnson, C.: Topics in Matrix Analysis. Cambridge University Press, Cambridge (1991)
9. Kahl, W.: CalcCheck: a proof checker for teaching the "logical approach to discrete math". In: Avigad, J., Mahboubi, A. (eds.) ITP 2018. LNCS, vol. 10895, pp. 324–341. Springer, Cham (2018). https://doi.org/10.1007/978-3-319-94821-8_19
10. Kahl, W.: CalcCheck—A proof-checker for Gries and Schneider's Logical Approach to Discrete Math. http://calccheck.mcmaster.ca/
11. Kanellakis, P.: Elements of relational database theory. In: van Leeuwen, J. (ed.) Handbook of Theoretical Computer Science. Volume B: Formal Models and Semantics, pp. 1073–1156. Elsevier (1990)
12. Kießling, W.: Preference queries with SV-semantics. In: International Conference on Management of Data (COMAD 2005), pp. 15–26 (2005)
13. Kießling, W., Endres, M., Wenzel, F.: The preference SQL system – an overview. Bull. Tech. Committee Data Eng. **34**(2), 11–18 (2011). http://www.markusendres.de/preferencesql/
14. Kießling, W., Hafenrichter, B.: Algebraic optimization of relational preference queries. Technical report No. 2003–01. University of Augsburg, Institute of Computer Science, February 2003
15. Maddux, R.: On the derivation of identities involving projection functions. In: Csirmaz, L., Gabbay, D., de Rijke, M. (eds.) Logic Colloquium '92. Studies in Logic, Languages, and Information, pp. 143–163. CSLI Publications (1995)
16. Möller, B., Roocks, P.: An algebra of database preferences. J. Logical Algebraic Methods Program. **84**(3), 456–481 (2015)
17. Schmidt, G., Ströhlein, T.: Relations and Graphs: Discrete Mathematics for Computer Scientists. EATCS Monographs on Theoretical Computer Science. Springer, Heidelberg (1993). https://doi.org/10.1007/978-3-642-77968-8
18. Zierer, H.: Programmierung mit Funktionsobjekten: Konstruktive Erzeugung semantischer Bereiche und Anwendung auf die partielle Auswertung. Institut für Informatik, Technische Universität München. Report TUM-I8803, February 1988

Bisimilarity of Diagrams

Jérémy Dubut[1,2(✉)]

[1] National Institute of Informatics, Tokyo, Japan
dubut@nii.ac.jp
[2] Japanese-French Laboratory of Informatics, CNRS IRL 3527, Tokyo, Japan

Abstract. In this paper, we investigate diagrams, namely functors from any small category to a fixed category, and more particularly, their bisimilarity. Initially defined using the theory of open maps of Joyal et al., we prove two characterisations of this bisimilarity: it is equivalent to the existence of a bisimulation-like relation and has a logical characterisation à la Hennessy and Milner. We then prove that we capture both path bisimilarity and strong path bisimilarity of any small open maps situation. We then look at the particular case of finitary diagrams with values in real or rational vector spaces. We prove that checking bisimilarity and satisfiability of a positive formula by a diagram are both decidable by reducing to a problem of existence of invertible matrices with linear conditions, which in turn reduces to the existential theory of the reals.

Keywords: Open maps · Diagrams · Path logic · Existential theories

1 Introduction

Diagrams in a category, namely functors from any small category to this specified category, are essential objects in category theory. Numerous basic constructions in category theory can be seen as a limit or colimit of a suitable diagram. However, their usefulness is not limited to those.

In the context of directed algebraic topology (see [7] for a textbook), Dubut et al. used diagrams with values in a category of modules to encode local geometric properties of a directed space and their evolution [5]. The domains of those diagrams are given by directed paths of the space and their extensions, while the diagrams themselves map such a path to some homology modules describing the default of directed homotopy of the space. It was then observed that a suitable notion of bisimilarity of diagrams, using the general theory of open maps from [10] was the right notion to compare such defaults of dihomotopy.

In the first part of this paper, we propose to look at the general theory of bisimilarity of diagrams, extending it to any category of observations. After describing the original definition using open morphisms (Sect. 2.2), we describe two equivalent characterisations. First (Sect. 2.3), it is equivalent to the existence

The author was supported by ERATO HASUO Metamathematics for Systems Design Project (No. JPMJER1603), JST and Grant-in-aid No. 19K20215, JSPS.

U. Fahrenberg et al. (Eds.): RAMiCS 2020, LNCS 12062, pp. 65–81, 2020.
https://doi.org/10.1007/978-3-030-43520-2_5

of a relation, similar to history preserving bisimulations of event structures from [14]. This result generalises a result from [5]. Secondly (Sect. 3), it has a logical characterisation, similar to a Hennessy-Milner theorem: two diagrams are bisimilar if and only if they both satisfy the same formulae of a path logic. We finally prove in Sect. 4 that we capture path and strong path bisimilarities of any open map situation [10] as the bisimilarity of a suitable notion of diagrams.

In a second part, we consider two decision problems for a class of diagrams with values in real or rational vector spaces, used in [5] for describing defaults of dihomotopy of geometric models of truly concurrent systems (See Sect. 5). For those diagrams, we prove in Sect. 7 that bisimilarity and the satisfaction of a positive formula are both decidable by reduction to a problem of invertible matrices, itself reduced to the existential theory of the reals (Sect. 6).

Existing Work: This theory of bisimilarity of diagrams is intimately related to categorical theories of bisimulations. If the relation with open maps is developed in Sects. 2.2 and 4, its relation with coalgebra is less clear. Relations between open maps and coalgebra are investigated in [11,19], however those cannot be applied in general to a category of diagrams. The main problem is that diagrams are not naturally coalgebras in general, and so there is no clear relationship between open maps as described in Sect. 2.2, and coalgebra homomorphisms (also called coverings of processes in [18]). Another important related line of work is the theory of bisimilarity of presheaves [4], which considers similar objects (presheaves are particular cases of diagrams), but from a very different point of view.

2 Bisimilarity and Bisimulations of Diagrams

Diagrams with values in a fixed category $\mathcal{A}$ are functors $F : \mathcal{C} \longrightarrow \mathcal{A}$ from any small category to $\mathcal{A}$. If $\mathcal{A}$ is thought as a category of "observations" and $\mathcal{C}$ as the category of executions of a system, a diagram encodes the trace of observations along every execution (typically, a label), and its actions on morphisms of $\mathcal{C}$ encodes how these observations change when the system evolves.

In this section, we describe the original form of the bisimilarity from [5], defined as the existence of a span of particular morphisms of diagrams having some lifting properties. We then develop an equivalent characterization using relations, similar to bisimulations of event structures as introduced in [14].

2.1 Category of Diagrams

The original definition of bisimilarity of diagrams was designed using particular morphisms of diagrams. Such a morphism, say from the diagram $F : \mathcal{C} \longrightarrow \mathcal{A}$ to the diagram $G : \mathcal{D} \longrightarrow \mathcal{A}$ is a pair (Φ, σ) with $\Phi : \mathcal{C} \longrightarrow \mathcal{D}$ a functor and σ is a natural *isomorphism* from F to $G \circ \Phi$. The composition $(\Psi, \tau) \circ (\Phi, \sigma)$ is defined as $(\Psi \circ \Phi, (\tau_{\Phi(c)} \circ \sigma_c)_{c \text{ object of } \mathcal{C}})$. We denote this category by $\mathbf{Diag}(\mathcal{A})$.

Example 1. Throughout the next two sections, we will develop a particular example of diagrams in which transition systems can be encoded. This example will allow us to relate constructions in diagrams to classical constructions in concurrency theory. From now, we fix a set L called the **alphabet**. Such a set induces a poset (which can be seen as a category) $\mathcal{A}_L$ whose elements are the finite words on L and whose order is the prefix order. A transition system T on L produces a diagram $F_T : \mathcal{C}_T \longrightarrow \mathcal{A}_L$ as follows. The category $\mathcal{C}_T$ is formed by considering as objects the runs of T, that is, sequences $i \xrightarrow{a_1} q_1 \xrightarrow{a_2} \ldots \xrightarrow{a_n} q_n$ of transitions of T where i is the initial state, and by ordering them by prefix. F_T then maps a run to its sequence of labels. This construction extends to a functor Π from the category **TS(L)** of transition systems on L to the category **Diag($\mathcal{A}_L$)**. Conversely, a diagram $F : \mathcal{C} \longrightarrow \mathcal{A}_L$ produces a transition system T as follows. First, such a diagram can be identified with a diagram with values in **TS(L)** by identifying a word $a_1.a_2.\ldots.a_n$ with the finite linear transition system $0 \xrightarrow{a_1} 1 \xrightarrow{a_2} \ldots \xrightarrow{a_n} n$. T is then obtained by forming the colimit of this diagram in **TS(L)**. This extends to a functor Γ from **Diag($\mathcal{A}_L$)** to **TS(L)**. Note that $\Gamma \circ \Pi$ is the unfolding of transition systems and that Γ is the left adjoint of Π.

The reason why we need natural isomorphisms in the definition of a morphism of diagram is not clear yet, as the only isomorphisms in the category $\mathcal{A}_L$ are the identities. This will be illustrated in the case where $\mathcal{A}$ is a category of vector spaces. Intuitively, two isomorphic vector spaces represent the same kind of observations (in the case of directed algebraic topology, the same kind and number of holes), which we do not want to discriminate.

2.2 Open Morphisms of Diagrams

The original idea from [5] was to compare diagrams similarly to transition systems using the theory of [10]. Let us call **branch** a diagram from $\mathbf{n}$ to $\mathcal{A}$ for $n \in \mathbb{N}$, where $\mathbf{n}$ is the poset (seen as a category) $\{1, \ldots, n\}$ with the usual ordering. An **evolution** of a diagram $F : \mathcal{C} \longrightarrow \mathcal{A}$ is then a morphism from any branch to F. Much as transition systems and executions, a morphism of diagrams (Φ, σ) from $F : \mathcal{C} \longrightarrow \mathcal{A}$ to $G : \mathcal{D} \longrightarrow \mathcal{A}$ maps evolutions of F to evolutions of G: if (Ψ, τ) is an evolution of F, i.e., a morphism from a branch to F, then $(\Phi, \sigma) \circ (\Psi, \tau)$ is an evolution of G. Then morphisms act as particular simulations of diagrams. The idea from [10] was to provide conditions on morphisms for them to act as particular bisimulations. The general idea is that a morphism induces a bisimulation if it lifts evolutions of G to evolutions of F. In the context of diagrams, this will be defined using **extensions of branches**. An extension of a branch $B : \mathbf{n} \longrightarrow \mathcal{A}$ is a morphism of diagrams (Π, θ) from $B : \mathbf{n} \longrightarrow \mathcal{A}$ to a branch $B' : \mathbf{n}' \longrightarrow \mathcal{A}$, with $n' \geq n$ such that:

- for every $i \leq n$, $B(i) = B'(i)$,
- for every $i \leq j \leq n$, the morphism $B'(i \leq j)$ of $\mathcal{A}$ is equal to $B(i \leq j)$,
- for every $i \leq n$, $\Pi(i) = i$,
- for every $i \leq n$, $\theta_i = id_{B(i)}$.

$$B(1) \xrightarrow{B(1 \leq 2)} \cdots \xrightarrow{B(n-1 \leq n)} B(n)$$

$$\text{id} \downarrow \qquad\qquad\qquad\qquad\qquad \downarrow \text{id}$$

$$B'(1) \xrightarrow[B'(1 \leq 2)]{} \cdots \xrightarrow[B'(n-1 \leq n)]{} B'(n) \xrightarrow[B'(n \leq n+1)]{} \cdots \xrightarrow[B'(n'-1 \leq n')]{} B'(n')$$

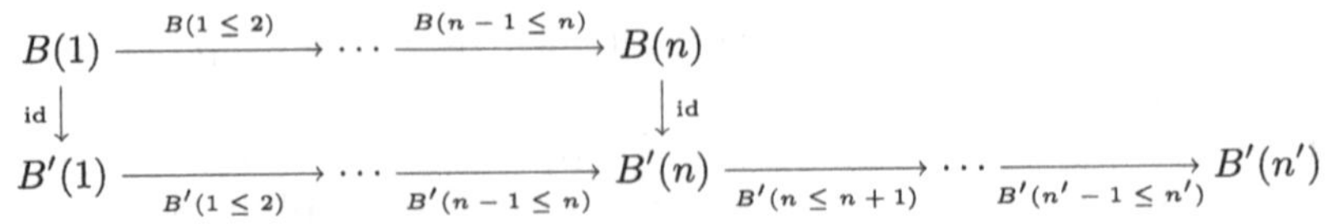

Fig. 1. Extension of branches

Those conditions mean that the restriction of B' to **n** is B and that the morphism (Π, θ) is the inclusion of B in B' (Fig. 1).

Following [10], we then say that a morphism (Φ, σ) from $F : C \longrightarrow A$ to $G : D \longrightarrow A$ is **open** if for every diagram of the form (in plain):

$$
\begin{array}{ccc}
 & (\Psi, \tau) & \\
B & \longrightarrow & F \\
{\scriptstyle (\Pi,\theta)} \downarrow & {\scriptstyle \exists} \nearrow & \downarrow {\scriptstyle (\Phi,\sigma)} \\
B' & \longrightarrow & G \\
 & (\Psi', \tau') &
\end{array}
$$

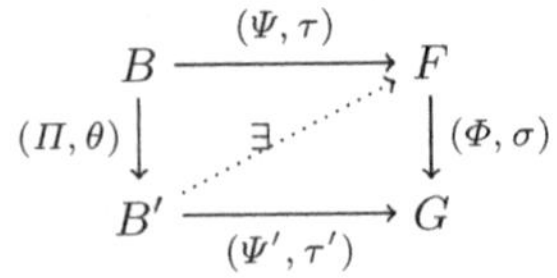

where (Π, θ) is an extension of branches, there is an evolution of F (in dots) which makes the two triangles commute. This means that if we can extend an evolution of F, mapped on an evolution of G by (Φ, σ), as a longer evolution of G, then we can extend it as a longer evolution of F that is mapped to this longer evolution of G. This means in particular that F and G have exactly the same evolutions. As observed in [5], the definition of an open map can be simplified as follows:

Theorem 1. *A morphism (Φ, σ) is open if and only if:*

- *Φ is surjective on objects, i.e., for every object d of D, there is an object c of C such that $\Phi(c) = d$,*
- *Φ is a fibration, i.e., for every morphism of D of the form $j : \Phi(c) \longrightarrow d'$, there is a morphism $i : c \longrightarrow c'$ of C such that $\Phi(i) = j$.*

Following [10], we say that two diagrams $F : C \longrightarrow A$ and $G : D \longrightarrow A$ are **bisimilar** if there is a span of open morphisms between them, that is, a diagram $H : \mathcal{E} \longrightarrow A$ and two open morphisms, one from H to F, one from H to G.

Example 2. In the case of diagrams in A_L, the notion of open morphisms is related to the notion of open morphisms of transition systems as defined in [10]. First, an open morphism $f : T \longrightarrow S$ between transition systems always induces an open morphism $\Pi(f) : \Pi(T) \longrightarrow \Pi(S)$ between the associated diagrams. In particular, if two transition systems are bisimilar then their diagrams are bisimilar. The converse also holds but proving it using open morphisms is hard (the reason will be explained later). For example, we may expect that an open morphism of diagrams of the form $\Phi : F \longrightarrow \Pi(T)$ induces an open morphism between transition systems $\Gamma(\Phi) : \Gamma(F) \longrightarrow \Gamma \circ \Pi(T)$, but that is not true in general.

2.3 Bisimulations of Diagrams

In this section, we generalise a notion of bisimulation relations from [5], which is equivalent to the existence of a span of open morphisms. This result is an equivalent of Theorem 3.1 in [18] in the context of open maps of diagrams.

A **bisimulation** R between two diagrams $F : \mathcal{C} \longrightarrow \mathcal{A}$ and $G : \mathcal{D} \longrightarrow \mathcal{A}$ is a set of triples (c, f, d) where c is an object of $\mathcal{C}$, d is an object of $\mathcal{D}$ and $f : F(c) \longrightarrow G(d)$ is an isomorphism of $\mathcal{A}$ such that:

- for every (c, f, d) in R and $i : c \longrightarrow c' \in \mathcal{C}$, there exist $j : d \longrightarrow d' \in \mathcal{D}$ and $g : F(c') \longrightarrow G(d') \in \mathcal{A}$ such that $g \circ F(i) = G(j) \circ f$ and $(c', g, d') \in R$,

$$\begin{array}{ccccc} c & F(c) & \xrightarrow{\quad f \quad} & G(d) & d \\ {\scriptstyle i}\downarrow & \downarrow{\scriptstyle F(i)} & & \vdots\, {\scriptstyle G(j)} & \vdots\, {\scriptstyle j} \\ c' & F(c') & \xrightarrow[\quad g \quad]{\cdots\cdots} & G(d') & d' \end{array}$$

- symmetrically, for every (c, f, d) in R and $j : d \longrightarrow d' \in \mathcal{D}$, there exist $i : c \longrightarrow c' \in \mathcal{C}$ and $g : F(c') \longrightarrow G(d') \in \mathcal{A}$ such that $g \circ F(i) = G(j) \circ f$ and $(c', g, d') \in R$,
- for all $c \in \mathcal{C}$, there exists d and f such that $(c, f, d) \in R$,
- for all $d \in \mathcal{D}$, there exists c and f such that $(c, f, d) \in R$.

Theorem 2. *Two diagrams are bisimilar if and only if there is a bisimulation between them.*

Example 3. In the case of diagrams in $\mathcal{A}_L$, a bisimulation between diagrams $\Pi(T)$ and $\Pi(S)$ is just a rephrasing for a path bisimulation in the sense of [10] between the transition systems T and S. In the particular case of transition systems, the existence of a path bisimulation is equivalent to the existence of a strong path bisimulation and is equivalent to the existence of a bisimulation. Consequently:

Proposition 1. *Two transition systems T and S are bisimilar if and only if the diagrams $\Pi(T)$ and $\Pi(S)$ are bisimilar.*

3 Diagrammatic Path Logic

In this section, we focus on a logical characterization of bisimilarity of diagrams. The logic used, which we call **diagrammatic path logic**, is similar to the logic introduced in [9] for transition systems, or to path logics developed in [10]. A formula in this logic allows one to express that a diagram has some kind of evolutions or not.

The formulae used are generated by the following grammar:

Object formulae: $S ::= [x]P \qquad x \in \mathrm{Ob}(\mathcal{A})$

Morphism formulae: $P ::= \langle f \rangle P \mid ?S \mid \neg P \mid \bigwedge_{i \in I} P_i$ $f \in \mathrm{Mor}(\mathcal{A})$ and I a set

where $\mathrm{Ob}(\mathcal{A})$ is the class of objects of $\mathcal{A}$ and $\mathrm{Mor}(\mathcal{A})$ is its class of morphisms.

Intuitively, the object formula $[x]P$ means that the current object is isomorphic to x, and the morphism formula $\langle f \rangle P$ means that from the current object, one can fire a transition labelled by a morphism equivalent (in the sense of matrices, or conjugate in the language of group theory) to f. Observe that we have arbitrary conjunctions, in particular infinite and empty (we will denote the empty conjunction by $\top$).

Example 4. In the case of diagrams in $\mathcal{A}_L$, $[w]\top$ means that the current run is labeled by the word w and $\langle w \leq w' \rangle \top$ means that the current run is labeled by w and that it can be extended to a run labeled by w'. The idea is very similar to the Hennessy-Milner logic [9] and the forward path logic [10]. The next theorem proves that, for two transition systems, satisfying the same Hennessy-Milner formulae, forward path formulae or path formulae is the same as their diagrams satisfying the same diagrammatic formulae.

For a diagram $F : \mathcal{C} \longrightarrow \mathcal{A}$, an object c of $\mathcal{C}$, and an isomorphism f of $\mathcal{A}$ of the form $f : F(d) \longrightarrow x$ for some d and x, we define $F, c \models S$ for an object formula S and $F, f, d \models P$ for a morphism formula P by induction on S (resp. P) as follows:

- $F, f, c \models \top$ always,
- more generally, $F, f, c \models \bigwedge_{i \in I} P_i$ iff for all $i \in I$, $F, f, c \models P_i$,
- $F, c \models [x]P$ iff there exists an isomorphism $f : F(c) \longrightarrow x$ of $\mathcal{A}$ such that $F, f, c \models P$,
- for every $g : x \longrightarrow x'$, $F, f, c \models \langle g \rangle P$ iff there exists $i : c \longrightarrow c'$ in $\mathcal{C}$ and an isomorphism $h : F(c') \longrightarrow x'$ such that $h \circ F(i) = g \circ f$ and $F, h, c' \models P$,

$$
\begin{array}{ccc}
F(c) & \xrightarrow{\;\;F(i)\;\;} & F(c') \\
{\scriptstyle f}\big\downarrow & & \big\downarrow{\scriptstyle h} \\
x & \xrightarrow{\;\;g\;\;} & x'
\end{array}
$$

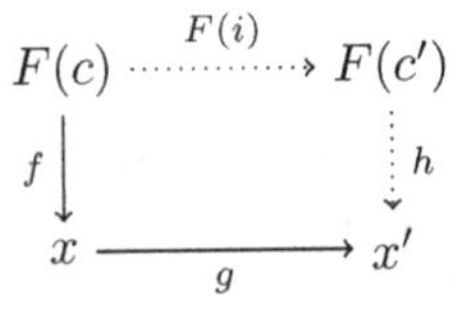

- $F, f, c \models ?S$ iff $F, c \models S$,
- $F, f, c \models \neg P$ iff $F, f, c \not\models P$.

We say that a diagram $F : \mathcal{C} \longrightarrow \mathcal{A}$ is **logically simulated** by another diagram $G : \mathcal{D} \longrightarrow \mathcal{A}$ if for every object c of $\mathcal{C}$, there exists an object d of $\mathcal{D}$ such that for all object formula S, $F, c \models S$ iff $G, d \models S$. Two diagrams F and G are **logically equivalent** if F is logically simulated by G and vice-versa.

Theorem 3. *Two diagrams are bisimilar iff they are logically equivalent.*

4 Relation to Path Bisimilarities of Open Maps Situations

In Sect. 2.2, we understood bisimilarity of diagrams using branches, as the existence of a span of open maps. In the context of [10], it means that diagrams, together with their subcategory of branches is an **open map situation**. Concretely, an open map situation is a category $\mathcal{M}$, called the category of **systems** (in our case $\mathbf{Diag(\mathcal{A})}$), together with a subcategory $\mathcal{P} \hookrightarrow \mathcal{M}$, said of **paths** (here the subcategory of branches). Another typical example is the category of transition systems $\mathbf{TS(L)}$ with its subcategory of finite linear systems.

In [10], two notions of bisimulations between objects of $\mathcal{M}$ are described: the **path bisimulations** and the **strong path bisimulations**. However, for them to make sense, some conditions on the open map situation are required: $\mathcal{P}$ need to be small and $\mathcal{P}$ and $\mathcal{M}$ must have a common initial object, which we denote by I. For example, the open maps situation of transition systems satisfies those requirements, while the one of diagrams does not in general (smallness is the issue).

Concretely, a path bisimulation R between objects X and Y of $\mathcal{M}$ is a set of pairs of morphisms of the form $(x : P \longrightarrow X, y : P \longrightarrow Y)$ for some object P of $\mathcal{P}$ such that:

- The pair of initial morphisms $(I \to X, I \to Y)$ belongs to R.
- For every $(x : P \longrightarrow X, y : P \longrightarrow Y)$ in R, and for every morphisms $p : P \longrightarrow Q$ of $\mathcal{P}$ and $x' : Q \longrightarrow X$ with $x' \circ p = x$, there is a morphism $y' : Q \longrightarrow Y$ such that $y' \circ p = y$ and $(x', y') \in R$.
- Symmetrically, swapping the roles of X and Y.

Furthermore, we say that R is strong if it additionally satisfies that for every $(x : P \longrightarrow X, y : P \longrightarrow Y)$ in R, and for every morphism $p : Q \longrightarrow P$ of $\mathcal{P}$, $(x \circ p, y \circ p) \in R$.

It has to be remarked that those bisimulations induce different notions of bisimilarity in general. Furthermore, they both are different to the existence of a span of open morphisms, although strong path bisimilarity coincide with this existence in many concrete cases [6]. In the case of transition systems and finite linear systems, those three notions coincide.

We propose now to characterise those two notions of bisimulations using diagrams, namely, we will now describe two functors $\mathbf{Ex} : \mathcal{M} \longrightarrow \mathbf{Diag}(\mathcal{P})$ and $\overline{\mathbf{Ex}} : \mathcal{M} \longrightarrow \mathbf{Diag}(\overline{\mathcal{P}})$ such that the existence of a path (resp. strong path) bisimulation between X and Y is equivalent to the fact that $\mathbf{Ex}(X)$ and $\mathbf{Ex}(Y)$ (resp. $\overline{\mathbf{Ex}(X)}$ and $\overline{\mathbf{Ex}(Y)}$) are bisimilar as diagrams.

First, $\mathbf{Ex}(X)$ is the functor from $\mathcal{P} \downarrow X$, the category of morphisms from any objects of $\mathcal{P}$ to X and commutative triangles, to the category $\mathcal{P}$ which maps any morphism $x : P \longrightarrow X$ to P and every commutative triangles $x' \circ p = x$ to p.

Theorem 4. *X and Y are path bisimilar if and only if $\boldsymbol{Ex}(X)$ and $\boldsymbol{Ex}(Y)$ are bisimilar.*

Given a category $\mathcal{C}$, we denote by $\mathbf{Zig}(\mathcal{C})$ the category whose objects are those of $\mathcal{C}$ and whose morphisms are generated by those of $\mathcal{C}$ and those of $\mathcal{C}^{\mathrm{op}}$. This naturally extends to an endofunctor $\mathbf{Zig} : \mathbf{Cat} \longrightarrow \mathbf{Cat}$. We then define $\overline{\mathbf{Ex}} = \mathbf{Zig}(\mathbf{Ex})$ from $\mathbf{Zig}(\mathbf{Ex}(X))$ to $\mathbf{Zig}(\mathcal{P})$.

Theorem 5. *X and Y are strong path bisimilar if and only if $\overline{\mathbf{Ex}}(X)$ and $\overline{\mathbf{Ex}}(Y)$ are bisimilar.*

Remark 1. This pattern of characterising a notion bisimilarity as bisimilarity of suitable diagrams whose domain is a category of "runs" and whose codomain is a category of "observations" is more general than for (strong) path bisimilarity of open maps situations, and can be pursued for Higher-Dimensional Automata [8,13] for example.

5 Interlude

In the first part of the paper, we focused on the general theory of bisimilarity of diagrams and its relationship with usual notions of bisimilarity of transition systems (and so of process algebra). In the second part of the paper, we would like to turn our attention to other kinds of diagrams that appeared in the theory of directed algebraic topology [5]. While diagrams in Sect. 4 were typically with values in a category of words, we will now consider diagrams with values in modules on a ring. More precisely, we will focus on the following two problems for such diagrams:

- **bisimilarity:** given two diagrams, are they bisimilar?
- **diagrammatic model-checking:** given a diagram F, an object c of its domain and a state formula S, does $F, c \models S$ hold?

The difficulty of those problems lies in the possibility to decide whether two modules are isomorphic, problem which does not appear in the context of process algebras and transition systems. Indeed, it is known that those problems are decidable in the category of transition systems (see [16] for a dynamic list of such (un)decidability results), while they would be undecidable for diagrams with values in groups and group morphisms because it is undecidable whether two groups are isomorphic. In this paper, we will focus on the category of finite dimensional real or rational vector spaces and matrices.

More precisely, we will stick to finitary diagrams and finitary positive formulae defined as follows. By a **finitary diagram** F, we mean the following data:

- a finite poset $(\mathcal{C}, \leq)$, the **domain**,
- for every element c of $\mathcal{C}$, a natural number $F(c)$ (which stands for the real vector space $\mathbb{R}^{F(c)}$),
- for every pair $c \leq c'$ of $\mathcal{C}$, a matrix $F(c \leq c')$ of size $F(c) \times F(c')$, with coefficients in rational numbers, presented as the list of all its elements, such that:

- $F(c \leq c)$ is the identity matrix,
- for every triple $c \leq c' \leq c''$, $F(c \leq c'') = F(c' \leq c'').F(c \leq c')$, where '.' denotes the matrix multiplication.

In short, a finitary diagram is a functor from a finite poset to the category of matrices with coefficients in rational numbers. One may argue that those assumptions are not reasonable, because they are not satisfied by the diagrams from Sect. 4 as soon as there is a loop. The reason is that when deciding this bisimilarity, there are two problems: finding out how to relate the executions and constructing the bisimulation, in particular, the isomorphism part. Loops make the first part difficult, because this relation is necessarily infinite in this case. In this paper, we want to focus on the second problem because: (1) reducing the problem of existence of a bisimulation to a problem of isomorphisms in a category is the main difference from existence of bisimulation for process algebra, (2) solving this question addresses the problem of comparing natural homologies of geometric models of true concurrency from [5].

We call **finitary formulae**, the formulae generated by the following grammar:

$$\textbf{Object formulae:} \quad S ::= [n]P \qquad n \in \mathbb{N}$$

$$\textbf{Morphism formulae:} \quad P ::= \langle M \rangle P \mid ?S \mid \neg P \mid \top \mid P_1 \wedge P_2$$

where M is a matrix with coefficients in rational numbers. Here, $[n]P$ stands for $[\mathbb{R}^n]P$ which makes finitary formulae diagrammatic formulae in real vector spaces. This time, since we only have finitely branching diagrams, we only consider finite conjunctions. We will more particularly consider **positive** formulae, i.e., formulae without any occurrences of the negation. For example, a formula of the form $\langle M_1 \rangle \ldots \langle M_k \rangle \top$ means that there is a sequence of matrices $N_1, \ldots, N_k$ in the diagrams where N_i is equivalent to M_i, and those equivalences are natural (in the categorical meaning).

In this case, bisimilarity and model checking problems become a problem of existence of invertible matrices satisfying some linear conditions, as we will see in Sect. 7. In Sect. 6, we will start by proving that this problem of matrices can be encoded in the existential theory of the reals, which is known to be decidable.

6 Existential Theory of Invertible Matrices

In the present section, we focus on an existential theory of matrices. We first recall the case of the existential theory of the reals, which is known to be decidable. We then introduce the existential theory of invertible matrices in $\mathbb{R}$ and $\mathbb{Q}$ and we finally prove the decidability of their satisfiability problems.

6.1 The Existential Theory of Some Rings

Designing algorithms for finding solutions of equations is an old problem in mathematics. The famous Hilbert's tenth problem posed the problem for polynomial equations in integers, but the question can be asked for other rings.

Tarski in [17] solved this question for real numbers: the first-order logic of real closed fields is decidable, although the solution is of non-elementary complexity. Several improvements have been made: it was proved to be in EXPSPACE in [2] and that the existential theory of the reals is in PSPACE in [3]. On the contrary, Matiyasevich's negative answer of the tenth problem [12], means that the existential theory of the integers is undecidable. In particular, since it is possible to express that a rational number is an integer (using possibly universal quantifiers), the full first-order logic of the rationals is undecidable. However, it is still an open question whether its existential fragment is decidable or not.

6.2 Theory of Matrices

In this section, we will consider a logic of matrices that will be expressible in the existential theory of the reals. It will be the main ingredient to decide some problems in diagrams with values in vector spaces. Namely, we consider formulae of the form:

$$\exists_{n_1} X_1.\ldots.\exists_{n_k} X_k.\ \bigwedge_{j=1}^{m} P_j(X_1, \ldots, X_k)$$

where:

- $n_i \geq 0$, is a natural number,
- X_i is a variable ranging over invertible matrices of dimension n_i,
- P_j is a predicate of the form $A.X_i = X_k.B$ for some i, k and matrices A, B with coefficients in rational numbers, A and B are of size $n_k \times n_i$, and . denotes the matrix multiplication. We call it the **existential theory of invertible matrices**.

We will consider the following decision problem: given such a formula, is it satisfiable, that is, are there matrices M_1, ..., M_k, with M_i of size $n_i \times n_i$, invertible such that for every j, $P_j(M_1, ..., M_k)$ is true?

We may ask this question for matrices M_i in coefficients in real or rational numbers. We will prove that both problems actually coincide and are decidable in PSPACE.

6.3 Decidability in $\mathbb{R}$

We stick here to the case of real numbers. We prove that we have a reduction to the existential theory of the reals. Given a formula

$$\Phi = \exists_{n_1} X_1.\ldots.\exists_{n_k} X_k.\ \bigwedge_{j=1}^{m} P_j(X_1, \ldots, X_k)$$

we will construct a formula Ψ in the existential theory of the reals which is satisfiable if and only if Φ is.

First, for every variable X_i, check if it appears in some P_j. If not, forget it. Indeed, if it does not appear in any predicate, then we can just choose the identity. Then, for every other quantifier $\exists_{n_i} X_i$, we fix $2.n_i^2$ fresh first-order variables $x_i^{r,s}$ and $y_i^{r,s}$ for $r, s \in \{1, ..., n_i\}$. Let X_i be the matrix of size $n_i \times n_i$ whose coefficients are $x_i^{r,s}$, and Y_i whose coefficients are $y_i^{r,s}$. Developing $A.X_i = X_j.B$ leads to $n_j n_i$ linear equations on the variables $x_i^{r,s}$ and $x_j^{r,s}$. So every predicate P_j induces a set L_j of linear equations. It remains to express that X_i is invertible in the first-order logic. The idea is to express that Y_i is its inverse. Developing $X_i.Y_i = \mathrm{Id}$ and $Y_i.X_i = \mathrm{Id}$, leads to $2.n_i^2$ polynomial equations on the variables $x_i^{r,s}$ and $y_i^{r,s}$. Let S_i be the set of these equations. We denote by Ψ the formula:

$$\exists x_1^{1,1}. \dots \exists x_k^{n_k,n_k}. \exists y_1^{1,1}. \dots \exists y_k^{n_k,n_k}. \bigwedge_{i=1}^{k} S_i \wedge \bigwedge_{j=1}^{m} L_j$$

Ψ is of polynomial size on the size of Φ: indeed, the only problem might be that we fix $2n_i^2$ variables while n_i is of size $\log(n_i)$, which may say that we fix an exponential number of variables. The point is that if we fixed those $2n_i^2$ variables, then it means that X_i appears in some P_j, and that the matrices appearing in P_j have a polynomial size in n_i. Consequently, we fix only a polynomial number of variables.

Theorem 6. *Ψ is satisfiable in the existential theory of the reals iff Φ is satisfiable in the existential theory of invertible matrices with coefficients in real numbers. Consequently, the existential theory of invertible matrices with coefficients in real numbers is decidable in PSPACE.*

6.4 The Rational Case

As we have seen previously, first-order theories of rationals are in general harder than those in reals. But there are some algebraic problems that are known to coincide when considering real and rational numbers. Given a linear system with coefficients in rational numbers, Gaussian elimination works independently of the coefficient field. Consequently, the real subspace $F_{\mathbb{R}}$ of solutions of this system has the same dimension as the rational subspace $F_{\mathbb{Q}}$ of solutions of the system. Actually, $F_{\mathbb{R}} \cap \mathbb{Q}^n = F_{\mathbb{Q}}$ and they have a common basis whose vectors are with coefficients in rational numbers. Similarly, the problem of equivalence of matrices coincides in the fields of real and rational numbers. Given two matrices A and B with coefficients in rational numbers, A and B are equivalent if there are two invertible matrices X and Y such that $A.X = Y.B$. This problem is also solvable using Gaussian elimination by computing the rank of A and B, which is independent of the coefficient field. Our problem is a generalization of the equivalence problem and it is not surprising that the same kind of results hold:

Theorem 7. *A formula Φ is satisfiable in the existential theory of invertible matrices with coefficients in real numbers if and only if it is satisfiable in the existential theory of invertible matrices with coefficients in rational numbers.*

7 Decidability in Diagrams

Finally, we prove two decidability results for bisimilarity of diagrams and diagrammatic logic using the existential theory of invertible matrices. In this section, we consider diagrams with values in real vector spaces (or rational, but as we have seen in the previous section, both theories will coincide). We prove the decidability of the following two problems:

- **bisimilarity:** given two finitary diagrams, are they bisimilar?
- **diagram model-checking:** given a finitary diagram F, an object c of its domain and a positive finitary state formula S, does $F, c \models S$ hold?

7.1 Decidability of Bisimilarity

We start with the bisimilarity problem. Assume given two finitary diagrams F and G, with domain $(\mathcal{C}, \leq)$ and $(\mathcal{D}, \preceq)$ respectively. The idea is to non-deterministically construct a bisimulation R, that is, a set of triples (c, M, d)

Algorithm 1. Bisimilarity of finitary diagrams

Require: Two finitary diagrams $F : \mathcal{C} \longrightarrow \mathcal{A}$ and $G : \mathcal{D} \longrightarrow \mathcal{A}$.
Ensure: Answer **Yes** iff F and G are bisimilar.

$S := \mathcal{C} \cup \mathcal{D}$; $R := \varnothing$; $lin := \varnothing$; $var := \varnothing$;
while S is non empty **do**
  Pick some $c \in S$. Let us assume that $c \in \mathcal{C}$, the other case is symmetric.
  Non-deterministically choose $d \in \mathcal{D}$ with $F(c) = G(d) = n$.
  if d does not exist **then FAIL end if**;
  Create a fresh variable X and add the pair (X, n) to var;
  Add (c, X, d) to R and do not mark it;
  while there is a non-marked element in R **do**
    Pick a non-marked element $(c, X, d) \in R$, with $F(c) = G(d) = n$;
    Mark (c, X, d);
    Non-deterministically choose a relation

$$Q \subseteq \{(c', d') \mid (c < c' \wedge d \preceq d') \vee (c \leq c' \wedge d \prec d')\}$$

  such that for every $c' > c$, there is $d' \succeq d$ with (c', d') in Q, and symmetrically;
    $S := S \setminus (\{c' \mid c' \geq c\} \cup \{d' \mid d' \succeq d\})$
    for all (c', d') in Q **do**
      Check if $F(c') = G(d') = m$, otherwise **FAIL**;
      Create a fresh variable X' and add the pair (X', m) to var;
      Add (c', X', d') to R and do not mark it;
      Add the equation $G(d \preceq d').X = X'.F(c \leq c')$ to lin;
    end for
  end while
end while
Let Φ be the formula of the theory of invertible matrices quantified by $\exists_n X$ for every $(X, n) \in var$ and whose predicates are the linear equations from lin.
return YES if Φ is valid, **FAIL** otherwise.

where M is a matrix with coefficients in real (or rational) numbers satisfying the properties of a bisimulation from Sect. 2. The only exception is that we will not guess explicitly the matrices M, but a formula in the existential theory of invertible matrices that encodes the fact that there exist some matrices M such that the bisimulation constructed satisfies those properties.

Consider the Algorithm 1 written in pseudo-code. It maintains the bisimulation R and two sets var, encoding the variables of the formula we are constructing and lin, encoding its predicates.

The algorithm always terminates. First, the innermost while loop terminates since after every loop an element (c, X, d) is marked and only elements of the form (c', X', d') with either $c < c'$ and $d \preceq d'$ or $c \leq c'$ and $d \prec d'$ are added. The outer loop terminates since after every loop at least one element of S is removed.

Assume that there is an execution of the algorithm that answers **Yes**. Let R and Φ constructed during this execution. Since the algorithm answers **Yes**, the formula Φ is satisfiable, that is, for every $(X, n) \in var$, there is an invertible matrix M_X of size $n \times n$ such that for every equation $A.X = X'.B$ in lin, $A.M_X = M_{X'}.B$ holds. Let R' be the set $\{(c, M_X, d) \mid (c, X, d) \in R\}$. Then by construction of R and Φ, R' is a bisimulation between F and G.

Assume that there is a bisimulation R' between F and G. We show that there are non-deterministic choices that lead to the answer **Yes**. The idea is to ensure that every (c, X, d) that belongs to R at some point corresponds to an element (c, f, d) of R'. To ensure this, we must:

1. when choosing d in line 7, choose it such that there is $(c, f, d) \in R'$. It exists by definition of a bisimulation.
2. when choosing Q in line 17, choose it in such a way that for every $(c', d') \in Q$, there is (c', f', d') in R' and that the element $(c, f, d) \in R'$ corresponding to (c, X, d) satisfies that $G(d \leq d') \circ f = f' \circ F(c \leq c')$. Such a Q always exists since R' is a bisimulation.

With this, the algorithm does not **FAIL** and the formula Φ is valid: the assignment that map X to the corresponding f satisfies Φ. Consequently, the algorithm answers **Yes**. Finally, this algorithm non-deterministically construct in exponential space a formula of exponential size in the size of the data. By Theorem 5, this algorithm is in NEXPSPACE. Consequently, by Savitch's theorem [15], since NEXPSPACE = EXPSPACE:

Theorem 8. *Knowing if two finitary diagrams are bisimilar in real or in rational numbers is decidable in EXPSPACE.*

Example 5. Consider the two finitary diagrams at the end of this Section, F on the left, G on the right. Let us apply a few steps of the algorithm on those two diagrams:

1. Pick a and choose 0. At this point $S = \{1, 2, b, c, d\}$, $var = [(X_1, 1)]$ and $R = [(0, X_1, a)]$ (we will only write the unmarked elements).

2. Pick $(0, X_1, a)$ and choose $Q = \{(1,c),(2,d),(0,b)\}$. At this point, $S = \varnothing$, $var = [(X_1,1);(X_2,2);(X_3,1);(X_4,1)]$, $R = [(1,X_2,c);(2,X_3,d);(0,X_4,b)]$ and $lin = [\begin{pmatrix}0\\2\end{pmatrix}.X_1 = X_2.\begin{pmatrix}1\\0\end{pmatrix}; 6.X_1 = X_3; 2X_1 = X_4]$.
3. Pick $(2, X_3, d)$ and choose $Q = \varnothing$. At this point, $R = [(1,X_2,c);(0,X_4,b)]$.
4. Pick $(1, X_2, c)$ and choose $Q = \{(2,d)\}$. At this point,

$$var = [(X_1,1);(X_2,2);(X_3,1);(X_4,1);(X_5,1)],$$

   $R = [(0,X_4,b),(2,X_5,d)]$ and

$$lin = [\begin{pmatrix}0\\2\end{pmatrix}.X_1 = X_2.\begin{pmatrix}1\\0\end{pmatrix}; 6.X_1 = X_3; 2X_1 = X_4; \begin{pmatrix}4 & 3\end{pmatrix}.X_2 = X_5.\begin{pmatrix}1 & 1\end{pmatrix}].$$

5. ...

At the end, the algorithm produces

$$var = [(X_1,1);(X_2,2);(X_3,1);(X_4,1);(X_5,1);(X_6,2);(X_7,1);(X_8,1)]$$

and their linear equations:

$$\begin{aligned} lin = [&\begin{pmatrix}0\\2\end{pmatrix}.X_1 = X_2.\begin{pmatrix}1\\0\end{pmatrix}; 6.X_1 = X_3; 2X_1 = X_4; \begin{pmatrix}4 & 3\end{pmatrix}.X_2 = X_5.\begin{pmatrix}1 & 1\end{pmatrix};\\ &\begin{pmatrix}0\\1\end{pmatrix}.X_4 = X_6.\begin{pmatrix}1\\0\end{pmatrix}; 3.X_4 = X_7; \begin{pmatrix}4 & 3\end{pmatrix}.X_6 = X_8.\begin{pmatrix}1 & 1\end{pmatrix}].\end{aligned}$$

The induced problem of invertible matrices is satisfiable, which means that both diagrams are bisimilar.

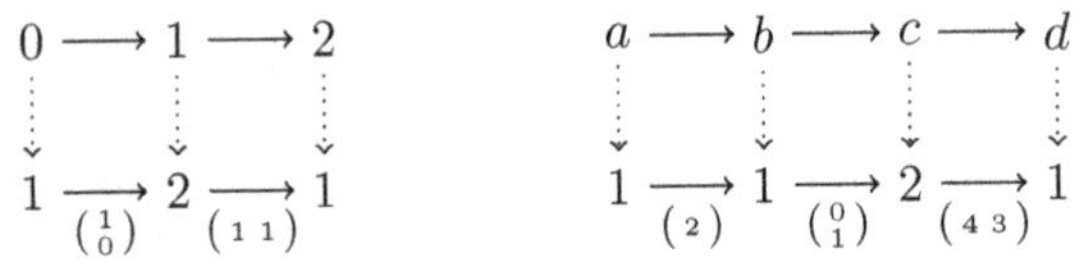

7.2 Decidability of the Model Checking

Starting with a finitary diagram F, an element c of its domain, and a positive finitary object formula S, we inductively construct two lists, initially empty, as previously:

- var of pairs (X, n) where X is a variable and n an integer. This will stand for $\exists_n X$.
- lin of equations $A.X = Y.B$ where X and Y are variables and A and B are matrices.

The formula S is of the form $[n]P$. We first check if $n = F(c)$. If it is not the case then we fail. Otherwise, let X be a fresh variable. Add the pair (X, n) to var. Continue with F, c, X, and P.

Now, assume that we consider the following data: a finitary diagram F, an element of its domain c, an X with (X, n) in var for some integer n and a positive finitary morphism formula P. Several cases:

- if $P = ?S'$, continue with F, c and S',
- if $P = \top$, stop,
- if $P = P_1 \wedge P_2$, first continue with F, c, X and P_1. When this part terminates, continue with F, c, X and P_2,
- if $P = \langle M \rangle P'$, with M of size $n_1 \times n_2$. If $n_1 \neq F(c)$, then we fail. Otherwise, non-deterministically choose an element $c' \geq c$, with $F(c') = n_2$. If such a c' does not exist, then we fail. Then, create a fresh variable X', add (X', n_2) to var and $M.X = X'.F(c \leq c')$ to lin. Finally, continue with F, c', X' and P'.

If the algorithm does not fail, construct a formula Φ from var and lin as previously and check if it is satisfiable using the existential theory of invertible matrices. The formula Φ is non-deterministically constructed in polynomial time and so is of polynomial size. So, this algorithm is in NPSPACE and again, by Savitch's theorem [15], since NPSPACE = PSPACE:

Theorem 9. *Knowing if a finitary diagram satisfies a positive finitary formula (either in real or in rational numbers) is decidable in PSPACE.*

Example 6. Let us consider the following positive finitary formula

$$\phi = [1]\langle\left(\begin{smallmatrix}1\\0\end{smallmatrix}\right)\rangle\langle\left(\begin{smallmatrix}1&1\end{smallmatrix}\right)\rangle\top.$$

It is not hard to check that $F, 0 \models \phi$, and so that $G, a \models \phi$ (you can unroll the algorithm, the identities will give a solution of the problem of matrices). Let H be the following diagram:

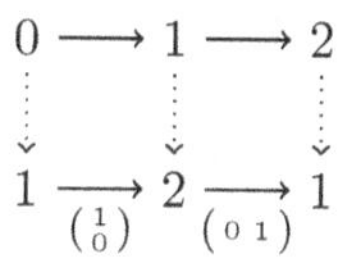

We will show that $H, 0 \not\models \phi$, and that H is not bisimilar to F and G. Let us unroll the algorithm on H, 0 and ϕ. We are in the first case, and we create a fresh variable X_1 and $var := [(X_1, 1)]$. We then continue the algorithm with H, 0, X_1 and $\langle\left(\begin{smallmatrix}1\\0\end{smallmatrix}\right)\rangle\langle\left(\begin{smallmatrix}1&1\end{smallmatrix}\right)\rangle\top$. We are then in the last case, and we can only choose 1 without failing. So, $var = [(X_1, 1); (X_2, 2)]$ and $lin = [\left(\begin{smallmatrix}1\\0\end{smallmatrix}\right).X_1 = X_2.\left(\begin{smallmatrix}1\\0\end{smallmatrix}\right)]$. We continue with H, 1, X_2 and $\langle\left(\begin{smallmatrix}1&1\end{smallmatrix}\right)\rangle\top$. We still are in the last case and we can only choose 2 without failing. So, $var = [(X_1, 1); (X_2, 2); (X_3, 1)]$ and $lin = [\left(\begin{smallmatrix}1\\0\end{smallmatrix}\right).X_1 = X_2.\left(\begin{smallmatrix}1\\0\end{smallmatrix}\right); \left(\begin{smallmatrix}1&1\end{smallmatrix}\right).X_2 = X_3.\left(\begin{smallmatrix}0&1\end{smallmatrix}\right)]$. Let us prove that we cannot solve this problem of invertible matrices. If we could, we would have that:

$$X_1 = \left(\begin{smallmatrix}1&1\end{smallmatrix}\right).\left(\begin{smallmatrix}1\\0\end{smallmatrix}\right).X_1 = \left(\begin{smallmatrix}1&1\end{smallmatrix}\right).X_2.\left(\begin{smallmatrix}1\\0\end{smallmatrix}\right) = X_3.\left(\begin{smallmatrix}0&1\end{smallmatrix}\right).\left(\begin{smallmatrix}1\\0\end{smallmatrix}\right) = 0$$

which is impossible since X_1 must be invertible.

8 Future Work

As a future work, we would like to investigate the case of diagrams with values in $\mathbb{Z}$-modules (that is, Abelian groups), i.e., diagrams with values in matrices whose coefficients are integers, for which the existential theory is undecidable, but for which we can still decide some problems of matrices. Another interesting direction is the relation between our algorithm of Sect. 7 to find a bisimulation and the final chain algorithm [1], which we let for a future work.

References

1. Adámek, J., Bonchi, F., Hülsbusch, M., König, B., Milius, S., Silva, A.: A coalgebraic perspective on minimization and determinization. In: Birkedal, L. (ed.) FoSSaCS 2012. LNCS, vol. 7213, pp. 58–73. Springer, Heidelberg (2012). https://doi.org/10.1007/978-3-642-28729-9_4
2. Ben-Or, M., Kozen, D., Reif, J.: The complexity of elementary algebra and geometry. J. Comput. Syst. Sci. **32**, 251–264 (1986)
3. Canny, J.: Some algebraic and geometric computations in PSPACE. In: Proceedings of the 20th Annual ACM Symposium on Theory of Computing (STOC), pp. 460–467 (1988)
4. Cattani, G.L., Stark, I., Winskel, G.: Presheaf models for the π-calculus. In: Moggi, E., Rosolini, G. (eds.) CTCS 1997. LNCS, vol. 1290, pp. 106–126. Springer, Heidelberg (1997). https://doi.org/10.1007/BFb0026984
5. Dubut, J., Goubault, É., Goubault-Larrecq, J.: Natural homology. In: Halldórsson, M.M., Iwama, K., Kobayashi, N., Speckmann, B. (eds.) ICALP 2015. LNCS, vol. 9135, pp. 171–183. Springer, Heidelberg (2015). https://doi.org/10.1007/978-3-662-47666-6_14
6. Dubut, J., Goubault, E., Goubault-Larrecq, J.: Bisimulations and unfolding in P-accessible categorical models. In: Proceedings of the 27th International Conference on Concurrency Theory (CONCUR 2016). Leibniz International Proceedings in Informatics, vol. 59, pp. 1–14. Schloss Dagstuhl-Leibniz-Zentrum fuer Informatik (2016)
7. Fajstrup, L., Goubault, E., Haucourt, E., Mimram, S., Raussen, M.: Directed Algebraic Topology and Concurrency. Springer, Cham (2016). https://doi.org/10.1007/978-3-319-15398-8
8. van Glabbeek, R.J.: On the expresiveness of higher dimensional automata. Electron. Notes Theoret. Comput. Sci. **128**(2), 5–34 (2005)
9. Hennessy, M., Milner, R.: On observing nondeterminism and concurrency. In: de Bakker, J., van Leeuwen, J. (eds.) ICALP 1980. LNCS, vol. 85, pp. 299–309. Springer, Heidelberg (1980). https://doi.org/10.1007/3-540-10003-2_79
10. Joyal, A., Nielsen, M., Winskel, G.: Bisimulation from open maps. Inf. Comput. **127**(2), 164–185 (1996)
11. Lasota, S.: Coalgebra morphisms subsume open maps. Theoret. Comput. Sci. **280**, 123–135 (2002)
12. Matiyasevitch, Y.: Hilbert's 10th Problem. MIT Press, Cambridge (1993)
13. Pratt, V.: Modeling concurrency with geometry. In: Proceedings of the 18th ACM SIGPLAN-SIGACT Symposium on Principles of Programming Languages (POPL), pp. 311–322 (1991)

14. Rabinovitch, A., Trakhtenbrot, B.: Behavior structures and nets. Fundamenta Informaticae **11**(4), 357–403 (1988)
15. Savitch, W.: Relationships between nondeterministic and deterministic tape complexities. J. Comput. Syst. Sci. **4**(2), 177–192 (1970)
16. Srba, J.: Roadmap of infinite results. Online
17. Tarski, A.: A Decision Method for Elementary Algebra and Geometry. University of California Press, Berkeley (1951)
18. Winter, M.: A relational algebraic theory of bisimulations. Fundamenta Informaticae **83**(4), 429–449 (2008)
19. Wißmann, T., Dubut, J., Katsumata, S., Hasuo, I.: Path category for free. In: Bojańczyk, M., Simpson, A. (eds.) FoSSaCS 2019. LNCS, vol. 11425, pp. 523–540. Springer, Cham (2019). https://doi.org/10.1007/978-3-030-17127-8_30

Generating Posets Beyond N

Uli Fahrenberg[1], Christian Johansen[2(✉)], Georg Struth[3], and Ratan Bahadur Thapa[2]

[1] École Polytechnique, Palaiseau, France
[2] University of Oslo, Oslo, Norway
cristi@ifi.uio.no
[3] University of Sheffield, Sheffield, UK

Abstract. We introduce iposets—posets with interfaces—equipped with a novel gluing composition along interfaces and the standard parallel composition. We study their basic algebraic properties as well as the hierarchy of gluing-parallel posets generated from singletons by finitary applications of the two compositions. We show that not only series-parallel posets, but also interval orders, which seem more interesting for modelling concurrent and distributed systems, can be generated, but not all posets. Generating posets is also important for constructing free algebras for concurrent semirings and Kleene algebras that allow compositional reasoning about such systems.

1 Introduction

This work is inspired by Tony Hoare's programme of building graph models of concurrent Kleene algebra (CKA) [14] for real-world applications. CKA extends the sequential compositions, nondeterministic choices and unbounded finite iterations of imperative programs modelled by Kleene algebra into concurrency, adding operations of parallel composition and iteration, and a weak interchange law for the sequential-parallel interaction. Such algebras have a long history in concurrency theory, dating back at least to Winkowski [38]. Commutative Kleene algebra—the parallel part of CKA—has been investigated by Pilling and Conway [3]. A double semiring with weak interchange—CKA without iteration—has been introduced by Gischer [10]; its free algebras have been studied by Bloom and Ésik [1]. CKA, like Gischer's concurrent semiring, has both interleaving and true concurrency models, that is, shuffle as well as pomset languages. Series-parallel pomset languages, which are generated from singletons by finitary applications

U. Fahrenberg—Author supported by the *Chaire ISC: Engineering Complex Systems – École polytechnique – Thales – FX – DGA – Dassault Aviation – Naval Group – ENSTA ParisTech – Télécom ParisTech*

C. Johansen—Author supported by the project IoTSec – Security in IoT for Smart Grids, with number 248113/O70, funded by the Norwegian Research Council.

G. Struth—Author supported by EPSRC grant EP/R032352/1 Verifiably Correct Transactional Memory.

U. Fahrenberg et al. (Eds.): RAMiCS 2020, LNCS 12062, pp. 82–99, 2020.
https://doi.org/10.1007/978-3-030-43520-2_6

of sequential and parallel compositions, form free algebras in this class [21,25] (at least when parallel iteration is ignored). The inherent compositionality of algebra is thus balanced by the generative properties of this model. Yet despite this and other theoretical work, applications of CKA remain rare.

One reason is that series-parallel pomsets are not expressive enough for many real-world applications: even simple producer-consumer examples cannot be modelled [27]. Tests, which are needed for the control structure of concurrent programs, and assertions are hard to capture in models of CKA (see [19] and its discussion in [20]). Finally, it remains unclear how modal operators could be defined over graph models akin to pomset languages, which is desirable for concurrent dynamic algebras and logics beyond alternating nondeterminism [9,31].

A natural approach to generating more expressive pomset languages is to "cut across" pomsets in more general ways when (de)composing them. This can be achieved by (de)composing along interfaces, and this idea can be traced back again to Winkowski [38]; see also [4,6,28] for interface-based compositions of graphs and posets, or [15,29,30] for recent interface-based graph models for CKA. As a side effect, interfaces may yield notions of tests, assertions or modalities. When they consist of events, cutting across them presumes that they extend in time and thus form intervals. Interval orders [7,37] of events with duration have been applied widely in partial order semantics of concurrent and distributed systems [17,22,23,33–36] and the verification of weak memory models [13], yet generating them remains an open problem [18].

Our main contribution lies in a new class and algebra of posets with interfaces (*iposets*) based on these ideas. We introduce a new gluing composition that acts like standard serial po(m)set composition outside of interfaces, yet glues together interface events, thus composing events that did not end in one component with those that did not start in the other one. Our definitions are categorical so that isomorphism classes of posets are considered ab initio. Their decoration with labels is then routine, so that we may focus on posets instead of pomsets.

Our technical results concern the hierarchy of gluing-parallel posets generated by finitary applications of this gluing composition and the standard parallel composition of po(m)sets, starting from singleton iposets.[1] Thus all series-parallel pomsets can be generated, but also all interval orders are captured at the second alternation level of the hierarchy. Beyond that, we show that the gluing-parallel hierarchy does not collapse and that posets with certain zigzag-shaped induced subposets are excluded. A precise characterisation of the generated (i) posets remains open. Series-parallel posets, by comparison, exclude precisely those posets with induced N-shaped subposets; interval orders those with induced subposets 2+2, which makes the two classes incomparable. Iposets thus retain the pleasant compositionality properties of series-parallel pomsets and the wide applicability of interval orders in concurrency and distributed computing.

In addition, we establish a bijection between isomorphism classes of interval orders and certain equivalence classes of interval sequences [33], and we study the basic algebraic properties of iposets, including weak interchange laws and a

[1] There is only one singleton poset, but with interfaces, there are four singleton iposets.

Levi lemma. The relationship between gluing-parallel ipo(m)set languages and CKA is left for another article.

2 Posets and Series-Parallel Posets

A *poset* $(P, \leq_P)$ is a set P equipped with a *partial order* $\leq_P$; a reflexive, transitive, antisymmetric relation on P (for which we often write $\leq$). A *morphism* of posets P and Q is an order-preserving function $f : P \to Q$, that is, $x \leq_P y$ implies $f(x) \leq_Q f(y)$. Posets and their morphisms define the category Pos.

A poset is *linear* if each pair of elements is comparable with respect to its order. We write $<$ for the strict part of $\leq$. We write $[n]$, for $n \geq 1$, for the *discrete n-poset* $(\{1, \dots, n\}, \leq)$, which satisfies $i \leq j \Leftrightarrow i = j$. Additionally, $[0] = \emptyset$.

The isomorphisms in Pos are *order bijections*: bijective functions $f : P \to Q$ for which $x \leq_P y \Leftrightarrow f(x) \leq_Q f(y)$. We write $P \cong Q$ if posets P and Q are isomorphic. We generally consider posets up-to isomorphism and assume that all posets are finite.

Concurrency theory often considers (isomorphism classes of) posets with points labelled by letters from some alphabet, which represent actions of some concurrent system. These are known as *partial words* or *pomsets*. As we are mainly interested in structural aspects of concurrency, we ignore such labels.

Series-parallel posets form a well investigated class that can be generated from the singleton poset by finitary applications of two compositions. Their labelled variants generalise rational languages into concurrency. For arbitrary posets, these compositions are defined as follows.

Definition 1. *Let $P_1 = (P_1, \leq_1)$ and $P_2 = (P_2, \leq_2)$ be posets.*

1. *Their* serial composition *is the poset $P_1; P_2 = (P_1 \sqcup P_2, \leq_1 \cup \leq_2 \cup P_1 \times P_2)$.*
2. *Their* parallel composition *is the poset $P_1 \otimes P_2 = (P_1 \sqcup P_2, \leq_1 \cup \leq_2)$.*

Here, $\sqcup$ means disjoint union (coproduct) of sets. We generalise serial composition to a gluing composition in Sect. 4, after equipping posets with interfaces.

Serial and parallel compositions respect isomorphism, and $[n+m]$ is isomorphic to $[n] \otimes [m]$ with isomorphism $\varphi_{n,m} : [n+m] \to [n] \otimes [m]$ given by

$$\varphi_{n,m}(i) = \begin{cases} i_{[n]} & \text{if } i \leq n, \\ (i-n)_{[m]} & \text{if } i > n. \end{cases}$$

Also note that parallel composition is the coproduct in Pos, hence $\otimes$ is also defined for morphisms.

By definition, a poset is *series-parallel* (an *sp-poset*) if it is either empty or can be obtained from the singleton poset by applying the serial and parallel compositions a finite number of times. It is well known [12,32] that a poset is series-parallel iff it does not contain the induced subposet $\mathsf{N} = \left(\begin{smallmatrix} \cdot & \to & \cdot \\ & \nearrow & \\ \cdot & \to & \cdot \end{smallmatrix}\right)$.[2]

[2] This means that there is no injection f from N satisfying $x \leq y \Leftrightarrow f(x) \leq f(y)$.

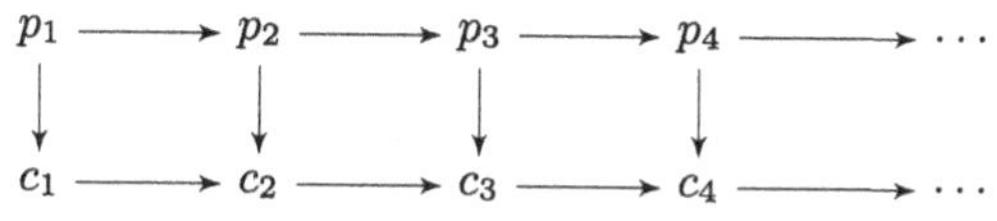

Fig. 1. The producer-consumer example.

Sp-po(m)sets form double monoids with respect to serial and parallel composition, and with the empty poset as shared unit—in fact the free algebras in this class. Compositionality of the recursive definition of sp-po(m)sets is thus reflected by the compositionality of their algebraic properties, which is often considered desirable for concurrent systems [36]. Yet sp-posets are, in fact, *too* compositional for many applications: even simple consumer-producer problems inevitably generate N's [27], as shown in Fig. 1 that contains the N spanned by c_1, c_2, p_2, and p_3 as an induced subposet among others.

3 Interval Orders and Interval Sequences

Interval orders [7,37] form another class of posets that are ubiquitous in concurrent and distributed computing. Intuitively, they are isomorphic to sets of intervals on the real line that are ordered whenever they do not overlap.

Definition 2. *An* interval order *is a relational structure $(P, <)$ with $<$ irreflexive such that $w < y$ and $x < z$ imply $w < z$ or $x < y$, for all $w, x, y, z \in P$.*

Transitivity of $<$ follows. An alternative geometric characterisation is that interval orders are precisely those posets that do not contain the induced subposet $2{+}2 = \left(\begin{smallmatrix}\cdot \longrightarrow \cdot\\ \cdot \longrightarrow \cdot\end{smallmatrix}\right)$.

The intuition is captured by Fishburn's theorem [7], which implies that a finite poset P is an interval order if and only if it has an *interval representation*: a pair of functions $b, e : P \to Q$ into some linear order $(Q, <_Q)$ such that $b(x) <_Q e(x)$ for all $x \in P$, and $x <_P y$ iff $e(x) <_Q b(y)$ for all $x, y \in P$. By the first condition, pairs $(b(x), e(x))$ correspond to intervals $I(x) = [b(x), e(x)]$ in Q; by the second condition, $x <_P y$ iff $I(x)$ lies entirely before $I(y)$ in Q.

We write $\rho_I(P)$ for the set of interval representations of P. Each representation can be rearranged so that all endpoints of intervals are distinct ([11], Lemma 1.5). We henceforth assume that all interval presentations have this property. It then holds that $|Q| = 2|P|$, and we can fix Q as the target type of any interval representation of P.

Finally, with relation $\sqsubset$ on the set of maximal antichains of poset P given by

$$A \sqsubset B \Leftrightarrow (\forall x \in A \setminus B. \forall y \in B \setminus A.\ x < y),$$

it is known that P is an interval order if and only if $\sqsubset$ is a strict linear order [8].

Interval orders occur implicitly in the ST-traces of Petri nets [33]. In a pure order-theoretic setting, these are *interval sequences*, that is, sequences of $b(x)$ and $e(x)$, with x from some finite set P, in which each $b(x)$ occurs exactly once and each $e(x)$ at most once and only after the corresponding $b(x)$. An interval sequence is *closed* if each $e(x)$ occurs exactly once [33,36]. An *interval trace* [18] is an equivalence class of interval sequences modulo the relations $b(x)b(y) \approx b(y)b(x)$ and $e(x)e(y) \approx e(y)e(x)$ for all $x, y \in P$. We write $\approx^*$ for the congruence generated by $\approx$ on interval sequences. We identify interval sequences and interval traces with the Hasse diagrams of their linear orders over Q.

Lemma 3. *Let P be an interval order and $(b, e) \in \rho_I(P)$. Then $(Q, <_Q)$ is a closed interval sequence.*

Proof. Trivial. □

We write $\sigma_{(b,e)}(P)$ for the interval sequence of interval order P and $(b, e) \in \rho_I(P)$, and $\Sigma(P)$ for the set of all interval sequences of interval representations of P.

Lemma 4. *If $\sigma \in \Sigma(P)$ and $\sigma \approx^* \sigma'$, then $\sigma' \in \Sigma(P)$.*

Proof. We show that $\sigma \in \Sigma(P)$ and $\sigma \approx \sigma'$ imply $\sigma' \in \Sigma(P)$. Suppose that $\sigma = \sigma_1 b(x)b(y)\sigma_2$ and $\sigma' = \sigma_1 b(y)b(x)\sigma_2$ and that $(b, e) \in \rho_I(P)$ generates σ. Then (b', e) with

$$b'(z) = \begin{cases} b(y), & \text{if } z = x, \\ b(x), & \text{if } z = y, \\ b(z), & \text{otherwise} \end{cases}$$

is in $\rho_I(P)$, as $b'(x) <_Q e(x)$, $b'(y) <_Q e(y)$ and, for all $v, w \in P$, $v <_P w \Leftrightarrow e(v) <_P b(w)$ still holds. In addition, (b', e) generates σ'. An analogous result for $\sigma = \sigma_1 e(x)e(y)\sigma_2$ and $\sigma' = \sigma_1 e(y)e(x)\sigma_2$ holds by opposition. The result for $\approx^*$ then follows by a simple induction. □

Lemma 5. *Let P be an interval order. If $(b, e), (b', e') \in \rho_I(P)$ assign b and e to elements of P in interval sequences, then $\sigma_{(b,e)}(P) \approx^* \sigma_{(b',e')}(P)$.*

Proof. Let $\prec_1$ and $\prec_2$ be the orderings of the interval sequences for (b, e) and (b', e') in Q. Then $b(x) \prec_1 e(x)$ and $b(x) \prec_2 e(x)$ for all $x \in X$, and, for all $x, y \in X$, $e(x) \prec_1 b(y) \Leftrightarrow e(x) \prec_2 b(y)$. It follows that there is no $b(z)$ in $\prec_1$ or $\prec_2$ between the positions of $e(x)$ in $\prec_1$ and $\prec_2$ and, by opposition, there is no $e(z)$ in $\prec_1$ or $\prec_2$ between the positions of $b(x)$ in $\prec_1$ and $\prec_2$. But this means that the positions of $e(x)$ and $b(x)$ can be rearranged by $\approx^*$. □

Proposition 6. *If P is an interval order and $(b, e) \in \rho_I(P)$, then $[\sigma_{(b,e)}(P)]_{\approx^*} = \Sigma(P)$. The mapping φ defined by $\varphi(P) = [\sigma_{(b,e)}(P)]_{\approx^*}$ is a bijection.*

Proof. By Lemmas 4 and 5, and by properties of interval representations. □

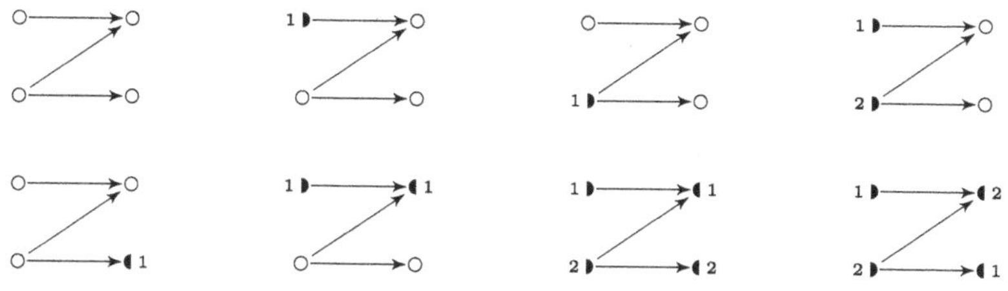

Fig. 2. Eight of 25 different iposets based on poset N.

4 Posets with Interfaces

An element s of poset $(P, \leq)$ is *minimal* (*maximal*) if $v \not< s$ ($v \not> s$) holds for all $v \in P$. We write $P_{\min}$ ($P_{\max}$) for the sets of minimal (maximal) elements of P.

Definition 7. *A* poset with interfaces (iposet) *consists of a poset P together with two injective morphisms*

$$[n] \xrightarrow{\ s\ } P \xleftarrow{\ t\ } [m]$$

such that $s[n] \subseteq P_{\min}$ and $t[m] \subseteq P_{\max}$.

Injection $s : [n] \to P$ represents the *source interface* of P and $t : [m] \to P$ its *target interface*. We write $(s, P, t) : n \to m$ for the iposet $s : [n] \to P \leftarrow [m] : t$.

Figure 2 shows some examples of iposets. Elements of source and target interfaces are depicted as filled half-circles to indicate the unfinished nature of the events they represent.

Next we define a sequential gluing composition on iposets whose interfaces agree and we adapt the standard parallel composition of posets to iposets.

Definition 8. *Let $(s_1, P_1, t_1) : n \to m$ and $(s_2, P_2, t_2) : \ell \to k$ be iposets.*

1. *For $m = \ell$, their* gluing composition *is the iposet $(s_1, P_1 \triangleright P_2, t_2) : n \to k$ with $P_1 \triangleright P_2 = \big((P_1 \sqcup P_2)_{/t_1(i)=s_2(i)}, \leq_1 \cup \leq_2 \cup (P_1 \setminus t_1[m]) \times (P_2 \setminus s_2[m])\big)$.*
2. *Their* parallel composition *is the iposet $(s, P_1 \otimes P_2, t) : n + \ell \to m + k$ with $s = (s_1 \otimes s_2) \circ \varphi_{n,l}$ and $t = (t_1 \otimes t_2) \circ \varphi_{m,k}$.*

Parallel composition of iposets thus puts components "side by side": it is the disjoint union of posets and interfaces. Gluing composition puts iposets "one after the other", P_1 before P_2, but glues their interfaces together (and adds arrows from all points in P_1 that are not in its target interface to all points in P_2 that are not in its source interface). As explained in the introduction, it thus glues events which did not end in P_1 with those that did not start in P_2. Figures 3 and 4 show examples. The filled half-circles in source and target interfaces are glued to unfilled circles in these diagrams.

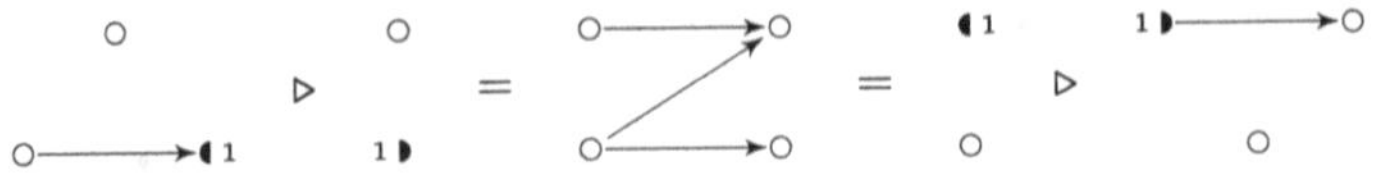

Fig. 3. Two different decompositions of the N.

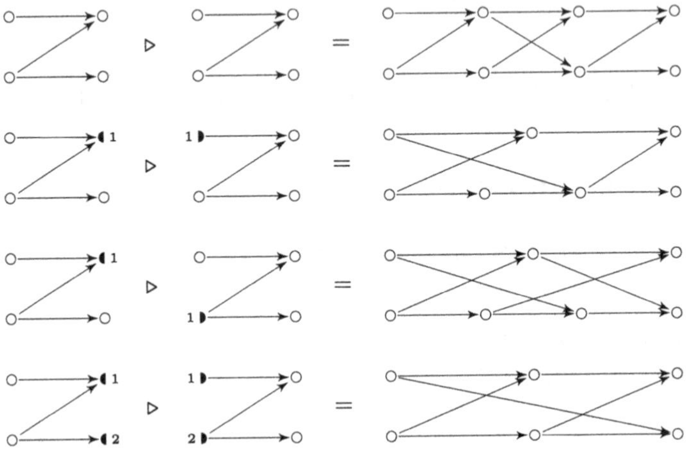

Fig. 4. Four gluings of different Ns with interfaces.

We define *identity iposets* $\mathsf{id}_n = (\mathsf{id}, [n], \mathsf{id}) : n \to n$, for $n \geq 0$. For convenience, we generalise this notation to other discrete posets with interfaces: for $k, \ell \leq n$, we write ${}^k\mathsf{id}_n^\ell$ for the iposet $(f_k^n, [n], f_\ell^n) : k \to \ell$, where $f_k^n : [k] \to [n]$ is the (identity) injection $x \mapsto x$ (similarly for f_ℓ^n). Hence $\mathsf{id}_n = {}^n\mathsf{id}_n^n$.

Proposition 9. *Iposets form a small category with natural numbers as objects, iposets* $(s, P, t) : n \to m$ *as morphisms,* $\rhd$ *as composition, and identities* id_n.

Checking the associativity and unit laws is routine. The following example shows that gluing composition is not commutative, as expected:

$$ {}^0\mathsf{id}_1^1 \rhd {}^1\mathsf{id}_1^0 = {}^0\mathsf{id}_1^0 = (\,\cdot\,) \neq (\,\cdot \longrightarrow \cdot\,) = {}^1\mathsf{id}_1^0 \rhd {}^0\mathsf{id}_1^1 \,. $$

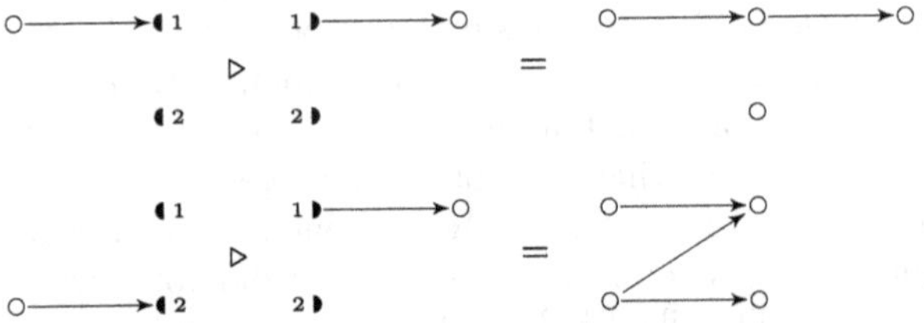

Fig. 5. Non-isomorphic gluings of symmetric parallel compositions.

Perhaps more unexpectedly, parallel composition need not be commutative, as the namings of interfaces in $P \otimes Q$ may differ from those in $Q \otimes P$. One can of course rename interfaces using *symmetries*: iposets $(s, [n], t) : n \to n$ with s and t bijective. Yet Fig. 5 shows two parallel compositions where renaming of interfaces and gluing with another iposet yields non-isomorphic posets.

The iposets $P : n \to m$, $Q : k \to \ell$ are mapped by $\otimes$ to $P \otimes Q : n+k \to m+\ell$, which has the signature of a tensor on the small category of Proposition 9. Yet gluing and parallel composition need *not* satisfy an interchange law:

$({}^0\mathsf{id}_1^0 \otimes {}^0\mathsf{id}_1^0) \rhd ({}^0\mathsf{id}_1^0 \otimes {}^0\mathsf{id}_1^0) = \Big($ 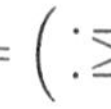 $\Big) \neq \Big($ 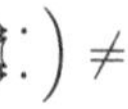 $\Big) = ({}^0\mathsf{id}_1^0 \rhd {}^0\mathsf{id}_1^0) \otimes ({}^0\mathsf{id}_1^0 \rhd {}^0\mathsf{id}_1^0)$.

Hence $\otimes$ is not a tensor, and iposets do *not* form a (strict) monoidal category (let alone a PROP). This situation differs from gluing compositions where interfaces of iposets are defined by *all* minimal and maximal elements [38], and also from sequential compositions with interfaces similar to ours but where interfaces disappear and no other order is induced [6,28]. All of these give rise to PROPs.

Instead of the interchange law above, we will state a *lax* interchange law in Proposition 12; the precise categorical relation between gluing and parallel composition is left open for future work. What we will need here is that iposets with $\otimes$ form a *graded monoid* over $\mathbb{N} \times \mathbb{N}$, *i.e.*, that there is a grading function (a monoid morphism) from iposets into $\mathbb{N} \times \mathbb{N}$.

Proposition 10. *Iposets form an $\mathbb{N} \times \mathbb{N}$-graded monoid with composition $\otimes$ and unit* id_0.

The grading function maps an iposet $P : n \to m$ to the tuple (n, m), with addition on tuples defined component-wise. The proof is routine.

A *morphism* of iposets is a commuting diagram

$$\begin{array}{ccccc} [n] & \xrightarrow{s} & P & \xleftarrow{t} & [m] \\ {\scriptstyle\nu}\downarrow & & {\scriptstyle f}\downarrow & & \downarrow{\scriptstyle\mu} \\ [n'] & \xrightarrow[s']{} & P' & \xleftarrow[t']{} & [m'] \end{array}$$

where ν and μ are strictly order preserving with respect to $<_{\mathbb{N}}$ and f is an order morphism between P and P'. Intuitively, iposet morphisms are thus order morphisms that also preserve interfaces and their order in $\mathbb{N}$. We write iPos for the so-defined category.

An iposet morphism (ν, f, μ) is an *isomorphism* if ν, f and μ are order isomorphisms. Hence $n = n'$, $m = m'$, $\nu = \mathsf{id} : n \to n$, and $\mu = \mathsf{id} : m \to m$ in the diagram above. In particular, iposets related by a symmetry $(s, [n], t) : n \to n$ need not be isomorphic. We write $P \cong Q$ if there exists an isomorphism $\varphi : P \to Q$. The following lemma shows $\otimes$ and $\rhd$ respect isomorphisms.

Lemma 11. *Let P, P', Q, Q' be iposets. Then $P \cong P'$ and $Q \cong Q'$ imply $P \otimes Q \cong P' \otimes Q'$ and $P \rhd Q \cong P' \rhd Q'$.*

Proof. Let $\varphi : P \to P'$ and $\psi : Q \to Q'$ be (the poset components of) isomorphisms. Define the functions $\varphi \otimes \psi : P \sqcup Q \to P' \sqcup Q'$ and $\varphi \triangleright \psi : (P \sqcup Q)_{/t_P(i)=s_Q(i)} \to (P' \sqcup Q')_{/t_{P'}(i)=s_{Q'}(i)}$ as

$$(\varphi \,\square\, \psi)(x) = \begin{cases} \varphi(x) & \text{if } x \in P, \\ \psi(x) & \text{if } x \in Q\,, \end{cases}$$

for $\square \in \{\otimes, \triangleright\}$. First, $\varphi \otimes \psi$ is obviously an isomorphism. Second, $\varphi \triangleright \psi$ is well-defined because $\varphi \circ t_P(i) = \psi \circ s_Q(i)$ for all $i \in [m]$, and easily seen to be an isomorphism as well. $\square$

We write $P \preceq Q$ if there is a bijective (on points) morphism $\varphi : Q \to P$ between iposets P and Q. Intuitively, $P \preceq Q$ if P has more arrows and is therefore less parallel than Q, while interfaces are preserved. Similar relations on posets and pomsets, sometimes called *subsumption*, are well studied [10,12]. In particular, $\preceq$ is a preorder on (finite) iposets and a partial order up-to isomorphism.

Proposition 12. *For iposets P, P', Q, Q', the following lax interchange law holds:*

$$(P \otimes P') \triangleright (Q \otimes Q') \preceq (P \triangleright Q) \otimes (P' \triangleright Q')\,.$$

Proof. Let $P_\ell = (P \otimes P') \triangleright (Q \otimes Q')$ and $P_r = (P \triangleright Q) \otimes (P' \triangleright Q')$. First, $P_\ell = (P \sqcup Q)_{/t_P \equiv s_Q} \sqcup (P' \sqcup Q')_{/t_{P'} \equiv s_{Q'}} = (P \sqcup Q \sqcup P' \sqcup Q')_{t_P \equiv s_Q, t_{P'} \equiv s_{Q'}} = P_r$, by definition of $\otimes$. Hence both posets have the same points, and we may choose $\varphi : P_r \to P_\ell$ to be the identity. It remains to show that φ is order preserving, which means that every arrow in P_r must be in P_ℓ.

Hence suppose $x \leq_{P_r} y$, that is, $x \leq_{P \triangleright Q} y$ or $x \leq_{P' \triangleright Q'} y$. In the first case, if $x \leq_P y$ or $x \leq_Q y$, then $x \leq_{P \otimes P'} y$ or $x \leq_{Q \otimes Q'} y$ and therefore $x \leq_{P_\ell} y$; and if $x \in P \setminus t_P$ and $y \in Q \setminus s_Q$, then $x \in P \sqcup P' \setminus t_{P \otimes P'}$ and $y \in Q \sqcup Q' \setminus s_{Q \otimes Q'}$ and therefore $x \leq_{P_\ell} y$, too. The second case is symmetric. Thus $x \leq_{P_\ell} y$ holds in any case. $\square$

In sum, the algebra of iposets is similar to a concurrent monoid [14], but $\triangleright$ is a partial operation with many units id_k. As $\otimes$ is not a tensor, the categorical structure of iposets is somewhat unusual and deserves further exploration.

Proposition 13. Pos *embeds into* iPos *as iposets with both interfaces* [0], *and likewise for morphisms. The so-defined inclusion functor J :* Pos $\to$ iPos *is fully faithful and the left adjoint to the forgetful functor F :* iPos $\to$ Pos *that maps (s, P, t) to P. The category* Pos *is therefore coreflective in* iPos. *Under J, serial (parallel) composition of posets becomes $\triangleright$ ($\otimes$) of iposets.*

Proof. It is clear that J is a functor. It is full because any morphism $\tilde{f}$ from $P : 0 \to 0$ to $Q : 0 \to 0$ in iPos must have the form $(\emptyset, f, \emptyset) = Jf$ for some f in Pos. It is faithful because $Jf = (\emptyset, f, \emptyset) = (\emptyset, g, \emptyset) = Jg$ implies $f = g$. For $P \in$ Pos and $\tilde{Q} \in$ iPos, J induces a natural bijection $J : \mathsf{Pos}(P, F\tilde{Q}) \cong \mathsf{iPos}(JP, \tilde{Q})$, hence J and F are indeed adjoint. The observations about $\triangleright$ and $\otimes$ are clear. $\square$

5 Further Properties of Iposets

We now derive further algebraic properties of iposets before turning to the iposets generated by gluing and parallel composition from singletons.

For an iposet P with order relation $\leq$ we write $\parallel = \not\leq \cap \not\geq$. Hence $x \parallel y$ if and only if x and y are incomparable and therefore *independent.*

First, in addition to the lax interchange in Proposition 12, we present an equational interchange law as a witness that the equational theory of iPos given by the laws in Propositions 9 and 10 is not free. The lemmas that follow then show that this law is the *only* non-trivial additional identity.

Proposition 14 (Interchange). *For all iposets P, Q and $k, \ell \in \{0,1\}$,*

$$({}^k\mathsf{id}_1^1 \otimes P) \rhd ({}^1\mathsf{id}_1^\ell \otimes Q) = {}^k\mathsf{id}_1^\ell \otimes (P \rhd Q).$$

Proof (sketch). The interface between ${}^k\mathsf{id}_1^1$ and ${}^1\mathsf{id}_1^\ell$ forces these iposets to be glued separately to the rest in the gluing composition $({}^k\mathsf{id}_1^1 \otimes P) \rhd ({}^1\mathsf{id}_1^\ell \otimes Q)$. □

On the one hand, the singleton iposets mentioned therefore do not interfere with compositions. On the other hand, Proposition 14 shows that some iposets can be decomposed both with respect to $\rhd$ and $\otimes$.

Let $\mathcal{S} = \{{}^k\mathsf{id}_1^\ell \mid k, \ell = 0, 1\}$ denote the set of singleton iposets and $\mathcal{C}_1 = \{P_1 \otimes \cdots \otimes P_n \mid P_1, \ldots, P_n \in \mathcal{S}\}$ the set of discrete iposets. The next lemma shows a kind of converse to the previous one: if an iposet is both $\rhd$-decomposable and $\otimes$-decomposable, then all components but one must be in $\mathcal{S}$.

Lemma 15 (Decomposition). *Let $P = P_1 \otimes P_2 = Q_1 \rhd Q_2$ such that $P_1 \neq \mathsf{id}_0$, $P_2 \neq \mathsf{id}_0$, and $Q_1 \neq {}^k\mathsf{id}_n^n$, $Q_2 \neq {}^n\mathsf{id}_n^k$ for any $k \leq n$. Then $P_1 \in \mathcal{C}_1$ or $P_2 \in \mathcal{C}_1$.*

Proof. Suppose $P_1 \notin \mathcal{C}_1$ and $P_2 \notin \mathcal{C}_1$. Then P contains a 2+2: there are $w, x \in P_1$ and $y, z \in P_2$ for which $w <_P x$, $y <_P z$, $w \parallel_P y$, $w \parallel_P z$, $x \parallel_P y$, and $x \parallel_P z$.

If $w, y \notin Q_2$, then $w, y \in Q_1 \setminus t_{Q_1}$. As $Q_2 \neq {}^n\mathsf{id}_n^k$ for any $k \leq n$, there must be an element $v \in Q_2 \setminus s_{Q_2}$. But then $w \leq_P v$ and $y \leq_P v$, which yields arrows between $w \in P_1$ and $y \in P_2$ that contradict $P = P_1 \otimes P_2$. A dual argument rules out that $x, z \notin Q_1$.

It follows that $w \in Q_2$ or $y \in Q_2$. Assume, without loss of generality, that $w \in Q_2$. Then $x \in Q_2 \setminus s_{Q_2}$ because $w <_{P_1} x$. Now if also $y \in Q_2$, then by the same argument, $z \in Q_2 \setminus s_{Q_2}$. Hence Q_2 contains two different points that are not in its starting interface; and as $Q_1 \setminus t_{Q_1}$ is non-empty, this again establishes a connection between $x \in P_1$ and $z \in P_2$ which cannot exist. Hence $y \notin Q_2$, but then $y \in Q_1 \setminus t_{Q_1}$, so that $y \leq_P x$, which contradicts $x \parallel_P y$. □

The next lemma generalises Levi's lemma for words [26].

Lemma 16 (Levi property). *Let $P \rhd Q = U \rhd V$. Then there is an R so that either $P = U \rhd R$ and $R \rhd Q = V$, or $U = P \rhd R$ and $R \rhd V = Q$.*

Proof (sketch). Obviously, $P \rhd Q$ and $U \rhd V$ have the same carrier set. By the assumption it is partitioned into three (disjoint) sets such that either $P = U \sqcup R$ and $R \sqcup Q = V$, or $U = P \sqcup R$ and $R \sqcup V = Q$. In the first case, it follows that $P \sqcup Q = U \sqcup R \sqcup Q$ and it remains to show that $P \rhd Q = U \rhd R \rhd Q$. If there are no interfaces, this is easy to see; otherwise, the proof is somewhat more tedious. The proof for the second case is similar. □

It is instructive to find the two cases in the decomposition of N in Fig. 3.

Levi's lemma is a factorisation property: every $P \rhd Q = U \rhd V$ factorises either as $U \rhd R \rhd Q$ or as $P \rhd R \rhd V$. Hence gluing decompositions are equal up-to associativity (and unit laws). For parallel composition, a Levi property as above does not hold, so we state a decomposition lemma directly:

Lemma 17. *Assume $P_1 \otimes \cdots \otimes P_n = Q_1 \otimes \cdots \otimes Q_m$, where each P_i and Q_j are (weakly) connected and not equal to id_0. Then $n = m$, and there exists a permutation $\sigma : [n] \to [n]$ such that $P_i = Q_{\sigma(i)}$ for each $i \in [n]$.*

Proof. Let $P = P_1 \otimes \cdots \otimes P_n$, then $P_1, \ldots, P_n$ are the connected components of P, but so are $Q_1, \ldots, Q_m$; the claim follows. □

The lemmas in this section are helpful for characterising the iposets generated by $\rhd$ and $\otimes$ from singletons. This is the subject of the next section.

6 Generating Iposets

Recall that $\mathcal{S}$ is the set of singleton iposets. It contains the four iposets ${}^0\mathsf{id}_1^0$, ${}^0\mathsf{id}_1^1$, ${}^1\mathsf{id}_1^0$ and ${}^1\mathsf{id}_1^1$, that is,

$$[0] \to [1] \leftarrow [0]\,, \qquad [0] \to [1] \leftarrow [1]\,, \qquad [1] \to [1] \leftarrow [0]\,, \qquad [1] \to [1] \leftarrow [1]\,,$$

with mappings uniquely determined. We are interested in the sets of iposets generated from singletons using $\rhd$ and $\otimes$. Strictly speaking, ${}^0\mathsf{id}_1^0$ should not count as a generator: it is equal to ${}^0\mathsf{id}_1^1 \rhd {}^1\mathsf{id}_1^0$ by Proposition 14. We may view $\mathcal{S}$ as a (directed) graph or quiver on vertices 0, 1.

Definition 18. *The set of* gluing-parallel *iposets (*gp-iposets*) is the smallest set that contains the empty iposet id_0 and the singleton iposets in $\mathcal{S}$ and that is closed under gluing and parallel composition.*

Theorem 19. *The gp-iposets are freely generated by the graph $\mathcal{S}$ in the variety of small categories (viewed as partial algebras) satisfying the equations of Propositions 9, 10 and 14.*

Proof (sketch). Suppose $(A, \rhd, \otimes, (1_i)_{i \geq 0})$ is any category on $\mathbb{N}$ satisfying the equations of Propositions 9, 10 and 14. Let $\varphi : \mathcal{S} \to A$ be any graph morphism that maps 0 to 0 and 1 to 1. We need to show that φ extends to a unique iposet morphism $\hat{\varphi}$.

We can generate any id_n from parallel compositions of id_1. For any $i \geq 0$, we map $\hat{\varphi}(\mathsf{id}_i) \mapsto 1_i$, and we map any other singleton $p \in \mathcal{S}$ as $\hat{\varphi}(p) = \varphi(p)$. For complex iposets we proceed by induction on the number of elements, assuming that homomorphism laws hold for iposets with n elements.

If the top composition of the size $n+1$ iposet P is $\triangleright$, then an inductive application of Lemma 16 implies that the $\triangleright$-decomposition of P is unique (up to associativity and unit laws), and we use associativity of $\triangleright$ to establish the homomorphism property of $\hat{\varphi}$. For $\otimes$ we proceed likewise, using the unique-decomposition property of Lemma 17. Finally, if the top composition is ambiguous, then Lemma 15 forces the configuration in which Proposition 14 can be applied, yielding a parallel composition of the same size. Finally, this extension is unique, as it was forced by the construction.

For a more detailed proof, care need to be taken because morphisms between partial algebras need to be compatible with the definedness conditions of partial operations. The notion of free partial algebra can be found in Burmeister's lecture notes [2]. Categories are single-sorted partial algebras through their object-free axiomatisation [24]. □

Next we define hierarchies of iposets generated from $\mathcal{S}$. (Removing ${}^0\mathsf{id}_1^0$ from $\mathcal{S}$ would change the hierarchy only for at most the first two alternations of $\triangleright$ and $\otimes$.) For any $\mathcal{Q} \subseteq \mathsf{iPos}$ and $\square \in \{\otimes, \triangleright\}$, let

$$\mathcal{Q}^{\square} = \{P_1 \square \cdots \square P_n \mid n \in \mathbb{N}, P_1, \ldots, P_n \in \mathcal{Q}\} .$$

Then define $\mathcal{C}_0 = \mathcal{D}_0 = \mathcal{S}$ and, for all $n \in \mathbb{N}$,

$$\mathcal{C}_{2n+1} = \mathcal{C}_{2n}^{\otimes}, \qquad \mathcal{D}_{2n+1} = \mathcal{D}_{2n}^{\triangleright}, \qquad \mathcal{C}_{2n+2} = \mathcal{C}_{2n+1}^{\triangleright}, \qquad \mathcal{D}_{2n+2} = \mathcal{D}_{2n+1}^{\otimes}$$

(this agrees with the $\mathcal{C}_1$ notation used earlier). Finally, let

$$\bar{\mathcal{S}} \stackrel{\text{def}}{=} \bigcup_{n\geq 0} \mathcal{C}_n = \bigcup_{n\geq 0} \mathcal{D}_n$$

be the set of all iposets generated from $\mathcal{S}$ by application of $\otimes$ and $\triangleright$.

Lemma 20. *For all $n \in \mathbb{N}$, $\mathcal{C}_n \cup \mathcal{D}_n \subseteq \mathcal{C}_{n+1} \cap \mathcal{D}_{n+1}$.*

Proof. We need to check the inclusions $\mathcal{C}_n \subseteq \mathcal{C}_{n+1}$, $\mathcal{D}_n \subseteq \mathcal{D}_{n+1}$, $\mathcal{C}_n \subseteq \mathcal{D}_{n+1}$ and $\mathcal{D}_n \subseteq \mathcal{C}_{n+1}$. The first two are trivial by construction. For the third one, note that $\mathcal{C}_0 \subseteq \mathcal{C}_0^{\triangleright} = \mathcal{S}^{\triangleright} = \mathcal{D}_0^{\triangleright} = \mathcal{D}_1$. Since $\mathcal{C}_n$ is constructed from $\mathcal{C}_0$ by the same alternations of $\otimes$ and $\triangleright$ as $\mathcal{D}_{n+1}$ is constructed from $\mathcal{D}_1$, the inclusion holds. The proof of the fourth inclusion is similar. □

Theorem 21. *An iposet is in $\mathcal{C}_2$ if and only if it is an interval order.*

Proof. For the forward direction, suppose $P \triangleright Q \in \mathcal{C}_2$. First it is clear that all elements of $\mathcal{C}_1$ are interval orders, so we will be done once we can show that the gluing composition of two interval orders is an interval order. This is precisely the

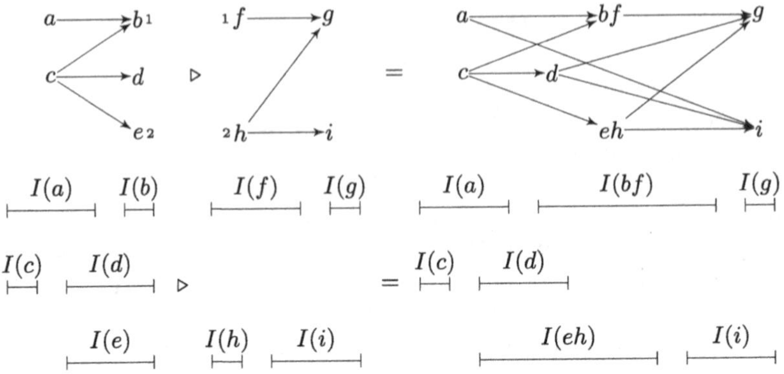

Fig. 6. Two interval orders and their concatenation: above as iposets, below using their interval representations. (Labels added for convenience.)

proof of Lemma 15: if $P \rhd Q$ contains a 2+2, then so do P or Q. Yet we also give a direct construction: Let σ_P be the interval sequence for interval representation (b_P, e_P) of $P : n \to m$ and σ_Q the interval sequence for interval representation (b_Q, e_Q) of $Q : m \to k$. Then concatenate σ_P and σ_Q, rename b_P, b_Q as b and e_P, e_Q as e, delete $e(t_P(i))$, $b(s_Q(i))$ and replace $e(s_Q(i))$ with $e(t_P(i))$ for each $i \in [m]$. This yields the interval sequence for interval representation (b, e) of $P \rhd Q$ and $P \rhd Q$ is therefore an interval order. Figure 6 gives an example.

For the backward direction, let P be an interval order and A_P its set of maximal antichains. Then A_P is totally ordered by the relation $\sqsubset$ defined in Sect. 3. Now write $A_P = \{P_1, \dots, P_k\}$ such that $P_i \sqsubset P_j$ for $i < j$. Then each P_i is an element of $\mathcal{S}^{\otimes}$. Write $s_1 : [n_1] \to P \leftarrow [n_{k+1}] : t_k$ for the sources and targets of P.

For $i = 2, \dots, k$, let $[n_i] = P_{i-1} \cap P_i$ be the overlap and $s_i : [n_i] \hookrightarrow P_i$, $t_{i-1} : [n_i] \hookrightarrow P_{i-1}$ the inclusions. Together with s_1 and t_k this defines iposets $s_i : [n_i] \to P_i \leftarrow [n_{i+1}] : t_i$. (Note that $s_1 : [n_1] \to P_1$ because P_1 is the minimal element in A_P; similarly for $t_k : [n_{k+1}] \to P_k$.) It is clear that $P = P_1 \rhd \dots \rhd P_k$; see also [16, Prop. 2]. □

In order to compare with series-parallel posets, we construct a similar hierarchy for these. Let $\mathcal{T}_0 = \mathcal{U}_0 = \mathcal{S}_0 = \{{}^0\mathrm{id}_1^0\}$ and, for all $n \in \mathbb{N}$,

$$\mathcal{T}_{2n+1} = \mathcal{T}_{2n}^{\otimes}, \qquad \mathcal{U}_{2n+1} = \mathcal{U}_{2n}^{\rhd}, \qquad \mathcal{T}_{2n+2} = \mathcal{T}_{2n+1}^{\rhd}, \qquad \mathcal{U}_{2n+2} = \mathcal{U}_{2n+1}^{\otimes}.$$

Then, because any element of any $\mathcal{T}_n$ or $\mathcal{U}_n$ has empty interfaces and because $\rhd$ is serial composition for all iposets with empty interfaces, we see that

$$\bar{\mathcal{S}}_0 \overset{\text{def}}{=} \bigcup_{n \ge 0} \mathcal{T}_n = \bigcup_{n \ge 0} \mathcal{U}_n$$

is the set of series-parallel posets. Note that $\mathcal{T}_n \subseteq \mathcal{C}_n$ and $\mathcal{U}_n \subseteq \mathcal{D}_n$ for all n, hence also $\bar{\mathcal{S}}_0 \subseteq \bar{\mathcal{S}}$. Now $\bar{\mathcal{S}}_0$ contains precisely the N-free posets whereas N is an interval order. Therefore $\mathsf{N} \in \mathcal{C}_2$, which implies the next lemma. On the other hand, we will show below that $\bar{\mathcal{S}}_0 \not\subseteq \mathcal{C}_n$ for any n.

Lemma 22. $\mathcal{C}_2 \not\subseteq \bar{\mathcal{S}}_0$.

Lemma 23. $\mathcal{C}_1 \cup \mathcal{D}_1 \subsetneq \mathcal{C}_2 \cap \mathcal{D}_2$, *i.e., there is an iposet with two non-trivial different decompositions.*

Proof. Directly from Proposition 14. □

Next we show that the $\mathcal{C}_n$ hierarchy is infinite, by exposing a sequence of witnesses for $\mathcal{C}_{2n-1} \subsetneq \mathcal{C}_{2n}$ for all $n \geq 1$.

Let $Q = {}^0\mathsf{id}_1^0$, $P_1 = Q \rhd Q$, and for $n \geq 1$, $P_{n+1} = Q \rhd (P_n \otimes P_n)$. Note that all these are series-parallel posets. Graphically:

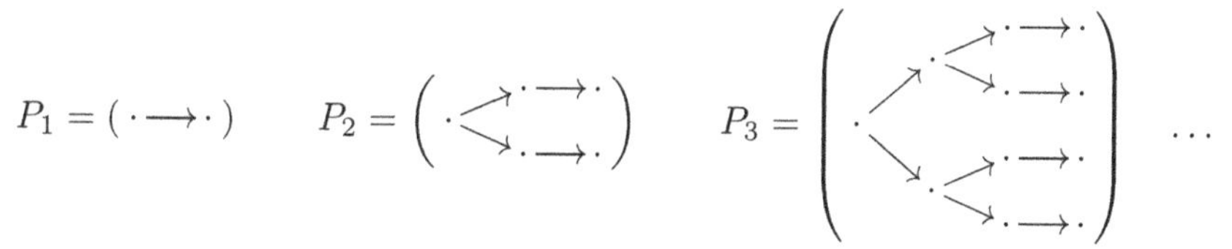

Lemma 24. $P_n \in \mathcal{C}_{2n} \setminus \mathcal{C}_{2n-1}$ *for all* $n \geq 1$.

Proof. By induction. For $n = 1$, $P_1 \notin \mathcal{C}_1$, but $Q \in \mathcal{C}_0 \subseteq \mathcal{C}_1$ and hence $P_1 = Q \rhd Q \in \mathcal{C}_2 = \mathcal{C}_1^{\rhd}$.

Now for $n \geq 1$, suppose $\mathcal{C}_{2n-1} \not\ni P_n \in \mathcal{C}_{2n}$. We use Lemma 15 to show that $P_n \otimes P_n \in \mathcal{C}_{2n+1} \setminus \mathcal{C}_{2n}$: Obviously $P_n \otimes P_n \in \mathcal{C}_{2n+1} = \mathcal{C}_{2n}^{\otimes}$. If $P_n \otimes P_n \in \mathcal{C}_{2n} = \mathcal{C}_{2n-1}^{\rhd}$, then $P_n \otimes P_n = Q_1 \rhd \cdots \rhd Q_k$ for some $Q_1, \dots, Q_k \in \mathcal{C}_{2n-1}$. Yet $P_n \notin \mathcal{C}_1$, which contradicts Lemma 15.

Now to $P_{n+1} = Q \rhd (P_n \otimes P_n)$. Trivially, $P_{n+1} \in \mathcal{C}_{2n+2} = \mathcal{C}_{2n+1}^{\rhd}$. Suppose $P_{n+1} \in \mathcal{C}_{2n+1} = \mathcal{C}_{2n}^{\otimes}$. P_{n+1} is connected, hence not a parallel product, so that P_{n+1} must already be in $\mathcal{C}_{2n} = \mathcal{C}_{2n-1}^{\rhd}$ and therefore $P_{n+1} = R_1 \rhd R_2$. Then, by Levi's lemma, there is an iposet S such that either $Q = R_1 \rhd S$ and $S \rhd (P_n \otimes P_n) = R_2$ or $R_1 = Q \rhd S$ and $S \rhd R_2 = P_n \otimes P_n$. In the second case, $S \rhd R_2 = P_n \otimes P_n$, which again contradicts Lemma 15; in the first case, both R_1 and S must be single points (with suitable interfaces), so that either $R_1 = {}^0\mathsf{id}_1^1$ and $R_2 = P_{n+1}$ (with an extra starting interface) or $R_1 = Q$ and $R_2 = P_n \otimes P_n$. This shows that $P_{n+1} = Q \rhd (P_n \otimes P_n)$ is the only non-trivial $\rhd$-decomposition of P_{n+1}. Thus $P_n \in \mathcal{C}_{2n-1}$, a contradiction, and therefore $P_{n+1} \notin \mathcal{C}_{2n+1}$. □

Corollary 25. $\mathcal{C}_{2n-1} \subsetneq \mathcal{C}_{2n}$ *for all* $n \geq 1$*, hence the* $\mathcal{C}_n$ *hierarchy does not collapse, and neither does the* $\mathcal{D}_n$ *hierarchy.*

Proof. The last statement follows from $\mathcal{D}_{2n-2} \subseteq \mathcal{C}_{2n-1} \subsetneq \mathcal{C}_{2n} \subseteq \mathcal{D}_{2n+1}$. □

Corollary 26. *For all* $n \in \mathbb{N}$, $\bar{\mathcal{S}}_0 \not\subseteq \mathcal{C}_n$ *and* $\bar{\mathcal{S}}_0 \not\subseteq \mathcal{D}_n$.

Proof. As explained already above, $P_n \in \bar{\mathcal{S}}_0$ for all n. This and Lemma 24 imply the first statement. The second one follows from $\mathcal{C}_n \subseteq \mathcal{D}_{n+1}$. □

We have seen that the $\mathcal{C}_n$ and $\mathcal{D}_n$ hierarchies are properly infinite and contain the set of sp-posets only in the limit $\bar{\mathcal{S}} = \bigcup_{n \geq 0} \mathcal{C}_n = \bigcup_{n \geq 0} \mathcal{D}_n$.

Finally, we turn to the question of characterising this limit $\bar{\mathcal{S}}$ geometrically. Given that a poset is series-parallel if and only if if it does not contain an induced subposet isomorphic to N, we aim for a similar characterisation of gp-(i)posets using forbidden subposets. We expose five such subposets, but leave the question of whether there are others to future work.

Define the following five posets on six points:

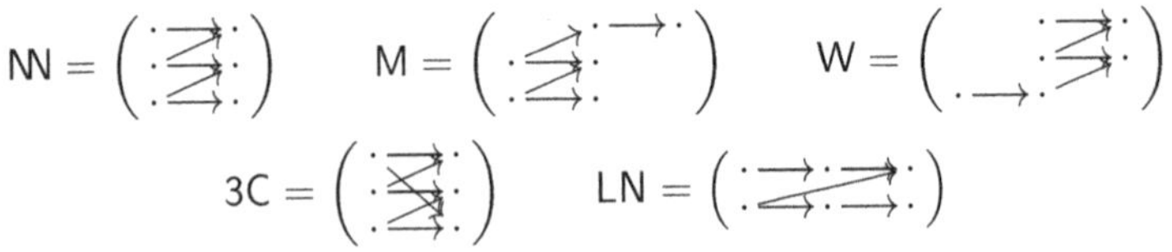

Proposition 27. *If* $P \in \bar{\mathcal{S}}$, *then* P *does not contain* NN, M, W, 3C, *or* LN *as an induced subposet.*

Proof. We only show the proof for NN; the others are very similar and are left to the reader. We can assume that P is connected. We use structural induction, noting that all $P \in \mathcal{S}$ are NN-free, so it remains to show that $P \triangleright Q$ is NN-free whenever P and Q are.

By contraposition, suppose $P \triangleright Q$ contains the induced sub-NN

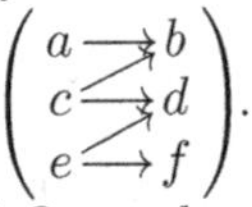

. Then we show that either P or Q also have an induced sub-NN.

Assume first that $a \in Q$. Then $a \leq_Q b$, hence also $b \in Q$, but $b \notin Q_{\min}$, that is, $b \notin s_Q$. Now $e \not\leq_{P \triangleright Q} b$, which forces $e \in t_P$ and therefore $e \in Q$. This in turn implies that $d, f \in Q$ and in particular $e \leq_Q f$. Thus $f \notin Q_{\min}$ and therefore $f \notin s_Q$, which forces $c \in t_P$ and therefore $c \in Q$. This shows that the NN lies entirely in Q.

Finally assume that $a \notin Q$. Then $a \in P \setminus t_P$, and as $a \not\leq_{P \triangleright Q} d$ and $a \not\leq_{P \triangleright Q} f$, we must have $d, f \in s_Q$ and therefore $d, f \in P$. This forces $c, e \in P$ and in particular $e \leq_P f$. Thus $e \notin P_{\max}$, whence $e \notin t_P$. This in turn forces $b \in s_Q$ and therefore $b \in P$. This shows that NN lies entirely in P. □

7 Experiments

We have encoded most of the constructions in this paper with Python to experiment with gluing-parallel (i)posets. Notably, Proposition 27 is, in part, a result of these experiments.[3] Our prototype is rather inefficient, which explains why some numbers are "n.a.", that is, not available in Table 1.

[3] Our software is available at http://www.lix.polytechnique.fr/~uli/posets/.

Table 1. Different types of posets with n points: all posets; sp-posets; gp-posets; (weakly) connected gp-posets; iposets with starting interfaces only; iposets; gp-iposets.

n	P(n)	SP(n)	GP(n)	GPC(n)	SIP(n)	IP(n)	GPI(n)
0	1	1	1	1	1	1	1
1	1	1	1	1	2	4	4
2	2	2	2	1	5	17	16
3	5	5	5	3	16	86	74
4	16	15	16	10	66	532	419
5	63	48	63	44	350	n.a	2980
6	318	167	313	233	n.a	n.a	26566

Using procedures to generate non-isomorphic posets of different types, we have used our software to verify that

1. all posets on five points are in $\bar{\mathcal{S}}$ and therefore gp-posets;
2. NN, M, W, 3C, and LN are the only six-point posets that are not in $\bar{\mathcal{S}}$.

We provide tables of gluing-parallel decompositions of posets in an extended version [5] to prove these claims.

We have also used our software to count non-isomorphic posets and iposets of different types, see Table 1. We note that P and SP are sequences no. A000112 and A003430, respectively, in the On-Line Encyclopedia of Integer Sequences (OEIS).[4] Sequences GPC, SIP, IP, and GPI are unknown to the OEIS.

The single iposet on two points which is not gluing-parallel is the symmetry $[2] : 2 \to 2$ with $s(1) = 1$, $s(2) = 2$, $t(1) = 2$, and $t(2) = 1$. The prefix of GP we were able to compute equals the corresponding prefix of sequence no. A079566 in the OEIS,(see footnote 4) which counts the number of connected (undirected) graphs which have no induced 4-cycle C_4. We leave it to the reader to ponder upon the relation between gp-posets and C_4-free connected graphs.

References

1. Bloom, S.L., Ésik, Z.: Free shuffle algebras in language varieties. Theoret. Comput. Sci. **163**(1&2), 55–98 (1996)
2. Burmeister, P.: Lecture notes on universal algebras, many sorted algebras (2002). https://www2.mathematik.tu-darmstadt.de/Math-Net/Lehrveranstaltungen/Lehrmaterial/SS2002/AllgemeineAlgebra/download/LNPartAlg.pdf
3. Conway, J.H.: Regular Algebra and Finite Machines. Chapman and Hall, Boca Raton (1971)
4. Courcelle, B., Engelfriet, J.: Graph Structure and Monadic Second-Order Logic. A Language-Theoretic Approach. Cambridge University Press, Cambridge (2012)
5. Fahrenberg, U., Johansen, C., Struth, G., Thapa, R.B.: Generating posets beyond N. CoRR, abs/1910.06162 (2019)

[4] See http://oeis.org/A000112, oeis.org/A003430, and oeis.org/A079566.

6. Fiore, M., Devesas Campos, M.: The algebra of directed acyclic graphs. In: Coecke, B., Ong, L., Panangaden, P. (eds.) Computation, Logic, Games, and Quantum Foundations. The Many Facets of Samson Abramsky. LNCS, vol. 7860, pp. 37–51. Springer, Heidelberg (2013). https://doi.org/10.1007/978-3-642-38164-5_4
7. Fishburn, P.C.: Intransitive indifference with unequal indifference intervals. J. Math. Psych. **7**(1), 144–149 (1970)
8. Fishburn, P.C.: Interval Orders and Interval Graphs: A Study of Partially Ordered Sets. Wiley, New York (1985)
9. Furusawa, H., Struth, G.: Concurrent dynamic algebra. ACM Trans. Comput. Log. **16**(4), 30:1–30:38 (2015)
10. Gischer, J.L.: The equational theory of pomsets. Theoret. Comput. Sci. **61**, 199–224 (1988)
11. Golumbic, M.C., Trenk, A.N.: Tolerance Graphs. Cambridge University Press, Cambridge (2004)
12. Grabowski, J.: On partial languages. Fund. Inf. **4**(2), 427 (1981)
13. Herlihy, M., Wing, J.M.: Linearizability: a correctness condition for concurrent objects. ACM Trans. Program. Lang. Syst. **12**(3), 463–492 (1990)
14. Hoare, T., Möller, B., Struth, G., Wehrman, I.: Concurrent Kleene algebra and its foundations. J. Log. Algebr. Program. **80**(6), 266–296 (2011)
15. Hoare, T., et al.: Developments in concurrent Kleene algebra. In: Höfner, P., Jipsen, P., Kahl, W., Müller, M.E. (eds.) RAMICS 2014. LNCS, vol. 8428, pp. 1–18. Springer, Cham (2014). https://doi.org/10.1007/978-3-319-06251-8_1
16. Janicki, R.: Modeling operational semantics with interval orders represented by sequences of antichains. In: Khomenko, V., Roux, O.H. (eds.) PETRI NETS 2018. LNCS, vol. 10877, pp. 251–271. Springer, Cham (2018). https://doi.org/10.1007/978-3-319-91268-4_13
17. Janicki, R., Koutny, M.: Structure of concurrency. Theoret. Comput. Sci. **112**(1), 5–52 (1993)
18. Janicki, R., Yin, X.: Modeling concurrency with interval traces. Inf. Comput. **253**, 78–108 (2017)
19. Jipsen, P., Moshier, M.A.: Concurrent Kleene algebra with tests and branching automata. J. Log. Algebr. Meth. Program. **85**(4), 637–652 (2016)
20. Kappé, T., Brunet, P., Rot, J., Silva, A., Wagemaker, J., Zanasi, F.: Kleene algebra with observations. In: CONCUR 2019, vol. 140 of LIPIcs. Schloss Dagstuhl - Leibniz-Zentrum für Informatik (2019)
21. Kappé, T., Brunet, P., Silva, A., Zanasi, F.: Concurrent Kleene algebra: free model and completeness. In: Ahmed, A. (ed.) ESOP 2018. LNCS, vol. 10801, pp. 856–882. Springer, Cham (2018). https://doi.org/10.1007/978-3-319-89884-1_30
22. Lamport, L.: The mutual exclusion problem: part I - a theory of interprocess communication. J. ACM **33**(2), 313–326 (1986)
23. Lamport, L.: On interprocess communication, part I: basic formalism. Distrib. Comput. **1**(2), 77–85 (1986)
24. Lane, S.M.: Categories for the Working Mathematician, 2nd edn. Springer, New York (1998)
25. Laurence, M.R., Struth, G.: Completeness theorems for pomset languages and concurrent Kleene algebras. CoRR, abs/1705.05896 (2017)
26. Levi, F.W.: On semigroups. Bull. Calcutta Math. Soc. **36**, 141–146 (1944)
27. Lodaya, K., Weil, P.: Series-parallel languages and the bounded-width property. Theoret. Comput. Sci. **237**(1–2), 347–380 (2000)
28. Mimram, S.: Presenting finite posets. In: TERMGRAPH 2014, vol. 183 of EPTCS (2014)

29. Möller, B., Hoare, T.: Exploring an interface model for CKA. In: Hinze, R., Voigtländer, J. (eds.) MPC 2015. LNCS, vol. 9129, pp. 1–29. Springer, Cham (2015). https://doi.org/10.1007/978-3-319-19797-5_1
30. Möller, B., Hoare, T., Müller, M.E., Struth, G.: A discrete geometric model of concurrent program execution. In: Bowen, J.P., Zhu, H. (eds.) UTP 2016. LNCS, vol. 10134, pp. 1–25. Springer, Cham (2017). https://doi.org/10.1007/978-3-319-52228-9_1
31. Peleg, D.: Concurrent dynamic logic. J. ACM **34**(2), 450–479 (1987)
32. Valdes, J., Tarjan, R.E., Lawler, E.L.: The recognition of series parallel digraphs. SIAM J. Comput. **11**(2), 298–313 (1982)
33. van Glabbeek, R.J.: The refinement theorem for ST-bisimulation semantics. In: IFIP TC2 Working Conference on Programming Concepts and Methods, North-Holland (1990)
34. van Glabbeek, R., Vaandrager, F.: Petri net models for algebraic theories of concurrency. In: de Bakker, J.W., Nijman, A.J., Treleaven, P.C. (eds.) PARLE 1987. LNCS, vol. 259, pp. 224–242. Springer, Heidelberg (1987). https://doi.org/10.1007/3-540-17945-3_13
35. Vogler, W.: Failures semantics based on interval semiwords is a congruence for refinement. Distrib. Comput. **4**, 139–162 (1991)
36. Vogler, W. (ed.): Modular Construction and Partial Order Semantics of Petri Nets. LNCS, vol. 625. Springer, Heidelberg (1992). https://doi.org/10.1007/3-540-55767-9
37. Wiener, N.: A contribution to the theory of relative position. Proc. Camb. Philos. Soc. **17**, 441–449 (1914)
38. Winkowski, J.: An algebraic characterization of the behaviour of non-sequential systems. Inf. Process. Lett. **6**(4), 105–109 (1977)

Automated Algebraic Reasoning for Collections and Local Variables with Lenses

Simon Foster(✉) and James Baxter

University of York, York, UK
{simon.foster,james.baxter}@york.ac.uk

Abstract. Lenses are a useful algebraic structure for giving a unifying semantics to program variables in a variety of store models. They support efficient automated proof in the Isabelle/UTP verification framework. In this paper, we expand our lens library with (1) dynamic lenses, that support mutable indexed collections, such as arrays, and (2) symmetric lenses, which allow partitioning of a state space into disjoint local and global regions to support variable scopes. From this basis, we provide an enriched program model in Isabelle/UTP for collection variables and variable blocks. For the latter, we adopt an approach first used by Back and von Wright, and derive weakest precondition and Hoare calculi. We demonstrate several examples, including verification of insertion sort.

1 Introduction

The use of algebraic structures for derivation of verification tools using theorem provers has been shown to be a successful and flexible approach [1–4]. It allows us to precisely and abstractly characterise the formal semantics of a spectrum of languages utilising different computational paradigms, including hybrid systems, concurrency, pointers, and probability. Once a suitable algebraic structure is fixed, a large array of axiomatic verification calculi can be generated, including Hoare logic [1], differential dynamic logic [4], separation logic [2], and rely-guarantee calculus [3]. This approach has significant advantages over concrete intermediate verification languages (IVLs) [5,6], since it allows us to unify languages and verification calculi at the algebraic level, and so promotes reuse.

Nevertheless, the underlying algebras for program verification largely focus on the *point-free* programming operators – those that do not explicitly characterise program variables – such as sequential composition ($\fatsemi$) and non-deterministic choice ($\sqcap$). Kleene Algebra with Tests (KAT) [3,7], for example, can characterise every operator of imperative while-programs, but is not sufficient to fully capture assignment, substitution, frames, and local variable blocks. Operators that manipulate the store via variables have to be defined in the model rather than the algebra [1,3]. This technically hampers the reuse of theorems across various languages. At the same time, an algebra of state should allow efficient use of automated proof facilities, so as to support scalable verification tools.

U. Fahrenberg et al. (Eds.): RAMiCS 2020, LNCS 12062, pp. 100–116, 2020.
https://doi.org/10.1007/978-3-030-43520-2_7

Lenses [8–10] allow us to characterise variables as abstract algebraic objects, which can be composed and manipulated. They provide a generic foundation for verification tools that can maximise proof automation in tools like Isabelle [11,12]. Although originating from a different intellectual stream [8], lenses are essentially Back and von Wright's variable manipulation functions [13]. However, lenses are also equipped with several operators that allow us to compose state space query operations in sequence and in parallel, for example. Lenses are the foundation for state modelling in our verification framework, Isabelle/UTP [10,14,15], which allows the use of UTP semantic models in developing program verification tools.

In previous work [10], we showed how lenses capture a variety of store models. In this paper, we extend our basic lens model in two ways. Firstly, we develop support for indexed collections, which requires the development of *dynamic lenses.* Secondly, we add support for local variables, for which we harness the work of Hoffmann et al. on *symmetric lenses* [16] that allow us to partition the state space into global and local variable scopes. This allows us to determine whether a particular assignment can be moved outside of a block. From this foundation, we adapt Back and von Wright's block operators [13], and prove Hoare logic theorems. Symmetric lenses allow us to unify a variety of variable block approaches, including extensible records [17] and list-based stacks [18].

In order to illustrate these features, consider the insertion sort algorithm:

Example 1.1 (Insertion Sort).

function $\mathit{insertion\text{-}sort}(\mathit{arr} : [\mathit{int}]\mathit{array})$
  var $i, j : \mathit{nat} \bullet$
  for $i := 1$ **to** $(\mathit{length}(\mathit{arr}) - 1)$ **do**
    $j := i \mathbin{;}$
    while $(0 < j \land \mathit{arr}[j] < \mathit{arr}[j-1])$ **do**
      $(\mathit{arr}[j-1], \mathit{arr}[j]) := (\mathit{arr}[j], \mathit{arr}[j-1]) \mathbin{;} j := j - 1$
    od od

It introduces two local variables, i and j, that are used to index into the array. The outer loop iterates through the list using i, and the inner loop inserts element i in the correct position into $\mathit{arr}[0...i-1]$, using j to count down. To give this a semantics, we need to (1) allow assignment to the indices of a collection, and (2) extend the state space to add i and j as local variables. Our goal is to support this abstract algorithmic presentation directly in our tool, through a shallow embedding, and provide syntax-directed reasoning support.

Our approach reduces reasoning about programs to proving properties of the state space. It is therefore applicable to any language semantics with an explicit state space model, including reactive [11] and hybrid languages [4,19]. The approach is therefore abstract, but also maximises Isabelle's proof automation.

The structure of this paper is as follows. After consideration of related work in Sect. 2, we describe how lenses give an algebraic semantics to variables in Sect. 3. In Sect. 4 we give an overview of Isabelle/UTP. In Sect. 5, we consider how a state space can be manipulated using lens operators. In Sect. 6 we describe dynamic lenses, which are needed for collections, like arrays. In Sect. 7, we describe our algebraic characterisation of symmetric lenses, and exhibit several models. In Sect. 8, we use symmetric lenses to implement local variable blocks. In Sect. 9 we use all the aforementioned results to verify insertion sort in Isabelle/UTP. Finally, in Sect. 10, we conclude.

All definitions and theorems are mechanised in Isabelle/UTP, and are often accompanied by an icon ( ) linking to the corresponding repository artefact.

2 Related Work

Isabelle/UTP [10,14,15,20] is a semantic framework for verification tools based in Hoare and He's Unifying Theories of Programming (UTP) [21]. It is broadly comparable to IVLs like Boogie [5] and Why3 [6], but harnesses algebraic and denotational semantic techniques, for application to languages of multiple paradigms. The UTP relational program model is built as a shallow embedding [20] in Isabelle/HOL, and so we compare with similar techniques in this prover.

Simpl is an IVL developed by Schirmer in Isabelle/HOL [17,22]. It is used in the AutoCorres verification platform [23] that was applied in the seL4 project[1]. It uses state monads augmented with exceptions to model low-level code. Our aim is to support the features and efficiency of Simpl, but using relational calculus and algebra to characterise language features abstractly so that they can be transferred between semantic models. Their work does not provide an algebraic semantics for variables, which we provide by lenses, but their comprehensive study of state space modelling techniques is a strong foundation for us [22].

Dongol et al. [24] characterise variables algebraically using Cylindric Kleene Lattices, which extend Kleene algebra with Cylindrification to support quantification. This, in turn, allows expression of both frames and local variable blocks. Their work is largely complementary to ours, since we focus on the algebraic semantics of the variables themselves. They use ordinals as indices into an implicit state space, whereas we characterise the state space explicitly. Our use of lenses also allows us to harness type checking and proof automation in Isabelle.

3 State Space Modelling with Lenses

Here, we review lenses [10], which give algebraic semantics to variables. Novelties include the *list-lens* and a more precise presentation compared to previous work [10]. We use the notation $X : V \Longrightarrow S$ when X is a lens that characterises a V-shaped subregion of a state space S. For instance, in a state space $A \times B$, we can define two lenses: $\mathbf{fst}_B^A : A \Longrightarrow A \times B$ and $\mathbf{snd}_B^A : B \Longrightarrow A \times B$, that select the respective components. As usual [8], we define lenses using two functions:

[1] The seL4 microkernel verification project: http://sel4.systems.

Definition 3.1. *A lens is a quadruple* $X \triangleq (V, S, \textsf{get}, \textsf{put})$, *where* V *and* S *are non-empty sets called the view type and state space, respectively, and* $\textsf{get} : S \to V$ *and* $\textsf{put} : S \to V \to S$ *are total functions. We often subscript* $\textsf{get}$ *and* $\textsf{put}$ *with the name of a lens. We define* $\textsf{create}_X\ v \triangleq \textsf{put}_X\ (\varepsilon s \bullet s \in S)\ v$, *which constructs an arbitrary, but fixed, state using Hilbert's choice* (ε) *and puts* v *into it.*

For example, $\textsf{fst}_B^A \triangleq (A, A \times B, \lambda(x, y) \bullet x, \lambda\, x'\,(x, y) \bullet (x', y))$, selects and updates the first element of a pair, leaving the second element unchanged. Lenses provide an intuitive and obvious way to model variables in a state space (cf. [13, 18]), which can be queried and updated using the two functions. Intuitively, we can think of them as pointers to distinct regions of a memory store modelled by S. As previously highlighted [10,22], there are a variety of possible memory models, and lenses provide a uniform algebraic interface for them.

Lenses provide the starting point for the UTP relational calculus [21], which has a model for imperative programs, including operators like sequential composition ($P \mathbin{⨟} Q$), conditional ($P \lhd b \rhd Q$), and assignment ($x := e$). Assignment is polymorphic over any lens x, provided that e matches its view type (see Sect. 4).

The behaviour of lenses is constrained by three intuitive axioms:

$$\textsf{get}\,(\textsf{put}\,s\,v) = v \quad (\text{L1}) \qquad \textsf{put}\,(\textsf{put}\,s\,v')\,v = \textsf{put}\,s\,v \quad (\text{L2}) \qquad \textsf{put}\,s\,(\textsf{get}\,s) = s \quad (\text{L3})$$

L1 states that a value put can be retrieved. L2 states that an earlier put is overwritten by a later put. L3 states that retrieving a value and then putting it back yields the original state. We distinguish *total* lenses, that obey all three axioms, from *partial* lenses that obey only L1 and L2. The **fst** and **snd** lenses are both total. A further example of a total lens is the total function lens:

$$\textsf{fun-lens}_B(x : A) \triangleq (B, A \to B, \lambda f \bullet f\,x, \lambda f\ v \bullet f(x := v))$$

It points to the value associated with a particular domain element x. It is useful, for instance, when the state space has type $Name \to Value$, which associates a named variable with a value in a given universe. The $\textsf{get}$ function simply applies f, and the $\textsf{put}$ function updates the value associated with x. It is clear that this lens obeys all three laws. We also define a relation called independence, $X \bowtie Y$, that characterises when two lenses view disjoint regions of the state space:

$$X \bowtie Y \triangleq \forall (s, u, v) \bullet \begin{pmatrix} \textsf{put}_X(\textsf{put}_Y\ s\ v)\ u = \textsf{put}_Y(\textsf{put}_X\ s\ u)\ v \\ \wedge\ \textsf{get}_X(\textsf{put}_Y\ s\ v) = \textsf{get}_X\ s \\ \wedge\ \textsf{get}_Y(\textsf{put}_X\ s\ u) = \textsf{get}_Y\ s \end{pmatrix} \quad \text{if } S_X = S_Y$$

It is defined only when the state spaces are the same: $S_X = S_Y$. X and Y are independent provided applications of $\textsf{put}$ commute, and each $\textsf{get}$ function is unaffected by the corresponding $\textsf{put}$ function. If X and Y are both total lenses, then the second and third conjuncts can be omitted. $X \bowtie Y$ means that X and Y do not interact, for example $\textsf{fun-lens}_B(i) \bowtie \textsf{fun-lens}_B(j)$, provided that $i \neq j$.

Partial lenses, which do not obey L3, are motivated by partial structures, such as arrays and heaps. The cells of an array can be modelled using list lenses:

$$\textit{list-lens}_A(i : \mathbb{N}) \triangleq (A, [A]\textit{list}, \lambda\, xs \bullet xs\,!\,i, \lambda\, xs\; v \bullet xs[i := v])$$

A lens $\textit{list-lens}_A(i) : A \Longrightarrow [A]\textit{list}$ points to the ith element of an inductive list of values drawn from A. The HOL operator $xs\,!\,i$ returns the ith element of xs, or an arbitrary element of A if $i \geq \#xs$, where $\#xs$ is the length. The operator $xs[i := v]$ updates the ith element to take value v. If xs is not long enough to hold v, it is first expanded by filling in the extra elements with arbitrary values. Here, *list-lens* and *fun-lens* are both examples of lenses indexed by a set. As for *fun-lens*, we have it that $\textit{list-lens}_B(i) \bowtie \textit{list-lens}_B(j)$ provided that $i \neq j$.

Clearly, *list-lens* satisfies both L1 and L2: we can always place a value in the ith component, potentially several times, and retrieve it. However, it does not satisfy L3. When a list is too short ($i \geq \#xs$), $\textit{list-lens}(i)$ returns an arbitrary value, which, if placed at i, alters the list structure and violates L3. Consequently, whilst *fun-lens* is a total lens, *list-lens* is only partial, and the same follows for data structures like partial functions. Nevertheless, as we shall see, partial lenses are sufficient to support most of the laws we need for verification calculi.

A useful class of state space is induced by records. In Isabelle/UTP, we can define a state type $\textit{rec-typ} \triangleq [x_1 : A_1 \cdots x_m : A_m]$ for m fields, each with a given type. Technically, it is isomorphic to a product type $A_1 \times \cdots \times A_m$, but with named lenses for manipulating each field. The ***alphabet*** command automates the creation of these lenses, and generates theorems that $x_i \bowtie x_j$ for any $i \neq j$.

State types can also be extended: $\textit{rec-typ}_2 \triangleq \textit{rec-typ} + [y_1 : B_1 \cdots y_n : B_n]$, which allows hierarchy. This approach is used in the IVL Impl [17] to represent local variables, and here we adopt a similar approach. The lenses are polymorphic: $x_i : A_i \Longrightarrow [\alpha]\textit{rec-typ-ext}$, where the parameter α allows application of x_i to both *rec-typ* and extensions thereof, such as $\textit{rec-typ}_2$ with $\alpha \cong A_{m+1} \times \cdots \times A_n$. This is important, as it means that the same lens name can be used in different state spaces: x_i can both have the type $A_i \Longrightarrow \textit{rec-typ}$ and $A_i \Longrightarrow \textit{rec-typ}_2$.

4 Relational Programs in Isabelle/UTP

In this section, we briefly introduce the foundations of Isabelle/UTP, which is a shallow embedding of UTP [21] in Isabelle/HOL. UTP is based on a variant of Tarski's relational calculus [25] where each relation is "alphabetised", meaning it is parameterised by the set of variables to which it can refer. In Isabelle/UTP, we instead opt to have relations parameterised by their state space type S, and variables are then lenses viewing this type. We can therefore use the Isabelle type system to ensure well-formedness: only relations and predicates with compatible alphabets can be composed using the Boolean and relational connectives.

Expressions are total functions: $[V, S]\textit{expr} \triangleq (S \to V)$, for some state space S and type V. Operators can be pointwise lifted and applied to them, for example, if $e, f : [\mathbb{N}, S]\textit{expr}$, then $e + f$ denotes $\lambda\, s : S \bullet e\, s + f\, s$. If x and y are lenses, then we can use them in expressions: $x + y$ denotes $\lambda\, s : S \bullet \textbf{get}_x\, s + \textbf{get}_y\, s$. We can determine whether e depends on part of the state using the unrestriction [10]:

Definition 4.1. $(x \sharp e) \triangleq (\forall (s, v) \bullet e\,(\mathit{put}_x\, s\, v) = e\, s)$

Lens x is unrestricted in e, written $x \sharp e$, when updating its value using *put* has no effect on the valuation of e. This can occur, for example, when x is a variable that e does not refer to. Unrestriction distributes through lifted functions [10].

Substitutions between two states spaces are modelled with functions, $\sigma : S_1 \to S_2$. A substitution can be updated using $\sigma(x \mapsto e)$. A heterogeneous substitution can be constructed using $(\!|x_1 \mapsto e_1, \cdots, x_n \mapsto e_n|\!)$, when $x_i : A_i \Longrightarrow S_2$ and $e_i : [A_i, S_1]\mathit{expr}$, which is a set of simultaneous updates. A homogeneous substitution, where $S_1 = S_2$, can be constructed similarly but using square brackets: $[x \mapsto e, \cdots]$. The difference between these two is that the former gives arbitrary values to unassigned variables, whereas the latter copies the original values. Substitutions can also be composed function-wise, $\sigma \circ \rho$, which corresponds to applications of the updates in ρ followed by those in σ.

We can apply a substitution to an expression using $\sigma \dagger e \triangleq e \circ \sigma$, which likewise composes the substitution and expression functions. Although this may seem redundant, it is useful to distinguish a separate operator to enable bespoke rewrite laws in Isabelle. Then, we can obtain the traditional substitution operator: $p[e/x] \triangleq [x \mapsto e] \dagger p$. Substitutions then obey a number of useful laws:

Theorem 4.2. *If x and y are partial lenses, then the following laws hold:*

$$\sigma(x \mapsto e, y \mapsto f) = \sigma(y \mapsto f, x \mapsto e) \qquad \text{if } x \bowtie y \qquad (1)$$
$$\sigma(x \mapsto e, x \mapsto f) = \sigma(x \mapsto f) \qquad (2)$$
$$\sigma(x \mapsto v) \circ \rho = (\sigma \circ \rho)(x \mapsto (\sigma \dagger v)) \qquad (3)$$
$$\sigma(x \mapsto e) \dagger x = e \qquad (4)$$
$$\sigma(x \mapsto v) \dagger e = \sigma \dagger e \qquad \text{if } x \sharp e \qquad (5)$$

Substitution updates commute when made to independent lenses (1), and can cancel earlier ones (2). Substitutions can be composed, and (3) shows how to pull out a variable update to the left-most substitution. These laws can be used to show that $[x \mapsto u] \circ [y \mapsto v]$ is equivalent to $[x \mapsto u[v/y], y \mapsto v]$, when $x \bowtie y$. Substitution application distributes through functions in the obvious way, and can be applied to variable expressions (4). If x is unrestricted in an expression, then any assignment to this variable can be dropped (5).

We define predicates, $[S]\mathit{pred} \triangleq [\mathit{bool}, S]\mathit{expr}$, relations, $[S_1, S_2]\mathit{rel} \triangleq [S_1 \times S_2]\mathit{pred}$, and the usual operators over them. Predicates and relations are ordered by refinement ($\sqsubseteq$). We import theorems for structures like complete lattices, quantales, and Kleene algebras [2,3,10]. With substitutions, it is easy to define a generalised assignment operator, in the style of Back and von Wright [13]: $\langle\sigma\rangle$, which lifts a substitution to a relation in the obvious way. This satisfies a useful law, $\langle\sigma\rangle \mathbin{;} \langle\rho\rangle = \langle\rho \circ \sigma\rangle$, which allows us to combine sequential assignments. Assignments can then be constructed with $x := e \triangleq \langle[x \mapsto e]\rangle$, and combining with non-deterministic choice ($\sqcap$) we define non-deterministic assignment: $x := * \triangleq \bigsqcap_{v \in V_x} x := v$. These definitions satisfy the laws of programming [26].

5 State Space Manipulation

Here, we show how to manipulate state spaces, and coerce variables and expressions between them. We use two additional relations, that are defined using get and put [10]: (1) $X \preceq Y$: the view of lens X is contained within the view of Y; (2) $X \approx Y$: the views of X and Y are isomorphic. These are both heterogeneous operators that can relate lenses with different view types. Relation $\preceq$ forms a preorder and $\approx$ is an equivalence relation. Ordering is needed because lenses can characterise both variables and sets thereof. We compose lenses, thus combining their respective views, using the pairing operator:

Definition 5.1 (Lens Pairing).

$$X \oplus Y \triangleq \begin{pmatrix} V_X \times V_Y, S_X, (\lambda\, s \bullet (\mathsf{get}_X\, s, \mathsf{get}_Y\, s)), \\ (\lambda\, s\, (u, v) \bullet \mathsf{put}_X\, (\mathsf{put}_Y\; s\; v)\; u) \end{pmatrix} \quad \mathit{when}\ S_X = S_Y, X \bowtie Y$$

Lens pairing combines two independent lenses with the same state space, creating a lens whose view type is $V_X \times V_Y$. The get function pairs the results of the get functions for X and Y, while its put function puts each element using the respective put. Using this, a set of variables, $\{x, y, z\}$ can be characterised by a lens, for example, by the summation $x \oplus y \oplus z$. Moreover, we have it that $X \preceq X \oplus Y$, since $X \oplus Y$ views more of the state space than X.

We define two basic total lenses: $\boldsymbol{1}_S \triangleq (S, S, \lambda\, s \bullet s, \lambda\, s\, v \bullet v)$ whose view and state space are identical, and $\boldsymbol{0}_S \triangleq (\{\emptyset\}, S, \lambda\, s \bullet \emptyset, \lambda\, s\, v \bullet s)$, whose view type is unitary. Intuitively, $\boldsymbol{1}$ characterises the entirety of S, and $\boldsymbol{0}$ characterises none of it, and cannot distinguish any states. Consequently, we have $\boldsymbol{0} \preceq X$ and $X \preceq \boldsymbol{1}$, since these are the least and most distinguishing lenses, respectively.

For variable blocks, we need expansion and contraction of the state space, for both lenses and expressions. For lenses, we define the composition and quotient:

Definition 5.2 (Lens Composition and Quotient).

$$X \,;\, Y \triangleq \begin{pmatrix} V_X, S_Y, \mathsf{get}_X \circ \mathsf{get}_Y, \\ (\lambda\, s\, v \bullet \mathsf{put}_Y\; s\, (\mathsf{put}_X\, (\mathsf{get}_Y\; s)\; v)) \end{pmatrix} \qquad \mathit{when}\ S_X = V_Y$$

$$X \,/\, Y \triangleq \begin{pmatrix} V_X, V_Y, \mathsf{get}_X \circ \mathsf{create}_Y, \\ (\lambda\, s\, v \bullet \mathsf{get}_Y\, (\mathsf{put}_X\, (\mathsf{create}_Y\; s)\; v)) \end{pmatrix} \qquad \mathit{when}\ S_X = S_Y$$

$X \,;\, Y$ has been previously defined [8]. It selects a subregion V_1, characterised by $X : V_1 \Longrightarrow V_2$, of a larger region V_2, characterised by $Y : V_2 \Longrightarrow S$. Intuitively, Y denotes a sub-space of S, X is a variable of this sub-space, and so $X \,;\, Y \preceq Y$. We sometimes write $\mathit{obj}\mathbf{:}\mathit{attr}$ for the composition $\mathit{attr} \,;\, \mathit{obj}$.

We believe the quotient operator, X / Y is novel[2]. Provided that $X : V_1 \Longrightarrow S$ is constructed by composition of $Y : V_2 \Longrightarrow S$ and $Z : V_1 \Longrightarrow V_2$, we have it that $X / Y = Z$. The get function first creates an arbitrary state and populates the V_2 region with the incoming state. It then uses the get_X function to obtain

[2] The similarly named *quotient lens* of Foster et al. [9] is a rather different concept.

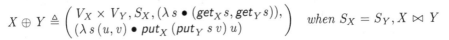

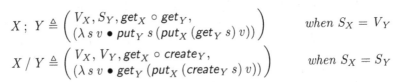

the V_1 element. The assumption is that all the information needed to construct a V_1 can be obtained from V_2. The put function creates an S element, uses put_X to update this with $v : V_1$, and finally applies get_Y to obtain a V_2 element. Again, the assumption is that put_X will only manipulate data within V_1.

Lens quotient gives rise to some useful, and intuitive, properties.

Theorem 5.3. $(X \,;\, Y)/Y = X \qquad (X/X) = \mathbf{1} \qquad X/\mathbf{1} = X$

The first identity gives the intuition of quotient: it removes the second element of a composition. The second identity shows that if we remove a lens from itself, then only a residual *1* remains. The third identity shows that removal of *1* has no effect, because of course $X \,;\, \mathbf{1} = X$.

In addition, we need to expand and contract the state space of expressions:

Definition 5.4. *We fix* $X : S_1 \Longrightarrow S_2$*, expressions* $e : [A, S_1]\mathsf{expr}$ *and* $f : [B, S_2]\mathsf{expr}$*, and define:* $e \uparrow X \triangleq e \circ \mathit{get}_X$ *and* $f \downarrow X \triangleq f \circ \mathit{create}_X$

Here, X is a lens that describes how S_1 is embedded into a larger space S_2. The first operator, $e \uparrow X$, extends the state space of e to be S_2, and the second, $f \downarrow X$, restricts it to be S_1. These operators coerce an expression to have a different type, for use in a context with a different state space. They satisfy several theorems.

Theorem 5.5 (State Space Extension and Restriction).

$$
\begin{aligned}
(f\, e_1 \cdots e_n) \uparrow A &= f\,(e_1 \uparrow A \cdots e_n \uparrow A) & \qquad (f\, e_1 \cdots e_n) \downarrow A &= f\,(e_1 \downarrow A \cdots e_n \downarrow A) \\
x \uparrow A &= A{:}x & x \downarrow A &= x/A \\
A \bowtie B &\Rightarrow B \sharp (e \uparrow A) & (e \uparrow A) \downarrow A &= e
\end{aligned}
$$

Both extension and restriction distribute through function application in the obvious way. Extension of a lens expression entails a lens composition, and restriction entails a lens quotient. If we extend an expression's state space, $e \uparrow A$, then the resulting expression does not depend on a lens B that is independent of A. The reason is that the original state space of e is characterised by A. Finally, we have it that restriction is the inverse of extension. The converse theorem does not hold, because restricting a state space may result in a loss of information.

We can define $e^{\blacktriangleleft} \triangleq e \uparrow \mathit{fst}$, $e_{\blacktriangleleft} \triangleq e \downarrow \mathit{fst}$, and $e^{\blacktriangleright} \triangleq a \uparrow \mathit{snd}$, that characterise relational preconditions and postconditions. Specifically, $e^{\blacktriangleleft}$ lifts an expression on S to one on $S \times S$, thus turning a predicate into a relation. We can characterise initial and final variables, $x^{\blacktriangleleft}$ and $x^{\blacktriangleright}$, in the style of notations like Z. We can also define weakest preconditions, $P \,\mathbf{wp}\, b \triangleq (P \fatsemi b^{\blacktriangleleft})_{\blacktriangleleft}$, and also the Hoare triple, $\{\, p \,\}\, Q \,\{\, r \,\} \triangleq (p^{\blacktriangleleft} \Rightarrow q^{\blacktriangleright}) \sqsubseteq Q$. These definitions admit, as theorems, the usual laws [27,28]. For example, we have the assignment law, $\{\, p[v/x] \,\}\, x := v \,\{\, p \,\}$, for any lens x, and the more general $\{\, \sigma \dagger p \,\}\, \langle \sigma \rangle \,\{\, p \,\}$.

6 Dynamic and Collection Lenses

In this section we give semantics to the notation $x[i]$, which refers to the ith element of a collection x. We model x with a lens that points to a collection, such as a list, and i with an index expression. The generality of the lens axioms means that we can define $x[i]$ itself to be a type of lens, which we call the collection lens. Consequently, we can manipulate it like any other lens, employing the theorems of Sect. 4. In order to define this, we first need to define dynamic lenses:

Definition 6.1. *We fix sets A and B that denote elements and collections, and a set I of indices. We assume a family of I-indexed lenses $F : I \to (A \Longrightarrow B)$ and an expression $e : B \to I$. A dynamic lens is defined as follows:*

$$\textsf{dyn-lens}\, F\ e \triangleq (A, B, \lambda\, s \bullet \textsf{get}_{F(e\ s)}\ s, \lambda\, s\ v \bullet \textsf{put}_{F(e\ s)}\ s\ v)$$

Intuitively, a dynamic lens points to the eth element of the indexed lens F. Since e is an expression, it can change value, and consequently the current index depends on the state space. The *get* and *put* function both instantiate the indexed lens with e applied to the current state, and then apply its respective *get* and *put* function. From this definition, we can prove the following closure theorem:

Theorem 6.2. *We assume that, for all $i : I$, e does not refer to $F\,i$, that is $(F\ i) \sharp\, e$, and $F\,i$ is a partial or total lens. We can then show that $\textsf{dyn-lens}\, F\ e$ is a partial or total lens, respectively.*

The intuition is that $F\,i$ must satisfy the lens axioms, for all indexes, and the index expression e should not itself refer to the $F\,i$, to avoid self references. From this definition, we can now define collection lenses:

Definition 6.3 (Collection Lenses). *We fix $F : I \to (A \Longrightarrow B)$, a lens indexed by the set I. Then, given a lens $x : B \Longrightarrow S$, for some state space S, and an index expression $e : S \to I$, a collection lens is defined as follows:*

$$x[e] \triangleq \textsf{dyn-lens}\,(\lambda\, i : I \bullet F\ i\ ;\ x)\ e$$

The collection lens, $x[e]$, is a dynamic lens where the underlying indexed lens F is applied after selection of the collection location in the lens x. It is clear that Theorem 6.2 can be applied here too, provided that x is also a total lens. The intuition of the collection lens is perhaps clearer if we consider a concrete example where $F = \textit{fun-lens}$. In this case, we can prove the following identity:

$$(x[i] := e) = (x := x(i := e))$$

An assignment to $x[i]$ frames the remainder of the state, and thus x takes its original value with the i index updated. We can also derive the identity:

$$(x[i] := e \,\text{⨾}\, x[j] := f) \;=\; (x[j] := f \,\text{⨾}\, x[i] := e)$$

whenever $i \neq j$, $x[i] \sharp f$, and $x[j] \sharp e$, by using the generalised assignment laws.

In Isabelle/UTP, we make F an overloaded polymorphic constant that associates a suitable indexed lens to a collection type. In many situations, $x[e]$ is a partial lens, since it is only meaningful when x is a collection where the key e is defined. For example, the assignment $(arr[j-1], arr[j]) := (arr[j], arr[j-1])$ in Example 1.1 is meaningful only when $j < \#arr$. Thus, when verifying programs with collection lenses, it is necessary to guard them with definedness predicates.

7 Symmetric Lenses

Symmetric lenses [16] stand in contrast to the lenses that were introduced in Sect. 3, which are "asymmetric" because, once a view has been extracted from a source, it is not possible to reconstruct the source from the view alone [16]. Symmetric lenses are effectively lenses of type $V \times C \Longrightarrow S$, where C is the "complement" of V with respect to S – the remainder of S once V is removed. In general, it is not possible to compute the complement of an asymmetric lens. Symmetric lenses thus capture the notion of partitioning the state into disjoint regions. These regions are represented by two lenses, which we refer to as the *view* and the *coview*, and for a given symmetric lens $\mathcal{X}$, we write $\mathcal{V}_{\mathcal{X}}$ and $\mathcal{C}_{\mathcal{X}}$ to represent them. Such a partitioning of the state space is fundamental to framing of certain variables, and allows us to distinguish the global and local store.

To characterise symmetric lenses, we must capture both the disjointness of the view and coview, and the fact that, taken together, they cover the state space. Coverage is captured by first combining the view and coview into a pairing $\mathcal{V}_{\mathcal{X}} \oplus \mathcal{C}_{\mathcal{X}}$, and requiring that this covers the state space. Such a definition is provided for by the concept of *bijective lenses*, defined below.

Definition 7.1. *A partial bijective lens satisfies L1, and also* $\mathsf{put}\, s\, v = \mathsf{put}\, s'\, v$*. A (total) bijective lens satisfies L1, and also* $\mathsf{put}\, s\, (\mathsf{get}\, s') = s'$*.*

For total bijective lenses, we require that getting the view of s' and putting it into s replaces the whole of s with s'. For a partial lens, get may return an incorrect value for states outside its domain, so a partial bijective lens is characterised by $\mathsf{put}\, s\, v = \mathsf{put}\, s'\, v$. This captures the property that put replaces the state space, without constraining get. A bijective lens fulfils all the axioms of a partial or total lens, but it is sufficient to require L1, so the overall definition of a bijective lens is as shown above. We can now define symmetric lenses:

Definition 7.2. *A (partial) symmetric lens* $\mathcal{X} \triangleq (\mathcal{V}, \mathcal{C})$ *over a state space* S *is a pair of (partial) total lenses,* $\mathcal{V} : V_1 \Longrightarrow S$ *and* $\mathcal{C} : V_2 \Longrightarrow S$ *such that (1)* $\mathcal{V} \bowtie \mathcal{C}$*, and (2)* $\mathcal{V} \oplus \mathcal{C}$ *is a (partial) bijective lens. We denote the set of symmetric lenses between* $V_1 \times V_2$ *and* S *with the notation* $[V_1, V_2] \Longleftrightarrow [S]$*.*

As an example of a symmetric lens, consider $\mathcal{X} \triangleq (\mathsf{fst}_A, \mathsf{snd}_B)$. These lenses are clearly independent, and $\mathsf{fst}_A \oplus \mathsf{snd}_B$ provides a view of the entire product, so it is a bijective lens. Thus, $\mathcal{X}$ is a (total) symmetric lens.

A more interesting example is the list symmetric lens, the view and coview of which are the head and tail of a list. Formally, they are the head lens, hd_A :

$A \Longrightarrow [A]list$, and the tail lens, $tl_A : [A]list \Longrightarrow [A]list$. The head lens is defined in terms of the list lens: $hd_A \triangleq list\text{-}lens_A(0)$. The tail lens is defined as $tl_A \triangleq ([A]list, [A]list, tl, \lambda\, xs\ v \bullet hd\ xs ^\frown v)$. It gets the tail of the list, and puts xs as the tail of the new list, preserving the old head. These lenses are independent, since they operate on different parts of a list. The head lens, as an instance of the list lens, is a partial lens, since it is not defined for an empty list. The list symmetric lens is thus an example of a partial symmetric lens, since putting a list head and tail replaces the whole list. We note that the tail lens has the same view and source types. This is an important property for allowing variable blocks based on such symmetric lenses to be recursed on [18], as we discuss in Sect. 8.

Another symmetric lens is induced by record state spaces, each of which induces two regions: the *base* region, which consists of the defined fields $(x_0 \cdots x_m)$, and the *extension* region, with any additional fields $(y_0 \cdots y_n)$. These can be characterised by $\mathbf{base} : rec\text{-}typ \Longrightarrow [\alpha]rec\text{-}typ\text{-}ext$ and $\mathbf{more} : \alpha \Longrightarrow [\alpha]rec\text{-}typ\text{-}ext$, where $\mathbf{base} \bowtie \mathbf{more}$ and $\mathbf{base} \oplus \mathbf{more}$ is a bijective lens. Consequently, for a record we define $\mathbf{all} \triangleq (\mathbf{base}, \mathbf{more})$, which forms a total symmetric lens. Moreover, we have it that $x_i \preceq \mathbf{base}$, for $0 \leq i \leq m$, and $y_j \preceq \mathbf{more}$, for $0 \leq j \leq n$.

The polymorphic nature of a record lens means that type coercions can be handled easily, as the following theorem shows.

Theorem 7.3. *$x_i \mathbin{;} \mathbf{base} = x_i$ and $x_i / \mathbf{base} = x_i$ whenever $x_i \preceq \mathbf{base}$*

Composition and quotient using the **base** lens corresponds to moving it into and out of an extended state space. Since $x_i : V_i \Longrightarrow [\alpha]rec\text{-}typ\text{-}ext$ is polymorphic, such a coercion yields the same lens but with a different type. These laws are important for when moving a global variable into a local scope in Sect. 8.

8 Variables Blocks

Having defined symmetric lenses, and demonstrated several models, we now use these to characterise local variable blocks. The basic idea is to implement operators analogous to begin and end from Back and von Wright [13, § 5.6], that grow and shrink the state space with additional variables.

Here, however, we fix a symmetric lens $\mathcal{X} : [S_2, C] \Longleftrightarrow [S_1]$ to give a concrete semantics to scope expansion and contraction. Intuitively, S_2 is the global state space, S_1 extends S_2 with local variables, and C is the complement of S_2 wrt. S_1. Then we have it that $\mathcal{V}_{\mathcal{X}}$ characterises the global state region of S_1, and $\mathcal{C}_{\mathcal{X}}$ the local state region. The symmetric lens allows us to distinguish global and local variables, so that we can determine whether an assignment can be moved outside a block or not. Unlike [13], where types are implicit, we have to explicitly handle type coercion when a variable and expression moves between state spaces.

We give the following program that swaps two variables as a running example:

Example 8.1. $swap(x, y : int) \triangleq \mathbf{var}\ z : int \bullet (z := y \mathbin{⨾} y := x \mathbin{⨾} x := z)$

It creates a third variable, z, and then uses this as a temporary store in which to place the value of y. We show how this can be modelled and verified in Isabelle/UTP, with the aim of supporting the larger insertion sort example in Sect. 9. We first define substitutions that extend and contract the state space.

Definition 8.2 (Extension and Contraction Substitutions).

$$\mathbf{ext}_{\mathcal{X}} \triangleq (\!|\mathcal{V}_{\mathcal{X}} \mapsto \mathbf{v}, \mathcal{C}_{\mathcal{X}} \mapsto \varepsilon v \bullet v \in V_2|\!) \qquad \mathbf{con}_{\mathcal{X}} \triangleq (\!|\mathbf{v} \mapsto \mathcal{V}_{\mathcal{X}}|\!)$$

Here, $\mathbf{ext}_{\mathcal{X}} : S_2 \rightarrow S_1$ and $\mathbf{con}_{\mathcal{X}} : S_1 \rightarrow S_2$ are heterogeneous substitutions. Extension assigns the original state ($\mathbf{v} : S_2$) to the view lens ($\mathcal{V}_{\mathcal{X}}$), and assigns an arbitrary but fixed element of C to the coview lens. Effectively this extends the state space, retaining the values for the global variables, and assigning an arbitrary value to the local ones. Contraction, conversely, assigns the view lens to the entire state lens, leading to the loss of the local state. Extension and contraction satisfy the theorem below:

Theorem 8.3. *Any symmetric lens $\mathcal{X}$ satisfies $\mathbf{con}_{\mathcal{X}} \circ \mathbf{ext}_{\mathcal{X}} = \mathbf{id}_{S_2}$*

Specifically, if we extend a state space and then contract it, we always get the original state space back. The converse of this law does not hold since contracting a state space, of course, loses the local state stored in the coview. Moreover, the law only follows for total symmetric lenses since extending and then contracting using a partial lens can alter the state. It is now straightforward to define relations that open and close a block using the substitutions:

Definition 8.4 (Blocks). $\mathbf{open}_{\mathcal{X}} \triangleq \langle \mathbf{ext}_{\mathcal{X}} \rangle \,;\, \mathcal{C}_{\mathcal{X}} := * \qquad \mathbf{close}_{\mathcal{X}} \triangleq \langle \mathbf{con}_{\mathcal{X}} \rangle$

Here, $\mathbf{open}_{\mathcal{X}} : [S_2, S_1]\mathit{rel}$ first extends the state space and then non-deterministically assigns a value to the coview, replacing the arbitrary but fixed value. Also, $\mathbf{close}_{\mathcal{X}}$ simply contracts the state space. We prove a useful law:

Theorem 8.5. *Any symmetric lens $\mathcal{X}$ satisfies $\mathbf{open}_{\mathcal{X}} \,;\, \mathbf{close}_{\mathcal{X}} = \mathit{II}$.*

Aside from being an important property of variable blocks, this law allows us to introduce a local variable block at any point in a program, which can facilitate step-wise refinement. We now prove three algebraic laws for variable blocks and assignments, which are adapted from [13, page 102].

Theorem 8.6 (Variable Block Laws).

$$x := v \,;\, \mathbf{open}_{\mathcal{X}} = \mathbf{open}_{\mathcal{X}} \,;\, \mathcal{V}_{\mathcal{X}}{:}x := (v \uparrow \mathcal{V}_{\mathcal{X}}) \tag{6}$$

$$y := v \,;\, \mathbf{close}_{\mathcal{X}} = \mathbf{close}_{\mathcal{X}} \qquad \text{when } y \preceq \mathcal{C}_{\mathcal{X}} \tag{7}$$

$$x := v \,;\, \mathbf{close}_{\mathcal{X}} = \mathbf{close}_{\mathcal{X}} \,;\, (x/\mathcal{V}_{\mathcal{X}}) := (v \downarrow \mathcal{V}_{\mathcal{X}}) \qquad \text{when } x \preceq \mathcal{V}_{\mathcal{X}}, \mathcal{C}_{\mathcal{X}} \,\sharp\, v \tag{8}$$

An assignment to a global variable x can be pushed into a variable block (6). We have to coerce both the variable and the assigned expression using lens composition and the state space extension operators, respectively. An assignment to a local variable $y \preceq \mathcal{C}_{\mathcal{X}}$ at the end of a block is lost (7). An assignment to a

```
alphabet global = x :: int  y :: int

abbreviation swap :: "global hrel" where
"swap ≡ var z :: int in allL • z := x ;; x := y ;; y := &z"

lemma swap_wp: "swap wp (x = Y ∧ y = X) = U(y = Y ∧ x = X)"
  by (simp add: vblock_def wp usubst unrest)

lemma swap_hoare: "{x = X ∧ y = Y} swap {x = Y ∧ y = X}" by (rel_auto)

lemma swap_alt_def: "swap = (x, y) := (y, x)" by (rel_auto)
```

Fig. 1. Modelling *swap*, and its properties in Isabelle/UTP

global variable in a block can be moved past the end (8). Again, it is necessary to coerce the variable and expression, using lens quotient and state space restriction, this time to contract the state space. Moreover, this law only applies when the expression does not refer to local variables, given by the condition $\mathcal{C}_{\mathcal{X}} \sharp v$. These latter two laws show how the symmetric lens allows us to distinguish local and global variables. We can also derive a Hoare logic law for variable blocks:

Theorem 8.7. If $\{\, p \uparrow \mathcal{V}_{\mathcal{X}} \,\}\, S \,\{\, q \uparrow \mathcal{V}_{\mathcal{X}} \,\}$ then $\{\, p \,\}\, \textbf{\textit{open}}_{\mathcal{X}} \,;\, S \,;\, \textbf{\textit{close}}_{\mathcal{X}} \,\{\, q \,\}$

The intuition is that p and q must be augmented with additional variables in the enlarged state space, and references to global variables must be type cast.

We can now model the algorithm in Example 8.1. First, we need to create global and local state spaces and a suitable symmetric lens. The global state space can be described by $global \triangleq [x : int, y : int]$, as explained in Sect. 3. This gives rise to ***base*** : $global \Longrightarrow [\alpha]global\text{-}ext$ and ***more*** : $\alpha \Longrightarrow [\alpha]global\text{-}ext$, which together form a symmetric lens ***all***. Moreover, the local state can be described by the record $local \triangleq global + [z : int]$, and so we specialise *global*'s base lens, ***base***, to have type $global \Longrightarrow local$. In Isabelle/UTP, *local* can be generated on-the-fly in a record block, to support the syntax given in Example 8.1. This approach, using extensible records for variable blocks, is used in Simpl [17].

An implementation is shown in Fig. 1, along with several theorems. We construct *global* using the ***alphabet*** command, and then define *swap*, with a near identical representation to Example 8.1. The decorations $\&z$ and $\boldsymbol{U}(\cdot)$ are hints to parser with no semantic content. The machinery for creating *local* and instantiating the symmetric lens is hidden behind the **var** construct, though we have to explicitly state that we are using the ***all*** symmetric lens from the global state space. We then prove three theorems. The first one calculates a weakest precondition, the second a Hoare triple, and the final one shows that *swap* can actually be replaced by a simultaneous assignment, assuming this is supported.

While the use of records in blocks provides strong typing, the fact that the ***all*** symmetric lens changes the type of the state space means it cannot be used in recursive functions. This was previously observed by Back and Preoteasa [18]. To handle recursion, the symmetric lens must describe a global state with the same type as the overall state space (which includes both global and local). As mentioned previously, (hd_A, tl_A) is such a lens, and so is its converse (tl_A, hd_A).

```
definition insert_elem :: "int list ⇒ local hrel" where
"insert_elem xs =
  while (0 < j ∧ arr!j < arr!(j-1)) invr @(I xs)
  do
    (arr[j-1], arr[j]) := (arr!j, arr!(j-1)) ;; j := j - 1
  od"

abbreviation insertion_sort :: "int list ⇒ global hrel" where
"insertion_sort xs ≡
  arr := xs ;;
  open⇘lv⇙ ;;
  (i := 1 ;;
  while (i < length arr)
  invr 0 < i ∧ i ≤ length arr ∧ sorted(nths arr {0..i-1}) ∧ perm arr xs
  do
    j := i ;; insert_elem xs ;; i := i + 1
  od) ;;
  close⇘lv⇙"
```

Fig. 2. Insertion sort in Isabelle/UTP

This symmetric lens creates variable blocks that push an arbitrary value onto the start of a list, creating a stack semantics.

The fact that the list symmetric lens forms a partial lens creates the need for domain checks when variables in the list are accessed. Such checks can be avoided by using a state space with a function from natural numbers to values instead of a list. We define head and tail lenses on such a state space as follows:

$$\mathit{hd}^f_A \triangleq (A, \mathbb{N} \to A, \lambda f \bullet f\,0, \lambda f\; v\; n \bullet v \lhd n = 0 \rhd f\; n)$$
$$\mathit{tl}^f_A \triangleq (\mathbb{N} \to A, \mathbb{N} \to A, \lambda f\; n \bullet f\,(n+1), \lambda f\; v\; n \bullet f\,0 \lhd n = 0 \rhd f\,(n-1))$$

These head and tail lenses are total lenses, since they are defined on a total function. We can thus define a total symmetric lens using them in a similar way to the list symmetric lens. These lenses can also be lifted to a state space with additional global state in a similar way to the list symmetric lens. We have mechanised these definitions in Isabelle/UTP and proved the resultant lens is indeed a total symmetric lens. We have also proved properties for a swap function using this symmetric lens as we did for the record symmetric lens. This shows the flexibility of our lens-based approach to local variables: list or function lenses can be used where support for recursion is required, while record lenses can be used where the added structure of Isabelle's type system is desired.

9 Insertion Sort

Here, we show how we have used the collected results of the previous sections to verify the insertion sort algorithm in Isabelle/UTP. We model the algorithm using both collection lenses and symmetric lens variable blocks, as shown in Fig. 2. In order to ease verification, we split the algorithm into two definitions: one for the inner loop (insert-elem), and one for the outer loop (insertion-sort). Both are specified as functions that take the list to be sorted, xs, as a parameter.

```
lemma insert_elem_correct: "
    {0 < i ∧ i < length arr ∧ j = i ∧ sorted(nths arr {0..i-1}) ∧ perm arr xs}
    insert_elem xs
    {0 < i ∧ i < length arr ∧ sorted(nths arr {0..i}) ∧ perm arr xs}"
proof -  [57 lines]

lemma insertion_sort_correct:
  "{true} insertion_sort xs {sorted(arr) ∧ perm arr xs}"
proof - [26 lines]
```

Fig. 3. Insertion sort verification

The syntax of the program broadly follows that given in Example 1.1. The only significant deviation is that we have manually constructed a symmetric lens *lv* that uses an explicit local state space. This is so that i and j can be referred to as global names in the Isabelle theory, to enable description of the invariants. The outer program itself operates on a state space where only *arr* is present, and the other variables are introduced by ***open*** and ***close***.

As usual [1], our loop construct supports invariant annotation, using the invr keyword. The invariant of the inner loop ($I\ xs$) is not shown due to its complexity. The outer loop invariant states that (1) $0 < i \leq \#arr$, it is within the array bounds; (2) the array in the range $0 \cdots i-1$ is sorted; and (3) *arr* is a permutation of the original list xs. The function $sorted : [A]list \rightarrow bool$ determines that a list is sorted by a predefined total order on A, and $perm : [A]list \rightarrow [A]list \rightarrow bool$ states two lists have the same elements, including repetitions. Both of the latter functions are provided as part of the Isabelle/HOL library. Function $nths : [A]list \rightarrow [nat]set \rightarrow [A]list$ gives the elements of a list described by an index set.

The program is verified using Hoare logic as shown in Fig. 3. The proof is quite long, due to the number of proof obligations, and some manual effort is required. This seems mainly due to missing lemmas, and so in future the proof should be more automated (cf. [12]). Nevertheless, for now we omit details of the proof steps. The first Hoare triple demonstrates that the inner loop preserves the invariant of the outer loop. The outer loop shows that, when provided with a non-empty list a sorted permutation of xs is returned in *arr*.

10 Conclusions

In this paper we have shown how lenses support modelling and verification of algorithms in the Isabelle/UTP tool [11]. We introduced dynamic lenses, that allow us to handle collections, and symmetric lenses, that allow partitioning of the state space into disjoint regions. Collection lenses allow us to generically characterise a variety of different indexed collection types in Isabelle/UTP, including arrays and maps. Symmetric lenses [16] allow us to characterise state partitioning, and we have used them here to distinguish local and global variables scopes. Due to typed nature of our state spaces, coercions are necessary when moving between scopes, which we can also handle using lenses. Our conclusion is that algebraic characterisation in this way is both flexible and practical, as our verification of insertion sort demonstrates. Moreover, since our characterisation sits at the state space level, our results are applicable to paradigms beyond imperative programming, such as reactive [11] and hybrid programming [4,19].

In future work, we will explore symmetric lenses and their properties further. We note that there are several different models for symmetric lenses, including extensible records, lists, and total functions, each with unique advantages. We can use extensible records to support variables with native type checking, but they cannot support recursion, as for example required by quicksort, due to *a priori* bounding of the state space. In contrast, list and function symmetric lenses overcome this limitation, but require a fixed element type. Our mechanisation thus allows us chose the best model for a particular circumstance. In the future, we will perform a detailed comparison of the different models.

Acknowledgements. This work is funded by the EPSRC projects CyPhyAssure (CyPhyAssure Project: https://www.cs.york.ac.uk/circus/CyPhyAssure/.) (Grant EP/S001190/1) and RoboTest (Grant EP/R025479/1).

References

1. Armstrong, A., Gomes, V.B.F., Struth, G.: Building program construction and verification tools from algebraic principles. Formal Aspects Comput. **28**(2), 265–293 (2015). https://doi.org/10.1007/s00165-015-0343-1
2. Dongol, B., Gomes, V.B.F., Struth, G.: A program construction and verification tool for separation logic. In: Hinze, R., Voigtländer, J. (eds.) MPC 2015. LNCS, vol. 9129, pp. 137–158. Springer, Cham (2015). https://doi.org/10.1007/978-3-319-19797-5_7
3. Gomes, V.B.F., Struth, G.: Modal Kleene algebra applied to program correctness. In: Fitzgerald, J., Heitmeyer, C., Gnesi, S., Philippou, A. (eds.) FM 2016. LNCS, vol. 9995, pp. 310–325. Springer, Cham (2016). https://doi.org/10.1007/978-3-319-48989-6_19
4. Huerta y Munive, J.J., Struth, G.: Verifying hybrid systems with modal Kleene algebra. In: Desharnais, J., Guttmann, W., Joosten, S. (eds.) RAMiCS 2018. LNCS, vol. 11194, pp. 225–243. Springer, Cham (2018). https://doi.org/10.1007/978-3-030-02149-8_14
5. Barnett, M., Chang, B.-Y.E., DeLine, R., Jacobs, B., Leino, K.R.M.: Boogie: a modular reusable verifier for object-oriented programs. In: de Boer, F.S., Bonsangue, M.M., Graf, S., de Roever, W.-P. (eds.) FMCO 2005. LNCS, vol. 4111, pp. 364–387. Springer, Heidelberg (2006). https://doi.org/10.1007/11804192_17
6. Filliâtre, J.-C., Paskevich, A.: Why3—where programs meet provers. In: Felleisen, M., Gardner, P. (eds.) ESOP 2013. LNCS, vol. 7792, pp. 125–128. Springer, Heidelberg (2013). https://doi.org/10.1007/978-3-642-37036-6_8
7. Kozen, D.: Kleene algebra with tests. ACM TOPLAS **19**(3), 427–443 (1992)
8. Foster, J., Greenwald, M., Moore, J., Pierce, B., Schmitt, A.: Combinators for bidirectional tree transformations: a linguistic approach to the view-update problem. ACM Trans. Program. Lang. Syst. **29**(3), 17-es (2007)
9. Foster, J., Pilkiewicz, A., Pierce, B.: Quotient lenses. In: Proceedings of the 13th International Conference on Functional Programming (ICFP). ACM (2008)
10. Foster, S., Zeyda, F., Woodcock, J.: Unifying heterogeneous state-spaces with lenses. In: Sampaio, A., Wang, F. (eds.) ICTAC 2016. LNCS, vol. 9965, pp. 295–314. Springer, Cham (2016). https://doi.org/10.1007/978-3-319-46750-4_17

11. Foster, S., Cavalcanti, A., Canham, S., Woodcock, J., Zeyda, F.: Unifying theories of reactive design contracts. Theor. Comput. Sci. **802**, 105–140 (2020)
12. Bockenek, J., Lammich, P., Nemouchi, Y., Wolff, B.: Using Isabelle/UTP for the verification of sorting algorithms. In: Proceedings of the Isabelle Workshop (FLoC) (2018)
13. Back, R.J., von Wright, J.: Refinement Calculus: A Systematic Introduction. Springer, New York (1998). https://doi.org/10.1007/978-1-4612-1674-2
14. Foster, S., Zeyda, F., Woodcock, J.: Isabelle/UTP: a mechanised theory engineering framework. In: Naumann, D. (ed.) UTP 2014. LNCS, vol. 8963, pp. 21–41. Springer, Cham (2015). https://doi.org/10.1007/978-3-319-14806-9_2
15. Foster, S., Zeyda, F., Nemouchi, Y., Ribeiro, P., Wolff, B.: Isabelle/UTP: mechanised theory engineering for unifying theories of programming. Arch. Formal Proofs (2019). https://www.isa-afp.org/entries/UTP.html
16. Hofmann, M., Pierce, B., Wagner, D.: Symmetric lenses. In: Proceedings of the 38th International Symposium on Principles of Programming Languages (POPL), pp. 371–384. IEEE (2011)
17. Alkassar, E., Hillebrand, M.A., Leinenbach, D., Schirmer, N.W., Starostin, A.: The Verisoft approach to systems verification. In: Shankar, N., Woodcock, J. (eds.) VSTTE 2008. LNCS, vol. 5295, pp. 209–224. Springer, Heidelberg (2008). https://doi.org/10.1007/978-3-540-87873-5_18
18. Back, R.J., Preoteasa, V.: An algebraic treatment of procedure refinement to support mechanical verification. Formal Aspects Comput. **17**(1), 69–90 (2005). https://doi.org/10.1007/s00165-004-0060-7
19. Foster, S.: Hybrid relations in Isabelle/UTP. In: Ribeiro, P., Sampaio, A. (eds.) UTP 2019. LNCS, vol. 11885, pp. 130–153. Springer, Cham (2019). https://doi.org/10.1007/978-3-030-31038-7_7
20. Feliachi, A., Gaudel, M.-C., Wolff, B.: Unifying theories in Isabelle/HOL. In: Qin, S. (ed.) UTP 2010. LNCS, vol. 6445, pp. 188–206. Springer, Heidelberg (2010). https://doi.org/10.1007/978-3-642-16690-7_9
21. Hoare, C.A.R., He, J.: Unifying Theories of Programming. Prentice-Hall, Upper Saddle River (1998)
22. Schirmer, N., Wenzel, M.: State spaces - the locale way. ENTCS **254**, 161–179 (2009). (SSV 2009)
23. Greenaway, G., Lim, J., Andronick, J., Klein, G.: Don't sweat the small stuff: formal verification of C code without the pain. In: Proceedings of the ACM SIGPLAN Conference on Programming Language Design and Implementation (PLDI). ACM, June 2014
24. Dongol, B., Hayes, I., Meinicke, L., Struth, G.: Cylindric Kleene lattices for program construction. In: Hutton, G. (ed.) MPC 2019. LNCS, vol. 11825, pp. 197–225. Springer, Cham (2019). https://doi.org/10.1007/978-3-030-33636-3_8
25. Tarski, A.: On the calculus of relations. J. Symb. Log. **6**(3), 73–89 (1941)
26. Hoare, C.A.R., et al.: The laws of programming. Commun. ACM **30**(8), 672–686 (1987)
27. Hoare, C.A.R.: An axiomatic basis for computer programming. Commun. ACM **12**(10), 576–580 (1969)
28. Dijkstra, E.W.: Guarded commands, nondeterminacy and formal derivation of programs. Commun. ACM **18**(8), 453–457 (1975)

Weakening Relation Algebras and FL^2-algebras

Nikolaos Galatos[1,2] and Peter Jipsen[1,2(✉)]

[1] University of Denver, Denver, CO, USA
[2] Chapman University, Orange, CA, USA
jipsen@chapman.edu

Abstract. FL^2-algebras are lattice-ordered algebras with two sets of residuated operators. The classes RA of relation algebras and GBI of generalized bunched implication algebras are subvarieties of FL^2-algebras. We prove that the congruences of FL^2-algebras are determined by the congruence class of the respective identity elements, and we characterize the subsets that correspond to this congruence class. For involutive GBI-algebras the characterization simplifies to a form similar to relation algebras.

For a positive idempotent element p in a relation algebra $\mathbf{A}$, the double division conucleus image $p\backslash\mathbf{A}/p$ is an (abstract) weakening relation algebra, and all representable weakening relation algebras (RWkRAs) are obtained in this way from representable relation algebras (RRAs). The class $S(\mathsf{dRA})$ of subalgebras of $\{p\backslash\mathbf{A}/p : A \in \mathsf{RA}, 1 \le p^2 = p \in A\}$ is a discriminator variety of cyclic involutive GBI-algebras that includes RA. We investigate $S(\mathsf{dRA})$ to find additional identities that are valid in all RWkRAs. A representable weakening relation algebra is determined by a chain if and only if it satisfies $0 \le 1$, and we prove that the identity $1 \le 0$ holds only in trivial members of $S(\mathsf{dRA})$.

Keywords: Relation algebras · Residuated lattices · Bunched implication algebras

1 Introduction

Tarski defined a relation algebra $(A, \wedge, \vee, \neg, \top, \bot, ;, ^{\smile}, 1, 0)$ to be an algebra that satisfies a short list of identities that hold in all algebras of binary relations on a set: $(A, \wedge, \vee, \neg, \top, \bot)$ is a Boolean algebra, $(A, ;, 1)$ is a monoid, ; and $^{\smile}$ distribute over $\vee$, $x;\bot = \bot = \bot;x$, $0 = \neg 1$, $x^{\smile\smile} = x$, $(xy)^{\smile} = y^{\smile}x^{\smile}$ and $x^{\smile} \cdot \neg(xy) \le \neg y$.

An interesting generalization is to consider algebras of weakening closed binary relations on partially ordered sets $\mathbf{P}=(P, \le)$. A relation $R \subseteq P^2$ is *weakening closed* or a *weakening relation* if $x' \le x\ R\ y \le y'$ implies $x'\ R\ y'$, or equivalently, $\le;R;\le\ \subseteq R$. The collection of all weakening relations on $\mathbf{P}$ is denoted $\mathrm{Wk}(\mathbf{P})$. If R is weakening closed, so is its *complement-converse* $R^{c\smile} = \{(y, x) \mid (x, y) \notin R\}$. This unary operation is denoted by $\sim R$.

U. Fahrenberg et al. (Eds.): RAMiCS 2020, LNCS 12062, pp. 117–133, 2020.
https://doi.org/10.1007/978-3-030-43520-2_8

Weakening relations are also closed under union, intersection, Heyting implication $\to$ (= residual of intersection), relation composition ; and residuals $\backslash, /$ of composition. The partial order relation $\leq$ is a weakening relation and, since it is the identity of composition, it is denoted by 1. The complement-converse of 1 is denoted by 0. The *full weakening relation algebra* on a poset $\mathbf{P}$ is

$$\mathbf{Wk}(\mathbf{P}) = (\mathrm{Wk}(\mathbf{P}), \cap, \cup, \to, P^2, \varnothing, ;, \sim, 1, 0).$$

The residuals $\backslash, /$ are omitted since they are definable via $x \backslash y = \sim(\sim y; x)$ and $x / y = \sim(y; \sim x)$. The variety RWkRA of representable weakening relation algebras is generated by the class $\{\mathbf{Wk}(\mathbf{P}) \mid \mathbf{P} \text{ is a poset}\}$. When the poset is an antichain, or equivalently, when $\leq$ is the identity relation then $\mathbf{Wk}(\mathbf{P})$ is the usual *full relation algebra* $\mathbf{Rel}(P)$ since in this case $R \to \varnothing = R^c$ is the complement of R, and $\sim(R^c) = R^{\smile}$ is the converse of R. Hence RWkRA contains the variety RRA of all representable relation algebras (which is generated by all full relation algebras).

Some applications of weakening relation algebras were given by Stell [22,23] in the area of image processing and hypergraphs. Since the lattice reducts of weakening relation algebras are Heyting algebras rather than Boolean algebras, weakening relations can be thought of as intuitionistic relations.

The variety RWkRA retains many of the algebraic properties of RRA, as shown in [8] and reviewed in Sect. 3. The aim of this paper is to investigate the identities that hold in RWkRA. We do this in the more general context of generalized bunched implication algebras, residuated lattices, and FL^2-algebras (defined below) in order to point out some of the syntactic symmetries of weakening relation algebras and to relate this variety to some other well-studied classes of algebras.

A *residuated lattice* is of the form $(A, \wedge, \vee, \cdot, \backslash, /, 1)$ such that $(A, \wedge, \vee)$ is a lattice, $(A, \cdot, 1)$ is a monoid, and for all $x, y, z \in A$ the *residuation property* holds:

$$xy \leq z \iff x \leq z/y \iff y \leq x \backslash z.$$

A *full Lambek algebra* or *FL-algebra* $(A, \wedge, \vee, \cdot, \backslash, /, 1, 0)$ is a residuated lattice with an additional constant 0, hence FL-algebras are also called *pointed residuated lattices*. The residuation property implies that $x(y \vee z) = xy \vee xz$ and $(x \vee y)z = xz \vee yz$ hence FL-algebras include idempotent semirings as reducts. In fact any finite idempotent semiring expands uniquely to an FL-algebra in which 0 is the bottom element. Hence FL-algebras are closely related to many computational algebraic theories, such as Kleene algebras, Kleene lattices and Pratt's action algebras.

FL-algebras and their reducts cover the algebraic semantics of a large number of logics, including classical propositional logic, intuitionistic logic, relevance logic, multi-valued logic, Hajek's basic logic, abelian logic, BCK-logic and many others. However they do not capture bunched implication logic or the logic of relation algebras (also known as arrow logic). Bunched implication logic is an integral part of separation logic, a Hoare logic for reasoning about pointer structures and concurrent programs [19–21]. Generalized bunched implication algebras were defined in [7] to provide a common algebraic version of bunched implication algebras and relation algebras.

In this paper we introduce FL^2-algebras in order to give a new definition of relation algebras and bunched implication algebras that exposes interesting symmetries of both algebraic theories. A *FL^2-algebra* is of the form $\mathbf{A} = (A, \wedge, \vee, \diamond, \to, \leftarrow, t, f, \cdot, \backslash, /, 1, 0)$ such that

$$\mathbf{A}_t = (A, \wedge, \vee, \diamond, \to, \leftarrow, t, f) \text{ and } \mathbf{A}_1 = (A, \wedge, \vee, \cdot, \backslash, /, 1, 0)$$

are both FL-algebras. We call $\mathbf{A}_t$ the *logical reduct* and $\mathbf{A}_1$ the *dynamic reduct* of $\mathbf{A}$. The class of all FL-algebras can be defined by identities, hence it and the class FL^2-algebras are varieties, denoted by FL and FL^2 respectively. To reduce the number of parentheses, we adopt the convention that $\cdot$ binds stronger than $\backslash, /$ followed by $\diamond$, $\wedge, \vee$ and $\to, \leftarrow$.

Define $\neg x = x \to f$, $\urcorner x = f \leftarrow x$, ${\sim}x = x\backslash 0$ and $-x = 0/x$. An FL^2-algebra is *involutive* if ${\sim}{-x} = x = -{\sim}x$, *$f$-involutive* if $\neg\urcorner x = x = \urcorner\neg x$, and *doubly involutive* if all four identities hold. An FL^2-algebra is *cyclic* if ${\sim}x = -x$, *f-cyclic* if $\neg x = \urcorner x$, and *doubly cyclic* if both hold.

Relation algebras are well known examples of doubly cyclic FL^2-algebras. In fact they are term-equivalent to the subvariety defined by the identities $x \wedge y = x \diamond y$ (hence $y \leftarrow x = x \to y$ and A_t is a Boolean algebra) and $\neg{\sim}(xy) = (\neg{\sim}y)(\neg{\sim}x)$. The operation $\neg{\sim}x$ is the *converse* of relation algebras, usually written $x^{\smile}$.

A *generalized bunched implication* (GBI-)*algebra* $(A, \wedge, \vee, \to, \top, \cdot, \backslash, /, 1)$ is defined as a Brouwerian algebra $(A, \wedge, \vee, \to, \top)$ such that $(A, \wedge, \vee, \cdot, \backslash, /, 1)$ is a residuated lattice. Equivalently a GBI-algebra is an FL^2-algebra that satisfies $x \wedge y = x \diamond y$, $t = f$ and $0 = 1$. A *bunched implication algebra*, or *BI-algebra*, is a commutative GBI-algebra (i.e., $xy = yx$) that has been expanded by a constant $\bot$ denoting the least element of the lattice. Alternatively, it is an FL^2-algebra that satisfies $x \wedge y = x \diamond y$, $f \leq x$, $0 = 1$ and $xy = yx$. Since the logical constants t, f are the top and bottom elements in this algebra they are usually denoted by $\top$, $\bot$.

An interesting subclass of cyclic GBI-algebras is the variety of *symmetric Heyting relation algebras* or SHRAs [23], defined by adding the identity ${\sim}\neg(xy) \leq ({\sim}\neg y)({\sim}\neg x)$. This identity holds in all representable weakening relation algebras, hence RWkRA is a subvariety of SHRA.

Another subvariety of FL^2-algebras is the variety of *skew relation algebras* [6], defined in this setting as Boolean involutive FL^2-algebras. As mentioned before, the variety of relation algebras is obtained by adding the identity $(xy)^{\smile} = y^{\smile}x^{\smile}$ where $x^{\smile} = \neg{\sim}x$ ([6], Cor. 29).

We provide a simpler characterization of the congruences of GBI-algebras in Sect. 3 (using congruence terms that have only *one* parameter), which also reveals hidden symmetries in the description given in [8]. Towards that goal, we first provide this description in the more natural and symmetric setting of FL^2 in Sect. 2, and this is our main reason for introducing FL^2. Equivalent characterizations are provided in Lemma 3 (in a fully symmetric setting), which then specialize to two distinct characterizations in Corollaries 9 and 10 (one for the congruence filters of 1 and one for the congruence filters of $\top$).

All FL^2-algebras can be constructed by selecting two pointed residuated lattices that have a common underlying lattice. If the lattice is non-distributive, then the resulting FL^2-algebra is outside the variety of GBI-algebra. One of the appeals of GBI-algebras in computer science is the fact that they provide the means to study both the logical and the dynamic aspect of situations. Of course in GBI the logical part is restricted to intuitionistic logic, but FL^2 allows for considering cases where the logical part is any substructural logic, such as linear logic, relevance logic or a particular fuzzy logic. Methods for combining logics have been studied extensively, and FL^2 is an example of fusion of logics as described by Gabbay in [5]. The results in the first half of this paper provide some insight into the algebraic structure of the fusion of two substructural logics.

The models of relevance logic RW, namely distributive cyclic involutive residuated lattices, are exactly the implication subreducts of de Morgan BI-algebras, namely the extension of BI where the dynamic part is involutive. In [3] it is shown that the addition of a Boolean negation to de Morgan BI-algebras results in a non-conservative extension called *classical* BI. A display calculus for this logic shows remarkable symmetry between the classical logic part and the involutive dynamic part of this logic. The setting of FL^2 is well suited to studying weaker versions of this logic that omit some rules like contraction and/or weakening. It is also worth noting that classical BI-algebras coincide with commutative skew relation algebras (defined in [6]).

In Sect. 4 we recall the definition of discriminator variety and some results about weakening relation algebras from [8]. Finally, Sect. 5 defines the double-division conucleus construction and shows that the image of the variety of relation algebras under this construction produces a class of GBI-algebras that is a non-Boolean analogue to Tarski's variety of abstract relation algebras.

Throughout the paper we make use of elementary properties of the residuals, such as $x(x\backslash y) \leq y$, $x \leq xy / y$, $x(y \vee z)w = xyw \vee xzw$ and that residuals are order-reversing in the "denominator" or antecedent and order-preserving in the "numerator" or consequent.

2 Congruences of FL^2-algebras

An algebraic theory determines a category in which all models of the theory are objects and the morphisms are homomorphisms between the algebraic models. The kernel $\{(x, y) : h(x) = h(y)\}$ of a homomorphism $h : \mathbf{A} \to \mathbf{B}$ is a congruence relation (i.e., an equivalence relation that is preserved by all algebraic operations) on $\mathbf{A}$, and an important step in understanding the structure of the category is to be able to describe the lattice of congruences $\mathrm{Con}(\mathbf{A})$ on each object $\mathbf{A}$.

An FL^2-algebra $\mathbf{A}$ has two residuated lattices as reducts, hence any congruence on $\mathbf{A}$ is a residuated lattice congruence. The description of congruences in residuated lattices is due to Blount and Tsinakis [2]. Here we use a version of this result that appears in [9].

An algebra is said to be *c-regular* if c is a constant in the algebra and each congruence of the algebra is determined by its c-congruence class. Residuated lattices are 1-regular since for a congruence θ on a residuated lattice $\mathbf{L}$

$$x\,\theta\,y \iff x/y \wedge y/x \wedge 1 \in [1]_\theta$$

where $[1]_\theta = \{z \in L : z\theta 1\}$ is the 1-congruence class of θ. If we define $x \leq_\theta y$ by $x \leq z$ and $z\,\theta\,y$ for some $z \in L$, or equivalently by $x\,\theta\,w$ and $w \leq y$ for some $w \in L$, then $x\,\theta\,y \iff x \leq_\theta y$ and $y \leq_\theta x$, hence the above equivalence follows from the observation that

$$x \leq_\theta y \iff 1 \leq_\theta y/x \iff y/x \wedge 1\,\theta\,1.$$

Instead of the right residual $/$ one could also use the left residual $\backslash$ for this equivalence. Rather than working with 1-congruence classes, it is convenient to use certain filters.

Recall that a filter of a lattice L is a subset F such that $x \wedge y, a \vee x \in F$ for all $x, y \in F$ and $a \in L$. For $x \in X \subseteq L$ let ${\uparrow}x = \{y \in L : x \leq y\}$ be the principal (lattice) filter generated by x and ${\uparrow}X = \bigcup_{x\in X}{\uparrow}x$.

A *congruence filter* of a residuated lattice or FL-algebra is a subset of the form $F = {\uparrow}([1]_\theta)$. This is a lattice filter since the congruence class of 1 is closed under meet. It is also a union of θ-classes since $\leq_\theta$ is transitive. The class $[1]_\theta$ can be recovered from F since $[1]_\theta = \{x : x, 1/x \in F\}$.

It is easy to check that

$$1,\ xy,\ \lambda_a(x) := a\backslash xa,\ \rho_a(x) := ax/a \in F \quad \text{for all } x, y \in F \text{ and } a \in L.$$

Note that the closure of F under the *conjugation terms* $ax/a, a\backslash xa$ is equivalent to the following *normality conditions* (where quantifiers range over F):

$$(\lambda_a)\ \forall x \exists x',\ ax' \leq xa \quad \text{and} \quad (\rho_a)\ \forall x \exists x',\ x'a \leq ax.$$

A filter F is said to satisfy (λ) if (λ_a) holds for all $a \in L$ and likewise for (ρ). The set of congruence-filters of $\mathbf{L}$ is denoted by $\mathrm{CF}(\mathbf{L})$.

Theorem 1 ([9]). *For a residuated lattice or FL-algebra* $\mathbf{A}$, *a subset* F *is a congruence-filter if and only if* F *is a lattice filter and a submonoid of* $\mathbf{A}$ *that satisfies* (λ) *and* (ρ).

Moreover, $\mathrm{Con}(\mathbf{A})$ *is isomorphic to the lattice* $\mathrm{CF}(\mathbf{A})$ *of congruence-filters via the bijection* $\theta \mapsto {\uparrow}([1]_\theta)$ *and* $F \mapsto \{(x, y) : x/y, y/x \in F\}$.

Since there are two signatures for FL-algebras, there are two ways to characterize the congruences of an FL^2-algebra, either by congruence 1-*filters* ${\uparrow}([1]_\theta)$ or by congruence t-*filters* ${\uparrow}([t]_\theta)$. We usually drop the prefix "congruence", and mostly work with 1-filters. However all results can be translated to t-filters by interchanging the operation symbols of the two signatures.

For FL^2 we need the following stronger t-*normality* conditions to determine the 1-filters (the quantifiers range over the filters). For any $a \in A$,

$$(U_a)\ \forall x \exists x_1,\ x_1 a \leq xt \diamond a, \qquad (U'_a)\ \forall x \exists x_2,\ a x_2 \leq a \diamond xt,$$

$$(V_a)\ \forall x \exists x_3,\ x_3 t \diamond a \leq ax, \qquad (V'_a)\ \forall x \exists x_4,\ a \diamond x_4 t \leq xa.$$

A filter satisfies (U) if (U_a) holds for all $a \in A$, and the same for (U'), (V) and (V'). The conjunction of these four conditions is referred to as (UV_a) or, if they hold for all a, as (UV).

Lemma 3 below shows that (UV_a) is indeed stronger than $(\lambda_a), (\rho_a)$. With the help of normality we can derive several other variants of the inequations in (UV_a) such as $\forall x \exists x', x'a \leq a \diamond tx$. We will use these variations occasionally in the following lemma about some useful two-parameter conditions. For $a, b \in A$ define

$$\forall x \exists x_1, x_2,\ x_1(a \diamond b) \leq xa \diamond b \quad \text{and} \quad x_2(a \diamond b) \leq a \diamond xb, \qquad (Q_{a,b})$$

$$\forall x \exists x_1, x_2,\ a \diamond x_1 b \leq xa \diamond b \quad \text{and} \quad x_2 a \diamond b \leq a \diamond xb, \qquad (R_{a,b})$$

$$\forall x \exists x_1, x_2,\ x_1(a \to b) \leq a \to xb \quad \text{and} \quad x_2(a \leftarrow b) \leq xa \leftarrow b. \qquad (S_{a,b})$$

Lemma 2. *The condition (UV) implies (Q), (R) and (S).*

Proof. We first derive $(R_{a,b})$ from (V_a') and (U_b). Given $x \in F$, there exist $x_1, x_4 \in F$ such that (reading from right to left)

$$a \diamond x_1 b \leq a \diamond (x_4 t \diamond b) = (a \diamond x_4 t) \diamond b \leq xa \diamond b.$$

By a similar calculation using (V_b') and (U_a'), there exist $x_2, x_4 \in F$ such that $x_2 a \diamond b \leq a \diamond x_4 t \diamond b \leq a \diamond xb$.

Next we derive $(Q_{a,b})$ from $(R_{t,a})$ and $(R_{t,a\diamond b})$. Given $x \in F$, there exist $x_1, x_2 \in F$ with

$$x_1(a \diamond b) = t \diamond x_1(a \diamond b) \leq x_2 t \diamond (a \diamond b) = (x_2 t \diamond a) \diamond b \leq (t \diamond xa) \diamond b = xa \diamond b.$$

For $(S_{a,b})$ the relevant calculation shows there exist $x_1, x_2, x_3, x_4 \in F$ such that

$$a \diamond x_4(a \to b) \leq a \diamond (a \to b)x_3 \diamond t \leq a \diamond (a \to b) \diamond x_1 t \leq b \diamond x_1 t \leq bx_2 \diamond t \leq xb,$$

hence for all $x \in F$ there exists $x_1 \in F$ such that $x_1(a \to b) \leq a \to xb$. The remaining inequalities are derived in a similar way. □

We also consider the conditions

$$(\lambda_a')\ \forall x \exists x',\ x't \diamond a \leq a \diamond xt \quad \text{and} \quad (\rho_a')\ \forall x \exists x',\ a \diamond x't \leq xt \diamond a.$$

As before, (λ') means that (λ_a') holds for all a, and likewise for (ρ').

Lemma 3. *We have the following implications between the above conditions.*

1. (U) *and* $(V) \Rightarrow (\rho)$ 2. (U') *and* $(V') \Rightarrow (\lambda)$
3. (U') *and* $(V) \Rightarrow (\rho')$ 4. (U) *and* $(V') \Rightarrow (\lambda')$.

Moreover, the following sets of conditions are equivalent:

5. $(U), (U'), (V), (V')$, *that is* (UV)
6. $(U), (V), (\lambda), (\lambda'), (\rho')$ 7. $(U'), (V'), (\rho), (\lambda'), (\rho')$
8. $(U'), (V), (\lambda), (\rho), (\rho')$ 9. $(U), (V'), (\lambda), (\rho), (\lambda')$.

Proof. For (1) we have for all $x \in F$ there exist $x_1, x_3 \in F$ such that $x_1 a \le x_3 t \diamond a \le ax$. For (2) there exist $x_2, x_4 \in F$ such that $ax_2 \le a \diamond x_4 t \le xa$. For (3), we have $x_3 t \diamond a \le ax_2 \le a \diamond xt$, while for (4) $a \diamond x_4 t \le x_1 a \le xt \diamond a$.

That (5) implies (6) follows from (1–4). For the converse, $(\lambda'), (U), (\lambda)$ imply $ax'' \le x_1 a \le x't \diamond a \le a \diamond xt$, giving (U'), and $(\lambda), (V), (\rho')$ imply $a \diamond x't \le x_3 t \diamond a \le ax \le xa$, yielding (V'). The equivalence of (5) and (7) is analogous.

That (5) implies (8) follows from (1–4). For the converse, $(\rho'), (U'), (\rho)$ show $x''a \le ax_2 \le a \diamond x't \le xt \diamond a$, yielding (U), and $(\lambda), (V), (\rho')$ show $a \diamond x't \le x_3 t \diamond a \le ax \le xa$ giving (V'). Likewise (5) and (9) are equivalent. □

Theorem 4. *For an FL^2-algebra* $\mathbf{A}$*, a subset* F *is the* 1*-filter of some congruence* θ *of* $\mathbf{A}$ *if and only if* F *is a lattice filter and a* $\cdot, 1$*-submonoid of* $\mathbf{A}$ *that satisfies* (UV)*, or any of the equivalent conditions 6.–9. of Lemma 3.*

Proof. Assume $F = \uparrow([1]_\theta)$ is the 1-filter of some FL^2-congruence θ. As observed earlier, F is a lattice filter that contains 1, so if $x, y \in F$ then there exist $u, v \in [1]_\theta$ with $u \le x, v \le y$ and $1 \cdot 1 \,\theta\, uv \le xy$, hence $xy \in F$ showing that F is a submonoid. Next we prove (U_a) $x_1 a \le xt \diamond a$. For $a \in A$ and $x \in F$ there exists $y \in F$ such that $y \le x$ and $y \in [1]_\theta$. Now

$$1 \,\theta\, y \;\Rightarrow\; t \,\theta\, ty \;\Rightarrow\; a = t \diamond a \,\theta\, yt \diamond a \;\Rightarrow\; 1 \le a/a \,\theta\, (yt \diamond a)/a \le (xt \diamond a)/a.$$

Hence $(xt \diamond a)/a \in F$ using the observation that if $1 \le u \theta v \le w$ then $w \in \uparrow([1]_\theta)$. Letting $x_1 = (xt \diamond a)/a$, we obtain $x_1 a \le xt \diamond a$.

For (V_a') $a \diamond x_4 t \le xa$ we use the following calculation.

$$1 \,\theta\, y \Rightarrow a \,\theta\, ya \Rightarrow t \le a \to a \,\theta\, a \to ya \Rightarrow 1 \le t/t \,\theta\, (a \to ya)/t \le (a \to xa)/t,$$

hence $(a \to xa)/t \in F$, and choosing $x_4 = (a \to xa)/t$ implies $a \diamond x_4 t \le xa$. The conditions (U_a') and (V) are proved in a similar way.

Conversely, assume F is a lattice filter with $1, xy \in F$ for all $x, y \in F$ and (UV) holds. Define $\theta = \{(a, b) : a/b, b/a \in F\}$. This relation is reflexive since $1 \in F$, transitive since $(x/y)(y/z) \le x/z$, and obviously symmetric. Assuming $a \,\theta\, b$, it suffices to show

$$(a \wedge c)/(b \wedge c), (a \vee c)/(b \vee c), (a \diamond c)/(b \diamond c), (c \diamond a)/(c \diamond b) \in F,$$
$$ac/bc, ca/cb, (a/c)/(b/c), (c/a)/(c/b), (a\backslash c)/(b\backslash c), (c\backslash a)/(c\backslash b) \in F \text{ and}$$
$$(a \to c)/(b \to c), (c \to a)/(c \to b), (a \leftarrow c)/(b \leftarrow c), (c \leftarrow a)/(c \leftarrow b) \in F$$

since interchanging a, b the same statements follow from $b \,\theta\, a$, hence θ is compatible with all FL^2 operations.

From $a \,\theta\, b$ we obtain $a/b, b/a \in F$ and since F is a filter $a/b \wedge 1 \in F$. The calculation for compatibility of meet is as follows:

$$(a/b \wedge 1)(b \wedge c) \le (a/b)b \wedge 1c \le a \wedge c$$

hence $a/b \wedge 1 \le (a \wedge c)/(b \wedge c) \in F$. The calculation for join is the same, using the distribution of $\cdot$ over $\vee$.

It is remarkable that all the remaining statements can be deduced from (UV). Lemma 3 shows that $(\lambda), (\rho)$ follow and, by Lemma 2 conditions (Q), (R), (S) also hold. The implication $a/b \in F \Rightarrow ac/bc \in F$ is easy since $(a/b)bc \le ac$ follows from $(a/b)b \le a$. However $a/b \in F \Rightarrow ca/cb \in F$ uses (ρ_c) with $x = a/b$, so there exists $x' \in F$ such that

$$x'cb = (x'c)b \le (c(a/b))b = c((a/b)b) \le ca$$

and therefore $x' \le ca/cb$ implies $ca/cb \in F$. Similarly the implication $a/b \in F \Rightarrow (a/c)/(b/c) \in F$ is easy since $a/b \le (a/c)/(b/c)$ holds, while for $b/a \in F \Rightarrow (c/a)/(c/b) \in F$ we use $(\rho_{c/b})$ with $x = b/a$ to get $x' \in F$ such that

$$x'(c/b)a \le (c/b)(b/a)a \le (c/b)b \le c$$

and then $x' \le (c/a)/(c/b)$ implies $(c/a)/(c/b) \in F$.

From $a/b \in F$ and $(Q_{b,c})$ we obtain $x_1 \in F$ such that $x_1(b \diamond c) \le (a\,/\,b)b \diamond c \le a \diamond c$ hence $(a \diamond c)/(b \diamond c) \in F$. Similarly $(S_{c,b})$ is used to find $x_1 \in F$ such that $x_1(c \to b) \le c \to (a/b)b \le c \to a$, which shows that $(c \to a)/(c \to b) \in F$. For $(a \to c)/(b \to c) \in F$ we use $(R_{a,b\to c})$ and $x = b/a \in F$ to get $x_1 \in F$ with

$$a \diamond x_1(b \to c) \le (b/a)a \diamond (b \to c) \le b \diamond (b \to c) \le c$$

hence $x_1 \le (a \to c)/(b \to c) \in F$. The remaining terms are shown to be in F by mirror-image arguments, so θ is a congruence for the FL^2-algebra $\mathbf{A}$.

It remains to show that $F = \uparrow([1]_\theta)$. If $a\,\theta\,1$ then by definition of θ, $a = a/1 \in F$ hence $\uparrow([1]_\theta) \subseteq F$. Conversely, given $a \in F$ we need to find $c \in F$ such that $1\,\theta\,c \le a$. By assumption $1 \in F$ so we can take $c = a \wedge 1 \in F$, in which case $1 \le 1/c$. It follows that $1/c$ and $c/1$ are in F, hence $1\,\theta\,c$. □

Note that join is only used to prove compatibility of join, hence the result generalizes to a meet-semilattice version of FL^2. The theorem also applies to the FL-algebra subvariety defined by $xy = x \diamond y$ (thus $1 = t$, $/ = \leftarrow$, $\backslash = \to$), hence the result implies Theorem 1. It is also possible to prove a congruence characterization for nonassociative FL^2-algebras using the techniques of [10].

Since relation algebras and bunched implication algebras are subvarieties of FL^2-algebras, the description of the congruence filters also applies to them. While the congruences of relation algebras have been well understood since the 1950s [16], for bunched implication algebras a description first appeared in [8]. However, the description and the proof given here are both simpler and more general. Because of the symmetry in the signature of FL^2-algebras, we immediately get the following result. Consider the conditions

$$(\bar{U}_a)\ \forall x \exists x_1,\ x_1 \diamond a \le (x \diamond 1)a, \qquad (\bar{U}'_a)\ \forall x \exists x_2,\ a \diamond x_2 \le a(x \diamond 1),$$
$$(\bar{V}_a)\ \forall x \exists x_3,\ (x_3 \diamond 1)a \le a \diamond x, \qquad (\bar{V}'_a)\ \forall x \exists x_4,\ a(x_4 \diamond 1) \le x \diamond a.$$

Collectively we refer to them as $(\overline{UV_a})$ or, if they hold for all a, as $(\overline{UV})$. Similarly we have conditions $(\bar{\lambda}), (\bar{\rho}), (\bar{\lambda}'), (\bar{\rho}')$.

Corollary 5. *For an FL^2-algebra* $\mathbf{A}$*, a subset* G *is the* t*-filter of some congruence* θ *of* $\mathbf{A}$ *if and only if* G *is a lattice filter and a* $\diamond, t$*-submonoid of* $\mathbf{A}$ *that satisfies* $(\overline{UV})$.

Solving (U_a) for x_1 yields $x_1 \le u_a(x) := (xt \diamond a)/a$. Given that F is assumed to be upward closed, demanding the existence of an element $x_1 \in F$ is equivalent to asking that $u_a(x)$ is in F. Translating the remaining three conditions, we obtain that they are equivalent to closure under the terms

$$u_a(x) = (xt \diamond a)/a, \qquad u'_a(x) = a\backslash(a \diamond xt),$$
$$v_a(x) = (ax \leftarrow a)/t, \qquad v'_a(x) = (a \rightarrow xa)/t.$$

As noted above, condition (UV_a) for a filter is equivalent to the filter being closed under the unary terms u_a, u'_a, v_a and v'_a. It is an interesting problem to determine if these terms can be applied in a specific order, and how they interact with submonoid generation and closure under meets. We leave this for future research.

The condition (λ_a) can be expressed in a concise way by noting that

$$\forall x \in F\ \exists x' \in F,\ ax' \le xa \iff \forall x \in F,\ xa \in {\uparrow}(aF) \iff Fa \subseteq {\uparrow}(aF).$$

Hence (λ_a), (ρ_a) are equivalent to ${\uparrow}(aF) = {\uparrow}(Fa)$. The same argument proves the following result.

Corollary 6. *A lattice filter* F *in an FL^2-algebra satisfies* (UV_a) *if and only if* ${\uparrow}(a \diamond Ft) = {\uparrow}(Fa) = {\uparrow}(aF) = {\uparrow}(Ft \diamond a)$.

Likewise, ${\uparrow}(a(F \diamond 1)) = {\uparrow}(F \diamond a) = {\uparrow}(a \diamond F) = {\uparrow}((F \diamond 1)a)$ is equivalent to the condition $(\overline{UV_a})$ holding for F.

The characterization of congruences by 1-filters simplifies a bit when applied to algebras where $[1]_\theta$ has a least element for all congruences, as is the case for finite algebras. An element $c \in A$ is *central* if $ca = ac$ for all $a \in A$, *negative* if $c \le 1$ and *idempotent* if $cc = c$. A *congruence element* is a central negative idempotent element. The join and the product of two congruence elements is again a congruence element. It is a well known corollary of Theorem 1 that for a finite residuated lattice or FL-algebra, the congruence lattice is dually isomorphic to the lattice $(\mathrm{CE}(\mathbf{A}), \cdot, \vee)$ of congruence elements [9]. The dual isomorphism between $\mathrm{CE}(\mathbf{A})$ and the filter lattice $\mathrm{CF}(\mathbf{A})$ is given by $a \mapsto {\uparrow}a$ and $F \mapsto \bigwedge F$.

In an FL^2-algebra an element c is t*-central* if $a \diamond ct = ac = ca = ct \diamond a$ for all $a \in A$ and 1*-central* if $a(c \diamond 1) = a \diamond c = c \diamond a = (c \diamond 1)a$. A 1*-congruence element* is a t-central negative idempotent element and a t*-congruence element* is a 1-central negative idempotent element.

Corollary 7. *For an FL^2-algebra* $\mathbf{A}$ *in which all 1-congruence classes have a least element, the lattice* $CE_1(\mathbf{A})$ *of* 1*-congruence elements is dually isomorphic to the lattice* $CF_1(\mathbf{A})$ *of* 1*-filter elements.*

Let $\mathbf{A}$ be an FL^2-algebra and $\mathbf{A}_1$ its FL-algebra reduct with $\wedge, \vee, \cdot, \backslash, /, 1, 0$. The isomorphism between the lattice $\mathrm{CF}(\mathbf{A}_1)$ of congruence filters and the lattice $\mathrm{Con}(\mathbf{A}_1)$ of congruences restricts to an isomorphism between lattice $\mathrm{CF}_1(\mathbf{A})$ of 1-filters of $\mathbf{A}$ and its congruence lattice $\mathrm{Con}(\mathbf{A})$. The above characterization also applies to the t-filters of $\mathbf{A}$, hence the lattice $\mathrm{CF}_t(\mathbf{A})$ of t-filters is isomorphic to $\mathrm{Con}(\mathbf{A})$ as well. The next result shows how to map between corresponding 1-filters and t-filters without having to construct the congruence relation. As in Lemma 2, the condition $(\overline{UV})$ has the following consequence:

$$\forall x \exists x_1, x_2,\ x_1 \diamond ab \le (x \diamond a)b \ \text{ and } \ x_2 \diamond ab \le a(x \diamond b). \qquad (Q'_{a,b})$$

Theorem 8. *For FL^2-algebras there is a one-one correspondence between 1-filters and t-filters via the mutually inverse lattice isomorphisms $F \mapsto {\uparrow}(Ft)$ and $G \mapsto {\uparrow}(G \diamond 1)$.*

Proof. Let G be a t-filter of an FL^2-algebra, and define $F = {\uparrow}(G \diamond 1)$. Then $1 \in F$ since $t \in G$, and for $u, v \in F$ there exist $x, y \in G$ such that $x \diamond 1 \le u$ and $y \diamond 1 \le v$. Using $(Q'_{1,y\diamond 1})$ there exists $x' \in G$ such that

$$x' \diamond y \diamond 1 = x' \diamond 1(y \diamond 1) \le (x \diamond 1)(y \diamond 1) \le uv,$$

and since G is closed under $\diamond$, $x' \diamond y \in G$ implies $uv \in F$. Next we show (U_a) holds for F. Since G is a t-filter, $(\overline{UV})$ holds for G. From $u \in F$ we obtain $x \in G$ such that $x \diamond 1 \le u$. By $(\bar{U}_t), (\bar{\lambda})$ and $(\bar{V}_a)$ there exist $x_1, x_3, x' \in G$ with

$$(x_3 \diamond 1)a \le a \diamond x' \le x_1 \diamond a = (x_1 \diamond t) \diamond a \le (x \diamond 1)t \diamond a \le ut \diamond a,$$

hence choosing $u_1 = x_3 \diamond 1$ we have found $u_1 \in F$ such that $u_1 a \le ut \diamond a$. The conditions $(U'), (V), (V')$ can be derived in a similar way. The proof that ${\uparrow}(Ft)$ is a t-filter for any 1-filter F follows by symmetry.

It remains to check that $F = {\uparrow}(G \diamond 1) \iff {\uparrow}(Ft) = G$. Assume $F = {\uparrow}(G \diamond 1)$ and let $x \in {\uparrow}(Ft)$. Then there exists $u \in F$ such that $ut \le x$. Since $u \in F$ we have $y \diamond 1 \le u$ for some $y \in G$. By $(\bar{U}_t)$ there exists $y_1 \in G$ such that $y_1 = y_1 \diamond t \le (y \diamond 1)t$. It follows that $(y \diamond 1)t \in G$, and since $(y \diamond 1)t \le ut \le x$ we have $x \in G$. This show ${\uparrow}(Ft) \subseteq G$. Now let $x \in G$, and note that by $(\bar{V}_t)$ there exists $x_3 \in G$ such that $(x_3 \diamond 1)t \le t \diamond x = x$. Taking $u = x_3 \diamond 1$ we have $u \in F$ and $ut \le x$, hence $x \in {\uparrow}(Ft)$. We conclude that ${\uparrow}(Ft) = G$. The reverse implication follows by symmetry of the signature. □

This correspondence restricts to a bijection between 1-congruence elements c and t-congruence elements d: $c \mapsto ct$ and $d \mapsto d \diamond 1$.

3 Congruences in GBI-algebras

The results in this section can be specialized to various subvarieties of FL^2. For example, for GBI-algebras, we can characterize the 1-filters by taking multiplication to be $\cdot$ and meet to be $\diamond$. Note that the constant t is denoted by $\top$ for

GBI-algebras because it is always the top element of the algebra. Since $\wedge$ is commutative, conditions $(\lambda'), (\rho')$ are automatically satisfied and (6) and (7) of Lemma 3 apply. We state the characterization explicitly.

Corollary 9. *The* 1*-filters of a GBI-algebra* $\mathbf{A}$ *are the filter submonoids that are closed under the terms*

$$u_a(x) = (x \top \wedge a)/a, \quad v_a(x) = (a \to ax)/\top \quad \text{and} \quad \lambda_a(x) = a\backslash xa,$$

or equivalently by the terms

$$u_a'(x) = a\backslash(a \wedge x\top), \quad v_a'(x) = (a \to xa)/\top \quad \text{and} \quad \rho_a(x) = ax/a.$$

Equivalently, they are the filter submonoids that satisfy, for all $a \in A$,

$$(U_a)\ \forall x \exists x_1,\ x_1 a \le x\top \wedge a, \quad (V_a)\ \forall x \exists x_3,\ x_3 \top \wedge a \le ax, \quad (\lambda_a)\ \forall x \exists x',\ ax' \le xa$$

or equivalently the conditions

$$(U_a')\ \forall x \exists x_2,\ ax_2 \le a \wedge x\top, \quad (V_a')\ \forall x \exists x_4\ a \wedge x_4 \top \le xa, \quad (\rho_a)\ \forall x \exists x',\ x'a \le ax.$$

Likewise, we can characterize the $\top$-filters by taking multiplication to be $\diamond$ and meet to be $\cdot$, in which case $(\lambda), (\rho)$ are automatically satisfied, the condition of F being a submonoids with respect to $\wedge$ holds, and (8) and (9) of Lemma 3 give short descriptions. To clarify that we are using a different interpretation of the operations $\cdot$ and $\diamond$, we place a bar over the terms and conditions. Conditions $(\bar\lambda), (\bar\rho)$ are satisfied by the commutativity of meet.

Note that translating the FL^2 condition $(\bar V_a)$ to a term produces $\bar v_a(x) = 1 \to (a \wedge x)/a$ (in the GBI language). This simplifies to $\bar v_a(x) = 1 \to (x/a)$ since $1 \to (a \wedge x)/a = 1 \to (a/a \wedge x/a) = (1 \to a/a) \wedge (1 \to x/a)$ and $\top \le 1 \to (a/a)$.

Corollary 10. *The* $\top$*-filters of a GBI-algebra* $\mathbf{A}$ *are the filters that are closed under the terms*

$$\bar u_a'(x) = a \to a(x \wedge 1), \quad \bar v_a(x) = 1 \to (x/a) \quad \text{and} \quad \bar\lambda_a'(x) = 1 \to a\backslash(x \wedge 1)a,$$

or equivalently by the terms

$$\bar u_a(x) = a \to (x \wedge 1)a, \quad \bar v_a'(x) = 1 \to (a\backslash x) \quad \text{and} \quad \bar\rho_a'(x) = 1 \to a(x \wedge 1)/a.$$

Equivalently, they are the filter submonoids that satisfy, for all $a \in A$,

$$(\bar U_a)\ \forall x \exists x_1,\ x_1 \wedge a \le (x \wedge 1)a, \qquad (\bar V_a')\ \forall x \exists x_4,\ a(x_4 \wedge 1) \le x \quad \text{and}$$

$$(\bar\rho_a')\ \forall x \exists x',\ (x' \wedge 1)a \le a(x \wedge 1),$$

or equivalently the conditions

$$(\bar U_a')\ \forall x \exists x_2,\ a \wedge x_2 \le a(x \wedge 1), \qquad (\bar V_a)\ \forall x \exists x_3,\ (x_3 \wedge 1)a \le x \quad \text{and}$$

$$(\bar\lambda_a')\ \forall x \exists x',\ a(x' \wedge 1) \le (x \wedge 1)a.$$

In a GBI-algebra, by Theorem 4 and Corollary 6 the 1-filters are the submonoid filters F satisfying $\uparrow(Fa) = \uparrow(aF) = \uparrow(a \wedge F\top)$, for all $a \in A$; an element c is a 1-congruence element iff it is negative, idempotent and $\top$-central: $ca = ac = a \wedge c\top$, for all $a \in A$. Likewise, $\top$-filters are the filters G satisfying $\uparrow(a(G \wedge 1)) = \uparrow((G \wedge 1)a) = \uparrow(G \wedge a)$, for all $a \in A$; an element d is a $\top$-congruence element iff it is 1-central: $a(c \wedge 1) = (c \wedge 1)a = c \wedge a$, for all $a \in A$.

For involutive GBI-algebras the characterization simplifies even further. The following result from [8] shows that the characterization of t-filters does not require any parameters in this case.

Theorem 11. *For an involutive GBI-algebra, a lattice filter F is a $\top$-filter if and only if for all $x \in F$ it follows that $\neg{\sim}x, \neg{-}x, {\sim}(\top(-x)\top) \in F$.*

Involutive GBI-algebras include all relation algebras and all representable weakening relation algebras. Several results from relation algebras generalize to the setting of involutive GBI-algebras and other varieties of bunched implication algebras. For example, the term ${\sim}(\top(-x)\top)$ in the previous result is the dual of Tarski's term $\top x \top$ that is used to characterize congruence ideals in relation algebras.

4 Discriminator Varieties of GBI-algebras

Recall that an algebra is *subdirectly irreducible* if it has a smallest nontrivial congruence. For FL^2-algebras Theorem 4 and Corollary 5 imply that this property is the same as having a smallest nontrivial 1-filter or, equivalently, a smallest nontrivial t-filter. For example, this makes it easy to compute all finite subdirectly irreducible bunched implication algebras. Since they have lattice reducts, Jónsson's Lemma [14] implies that two nonisomorphic finite subdirectly irreducible BI-algebras generate distinct subvarieties, i.e., there exists an identity that holds in one of them and fails in the other. The same observations apply to finite FL^2-algebras.

Relation algebras form a discriminator variety, which means that the variety is generated by a class of algebras which have a *ternary discriminator term* $t(x, y, z)$ such that for all algebras in this generating class

$$t(x, y, z) = \begin{cases} x & \text{if } x \neq y \\ z & \text{otherwise.} \end{cases}$$

For relation algebras such a term is given by

$$t(x, y, z) = ((\top; (x \oplus y); \top) \wedge x) \vee (\neg(\top; (x \oplus y); \top) \wedge z)$$

where $x \oplus y = (x \vee y) \wedge \neg(x \wedge y)$ is the symmetric difference operation.

Discriminator varieties are well behaved in the sense that all subvarieties are also discriminator varieties (with the same term t) and all their subdirectly irreducible members are *simple*, i.e., the congruence lattice has only two elements, namely the identity congruence and the top congruence that relates all pairs. In addition, every subalgebra of a simple member is simple, and every finite member is a direct product of simple members. For relation algebras, simplicity is characterized by the Tarski rule $x \neq \bot \ \Rightarrow\ \top x \top = \top$.

An interesting question is whether there are other prominent subvarieties of FL^2-algebras that are discriminator varieties. This is not the case for the variety of BI-algebras since it contains the subvariety of Heyting algebras, defined relative to FL^2 by $x \wedge y = xy = x \diamond y$. Heyting algebras are not a discriminator variety because, e.g., the 3-element Heyting algebra is not simple.

The full weakening relation algebras $\mathbf{Wk}(\mathbf{P})$ for any poset $\mathbf{P}$ satisfy the Tarski rule (since composition and $\top$ are the same as for relation algebras), but the term $t(x, y, z)$ has to be constructed differently since negation is not classical. The following dual form has the required property:

$$t'(x, y, z) = (c(x \leftrightarrow y) \wedge z) \vee (\neg c(x \leftrightarrow y) \wedge x)$$

where $x \leftrightarrow y = (x \to y) \wedge (y \to x)$ and $c(x) = \top \backslash x / \top$. The term c is known as a (dual) unary discriminator [12] since it satisfies $c(\top) = \top$ and for $x \neq \top$, $c(x) = \bot$, i.e., it behaves like a dual Tarski rule, also known as a Baaz Delta [1] in fuzzy logic. Some concepts from relation algebra need to be dualized since in the theory of relation algebras ideals and atoms are more suitable concepts, but in the weaker (noninvolutive) theories of BI-algebras and FL^2-algebras, filters are needed to characterize the congruences. It is easy to check that t' is a discriminator in all full weakening relation algebras, hence the variety RWkRA generated by them is a discriminator variety.

In [8] it is shown that $\mathsf{RWkRA} = SP(\{\mathbf{Wk}(\mathbf{P}) : \mathbf{P} \text{ is a poset}\})$, hence every member of RWkRA is embedded in an algebra of relations and deserves to be called representable. In other words, RWkRA is analogous to the variety RRA of representable relation algebras. Since RRA is not finitely axiomatizable and can be defined from RWkRA by adding a single equation, it follows that RWkRA is also not finitely axiomatizable. A natural problem is to define a finitely based variety WkRA analogous to Tarski's variety RA of relation algebras. The variety SHRA defined in [23] is too large since it fails some short identities that hold in all full weakening relation algebras. It is also not known whether SHRA is a discriminator variety.

In the next section we recall a construction from [8] that generalizes the double coset construction of relation algebras. Applying this construction to RA leads to a variety $S(\mathsf{dRA})$ of cyclic involutive GBI-algebras that contains $\mathsf{RA} \cup \mathsf{RWkRA}$ and is properly contained in SHRA. Currently no (finite) axiomatization is known for $S(\mathsf{dRA})$ but we obtain several identities that hold in all its members.

5 The Double Division Conuclei Construction

The process of factoring a set by an equivalence relation is captured at the level of relation algebras by a construction described in [15]. In a relation algebra $\mathbf{A}$, let e be an idempotent ($ee = e$) symmetric ($e = e^{\smile}$) element and $eAe = \{exe : x \in A\}$.

Then $e\mathbf{A}e = (eAe, \wedge, \vee, \neg_e, e\top e, \bot, \cdot, {}^\smile, e, \neg_e e)$ is a relation algebra, where $\neg_e x = \neg x \wedge e\top e$. For group relation algebras this construction is known as a double coset relation algebra, and in this case $e \geq 1$. In [8] this construction is generalized to residuated lattices and GBI-algebras for arbitrary positive idempotents $p{=}p^2 \geq 1$. Given such an element p, let $\delta_p(x) = p\backslash x/p$ and note that this *double division* operation is a *conucleus*, i.e., an interior operator that satisfies $\delta_p(x)\delta_p(y) \leq \delta_p(xy)$. This holds because $\delta_p(x) = \delta'(\delta''(x))$ where $\delta'(x) = p\backslash x$ and $\delta''(x) = x/p$, both of which are conuclei, and this property is preserved under composition. By a version of [9, Prop. 3.41] without the identity, the conucleus image $\delta(\mathbf{A})$ of a residuated lattice is a residuated lattice $(\delta(A), \wedge_\delta, \vee, \cdot, \backslash_\delta, /_\delta)$ possibly without an identity, where $x *_\delta y = \delta(x * y)$ for $* \in \{\wedge, \backslash, /\}$. For the conucleus image $\delta_p(\mathbf{A})$, the element p is the identity element: $p\backslash x/p \leq (p\backslash x/p)p$ since p is positive, and $(p\backslash x/p)pp = (p\backslash x/p)p \leq p\backslash x$ hence $(p\backslash x/p)p \leq p\backslash x/p$. An even easier way to show this is to make use of the result from [8] that $\delta_p(A) = \{pxp : x \in A\}$.

The double division conucleus δ_p is of special interest for relation algebras since a positive idempotent p in a full relation algebra $\mathrm{Rel}(P)$ on a set P is a *preorder* $\mathbf{P} = (P, \sqsubseteq)$ (i.e., $p = \sqsubseteq$ is reflexive and transitive). If we assume $p \wedge p^\smile = 1$, then $\mathbf{P}$ is a poset and it follows that the full weakening relation algebra $\mathbf{Wk}(\mathbf{P})$ is equal to $\delta_p(\mathrm{Rel}(P))$. This shows that the variety RWkRA contains all double division conucleus images of members of RRA. For any class $\mathcal{K}$ of GBI-algebras we define $\mathsf{d}\mathcal{K} = \{\delta_p(\mathbf{A}) : \mathbf{A} \in \mathcal{K}, 1 \leq p^2 = p \in A\}$. In [8] it is proved that if $\mathcal{V}$ is a variety of bounded GBI-algebras with $\top\backslash x/\top$ as unary discriminator on the subdirectly irreducible members then $S(\mathsf{d}\mathcal{V})$ is a discriminator variety with the same unary discriminator term. Applying this result to the variety RA results in the discriminator variety $S(\mathsf{dRA})$ that contains both RA and RWkRA.

For an element x in a GBI-algebra, define the domain $d(x) = x\top \wedge 1$ and the range $r(x) = \top x \wedge 1$. In [13] it was shown that RWkRA satisfies the standard domain and range identities $d(x)x = x$ and $xr(x) = x$, as well as the identity $\top x \top x \top = \top x \top$. Stell's results in [23] about SHRA, together with the fact that the latter contains RWkRA, imply that RWkRA satisfies the inequality ${\sim}\neg(xy) \leq ({\sim}\neg y)({\sim}\neg x)$. Mace4 [18] shows that these identities do not hold in all cyclic involutive GBI-algebras. The 3-element Łukasiewicz algebra $L = \{0 < a < 1\}$ with $aa{=}0$ is a commutative (hence cyclic) involutive BI-algebra and taking $x{=}a$ gives counterexamples for the first three identities below. The last identity fails with $x{=}1$ in the 4-element Boolean commutative involutive BI-algebra $B{=}\{0{<}1, \neg 1{<}\top\}$ where $(\neg 1)(\neg 1) = 1$ and ${\sim}1 = 1$, ${\sim}\neg 1 = \neg 1$.

Theorem 12. *The identities*

$$d(x)x = x, \quad xr(x) = x, \quad \top x \top x \top = \top x \top \quad \text{and} \quad {\sim}\neg(xy) \leq ({\sim}\neg y)({\sim}\neg x)$$

hold in $S(\mathsf{dRA})$.

Proof. Let x be an element in $\delta_p(A)$ for some relation algebra $\mathbf{A}$ and positive idempotent $p \in A$. The identity element of $\delta_p(\mathbf{A})$ is p, hence $d(x) = x\top \wedge p$. Since $p \geq 1$, $d(x)x \geq (x\top \wedge 1)x = x$ where the last equality holds because it already holds in RA. The opposite inequality holds since $d(x) \leq p$, and p is the identity element. The proof for $r(x)$ is similar.

Although a conucleus can in principle map the top element of **A** to a smaller element, this is not the case for δ_p since $p \top p = \top$ is in the image of δ_p. Hence the third identity is true since it evaluates the same way in **A** as in $\delta_p(\mathbf{A})$.

For the fourth identity, let $x, y \in \delta_p(A)$. Applying $\sim$ on both sides and reversing the inequality $\sim\neg(xy) \leq (\sim\neg y)(\sim\neg x)$ we get the equivalent version $\neg x + \neg y \leq \neg(xy)$, where $x+y = \sim((\sim y)(\sim x))$ is the *dual product*. Since $x \leq \neg y \Leftrightarrow y \leq \neg x$ holds in Heyting algebras, the inequality becomes $xy \leq \neg(\neg x + \neg y)$. The definition of $\neg x$ in $\delta_p(\mathbf{A})$ is $\delta_p(x^c)$ where x^c is the complement in **A**. Hence we get the equivalent version $xy \leq \delta_p((\neg x + \neg y)^c) = p\backslash(\neg x + \neg y)^c/p$. Using residuation this is equivalent to $pxyp \leq (\neg x + \neg y)^c$ and to $xy \leq (\delta_p(x^c) + \delta_p(y^c))^c$ since $xy \in \delta_p(A)$, hence $pxyp = xy$. Using de Morgan's law $xy = (x^c + y^c)^c$ in RA and applying complements on both sides the equation is equivalent to $\delta_p(x^c) + \delta_p(y^c) \leq x^c + y^c$, where the last inequality holds because δ_p is decreasing. □

These identities are easily derived from the equational basis of RA, but some of these derivations make use of identities that do not hold in all algebras of weakening relations. It would be interesting to find an equational basis for S(dRA). The inequality in Theorem 12 might be part of such a basis, while the other three identities are perhaps derivable from other identities that still need to be discovered.

An example of an identity that holds in all relation algebras but is not preserved by double division conuclei is $(x \wedge 1)(y \wedge 1) = x \wedge y \wedge 1$. Some new identities have nontrivial models RWkRA. For example it is proved in [8] that $0 \leq 1$ holds in $\mathbf{Wk}(\mathbf{P})$ if and only if **P** is a chain. Here we note that the opposite inequality cannot hold in S(dRA).

Lemma 13. *If* $\mathbf{A} \in S$(dRA) *satisfies* $1 \leq 0$ *then* **A** *is trivial.*

Proof. Suppose $1 \leq 0$ holds in $\delta_p(\mathbf{A})$ for some relation algebra **A** and positive idempotent $p \in A$. Then $p \leq \sim p$ in **A**. Applying complementation on both sides we get $p^{\smile} \leq p^c$, or equivalently $p^{\smile} \wedge p = \bot$. Since p is positive, $1 \leq p^{\smile}$ hence it follows that $1 = \bot$, forcing **A** to be trivial. □

This shows that the 3-element Sugihara chain [9] is not in S(dRA) since it satisfies $0 = 1$. However the 4-element Sugihara chain is representable by the following 4 relations on the rationals $\mathbf{Q}$: $\{\varnothing, <, \leq, \mathbf{Q}^2\}$.

Another problem is to find small algebras that are in S(dRA) but not in RWkRA. Of course many small nonrepresentable relation algebras are known, but they must have at least 16 elements. It is currently not known if there are smaller examples in S(dRA).

6 Conclusion

We have shown that several concepts from relation algebras can be lifted to more general settings where they apply to other classes of algebras that occur in logic and computer science. While the variety of FL^2-algebras is somewhat general, it is a convenient setting for results about congruences since the symmetry of the two sets of connectives allows for shorter proofs. Adapting the characterization of FL^2 congruences to GBI-algebras produces a description that is significantly simpler than the previous results in [8]. The variety RWkRA of representable weakening relation algebras is a subvariety of cyclic involutive FL^2 and generalizes RRA from relations over sets to weakening relations over posets. We defined a discriminator variety $S(\mathsf{dRA})$ of cyclic involutive GBI-algebras that contains $\mathsf{RA} \cup \mathsf{RWkRA}$ and showed that it satisfies some identities that hold in both relation algebras and weakening relation algebras.

We thank the referees for the interesting comments and corrections that have substantially improved this paper.

References

1. Baaz, M.: Infinite-valued Gödel logics with 0-1-projections and relativizations. In: Hájek, P. (ed.) Lecture Notes Logic, vol. 6, pp. 23–33. Springer, Berlin (1996)
2. Blount, K., Tsinakis, C.: The structure of residuated lattices. Int. J. Algebra Comput. **13**(4), 437–461 (2003)
3. Brotherston, J., Calcagno, C.: Classical BI: its semantics and proof theory. Log. Methods Comput. Sci. **6**(3), 42 (2010)
4. Di Nola, A., Grigolia, R., Vitale, G.: On the variety of Gödel MV-algebras. Soft Comput. (2019). https://doi.org/10.1007/s00500-019-04235-5
5. Gabbay, D.M.: Fibring Logics. Oxford Logic Guides, vol. 38. The Clarendon Press, Oxford University Press, New York (1999)
6. Galatos, N., Jipsen, P.: Relation algebras as expanded FL-algebras. Algebra Univers. **69**(1), 1–21 (2013)
7. Galatos, N., Jipsen, P.: Distributive residuated frames and generalized bunched implication algebras. Algebra Univers. **78**(3), 303–336 (2017)
8. Galatos, N., Jipsen, P.: The structure of generalized BI-algebras and weakening relation algebras (2019, preprint). http://math.chapman.edu/jipsen/preprints/GalatosJipsenGBIsubmitted.pdf
9. Galatos, N., Jipsen, P., Kowalski, T., Ono, H.: Residuated Lattices: An Algebraic Glimpse at Substructural Logics. Studies in Logic and the Foundations of Mathematics, vol. 151. Elsevier, Amsterdam (2007)
10. Galatos, N., Ono, H.: Cut elimination and strong separation for substructural logics: an algebraic approach. Ann. Pure Appl. Logic **161**(9), 1097–1133 (2010)
11. Gil-Férez, J., Ledda, A., Paoli, F., Tsinakis, C.: Projectable l-groups and algebras of logic: categorical and algebraic connections. J. Pure Appl. Algebra **220**(10), 3514–3532 (2016)
12. Jipsen, P.: Discriminator Varieties of Boolean Algebras with Residuated Operators. Algebraic Methods in Logic and in Computer Science, Warsaw 1991, Polish Academy of Science, Institute of Mathematics, Warsaw, vol. 28, pp. 239–252. Banach Center Publications (1993)
13. Jipsen, P.: Relation algebras, idempotent semirings and generalized bunched implication algebras. In: Höfner, P., Pous, D., Struth, G. (eds.) RAMICS 2017. LNCS, vol. 10226, pp. 144–158. Springer, Cham (2017). https://doi.org/10.1007/978-3-319-57418-9_9

14. Jónsson, B.: Algebras whose congruence lattices are distributive. Math. Scand. **21**, 110–121 (1967)
15. Jónsson, B.: Varieties of relation algebras. Algebra Univers. **15**(3), 273–298 (1982)
16. Jónsson, B., Tarski, A.: Boolean algebras with operators. II. Amer. J. Math. **74**, 127–162 (1952)
17. Jónsson, B., Tsinakis, C.: Relation algebras as residuated Boolean algebras. Algebra Univers. **30**(4), 469–478 (1993)
18. McCune, W.: Prover9 and Mace4 (2005). https://www.cs.unm.edu/mccune/mace4/
19. O'Hearn, P.W.: Resources, concurrency, and local reasoning. Theor. Comput. Sci. **375**(1–3), 271–307 (2007)
20. Pym, D.J.: The Semantics and Proof Theory of the Logic of Bunched Implications. APLS, vol. 26. Springer, Dordrecht (2002). https://doi.org/10.1007/978-94-017-0091-7
21. Reynolds, J.C.: Separation logic: a logic for shared mutable data structures. In: Proceedings of the 17th IEEE Symposium on Logic in Computer Science (LICS 2002), 22–25 July 2002, Copenhagen, Denmark, pp. 55-74 (2002)
22. Stell, J.G.: Relations on hypergraphs. In: Kahl, W., Griffin, T.G. (eds.) RAMICS 2012. LNCS, vol. 7560, pp. 326–341. Springer, Heidelberg (2012). https://doi.org/10.1007/978-3-642-33314-9_22
23. Stell, J.G.: Symmetric Heyting relation algebras with applications to hypergraphs. J. Log. Algebr. Methods Program. **84**(3), 440–455 (2015)

Verifying the Correctness of Disjoint-Set Forests with Kleene Relation Algebras

Walter Guttmann(✉)

Department of Computer Science and Software Engineering,
University of Canterbury, Christchurch, New Zealand
walter.guttmann@canterbury.ac.nz

Abstract. We give a simple relation-algebraic semantics of read and write operations on associative arrays. The array operations seamlessly integrate with assignments in computation models supporting while-programs. As a result, relation algebras can be used for verifying programs with associative arrays. We verify the correctness of an array-based implementation of disjoint-set forests with a naive union operation and a find operation with path compression. All results are formally proved in Isabelle/HOL.

1 Introduction

Relations, relation algebras, Kleene algebras and similar structures have been used for various aspects of program semantics, in particular, to model control flow, refinement and data structures [1,18,23,32]. For example, the control-flow of while-programs can be modelled in Kleene algebras with tests, where the Kleene star is used to define the semantics of while-loops [2,21]. Program transformations and refinements can be carried out algebraically; for example, see [2,20]. On the data side, relations are intimately connected with graphs through their adjacency matrices, whence the data-flow of graph algorithms can be modelled using relation algebras, frequently extended by a Kleene star to describe transitive closure [3–6,14,19,25]. Relations as a generalisation of functions are also useful for the specification and derivation of functional programs [7].

Hoare logic [16] is commonly used for verifying programs. A verification condition generator automatically derives from the structure of the program a collection of statements whose proof implies correctness of the program. When applied to graph algorithms using a relation algebra to represent graphs, the verification conditions are simply relation-algebraic formulas. They can be discharged by a combination of manual and automated reasoning in relation algebras [6].

When modelling graphs, the operations of relation algebras work on entire relations. This abstract view is useful for specification and verification, but typically not intended directly for implementation. Efficient algorithms are often expressed at a lower level, in particular, using arrays. For example, the pseudocode for disjoint-set forests in [10] uses two arrays: one for the rank of a node

U. Fahrenberg et al. (Eds.): RAMiCS 2020, LNCS 12062, pp. 134–151, 2020.
https://doi.org/10.1007/978-3-030-43520-2_9

and one for its parent. A difference between these arrays is that the rank of a node is a natural number while the parent of a node is also a node.

An associative array is just a finite mapping from a set of indices to a set of values, hence a relation. The term 'array' often implies that the set of indices is an interval of integer numbers, but it can be an arbitrary finite set for associative arrays. The rank array of a disjoint-set forest maps nodes to natural numbers, making it a heterogeneous relation. The parent array maps nodes to nodes, which gives a homogeneous relation.

In the present paper, we focus on associative arrays with the same index and value sets. We do not assume any specific structure on the index/value set. In this context we give a simple relation-algebraic semantics of reading from and writing to an array. These access operations can occur in assignments in while-programs, and are therefore amenable to the usual verification techniques. The generated verification conditions are relation-algebraic formulas using the semantics of the array operations.

As a case study, we implement disjoint-set forests in a way that is close to the pseudo-code in [10] and verify their correctness in Isabelle/HOL. This facilitates the use of relation-algebraic reasoning about algorithms expressed at a low level.

The contributions of this paper are:

- A simple relation-algebraic semantics of selective read and write in associative arrays.
- Verification of the correctness of disjoint-set forests in Kleene relation algebras.
- Constructive proof of a theorem of Kleene relation algebras using an imperative program.
- Formalisation of the above and all other results in Isabelle/HOL.

Proofs are omitted in this paper and can be found in the Isabelle/HOL theory file available at http://www.csse.canterbury.ac.nz/walter.guttmann/algebra/.

In Sect. 2 we discuss related approaches. Section 3 introduces the algebraic framework for the remainder of this paper including relation algebras and Kleene algebras. We give a simple semantics of read and write access to associative arrays in Sect. 4 and discuss basic properties. The semantics of disjoint-set forests is provided in Sect. 5. Forming the main part of this paper, Sect. 6 describes our Isabelle/HOL verification of the total correctness of the make-set, find-set and union-sets operations on disjoint-set forests.

2 Related Work

The semantics of array or general state access is well understood and has been described in many different formalisms. We discuss a selection of these related works. An early example are the a and c functions in [24] for updating and reading state vectors, which map variables to values. Arrays are modelled as mappings in [17] and selective array updates are defined as updates of mappings. Such updates are more formally defined in [31]. A relational definition of functional overriding is given in [33] and extended to override relations in the second edition of this book. Overwriting one relation with another also appears in [26] where it is used for pointer structures. Axioms for state attributes and array access are given in [1]; some of these are used for lenses [12]. A definition of general updates in Kleene algebras with domain is given in [11]. The relation-algebraic semantics given in the present paper specialises definitions given in the last five references to selective array updates studied in the first three references.

Relation-algebraic methods have been used for the description and verification of numerous algorithms, in particular, on graphs as mentioned in the introduction. Especially relevant to the present work on disjoint sets are relational formalisations of forests and reachability; for example, see [4,25,32]. Also relevant are relational models of stores modelling pointer structures and using relational overwrite operations [25–28].

There are several formally verified implementations of disjoint-set forests. A persistent version of the data structure is verified in Coq by [9]. The specification is in terms of predicate logic and the implementation is based on a mathematical model of ML including references. See [22] for a verification using separation logic in Isabelle/HOL also based on a logical specification. Program complexity and correctness of an OCaml implementation is proved in [8] using separation logic in Coq based on a predicate-logic specification. See the latter paper for an overview of other formal verifications and further related works. The present paper gives a relation-algebraic specification and proof, which does not cover complexity of the union and find operations.

3 Relation Algebras and Kleene Algebras

This section presents the algebraic structures used in this development including relation algebras and Kleene algebras and basic properties [20,32,34].

A *semilattice* $(S, \sqcup)$ is a set S with a binary operation $\sqcup$ that is associative, commutative and idempotent. In a semilattice the binary relation $\sqsubseteq$ defined by $x \sqsubseteq y \Leftrightarrow x \sqcup y = y$ is a partial order called the *semilattice order*. The operation $\sqcup$ is $\sqsubseteq$-isotone and gives the $\sqsubseteq$-least upper bound or join of two elements.

A *bounded semilattice* $(S, \sqcup, \bot)$ is a semilattice $(S, \sqcup)$ with a constant $\bot$ that is a unit of $\sqcup$. It follows that $\bot$ is the $\sqsubseteq$-least element of S.

A *lattice* $(S, \sqcup, \sqcap)$ comprises two semilattices $(S, \sqcup)$ and $(S, \sqcap)$ such that the absorption laws $x \sqcup (x \sqcap y) = x = x \sqcap (x \sqcup y)$ hold. The operation $\sqcap$ is $\sqsubseteq$-isotone and gives the $\sqsubseteq$-greatest lower bound or meet of two elements.

A *bounded lattice* $(S, \sqcup, \sqcap, \bot, \top)$ comprises two bounded semilattices $(S, \sqcup, \bot)$ and $(S, \sqcap, \top)$ such that $(S, \sqcup, \sqcap)$ is a lattice. It follows that $\top$ is the $\sqsubseteq$-greatest element of S and a zero of $\sqcup$, and that $\bot$ is a zero of $\sqcap$.

A lattice is *distributive* if the law $x \sqcup (y \sqcap z) = (x \sqcup y) \sqcap (x \sqcup z)$ holds. In a lattice this law is equivalent to its dual $x \sqcap (y \sqcup z) = (x \sqcap y) \sqcup (x \sqcap z)$.

A *Boolean algebra* $(S, \sqcup, \sqcap, \overline{}, \bot, \top)$ is a bounded lattice $(S, \sqcup, \sqcap, \bot, \top)$ that is distributive with a unary operation $\overline{}$ satisfying the laws $x \sqcup \overline{x} = \top$ and $x \sqcap \overline{x} = \bot$. The operation $\overline{}$ is $\sqsubseteq$-antitone.

A *monoid* $(S, \cdot, 1)$ is a set S with a binary composition operation $\cdot$ that is associative and a constant 1 that is a left unit and a right unit of $\cdot$.

An *idempotent semiring* $(S, \sqcup, \cdot, \bot, 1)$ is a bounded semilattice $(S, \sqcup, \bot)$ and a monoid $(S, \cdot, 1)$ such that $\cdot$ distributes over $\sqcup$ and $\bot$ is a left zero and a right zero of $\cdot$. The operation $\cdot$ is $\sqsubseteq$-isotone.

A *relation algebra* $(S, \sqcup, \sqcap, \cdot, ^{-}, ^{\mathsf{T}}, \bot, \top, 1)$ is a Boolean algebra $(S, \sqcup, \sqcap, ^{-}, \bot, \top)$ and an idempotent semiring $(S, \sqcup, \cdot, \bot, 1)$ with a unary transposition operation ^{T} satisfying the laws:

$$(x \sqcup y)^{\mathsf{T}} = x^{\mathsf{T}} \sqcup y^{\mathsf{T}} \qquad x^{\mathsf{T}\mathsf{T}} = x$$

$$(x \cdot y)^{\mathsf{T}} = y^{\mathsf{T}} \cdot x^{\mathsf{T}} \qquad (x \cdot y) \sqcap z \sqsubseteq x \cdot (y \sqcap (x^{\mathsf{T}} \cdot z))$$

It follows that the operation ^{T} is $\sqsubseteq$-isotone. A relation algebra satisfies the *Tarski rule* if $\top \cdot x \cdot \top = \top$ for each $x \neq \bot$.

A *Kleene algebra* $(S, \sqcup, \cdot, ^{*}, \bot, 1)$ is an idempotent semiring $(S, \sqcup, \cdot, \bot, 1)$ with a unary iteration operation * satisfying the laws:

$$1 \sqcup (y \cdot y^{*}) = y^{*} \qquad z \sqcup (y \cdot x) \sqsubseteq x \Rightarrow y^{*} \cdot z \sqsubseteq x$$

$$1 \sqcup (y^{*} \cdot y) = y^{*} \qquad z \sqcup (x \cdot y) \sqsubseteq x \Rightarrow z \cdot y^{*} \sqsubseteq x$$

The operation * is $\sqsubseteq$-isotone. It describes iterations with zero or more steps; the related operation $x^{+} = x \cdot x^{*}$ describes iterations with one or more steps.

A *Kleene relation algebra* $(S, \sqcup, \sqcap, \cdot, ^{-}, ^{\mathsf{T}}, ^{*}, \bot, \top, 1)$ comprises a relation algebra $(S, \sqcup, \sqcap, \cdot, ^{-}, ^{\mathsf{T}}, \bot, \top, 1)$ and a Kleene algebra $(S, \sqcup, \cdot, ^{*}, \bot, 1)$.

An element $x \in S$ of a relation algebra S is called *reflexive* if $1 \sqsubseteq x$, *transitive* if $x \cdot x \sqsubseteq x$, *symmetric* if $x^{\mathsf{T}} = x$, an *equivalence* if x is reflexive and transitive and symmetric, *total* if $1 \sqsubseteq x \cdot x^{\mathsf{T}}$, *surjective* if $1 \sqsubseteq x^{\mathsf{T}} \cdot x$, *univalent* if $x^{\mathsf{T}} \cdot x \sqsubseteq 1$, *injective* if $x \cdot x^{\mathsf{T}} \sqsubseteq 1$, *bijective* if x is injective and surjective, a *mapping* if x is univalent and total, a *vector* if $x \cdot \top = x$, a *point* if x is a vector and bijective, and an *arc* if $x \cdot \top$ and $x^{\mathsf{T}} \cdot \top$ are bijective.

An element $x \in S$ of a Kleene relation algebra S is called *acyclic* if $x^{+} \sqsubseteq \overline{1}$.

In this paper we work in a Kleene relation algebra S that satisfies the Tarski rule. For proving termination of programs we assume that S is finite.

The main model of Kleene relation algebras are binary relations over a set A, that is, subsets of $A \times A$. In this model $\sqcup$ is union, $\sqcap$ is intersection, $^{-}$ is complement, $\sqsubseteq$ is subset, $\bot$ is the empty set, $\top$ is $A \times A$, $\cdot$ is relational composition, ^{T} is relational transposition, 1 is the identity relation, * is reflexive transitive closure, $^{+}$ is transitive closure, and the Tarski rule holds.

We finally characterise vectors, points and arcs among the binary relations over A. A vector is a relation $B \times A$ for a subset $B \subseteq A$; hence vectors represent subsets of the base set such as a set of nodes in a graph. A point is a relation $\{a\} \times A$ for an element $a \in A$; hence points represent elements of the base set such as nodes in a graph. An arc is a relation $\{(a, b)\}$ for elements $a, b \in A$; hence arcs represent pairs of elements from the base set such as edges in a graph.

4 Associative Array Access

An array maps indices to values and therefore can be modelled as a binary relation between two sets. Under our assumption that indices and values come from the same set A, we can use binary relations on A and work with these using relation algebra. Because an array associates exactly one value to every index, the relation is a mapping in the relation-algebraic sense, that is, univalent and total. A relation that is just univalent corresponds to a partially defined array which associates at most one value to every index. An index or a value is an element of A, which can be modelled in relation algebras as a point. A relation that is just a vector corresponds to a set of indices or values.

These observations underlie the following simple semantics of array access. Let x, y and z be elements of a relation algebra such that y and z are points. The element x models the associative array, y corresponds to an index and z corresponds to a value. The array $x[y \mapsto z]$ obtained by updating array x at index y to new value z is:

$$x[y \mapsto z] = (y \sqcap z^\mathsf{T}) \sqcup (\overline{y} \sqcap x)$$

To understand this definition it is helpful to consider the matrix representation of the relation modelling the array x. A vector describes a set of rows of the matrix and a point describes a single row. The point y refers to the row at the corresponding index. Its complement $\overline{y}$ refers to all the other rows. The formula $\overline{y} \sqcap x$ specifies that in all other rows x is left unchanged. The formula $y \sqcap z^\mathsf{T}$ specifies that row y is updated to value z. Since z is a point, which refers to row, we take its transposition z^T, which refers to the column of the matrix at the corresponding value. In terms of binary relations, $y \sqcap z^\mathsf{T}$ constructs a relation containing a single pair of the index y and the value z. In relation algebras $y \sqcap z^\mathsf{T} = y \cdot z^\mathsf{T}$ is an arc for points y and z.

For example, consider the following relations x, y and z on $A = \{1, 2, 3\}$ given as Boolean matrices:

$$x = \begin{pmatrix} 0 & 0 & 1 \\ 0 & 1 & 0 \\ 0 & 0 & 0 \end{pmatrix} \qquad y = \begin{pmatrix} 0 & 0 & 0 \\ 1 & 1 & 1 \\ 0 & 0 & 0 \end{pmatrix} \qquad z = \begin{pmatrix} 1 & 1 & 1 \\ 0 & 0 & 0 \\ 0 & 0 & 0 \end{pmatrix}$$

Relation x represents a partially defined array that maps index 1 to value 3 and index 2 to value 2, point y represents index 2 and point z represents value 1. The updated array still maps index 1 to value 3, but maps index 2 to value 1:

$$y \sqcap z^\mathsf{T} = \begin{pmatrix} 0 & 0 & 0 \\ 1 & 0 & 0 \\ 0 & 0 & 0 \end{pmatrix} \qquad \overline{y} \sqcap x = \begin{pmatrix} 0 & 0 & 1 \\ 0 & 0 & 0 \\ 0 & 0 & 0 \end{pmatrix} \qquad x[y \mapsto z] = \begin{pmatrix} 0 & 0 & 1 \\ 1 & 0 & 0 \\ 0 & 0 & 0 \end{pmatrix}$$

Reading the value $x[y]$ of the associative array x at index y is done by:

$$x[y] = x^\mathsf{T} \cdot y$$

The composition of a relation with a vector always gives a vector. If x is interpreted as a transition relation, $x^\mathsf{T} \cdot y$ is a vector corresponding to the successors of the point y under a transition step according to x. In the matrix representation of an array, this is just the value of x at row y. If the array associates exactly one value to every index, the result is the unique value associated with index y, represented as a point.

Continuing the previous example, the value of x at index y is 2 and the value of x at index z is 3:

$$x^\mathsf{T} = \begin{pmatrix} 0 & 0 & 0 \\ 0 & 1 & 0 \\ 1 & 0 & 0 \end{pmatrix} \qquad x[y] = \begin{pmatrix} 0 & 0 & 0 \\ 1 & 1 & 1 \\ 0 & 0 & 0 \end{pmatrix} \qquad x[z] = \begin{pmatrix} 0 & 0 & 0 \\ 0 & 0 & 0 \\ 1 & 1 & 1 \end{pmatrix}$$

The following result shows basic preservation properties of these write and read operations on arrays. It uses the above equational definitions without implicitly assuming that y and z are points. Part 1 is similar to [26, Lemma 2.7].

Theorem 1.

1. $x[y \mapsto z]$ is univalent if x is univalent, y is a vector and z is injective.
2. $x[y \mapsto z]$ is total if x is total, y is a vector and z is surjective.
3. $x[y \mapsto z]$ is a mapping if x is a mapping, y is a vector and z is bijective.
4. $x[y]$ is injective if x is univalent and y is injective.
5. $x[y]$ is surjective if x is total and y is surjective.
6. $x[y]$ is bijective if x is a mapping and y is bijective.
7. $x[y]$ is a point if x is a mapping and y is a point.
8. $x[y] = z \Leftrightarrow y \sqcap x = y \cdot z^\mathsf{T}$ if y and z are points.

5 Disjoint Sets

A disjoint-set data structure keeps track of a set of elements that is partitioned into disjoint sets [13]. The basic operations are to initialise elements to be in their own singleton sets, to form the union of two sets and to look up which set an element belongs to.

The semantics of a disjoint-set data structure with elements from A is an equivalence relation on A. The disjoint sets are just the equivalence classes of the relation. A particular representative from each class identifies a set.

An element of a relation algebra is an equivalence if it is reflexive, transitive and symmetric. The $\sqsubseteq$-least equivalence is the identity relation 1. The $\sqsubseteq$-greatest equivalence is the universal relation $\top$. Equivalences are closed under the operations $\sqcap$ and $^\mathsf{T}$ and, in Kleene relation algebras, under * and $^+$.

Following [10] we implement the data structure as a disjoint-set forest. Each equivalence class corresponds to a tree in the forest. Singleton sets correspond to empty trees, which contain one node. Each tree in the forest has a root and is directed. Each node in a tree has a unique parent node; the root is its own parent. The root of a tree represents the corresponding equivalence class. An edge from a node to its parent points towards the root of the tree, which can be reached by successively following parents.

Disjoint-set forests can be modelled in Kleene relation algebras as follows. An element $x \in S$ of a Kleene relation algebra S is called a *forest* if x is a mapping and $x \sqcap \overline{1}$ is acyclic. Requiring x to be a mapping ensures that each node has a unique parent. It remains to ensure that there are no cycles. We cannot require that x is acyclic because every root has itself as its parent, which corresponds to a loop in the graph. However, $x \sqcap \overline{1}$ removes all loops, so we require that the result is acyclic. Related helpful lemmas are $x^* = (x \sqcap \overline{1})^*$ and $x^* \sqcap \overline{1} = x^+ \sqcap \overline{1}$.

In a forest x, it is possible to reach from a node every other node in the same component tree by going towards its root and then back to the desired node. This defines a relation $\mathrm{fc}(x)$ on the nodes of the forest, namely the relation of being in the same component:

$$\mathrm{fc}(x) = x^* \cdot x^{\mathsf{T}*}$$

Properties of this construction are given in the following result.

Theorem 2.

1. $\mathrm{fc}(x)$ is an equivalence if x is univalent.
2. fc is $\sqsubseteq$-increasing, that is, $x \sqsubseteq \mathrm{fc}(x)$.
3. fc is $\sqsubseteq$-isotone.
4. $\mathrm{fc}(\mathrm{fc}(x)) = \mathrm{fc}(x)$ if x is univalent.
5. $\mathrm{fc}(x)^* = \mathrm{fc}(x)^+ = \mathrm{fc}(x)$ if x is univalent.
6. $\mathrm{fc}(\bot) = \mathrm{fc}(1) = 1$.
7. $\mathrm{fc}(\top) = \top$.

6 Verifying Disjoint-Set Forests in Isabelle/HOL

For implementing the operations on disjoint-set forests and verifying their correctness we use a Hoare-logic library of Isabelle/HOL [29,30], which we have extended from partial correctness to total correctness [15]. The library supports while-programs, which have to be annotated with a precondition, a postcondition, and an invariant and a variant for each while-loop. From this, verification conditions are automatically generated.

Program variables can range over arbitrary HOL types. We write programs in the context of a class specifying the axioms of Kleene relation algebras, the Tarski rule and a finite universe for total correctness. Hence program variables range over elements from the universe of the class, which models the corresponding algebraic structure. Reasoning about these variables to discharge verification conditions is performed in the same context using existing libraries for Kleene algebras and relation algebras and newly derived theorems.

While-programs supported by the Hoare-logic library feature while-loops, conditionals, sequential composition and assignments as basic statements. We introduce new notation for array read and write operations, which are automatically translated to basic relation-algebraic expressions according to Sect. 4. The assignment $x[y] := z$ is translated to the assignment $x := x[y \mapsto z]$. The read expression $x[y]$ can be used directly on the right-hand side of assignments and in conditions, except we modify its syntax to $x[[y]]$ to avoid ambiguity with list syntax. This paper uses $x[y]$ except in Isabelle/HOL code which uses $x[[y]]$.

6.1 The Make-Set Operation

As a warm-up we implement the make-set operation of disjoint-set forests and prove its correctness. It is usually applied to each element when the data structure is initialised. Until the initialisation is complete, the underlying associative array is partial. Make-set puts an element x into its own singleton equivalence class by setting the parent of x to itself which creates an empty tree:

```
1 theorem make_set:
2    "VARS p
3       [ point x ∧ p0 = p ]
4       p[x] := x
5       [ make_set_postcondition p x p0 ]"
6    apply vcg_tc_simp
7    by (simp add: ...) – names of four lemmas omitted
```

Line 2 declares variables that are changed by the program and therefore need to be part of the state, in this case only p which contains the parent array. The variables x and p_0 are universally quantified variables of the theorem; because they are not changed they do not need to be part of the state. The variable p_0 transports the initial value of p to the postcondition, where it can related to the final value of p. Line 3 gives the precondition, which requires x to be a point, representing an element of the set partitioned by the data structure. Line 4 updates the parent array to make x the root of a tree. Line 5 gives the postcondition, which is discussed below. Line 6 generates the verification condition, which for this small program is a single goal, and applies some simplifications to it:

$$\text{point } x \wedge p_0 = p \Rightarrow \text{make\_set\_postcondition } p[x \mapsto x]\ x\ p$$

Line 7 proves this goal by invoking the simplifier with additional lemmas. The postcondition has two parts:

$$\text{make\_set\_postcondition } p\ x\ p_0 \Leftrightarrow x \sqcap p = x \cdot x^{\mathsf{T}} \wedge \overline{x} \sqcap p = \overline{x} \sqcap p_0$$

The first condition $x \sqcap p = x \cdot x^{\mathsf{T}}$ states that the parent array contains x at index x. It is equivalent to $p[x] = x$ by Theorem 1.8. The second condition $\overline{x} \sqcap p = \overline{x} \sqcap p_0$ states that the parent array remains unchanged at all indices different from x.

The precondition and postcondition can be strengthened by adding $p \sqsubseteq 1$. As a consequence, when a disjoint-set forest is initialised each equivalence class constructed by make-set is a singleton.

The method vcg_tc_simp generates conditions that prove total correctness. Since the above program does not contain any while-loops, there are no conditions related to its termination.

We use a basic Hoare-logic library which does not support the definition of procedures. So that other programs can use a disjoint-set operation such as make_set, we extract an Isabelle/HOL function from the above proof using a technique of [15]. Specifically, the above total-correctness theorem implies:

lemma make_set_exists: "point $x \Rightarrow \exists p'$. make_set_postcondition p' x p"
  using tc_extract_function make_set **by** blast

This is a consequence of how total correctness is defined on the underlying operational semantics. Hence we can introduce the following Isabelle/HOL function:

definition "make_set p x = (SOME p' . make_set_postcondition p' x p)"

The construct SOME y. $P(y)$ yields some element y that satisfies $P(y)$. In order to reason about this function in other programs we derive the following property:

lemma make_set_function:
  assumes "point x" **and** "p' = make_set p x"
  shows "make_set_postcondition p' x p"
  – proof omitted

6.2 The Find-Set Operation

We next implement the find-set operation of disjoint-set forests and verify its correctness. The find-set operation computes the representative of the equivalence class an element belongs to. We first demonstrate a basic implementation of find-set and then extend it by path compression. The pseudo-code in [10] uses recursion whereas we use a while-loop. The find-set operation follows the chain of parents from a node x to the root of its tree:

1 **theorem** find_set:
2   "VARS y
3   [find_set_precondition p x]
4     $y := x$;
5   WHILE $y \neq p[[y]]$
6   INV { find_set_invariant p x y }
7   VAR { card $\{z \,.\, z \sqsubseteq p^{\mathsf{T}*} \cdot y\}$ }
8   DO $y := p[[y]]$
9   OD
10   [find_set_postcondition p x y]"
11   **apply** vcg_tc_simp
12   – proof of three verification conditions omitted

In line 4, variable y is initialised with the start node x. The while-loop stops when it finds a node that is its own parent in line 5. Otherwise it continues with the parent of the current node in line 8.

The precondition requires that p is a forest (representing the disjoint sets) and x is a point (representing a node in the forest):

$$\text{find\_set\_precondition}\ p\ x \Leftrightarrow \text{forest}\ p \wedge \text{point}\ x$$

Every while-loop in the program needs to be annotated with an invariant. In this case, the invariant requires the precondition and that y is a point reachable from x along a chain of parents:

$$\text{find\_set\_invariant}\ p\ x\ y \Leftrightarrow \text{find\_set\_precondition}\ p\ x \wedge \text{point}\ y \wedge y \sqsubseteq p^{\top *} \cdot x$$

Vector $p^{\top *} \cdot x$ contains all successors of x under zero or more transitions of p. The postcondition states that y is a point and the root of the tree containing x:

$$\text{find\_set\_postcondition}\ p\ x\ y \Leftrightarrow \text{point}\ y \wedge y = \text{root}\ p\ x$$

The root of a node x in the disjoint-set forest represented by p is the unique node that has a loop and is reachable from x along a chain of parents:

$$\text{root}\ p\ x = (p^{\top *} \cdot x) \sqcap ((p \sqcap 1) \cdot \top)$$

The vector $(p \sqcap 1) \cdot \top$ contains all roots of the forest p, constructed from the relation $p \sqcap 1$ containing all loops of p. Part 1 of the following result gives an equivalent characterisation. Part 2 shows that following the parents of roots one or several times gives the roots again. We discuss part 3 below.

Theorem 3.

1. $\text{root}\ p\ x = (p \sqcap 1) \cdot p^{\top *} \cdot x$.
2. $\text{root}\ p\ x = p[\text{root}\ p\ x] = p^{\top *} \cdot (\text{root}\ p\ x)$ if p is univalent.
3. $\text{root}\ p\ x$ is a point if p is a forest and x is a point.

Because the above program contains one while-loop, three verification conditions are generated: one to establish the loop invariant before execution of the while-loop, one to maintain the loop invariant across execution of the body of the while-loop, and one to show the postcondition at the end of the while-loop. For partial correctness, the generated conditions are:

1. $\text{find\_set\_precondition}\ p\ x \Rightarrow \text{find\_set\_invariant}\ p\ x\ x$
2. $\text{find\_set\_invariant}\ p\ x\ y \wedge y \neq p[[y]] \Rightarrow \text{find\_set\_invariant}\ p\ x\ p[[y]]$
3. $\text{find\_set\_invariant}\ p\ x\ y \wedge y = p[[y]] \Rightarrow \text{find\_set\_postcondition}\ p\ x\ y$

To maintain the invariant we can assume that the condition of the while-loop holds. To show the postcondition we can assume that the condition of the while-loop does not hold. For total correctness, the first and third verification conditions are the same but maintenance of the invariant is modified taking into account the variant of the while-loop:

2. $\text{find\_set\_invariant}\ p\ x\ y \wedge y \neq p[[y]] \wedge \text{card}\ \{z\ .\ z \sqsubseteq p^{\top *} \cdot y\} = n \Rightarrow$ $\text{find\_set\_invariant}\ p\ x\ p[[y]] \wedge \text{card}\ \{z\ .\ z \sqsubseteq p^{\top *} \cdot p^{\top} \cdot y\} < n$

Every while-loop in the program needs to be annotated with a variant. The variant is an expression that yields a natural number depending on the program variables. The value of this expression decreases after execution of the body of the loop. Because it is a natural number, it will decrease only a finite number of times which ensures termination of the while-loop. The variable n transports the initial value of the variant from the assumption to the conclusion, where it is compared with the final value of the variant.

For the above program, the variant is the number of elements in the algebra below $p^{\mathsf{T}*} \cdot y$. The expression $p^{\mathsf{T}*} \cdot y$ is a vector representing the set of nodes reachable from y by successively following parents. The variant is an order-preserving expression that turns this vector into a natural number. This works because the algebra is finite.

We now discuss Theorem 3.3, which states that the root of the tree containing point x in the forest p is a point, that is, a vector representing a single node. This result could be proved by working with the definitions of roots, points and forests. We give a different proof based on find-set. Observe that this operation computes the desired root and the postcondition states it is a point. Moreover the precondition of find-set contains just the assumptions of Theorem 3.3. Hence this result immediately follows from total correctness of find-set. In Isabelle/HOL, similarly to make-set discussed in Sect. 6.1 we obtain:

lemma find_set_exists:
  "find_set_precondition p x $\Rightarrow$ $\exists y$. find_set_postcondition p x y"
  using tc_extract_function find_set **by** blast

Theorem 3.3 then is a simple consequence:

lemma root_point: "forest p $\wedge$ point x $\Rightarrow$ point (root p x)"
  using find_set_exists find_set_precondition_def find_set_postcondition_def
  by simp

Essentially this is a constructive proof using the imperative programs supported by the Hoare-logic library. This method does not necessarily reduce the amount of work needed for proving a result but shifts the work to the correctness proof of a program. However, once the correctness proof is established it saves additional work. Moreover, this approach facilitates computational reasoning.

6.3 Path Compression

Path compression is a technique to decrease the depth of the disjoint-set forest, which makes subsequent find-set operations faster. The idea is to change the parent of every node encountered during the execution of find-set to the root of the tree. Because the root is known only after the chain of parents has been traversed, modifying the parents takes place in a separate traversal. In a recursive implementation of find-set, these modifications would take place on the way out from the recursion. We use two while-loops for the same purpose. The first loop is the find-set operation described in Sect. 6.2 to find the root y of the tree.

As shown here, the second loop traverses the same sequence of nodes and adjusts the parents on the way:

```
1  theorem path_compression:
2    "VARS p t w
3      [ path_compression_precondition p x y ∧ p0 = p ]
4      w := x;
5      WHILE y ≠ p[[w]]
6        INV { path_compression_invariant p x y p0 w }
7        VAR { card {z . z ⊑ p^T* · w} }
8        DO
9          t := w;
10         w := p[[w]];
11         p[t] := y
12       OD
13     [ path_compression_postcondition p x y p0 ]"
14   apply vcg_tc_simp
15   – proof of three verification conditions omitted
```

This program is executed immediately after the while-loop of find-set, where p is the parent array, x is the original node and y is its representative computed by find-set, which is the root of the tree that contains x. The assignments in lines 4 and 10 traverse the same sequence of nodes as find-set. According to line 5 this finishes when the root is reached. Lines 9 and 11 set the parent of the current node to the root. Temporary variable t is used to save the current node w, which is changed by line 10.

The variant in line 7 is the same as the one used for find-set, except the current node is now stored in w. Also the generated verification conditions have the same structure as in the proof for find-set. It remains to discuss the actual precondition, invariant and postcondition. The precondition is:

$$\begin{aligned}&\text{path\_compression\_precondition } p\ x\ y \Leftrightarrow\\&\quad \text{forest } p \wedge \text{point } x \wedge \text{point } y \wedge y = \text{root } p\ x\end{aligned}$$

It extends the precondition of find-set by two conditions, which are just the postcondition of find-set. This ensures the two loops can be composed sequentially. The invariant significantly extends the precondition:

$$\begin{aligned}&\text{path\_compression\_invariant } p\ x\ y\ p_0\ w \Leftrightarrow\\&\quad \text{path\_compression\_precondition } p\ x\ y \wedge \text{fc}(p) = \text{fc}(p_0) \wedge p \sqcap 1 = p_0 \sqcap 1\\&\quad \wedge \text{point } w \wedge y \sqsubseteq p^{\mathsf{T}*} \cdot w \wedge (w \neq x \Rightarrow (y \neq x \wedge p[[x]] = y \wedge p^{\mathsf{T}+} \cdot w \sqsubseteq \overline{x}))\end{aligned}$$

First, $\text{fc}(p) = \text{fc}(p_0)$ states that the components of p do not change, that is, p represents the same disjoint sets. Second, $p \sqcap 1 = p_0 \sqcap 1$ states that the roots of the component trees of p do not change. Third, the invariant requires that w is a point. Fourth, $y \sqsubseteq p^{\mathsf{T}*} \cdot w$ states that the root y is reachable from w by following the chain of parents. The last part of the invariant only applies if $w \neq x$, that is, in the second or later iterations of the while-loop. In these iterations, the start node x and the root y are different, the parent of x is y, and any node reachable from w by one or more steps along the chain of parents is different from x. The postcondition is part of the invariant:

$$\begin{aligned}&\text{path\_compression\_postcondition } p\ x\ y\ p_0 \Leftrightarrow \\ &\quad \text{path\_compression\_precondition } p\ x\ y \wedge \text{fc}(p) = \text{fc}(p_0) \wedge p \sqcap 1 = p_0 \sqcap 1\end{aligned}$$

For correctness we only require that path compression does not change the disjoint sets represented by the forest. We also get that the roots do not change.

We discuss a selection of results used for maintaining the invariant. Part of the maintenance is to show that the parent relation (without loops) remains acyclic. Path compression updates the parent relation by letting the parents of visited nodes point to the root of the tree. Part 1 of the following theorem shows that updating the parent of a node w to any node y reachable from w along the chain of parents does not introduce cycles (ignoring loops).

Theorem 4.

1. $p[w \mapsto y] \sqcap \overline{1}$ is acyclic if $p \sqcap \overline{1}$ is acyclic, w and y are points and $y \sqsubseteq p^{\mathsf{T}*} \cdot w$.
2. $x \sqcap p^* = (x \sqcap 1) \sqcup ((x \sqcap p) \cdot (\overline{x} \sqcap p)^*)$ if x is a point.
3. $x \sqcap y = \bot$ if x and y are points such that $x \neq y$.

Part 2 optimises iterations similar to [25, Lemma 4]; for related techniques see also [3]. The element $x \sqcap p^*$ on the left-hand side relates the node x to all nodes reachable from it by zero or more steps in the graph p. The right-hand side contains $x \sqcap 1$, which relates x to itself, and $(x \sqcap p) \cdot (\overline{x} \sqcap p)^*$, which relates x to nodes reachable from it by one step in p followed by zero or more steps in $\overline{x} \sqcap p$. This means that edges starting in x have to be considered at most in the first step and can be omitted in the remaining steps. In maintaining the invariant, this is applied with $x = w$, so that the remaining steps only use edges not starting in w, which is important since these edges are not affected by the update to the forest.

Part 3 of the previous theorem ultimately derives from the Tarski rule and states that different points are disjoint as relations. This is a general result used in several arguments; we explain one of them. In maintaining the invariant, we need to show that updating p does not change the set of its roots. The update changes p at index w to the new value y, so this part of p changes from $w \sqcap p$ to $w \sqcap y^{\mathsf{T}}$. The roots in this part are $w \sqcap p \sqcap 1$ and $w \sqcap y^{\mathsf{T}} \sqcap 1$ and we show that both expressions are $\bot$. To this end, observe that $y \neq w$ since y is a root according to the precondition, but the parent of w is different from y according to the condition of the while-loop. First, $w \sqcap p \sqcap 1 \sqsubseteq w \sqcap 1 = \bot$ because the node w does not have a loop; otherwise $y = w$ would hold since y is reachable from w according to the loop invariant. Second, $w \sqcap y^{\mathsf{T}} \sqcap 1 = w \sqcap y \sqcap 1 \sqsubseteq w \sqcap y = \bot$ by a general property of relation algebras and part 3 of the previous theorem.

6.4 The Find-Set Operation with Path Compression

Using the technique of Sect. 6.1 we extract function definitions for the find-set operation of Sect. 6.2 and the path-compression operation of Sect. 6.3. This allows us to combine the two programs into the following one with a simple correctness proof:

```
theorem find_set_path_compression:
  "VARS p y
    [ find_set_precondition p x ∧ p0 = p ]
    y := find_set p x;
    p := path_compression p x y
    [ path_compression_postcondition p x y p0 ]"
  apply vcg_tc_simp
  using find_set_function find_set_postcondition_def
    find_set_precondition_def path_compression_function
    path_compression_precondition_def by fastforce
```

We can also extract a function for this program, but this function returns a pair of values as the find-set operation with path compression both modifies the disjoint-set forest and returns the root of the tree containing node x:

```
definition "find_set_path_compression p x =
  (SOME (p', y) . path_compression_postcondition p' x y p)"
```

6.5 The Union-Sets Operation

We finally consider the union-sets operation, which takes two elements and joins the corresponding disjoint sets into a single set. To this end it finds the representatives of the equivalence classes of the elements and links one to the other:

```
1  theorem union_sets:
2    "VARS p r s t
3      [ union_sets_precondition p x y ∧ p0 = p ]
4      t := find_set_path_compression p x;
5      p := fst t;
6      r := snd t;
7      t := find_set_path_compression p y;
8      p := fst t;
9      s := snd t;
10     p[r] := s
11     [ union_sets_postcondition p x y p0 ]"
12   apply vcg_tc_simp
13   – proof of one verification condition omitted
```

Because the Hoare-logic library does not support parallel assignments, we assign the resulting pair of find-set to a temporary variable in lines 4 and 7 and separate the components in lines 5–6 and 8–9, respectively. Note how the forest p is threaded through both occurrences of find-set, where it may be modified by path compression, before line 10 adds the link from the root r of the tree containing x to the root s of the tree containing y.

The precondition of union-sets requires that p is a forest and x and y are single nodes:

$$\text{union\_sets\_precondition}\ p\ x\ y \Leftrightarrow \text{forest}\ p \wedge \text{point}\ x \wedge \text{point}\ y$$

The postcondition also requires that the final value of p represents the equivalence relation where the sets containing x and y have been merged into one:

$$\begin{aligned}&\text{union\_sets\_postcondition}\ p\ x\ y\ p_0 \Leftrightarrow\\ &\quad \text{union\_sets\_precondition}\ p\ x\ y \wedge \text{fc}(p) = \text{wcc}(p_0 \sqcup (x \cdot y^\mathsf{T}))\end{aligned}$$

To get the latter equivalence relation, we add the pair (x, y) to the initial equivalence relation p_0 and compute its equivalence closure, that is, the smallest equivalence relation containing p_0 and the pair (x, y). The pair (x, y) is described by $x \cdot y^\mathsf{T}$ and according to [32] the equivalence closure is given by:

$$\text{wcc}(x) = (x \sqcup x^\mathsf{T})^*$$

Interpreting the relation x as a directed graph, the equivalence closure represents the weakly-connected components of x, which are obtained by reachability while ignoring the direction of edges. Properties of wcc are given in the following result.

Theorem 5.

1. $\text{wcc}(x)$ is an equivalence.
2. wcc is a closure operation, that is, idempotent, $\sqsubseteq$-isotone and $\sqsubseteq$-increasing.
3. $\text{wcc}(x) \sqsubseteq \text{wcc}(y)$ if $x \sqsubseteq \text{wcc}(y)$.
4. $\text{wcc}(\bot) = \text{wcc}(1) = 1$.
5. $\text{wcc}(\top) = \top$.
6. $\text{wcc}(x \sqcup 1) = \text{wcc}(x \sqcap \overline{1}) = \text{wcc}(x)$.
7. $\text{wcc}(x) = \text{fc}(x)$ if x is univalent.

We further discuss a selection of results used for proving the correctness of union-sets. Part 1 of the following result is similar to [6, Proposition 3]. It considers reachability under the union of two relations x and y, where x is an arc containing just one edge. It then suffices to use the edge x at most once: y^+ describes the case where x is not needed and $y^* \cdot x \cdot y^*$ describes the case where x is used once, preceded and followed by any number of edges in y.

Theorem 6.

1. $(x \sqcup y)^+ = y^+ \sqcup (y^* \cdot x \cdot y^*)$ if x is an arc.
2. $p[w \mapsto y] \sqcap \overline{1}$ is acyclic if $p \sqcap \overline{1}$ is acyclic, w and y are points, and $y \sqcap p^* \cdot w = \bot$.
3. $p[w \mapsto w] \sqcap \overline{1}$ is acyclic if $p \sqcap \overline{1}$ is acyclic and w is a point.

Parts 2 and 3 are similar to Theorem 4.1. In part 2 the parent of w is updated to a node y from which w is not reachable in p. This does not introduce a cycle (ignoring loops). Part 3 shows that creating a loop on w does not introduce a cycle (ignoring loops).

These results are used to show that the assignment in line 10 of union-sets maintains the forest property. If the arguments x and y of union-sets are in the same tree, the roots r and s will be equal, so line 10 creates a loop and the correctness proof uses part 3 of the preceding result. Alternatively, it could be proved that the assignment in line 10 does not change the forest in this case. Part 2 of the preceding result is used if nodes x and y are in different trees.

7 Conclusion

This paper has given a simple relation-algebraic semantics for read and write operations on associative arrays. Based on this semantics, we added such operations to a sequential programming language used for specifying and verifying programs in Isabelle/HOL. We implemented disjoint-set forests with path compression this way and proved their correctness.

Correctness of the union-sets operation would not be affected if the assignment in line 10 of the program in Sect. 6.5 was replaced with $p[s] := r$. More efficient implementations of union-sets therefore decide which of these two assignments to use based on heuristics such as union by rank. The rank of a node is a natural number giving an upper bound on the depth of the subtree at the node. It is more efficient to add a link from the root with smaller rank to the other. Using ranks in a disjoint-set forest implementation requires comparisons and simple arithmetic operations. In future work we will consider how to implement this extension using relation-algebraic methods.

Another task is to integrate the implementation given in this paper with relation-algebraic implementations of Kruskal's minimum spanning tree algorithm. For this reason our Isabelle/HOL theory uses Stone-Kleene relation algebras, which are weaker than Kleene relation algebras and can represent weighted graphs [15]. A further direction of research is to consider how relation-algebraic methods can support complexity analysis of algorithms.

Acknowledgement. I thank the anonymous referees for their helpful comments.

References

1. Back, R.J., von Wright, J.: Refinement Calculus. Springer, New York (1998). https://doi.org/10.1007/978-1-4612-1674-2
2. Back, R.J.R., von Wright, J.: Reasoning algebraically about loops. Acta Inf. **36**(4), 295–334 (1999)
3. Backhouse, R.C., Carré, B.A.: Regular algebra applied to path-finding problems. J. Inst. Math. Appl. **15**(2), 161–186 (1975)
4. Berghammer, R.: Combining relational calculus and the Dijkstra-Gries method for deriving relational programs. Inf. Sci. **119**(3–4), 155–171 (1999)
5. Berghammer, R., von Karger, B., Wolf, A.: Relation-algebraic derivation of spanning tree algorithms. In: Jeuring, J. (ed.) MPC 1998. LNCS, vol. 1422, pp. 23–43. Springer, Heidelberg (1998). https://doi.org/10.1007/BFb0054283
6. Berghammer, R., Struth, G.: On automated program construction and verification. In: Bolduc, C., Desharnais, J., Ktari, B. (eds.) MPC 2010. LNCS, vol. 6120, pp. 22–41. Springer, Heidelberg (2010). https://doi.org/10.1007/978-3-642-13321-3_4
7. Bird, R., de Moor, O.: Algebra of Programming. Prentice Hall, Englewood Cliffs (1997)
8. Charguéraud, A., Pottier, F.: Verifying the correctness and amortized complexity of a union-find implementation in separation logic with time credits. J. Autom. Reason. **62**(3), 331–365 (2019)

9. Conchon, S., Filliâtre, J.C.: A persistent union-find data structure. In: Dreyer, D., Russo, C. (eds.) ML 2007, pp. 37–45. ACM (2007)
10. Cormen, T.H., Leiserson, C.E., Rivest, R.L.: Introduction to Algorithms. MIT Press, Cambridge (1990)
11. Ehm, T.: Pointer Kleene algebra. In: Berghammer, R., Möller, B., Struth, G. (eds.) RelMiCS 2003. LNCS, vol. 3051, pp. 99–111. Springer, Heidelberg (2004). https://doi.org/10.1007/978-3-540-24771-5_9
12. Foster, J.N., Greenwald, M.B., Moore, J.T., Pierce, B.C., Schmitt, A.: Combinators for bidirectional tree transformations: a linguistic approach to the view-update problem. ACM Trans. Program. Lang. Syst. **29**(3:17), 1–65 (2007)
13. Galler, B.A., Fisher, M.J.: An improved equivalence algorithm. Commun. ACM **7**(5), 301–303 (1964)
14. Gondran, M., Minoux, M.: Graphs, Dioids and Semirings. Springer, Boston (2008). https://doi.org/10.1007/978-0-387-75450-5
15. Guttmann, W.: Verifying minimum spanning tree algorithms with Stone relation algebras. J. Log. Algebraic Methods Program. **101**, 132–150 (2018)
16. Hoare, C.A.R.: An axiomatic basis for computer programming. Commun. ACM **12**(10), 576–580/583 (1969)
17. Hoare, C.A.R.: Notes on data structuring. In: Dahl, O.J., Dijkstra, E.W., Hoare, C.A.R. (eds.) Structured Programming (Chapter 2), pp. 83–174. Academic Press, Cambridge (1972)
18. Hoare, C.A.R., He, J.: Unifying Theories of Programming. Prentice Hall Europe (1998)
19. Höfner, P., Möller, B.: Dijkstra, Floyd and Warshall Meet Kleene. Formal Aspects Comput. **24**(4), 459–476 (2012)
20. Kozen, D.: A completeness theorem for Kleene algebras and the algebra of regular events. Inf. Comput. **110**(2), 366–390 (1994)
21. Kozen, D.: Kleene algebra with tests. ACM Trans. Program. Lang. Syst. **19**(3), 427–443 (1997)
22. Lammich, P., Meis, R.: A separation logic framework for Imperative HOL. Archive of Formal Proofs (2012)
23. Maddux, R.D.: Relation-algebraic semantics. Theor. Comput. Sci. **160**(1–2), 1–85 (1996)
24. McCarthy, J.: Towards a mathematical science of computation. In: Popplewell, C.M. (ed.) IFIP 1962. IFIP Congress Series, vol. 2, pp. 21–28. North-Holland Publishing Company, Amsterdam (1963)
25. Möller, B.: Derivation of graph and pointer algorithms. In: Möller, B., Partsch, H., Schuman, S. (eds.) Formal Program Development. LNCS, vol. 755, pp. 123–160. Springer, Heidelberg (1993). https://doi.org/10.1007/3-540-57499-9_19
26. Möller, B.: Towards pointer algebra. Sci. Comput. Program. **21**(1), 57–90 (1993)
27. Möller, B.: Calculating with pointer structures. In: Bird, R., Meertens, L.G.L.T. (eds.) Algorithmic Languages and Calculi 1997. IFIP Conference Proceedings, vol. 95, pp. 24–48. Chapman and Hall, London (1997)
28. Möller, B.: Calculating with acyclic and cyclic lists. Inf. Sci. **119**(3–4), 135–154 (1999)
29. Nipkow, T.: Winskel is (almost) right: towards a mechanized semantics textbook. Formal Aspects Comput. **10**(2), 171–186 (1998)
30. Nipkow, T.: Hoare logics in Isabelle/HOL. In: Schwichtenberg, H., Steinbrüggen, R. (eds.) Proof and System-Reliability, pp. 341–367. Kluwer Academic Publishers, Dordrecht (2002)

31. Reynolds, J.C.: Reasoning about arrays. Commun. ACM **22**(5), 290–299 (1979)
32. Schmidt, G., Ströhlein, T.: Relationen und Graphen. Springer, Heidelberg (1989). https://doi.org/10.1007/978-3-642-83608-4
33. Spivey, J.M.: The Z Notation: A Reference Manual. Prentice Hall, Englewood Cliffs (1989)
34. Tarski, A.: On the calculus of relations. J. Symb. Logic **6**(3), 73–89 (1941)

A Hierarchy of Algebras for Boolean Subsets

Walter Guttmann[1](✉) and Bernhard Möller[2]

[1] Department of Computer Science and Software Engineering, University of Canterbury, Christchurch, New Zealand
walter.guttmann@canterbury.ac.nz

[2] Institut für Informatik, Universität Augsburg, Augsburg, Germany
bernhard.moeller@informatik.uni-augsburg.de

Abstract. We present a collection of axiom systems for the construction of Boolean subalgebras of larger overall algebras. The subalgebras are defined as the range of a complement-like operation on a semilattice. This technique has been used, for example, with the antidomain operation, dynamic negation and Stone algebras. We present a common ground for these constructions based on a new equational axiomatisation of Boolean algebras. All results are formally proved in Isabelle/HOL.

1 Introduction

Boolean algebras abound in formal approaches to program semantics as well as algorithm derivation and verification. Often such an algebra arises as a subalgebra of some overall algebra for the problem at hand. There are various methods of defining a Boolean substructure, for example, introducing a special type or sort for the subalgebra and then stipulating one of the standard Boolean algebra axiom sets for it. However, the extra type may get into the way of automatic verification with tools that only support a single sort. Then the Boolean sort has to be simulated by a characterising predicate, and many otherwise equational formulas need to be enriched by a premise involving that predicate. This complicates specifications and may hamper efficient automatic treatment.

Therefore a different approach has been studied: enrich the algebra with a special operation leading into the intended subalgebra and add sufficiently many axioms to guarantee that the range of that operation has a Boolean structure. Examples for this are the antidomain operation in idempotent (left) semirings [10–12], dynamic negation [21], the operation yielding tests in [17,19], and the pseudocomplement operation in Stone algebras [13,16,18].

The axiomatisations in these examples are all similar since they follow the same goal. The aim of the present paper is to exhibit a ground pattern for them and so allow a more unified treatment. For instance, the common structure of the seemingly disparate topics of Stone algebras and antidomain semirings is exhibited. To this end we first propose a succinct yet understandable set of axioms for Boolean algebras. Imposing these on the range of the complement operation, we

U. Fahrenberg et al. (Eds.): RAMiCS 2020, LNCS 12062, pp. 152–168, 2020.
https://doi.org/10.1007/978-3-030-43520-2_10

develop a hierarchy of algebras with a Boolean subalgebra and further structure overall. The hierarchy ultimately specialises to antidomain semirings and Stone algebras.

The contributions of this paper are as follows:

- Formally verified proofs of Byrne's axiomatisations of Boolean algebras in Sects. 4.1 and 4.2.
- A new and formally verified axiomatisation of Boolean algebras, which is equational and based on join and complement, in Sect. 4.3.
- A hierarchy of algebras each with a subset that forms a Boolean algebra and successively stronger assumptions for the overall set in Sect. 6. Stone algebras arise as a specialisation of this hierarchy in Sect. 7. One of the algebras corresponds to antidomain semirings as shown in Sect. 8.

All results have been formally verified in Isabelle/HOL [31]. Due to their extent the proofs are omitted in this paper. They can be found in the Isabelle/HOL theory file at http://www.csse.canterbury.ac.nz/walter.guttmann/algebra/.

In Sects. 3 and 4 we review various axiomatisations of Boolean algebras from the literature and present a new equational one tailored to our needs. Section 5 adapts this for the above-mentioned construction of Boolean subalgebras of larger overall algebras. In Sect. 6 we add successively stronger assumptions to the overall algebra. Sections 7 and 8 show how Stone algebras and antidomain semirings fit into this hierarchy.

2 Related Work

Boolean algebras have been extensively studied in the literature. In the following we discuss a selection of related works.

Some approaches build Boolean algebras on a hierarchy of more basic algebraic structures, for example, as complemented distributive lattices [2]. Other approaches are based on fewer operations and axioms, and introduce further operations of Boolean algebras by definitions. For example, one of Huntington's axiomatisations uses just the operations of join and complement with three equational axioms [22].

Huntington postulates that join is associative and commutative, but the third axiom is quite complex and not handy for manual proofs. There have been attempts to replace this axiom. Byrne [5] substitutes an equivalence, as detailed in Sect. 4, and also combines associativity and commutativity into one equational axiom. A related axiomatisation was proposed by Frink [14]. A later axiomatisation based on join and complement [28] uses the following two equations:

$$\overline{(\overline{x} \sqcup y)} \sqcup x = x \qquad\qquad \overline{(\overline{x} \sqcup y)} \sqcup (z \sqcup y) = y \sqcup (z \sqcup x)$$

Here again the second axiom is not easy to explain. A single-equation axiomatisation in terms of the Sheffer stroke or NAND operation $|$ was given in [27]:

$$(x|((y|x)|x))|(y|(z|x)) = y$$

However, it seems too complex for practical purposes.

In Sect. 4.3 we present an axiomatisation in which we try to strike a balance between simplicity/understandability and small number of axioms.

Axioms for domain and antidomain in idempotent semirings and weaker semiring structures have been studied, for example, in [9–12]. Axioms for these operations in semigroups and monoids have been studied, for example, in [7,23].

3 Boolean Algebras

In this section we present Huntington's axioms for Boolean algebras and discuss how Boolean algebras are implemented in Isabelle/HOL.

3.1 Huntington's Axioms

Huntington gave the following axiomatisation of Boolean algebras [22]. It is based only on join and complement.

Definition 1. *A* Boolean algebra *is a set $S \neq \emptyset$ with a binary operation $\sqcup$ and a unary operation $\overline{}$ such that, for all $x, y, z \in S$,*

$$x \sqcup (y \sqcup z) = (x \sqcup y) \sqcup z$$
$$x \sqcup y = y \sqcup x$$
$$x = \overline{\overline{x} \sqcup y} \sqcup \overline{\overline{x} \sqcup \overline{y}}$$

The operation $\sqcup$ is called join and the operation $\overline{}$ is called complement. In a Boolean algebra, $x \sqcup \overline{x} = y \sqcup \overline{y}$ for all $x, y \in S$. Hence the order $\sqsubseteq$, the strict order $\sqsubset$, the meet operation $\sqcap$, the difference $-$, the greatest element $\top$ and the least element $\bot$ can be defined as follows.

Definition 2. *An* extended Boolean algebra *is a Boolean algebra S with relations $\sqsubseteq$ and $\sqsubset$, binary operations $\sqcap$ and $-$, and constants $\top$ and $\bot$ such that, for all $x, y \in S$,*

$$x \sqsubseteq y \Leftrightarrow x \sqcup y = y \qquad x \sqcap y = \overline{\overline{x} \sqcup \overline{y}} \qquad \top = x \sqcup \overline{x}$$
$$x \sqsubset y \Leftrightarrow x \sqsubseteq y \wedge \neg(y \sqsubseteq x) \qquad x - y = \overline{\overline{x} \sqcup y} \qquad \bot = \overline{\top}$$

3.2 Boolean Algebras in Isabelle/HOL

We explain the hierarchy of orders and lattices in Isabelle/HOL up to Boolean algebras. These structures are implemented as type classes, which offer means to group operations and axioms, arrange them in hierarchies, dynamically inherit results, and exhibit multiple instances [20]. Every class has a single type parameter, which can be instantiated with a HOL type. Types in HOL must not be empty.

A *partial order* $\sqsubseteq$ on a set $S \neq \emptyset$ is a reflexive, transitive and antisymmetric relation on S with associated strict order $\sqsubset$. This means, for all $x, y, z \in S$:

$$
\begin{aligned}
x &\sqsubseteq x \\
x \sqsubseteq y \wedge y \sqsubseteq z &\Rightarrow x \sqsubseteq z \\
x \sqsubseteq y \wedge y \sqsubseteq x &\Rightarrow x = y \\
x \sqsubset y &\Leftrightarrow x \sqsubseteq y \wedge \neg(y \sqsubseteq x)
\end{aligned}
$$

A *lattice* is a set S partially ordered by $\sqsubseteq$ where any two elements $x, y \in S$ have a least upper bound or join $x \sqcup y$ and a greatest lower bound or meet $x \sqcap y$. This means, for all $x, y, z \in S$:

$$
\begin{aligned}
x &\sqsubseteq x \sqcup y & x \sqcap y &\sqsubseteq x \\
y &\sqsubseteq x \sqcup y & x \sqcap y &\sqsubseteq y \\
x \sqsubseteq z \wedge y \sqsubseteq z \Rightarrow x \sqcup y &\sqsubseteq z & z \sqsubseteq x \wedge z \sqsubseteq y \Rightarrow z &\sqsubseteq x \sqcap y
\end{aligned}
$$

A *bounded lattice* is a lattice S with a least element $\bot$ and a greatest element $\top$. This means, for all $x \in S$:

$$
\bot \sqsubseteq x \qquad\qquad x \sqsubseteq \top
$$

A lattice S is *distributive* if the following axiom holds for all $x, y, z \in S$:

$$
x \sqcup (y \sqcap z) = (x \sqcup y) \sqcap (x \sqcup z)
$$

A *Boolean algebra* is a bounded distributive lattice S with a complement $\overline{}$ and a difference $-$ satisfying, for all $x, y \in S$:

$$
\begin{aligned}
x \sqcup \overline{x} &= \top \\
x \sqcap \overline{x} &= \bot \\
x - y &= x \sqcap \overline{y}
\end{aligned}
$$

The above axiomatisation is equivalent to the extended Boolean algebras based on Huntington's axioms. This has been proved in Isabelle/HOL in [33], which also shows the equivalence to Robbins algebras and to an axiomatisation basing the lattice structure on $\sqcup$ and $\sqcap$ rather than $\sqsubseteq$.

Next we describe Stone algebras. Previous work extended the Isabelle/HOL hierarchy by various pseudocomplemented algebras [18]. Their place is between bounded (distributive) lattices and Boolean algebras.

A *(distributive) p-algebra* is a bounded (distributive) lattice S with a unary pseudocomplement $\overline{}$ satisfying, for all $x, y \in S$:

$$
x \sqcap y = \bot \Leftrightarrow x \sqsubseteq \overline{y}
$$

A *Stone algebra* is a distributive p-algebra S satisfying the following equation for all $x \in S$:

$$
\overline{x} \sqcup \overline{\overline{x}} = \top
$$

An *extended Stone algebra* adds to a Stone algebra S a difference $-$ satisfying, for all $x, y \in S$:

$$x - y = x \sqcap \overline{y}$$

To simplify comparisons, we provide this and similar extensions of algebras to obtain the signature $(S, \sqsubseteq, \sqsubset, \sqcup, \sqcap, -, \overline{}, \bot, \top)$ used by Isabelle/HOL. Adding the axiom $x = \overline{\overline{x}}$ to extended Stone algebras gives extended Boolean algebras.

4 Alternative Axiomatisations of Boolean Algebras

In this section we consider three axiomatisations of Boolean algebras, which are based only on join and complement, as are Huntington's axioms. The first two are from the literature and the third is new. A motivation for these versions is that the axioms are easier to understand than Huntington's third axiom.

4.1 Lee Byrne's Formulation A

The following axiomatisation is from [5, Formulation A]; see also [14]. It replaces Huntington's third axiom with an equivalence. The formulas in the equivalence express $y \sqsubseteq x$ in two different ways, noting that $z \sqcup \overline{z}$ represents $\top$.

Theorem 3. *The structure $(S, \sqcup, \overline{})$ is a Boolean algebra if and only if, for all $x, y, z \in S$,*

$$x \sqcup (y \sqcup z) = (x \sqcup y) \sqcup z$$
$$x \sqcup y = y \sqcup x$$
$$x \sqcup \overline{y} = z \sqcup \overline{z} \Leftrightarrow x \sqcup y = x$$

□

4.2 Lee Byrne's Formulation B

The following axiomatisation is from [5, Formulation B]. It combines associativity and commutativity into one axiom.

Theorem 4. *The structure $(S, \sqcup, \overline{})$ is a Boolean algebra if and only if, for all $x, y, z \in S$,*

$$(x \sqcup y) \sqcup z = (y \sqcup z) \sqcup x$$
$$x \sqcup \overline{y} = z \sqcup \overline{z} \Leftrightarrow x \sqcup y = x$$

□

4.3 An Equational Axiomatisation Based on Semilattices

The following new axiomatisation is based on semilattices, that is, sets with an associative, commutative and idempotent $\sqcup$ operation. We add the double complement rule and that $\top$ is unique. The final axiom is similar to the logical statement $P \vee Q = P \vee (\neg P \wedge Q)$. The dual of the final axiom is used in [1] for an axiomatisation of pseudocomplemented semilattices.

Theorem 5. *The structure* $(S, \sqcup, ^{-})$ *is a Boolean algebra if and only if, for all* $x, y, z \in S$,

$$x \sqcup (y \sqcup z) = (x \sqcup y) \sqcup z$$
$$x \sqcup y = y \sqcup x$$
$$x \sqcup x = x$$
$$\overline{\overline{x}} = x$$
$$x \sqcup \overline{x} = y \sqcup \overline{y}$$
$$x \sqcup \overline{\overline{x} \sqcup y} = x \sqcup \overline{y}$$

□

This axiomatisation is equational with few and simple axioms, which is useful for both manual and automated proofs. Counterexamples generated by Nitpick [4] witness that the axioms are independent of each other. The smallest counterexample for independence of associativity the tool found has 16 elements.

5 Subset Boolean Algebras

In a number of situations a subset of the elements under consideration forms a Boolean algebra, whereas a more general structure is desired for the overall set. An example is that of Kleene algebras with tests [24] where the overall structure forms a Kleene algebra (with operations for join, composition and iteration) and a designated subset of tests forms a Boolean algebra (in which meet coincides with composition). In computation models, elements of the Kleene algebra model state changes while tests model conditions on states. Another example is that of weighted graphs [18] where the overall structure forms a Stone relation algebra and a subset forms a relation algebra. It uses the well-known fact that the elements of a Stone algebra satisfying $x = \overline{\overline{x}}$ form a Boolean subalgebra [16]. Elements of the Stone relation algebra model graphs with edge weights while elements of the Boolean subset model unweighted graphs. In both examples it is convenient to have a single-sorted structure, where the Boolean algebra axioms hold only for a subset of elements of the overall algebra.

In the remainder of this paper we study axiomatisations describing the common structure underlying these situations. Our most general setting, taken from [19], is a set S with a subset $S' \subseteq S$ of elements that forms a Boolean algebra. We axiomatise that Boolean algebra structure using the $\sqcup$ and $^{-}$ operations. To obtain a single-sorted structure in Isabelle/HOL these operations are introduced on the overall set S, however their axioms are restricted to the subset S'.

This first building block B_0 in our hierarchy of structures results by applying Huntington's axioms [22] to the range S' of operation $^{-}$, which serves as complement on the range. It provides a Boolean algebra structure on S' without imposing any further constraints on the overall set. Building block B_0 is used as a reference in the subsequent development and to prove results to be inherited by further, more special structures. Results that hold in Boolean algebras can be stated for the subset S' by using elements from the range of $^{-}$ instead of arbitrary elements; they are derived in the order used by [25].

Given a set S with a unary operation $\overline{}$ we write $S' = \{\overline{x} \mid x \in S\}$ for the range of $\overline{}$. The first three equations are Huntington's axioms for Boolean algebras applied to the range of $\overline{}$. The last equation states that S' is closed under $\sqcup$. Note that the behaviour of the operations on elements in $S \setminus S'$ is left unspecified by the axioms.

Definition 6. *A* B_0-algebra *is a set* $S \neq \emptyset$ *with a binary operation* $\sqcup$ *and a unary operation* $\overline{}$ *such that, for all* $x, y, z \in S$,

$$\overline{x} \sqcup (\overline{y} \sqcup \overline{z}) = (\overline{x} \sqcup \overline{y}) \sqcup \overline{z}$$
$$\overline{x} \sqcup \overline{y} = \overline{y} \sqcup \overline{x}$$
$$\overline{x} = \overline{\overline{\overline{x}} \sqcup \overline{y}} \sqcup \overline{\overline{\overline{x}} \sqcup \overline{\overline{y}}}$$
$$\overline{x} \sqcup \overline{y} = \overline{\overline{\overline{x} \sqcup \overline{y}}}$$

The remaining operations of Boolean algebras can be defined in terms of $\sqcup$ and $\overline{}$ on S'.

Definition 7. *An* extended B_0-algebra *is a* B_0*-algebra* S *with relations* $\sqsubseteq$ *and* $\sqsubset$*, binary operations* $\sqcap$ *and* $-$*, and constants* $\top$ *and* $\bot$ *such that, for all* $x, y \in S$,

$$\overline{x} \sqsubseteq \overline{y} \Leftrightarrow \overline{x} \sqcup \overline{y} = \overline{y} \qquad \overline{x} \sqcap \overline{y} = \overline{\overline{\overline{x}} \sqcup \overline{\overline{y}}} \qquad \top = \overline{x} \sqcup \overline{\overline{x}}$$
$$\overline{x} \sqsubset \overline{y} \Leftrightarrow \overline{x} \sqsubseteq \overline{y} \wedge \neg(\overline{y} \sqsubseteq \overline{x}) \qquad \overline{x} - \overline{y} = \overline{\overline{\overline{x}} \sqcup \overline{y}} \qquad \bot = \overline{\top}$$

The following result confirms that we obtain the desired Boolean algebra structure on S'.

Theorem 8.

1. *Let* $(S, \sqcup, \overline{})$ *be a* B_0*-algebra. Then* $(S', \sqcup, \overline{})$ *is a Boolean algebra.*
2. *Let* $(S, \sqsubseteq, \sqsubset, \sqcup, \sqcap, -, \overline{}, \bot, \top)$ *be an extended* B_0*-algebra.*
   Then $(S', \sqsubseteq, \sqsubset, \sqcup, \sqcap, -, \overline{}, \bot, \top)$ *is an extended Boolean algebra.* □

Structural results about extended algebras, such as part 2 of Theorem 8, enable the use of existing Isabelle/HOL theories for Boolean algebras.

6 Subset Boolean Algebras with Additional Structure

We now discuss axioms that make the range of $\overline{}$ a Boolean algebra, but add further properties that are common to the intended models. In these models, the unary operation can be a complement, a pseudocomplement or the antidomain operation. For simplicity, we mostly call $\overline{}$ the 'complement'.

We first look at structures based only on join and complement, and then add axioms for the remaining operations of Boolean algebras. In the intended models, the operation $\sqcap$, which is the meet on the range of the operation $\overline{}$, can be the meet in the overall algebra or the composition operation of a (left) semiring. For simplicity, we mostly call $\sqcap$ the 'meet'.

6.1 Assumptions Derived from the New Axiomatisation

The axioms of building block B_1 are based on the ones in Sect. 4.3. We follow the idea of applying the Boolean algebra axioms to the range of the operation $^-$, but we only do this where necessary for the intended models. For example, the intended models have a semilattice structure on the overall algebra, not just on the Boolean subset. In contrast, the double complement axiom only applies to the subset, not to the overall algebra.

Definition 9. *A* B_1-algebra *is a set* $S \neq \emptyset$ *with a binary operation* $\sqcup$ *and a unary operation* $^-$ *such that, for all* $x, y, z \in S$,

$$
\begin{aligned}
x \sqcup (y \sqcup z) &= (x \sqcup y) \sqcup z \\
x \sqcup y &= y \sqcup x \\
x \sqcup x &= x \\
\overline{\overline{\overline{x}}} &= \overline{x} \\
\overline{x \sqcup \overline{x}} &= \overline{y \sqcup \overline{y}} \\
\overline{x} \sqcup \overline{\overline{x} \sqcup y} &= \overline{x} \sqcup \overline{y}
\end{aligned}
$$

Using a similar approach, the remaining operations of Boolean algebras are introduced as follows.

Definition 10. *An* extended B_1-algebra *is a* B_1*-algebra* S *with relations* $\sqsubseteq$ *and* $\sqsubset$*, binary operations* $\sqcap$ *and* $-$*, and constants* $\top$ *and* $\bot$ *such that, for all* $x, y \in S$,

$$
\begin{aligned}
&x \sqsubseteq y \Leftrightarrow x \sqcup y = y & \qquad \overline{x} \sqcap \overline{y} &= \overline{\overline{\overline{x}} \sqcup \overline{\overline{y}}} & \qquad \bot &= \overline{x \sqcup \overline{x}} \\
&x \sqsubset y \Leftrightarrow x \sqsubseteq y \wedge \neg(y \sqsubseteq x) & \qquad \overline{x} - \overline{y} &= \overline{\overline{\overline{x}} \sqcup \overline{y}} & \qquad \top &= \overline{\bot}
\end{aligned}
$$

The following result shows that B_1-algebras specialise B_0-algebras. Hence we again obtain the Boolean algebra structure on S'.

Theorem 11.

1. *Every* B_1*-algebra is a* B_0*-algebra.*
2. *Every extended* B_1*-algebra is an extended* B_0*-algebra.* □

6.2 Stronger Assumptions Based on Join and Complement

In building block B_2 we add axioms covering further properties common to structures with antidomain or (pseudo)complement. In particular, they allow us to derive that $^-$ is antitone and satisfies one of De Morgan's laws in the overall algebra. Moreover, double complement distributes over $\sqcup$ in the overall algebra.

Definition 12. *A* B_2-algebra *is a set* $S \neq \emptyset$ *with a binary operation* $\sqcup$ *and a unary operation* $^{-}$ *such that, for all* $x, y, z \in S$,

$$x \sqcup (y \sqcup z) = (x \sqcup y) \sqcup z$$
$$x \sqcup y = y \sqcup x$$
$$x \sqcup x = x$$
$$x \sqcup \overline{y \sqcup \overline{\overline{y}}} = x$$
$$\overline{x \sqcup y} = \overline{\overline{\overline{x}} \sqcup \overline{\overline{y}}}$$
$$\overline{x} \sqcup \overline{\overline{\overline{x}} \sqcup y} = \overline{x} \sqcup \overline{y}$$

An extended B_2-algebra is obtained from this by adding the operations and axioms given in Definition 10. The following result shows consequences.

Theorem 13.

1. *Every (extended)* B_2*-algebra is an (extended)* B_1*-algebra.*
2. *Let* S *be a* B_2*-algebra. Then, for all* $x, y \in S$,

$$\overline{x \sqcup y} \sqcup \overline{x \sqcup \overline{y}} = \overline{x} \qquad \overline{\overline{x \sqcup y}} = \overline{\overline{x}} \sqcup \overline{\overline{y}}$$
$$\overline{\overline{x} \sqcup \overline{y}} \sqcup \overline{\overline{x} \sqcup y} = \overline{\overline{x}}$$

3. *Let* S *be an extended* B_2*-algebra. Then, for all* $x, y \in S$,

$$x \sqsubseteq y \Rightarrow \overline{y} \sqsubseteq \overline{x} \qquad \overline{x \sqcup y} = \overline{x} \sqcap \overline{y}$$
$$x \sqsubseteq y \Rightarrow \overline{\overline{x}} \sqsubseteq \overline{\overline{y}}$$

□

6.3 Axioms for Meet

In building block B_3 we add axioms of $\sqcap$ covering further properties common to the antidomain and pseudocomplement instances. We omit the left distributivity rule and the right zero rule as they do not hold in some models. For the same reason, the operation $\sqcap$ does not have to be commutative.

To simplify comparison with the antidomain model we supply a translation table for the operations and relations, where $+$, $\cdot$, 0 and 1 are operations known from semirings, a stands for antidomain and d for domain:

extended B_0-algebra	antidomain model
$\sqcup$	$+$
$\sqcap$	$\cdot$
$\overline{}$	a
$\overline{\overline{}}$	d
$\bot$	0
$\top$	1
$\sqsubseteq$	$\leq$
$\sqsubset$	$<$

We frequently write xy instead of $x \cdot y$. The additional equations in the following definition are just translations of the formulas on the left and not part of the axiomatisation. We translate results similarly in the remainder of this paper.

Definition 14. *An* extended B_3-algebra *is an extended B_2-algebra S such that, for all $x, y, z \in S$,*

$$x \sqcap (y \sqcap z) = (x \sqcap y) \sqcap z \qquad x(yz) = (xy)z$$

$$(x \sqcup y) \sqcap z = (x \sqcap z) \sqcup (y \sqcap z) \qquad (x + y)z = (xz) + (yz)$$

$$\overline{x} \sqcap x = \bot \qquad a(x)x = 0$$

$$\top \sqcap x = x \qquad 1x = x$$

$$\overline{x \sqcap \overline{\overline{y}}} = \overline{x \sqcap y} \qquad a(x \cdot d(y)) = a(xy)$$

The following result gives derived properties of $\sqcap$.

Theorem 15. *Let S be an extended B_3-algebra. Then, for all $x, y, z \in S$,*

$$x \sqsubseteq y \Rightarrow x \sqcap z \sqsubseteq y \sqcap z \qquad x \leq y \Rightarrow xz \leq yz$$

$$\bot \sqcap x = \bot \qquad 0x = 0$$

$$\overline{\overline{x}} \sqcap x = x \qquad d(x)x = x$$

$$\overline{\overline{\overline{x} \sqcap y}} = \overline{x} \sqcap \overline{\overline{y}} \qquad d(a(x)y) = a(x)d(y)$$

$$\overline{\overline{\overline{\overline{x}} \sqcap y}} = \overline{\overline{x}} \sqcap \overline{\overline{y}} \qquad d(d(x)y) = d(x)d(y)$$

□

Counterexamples generated by Nitpick witness that

$$x \sqcap \top = x \qquad x1 = x$$

$$x \sqcap y = y \sqcap x \qquad xy = yx$$

$$x \sqsubseteq y \Rightarrow z \sqcap x \sqsubseteq z \sqcap y \qquad x \leq y \Rightarrow zx \leq zy$$

do not hold for some extended B_3-algebra S and some $x, y, z \in S$. Hence our axiomatisation also covers structures weaker than idempotent left semirings (where the first and third of these properties are required).

6.4 Stronger Assumptions for Meet

The following axioms of building block B_4 also hold in the pseudocomplement and antidomain models, but follow from the axioms of B_5-algebras introduced below.

Definition 16. *An* extended B_4-algebra *is an extended* B_3*-algebra* S *such that, for all* $x, y, z \in S$,

$$x \sqcap \top = x \qquad\qquad x1 = x$$
$$x \sqsubseteq y \Rightarrow z \sqcap x \sqsubseteq z \sqcap y \qquad\qquad x \leq y \Rightarrow zx \leq zy$$

Counterexamples generated by Nitpick witness that

$$x \sqcup \top = \top \qquad\qquad x + 1 = 1$$
$$x \sqcap \bot = \bot \qquad\qquad x0 = 0$$
$$x \sqcap (y \sqcup z) = (x \sqcap z) \sqcup (y \sqcap z) \qquad\qquad x(y + z) = (xz) + (yz)$$
$$x \sqcap y = \bot \Leftrightarrow x \sqsubseteq \overline{y} \qquad\qquad xy = 0 \Leftrightarrow x \leq a(y)$$

do not hold for some extended B_4-algebra S and some $x, y, z \in S$.

We will come back to B_4-algebras when we study the antidomain model in more detail in Sect. 8.

7 Subset Boolean Algebras in Stone Algebras

In building block B_5 we specialise $\sqcap$ to meet and $\overline{}$ to pseudocomplement.

Definition 17. *An* extended B_5-algebra *is an extended* B_3*-algebra* S *such that, for all* $x, y \in S$,

$$x \sqcap y = y \sqcap x$$
$$x \sqcap (x \sqcup y) = x$$

The following result shows that B_5-algebras correspond to Stone algebras. Parts 2 and 3 do not combine to an equivalence because the difference operation $-$ is axiomatised only on S' in B_5-algebras but on S in Stone algebras.

Theorem 18.

1. *Every extended* B_5*-algebra is an extended* B_4*-algebra.*
2. *Every extended* B_5*-algebra is a Stone algebra.*
3. *Every extended Stone algebra is an extended* B_5*-algebra.* □

8 Antidomain Semirings

In this section we study the connection to antidomain semirings, which, in particular, are semilattices. We show that they correspond to extended B_4-algebras. We start by introducing idempotent left semirings (IL-semirings).

Definition 19. *An* IL-semiring *is a set $S \neq \emptyset$ with relations $\sqsubseteq$ and $\sqsubset$, binary operations $\sqcup$ and $\sqcap$, and constants $\top$ and $\bot$ such that, for all $x, y, z \in S$,*

$$\begin{array}{ccc} x \sqcup (y \sqcup z) = (x \sqcup y) \sqcup z & x \sqsubseteq y \Leftrightarrow x \sqcup y = y & x \sqcap (y \sqcap z) = (x \sqcap y) \sqcap z \\ x \sqcup y = y \sqcup x & x \sqsubset y \Leftrightarrow x \sqsubseteq y \wedge \neg(y \sqsubseteq x) & \top \sqcap x = x \\ x \sqcup x = x & x \sqsubseteq y \Rightarrow z \sqcap x \sqsubseteq z \sqcap y & x \sqcap \top = x \\ x \sqcup \bot = x & (x \sqcup y) \sqcap z = (x \sqcap z) \sqcup (y \sqcap z) & \bot \sqcap x = \bot \end{array}$$

$$\begin{array}{ccc} x + (y + z) = (x + y) + z & x \leq y \Leftrightarrow x + y = y & x(yz) = (xy)z \\ x + y = y + x & x < y \Leftrightarrow x \leq y \wedge \neg(y \leq x) & 1x = x \\ x + x = x & x \leq y \Rightarrow zx \leq zy & x1 = x \\ x + 0 = x & (x + y)z = (xz) + (yz) & 0x = 0 \end{array}$$

An IL-semiring S is partially ordered by $\sqsubseteq$.

We now introduce the notion of tests, using semiring notation for ease of reference. Our presentation follows [29]. Tests algebraically represent conditions in programs and can be used to construct conditionals, while-loops, assertions and related statements. All these statements have in common that they check if a condition is satisfied in the current state, but this check does not modify the state. A condition p acts as an identity on states that satisfy p, so it is reasonable to model it algebraically by an element below 1 which represents 'do nothing'.

In an IL-semiring a *test* is an element p that has a *complement* q relative to 1, that is, $p + q = 1$ and $p \cdot q = 0 = q \cdot p$. In particular, 0 and 1 are tests. By the requirement $p + q = 1$ every test is a *sub-identity*, that is, satisfies $p \leq 1$. The set of all tests of an IL-semiring S is denoted by $test(S)$. It is not hard to show that a complement of p is unique if it exists; we will denote it by $\neg p$.

Next we introduce an abstract domain operation d that assigns to a semiring element, which represents a set of transitions from states to states, the test that describes precisely its possible starting states.

As a motivation, consider the IL-semiring of binary relations over a set M, with union as $+$, relational composition as $\cdot$, the identity relation as 1 and the empty relation as 0. Then the domain $d(R)$ of a binary relation $R \subseteq M \times M$ is the set $\{u \in M \mid \exists v \in M : (u, v) \in R\}$. In the semiring setting, this set should be represented as a test in the IL-semiring of binary relations, that is, as the sub-identity $d(R) = \{(u, u) \in M \times M \mid \exists v \in M : (u, v) \in R\}$.

Abstracting from the relational IL-semiring to a general one, we arrive at the following definitions [8,29]. A *left prepredomain semiring* is an IL-semiring S with an additional *prepredomain operation* $d : S \to test(S)$ satisfying

$$x \leq d(x) \cdot x \qquad \text{(d1)}$$

for all $x \in S$. We call d a *predomain operation* if additionally

$$d(p \cdot x) \leq p \tag{d2}$$

for all $x \in S$ and $p \in test(S)$. Finally, a predomain operation d is called a *domain operation* if it satisfies the *locality* axiom

$$d(x \cdot d(y)) \leq d(x \cdot y) \tag{d3}$$

for all $x, y \in S$. See [9] for axioms (d1), (d2) and (d3) in idempotent semirings.

In IL-semirings, axioms (d1), (d2) and (d3) are independent of each other. However, (d1) and (d3) together with the assumption $d(0) = 0$ imply (d2). Moreover, having a predomain operation d implies that d is surjective and $test(S)$ forms a Boolean algebra [29, Theorem 2.4.6 items 1 and 8]. Predomain is studied since in a number of cases it already suffices for the purpose at hand. For example, the algebraic soundness proof of Hoare logic in [30] does not need (d3); that axiom is only used in the proof of relative completeness of the logic. Therefore we give an antidomain analogue of (d2) below.

Technically, by referring to $test(S)$ the above axioms have a 'two-sorted' flavour. So there have been approaches [10–12] to give a different axiomatisation in terms of a combination of d and $\neg$, namely the *antidomain operation* $a(x) = \neg d(x)$, and to leave $test(S)$ unmentioned in the axioms. Originally there were three axioms for antidomain corresponding roughly to the test property, (d1) and (d3). In the present paper we also discuss the role of a further axiom corresponding to (d2); here we can show that the original antidomain axioms imply that without an additional assumption corresponding to $d(0) = 0$.

To do this we first introduce *prepreantidomain* in PPA-semirings, *preantidomain* in PA-semirings and *antidomain* in A-semirings, and afterwards relate them to our general treatment of sets with a Boolean subset.

We start with prepreantidomain using axioms that correspond to (d1) and the test property. These are axioms (BD1) and (BD3) of [11]. In the antidomain model, $d(x) = a(a(x))$.

Definition 20. *A* PPA-semiring *is an IL-semiring S with a unary operation $\overline{}$ such that, for all $x \in S$,*

$$\overline{x} \sqcap x = \bot \qquad\qquad a(x)x = 0$$
$$\overline{x} \sqcup \overline{\overline{x}} = \top \qquad\qquad a(x) + d(x) = 1$$

It is somewhat unexpected that the simple PPA-semiring axioms already imply a rich set of consequences shown in the following result. Many of them are concerned with how tests interact with each other and general elements under meet/composition.

Theorem 21. *Let S be a PPA-semiring. Then, for all $x, y \in S$,*

$$\begin{array}{cccc} \overline{\bot} = \top & \overline{\overline{\overline{x}}} = \overline{x} & x \sqsubseteq \overline{\overline{x}} \sqcap x & \overline{\overline{x}} \sqsubseteq \overline{\overline{y}} \Rightarrow \overline{y} \sqsubseteq \overline{x} \\ \overline{\top} = \bot & \overline{x} \sqcap \overline{\overline{x}} = \bot & \overline{x} \sqcap \overline{y} = \overline{y} \sqcap \overline{x} & \overline{x} \sqsubseteq \overline{y} \Rightarrow \overline{x} \sqcap y = \bot \\ a(0) = 1 & a(d(x)) = a(x) & x \leq d(x)x & d(x) \leq d(y) \Rightarrow a(y) \leq a(x) \\ a(1) = 0 & a(x)d(x) = 0 & a(x)a(y) = a(y)a(x) & a(x) \leq a(y) \Rightarrow a(x)y = 0 \end{array}$$

□

To obtain preantidomain we add an axiom that corresponds to (d2). This axiom facilitates the import/export of composition with a test under a domain.

Definition 22. *A* PA-semiring *is a PPA-semiring S such that, for all $x, y \in S$,*

$$\overline{\overline{x}} \sqsubseteq \overline{\overline{x} \sqcap y} \qquad\qquad d(x) \leq a(a(x)y)$$

Consequences of this additional axiom are given in the following result. They are mostly concerned with the (anti)domain of joins and the (anti)domain of meets/compositions where the first component is a test.

Theorem 23. *Let S be a PA-semiring. Then, for all $x, y \in S$,*

$$\begin{array}{ccc} \overline{x} \sqcap \overline{y} = \overline{x \sqcup y} & \overline{\overline{\overline{\overline{x}} \sqcap y}} \sqsubseteq \overline{\overline{x}} & x \sqsubseteq y \Rightarrow \overline{y} \sqsubseteq \overline{x} \\ \overline{\overline{x \sqcup y}} = \overline{\overline{x}} \sqcup \overline{\overline{y}} & \overline{x \sqcap \overline{\overline{y}}} \sqsubseteq \overline{x \sqcap y} & \overline{x} \sqsubseteq \overline{y} \Leftrightarrow \overline{x} \sqcap y = \bot \\ \overline{x} \sqcup \overline{y} = \overline{\overline{\overline{x}} \sqcap \overline{\overline{y}}} & \overline{\overline{\overline{x} \sqcap y}} = \overline{x} \sqcap \overline{\overline{y}} & \end{array}$$

$$\begin{array}{ccc} a(x)a(y) = a(x+y) & d(d(x)y) \leq d(x) & x \leq y \Rightarrow a(y) \leq a(x) \\ d(x+y) = d(x) + d(y) & a(x \cdot d(y)) \leq a(xy) & a(x) \leq a(y) \Leftrightarrow a(x)y = 0 \\ a(x) + a(y) = a(d(x)d(y)) & d(a(x)y) = a(x)d(y) & \end{array}$$

□

To obtain antidomain, we finally add a version of (d3), called (BD2) in [11]. This axiom is concerned with the (anti)domain of meets/compositions where the second component is a test. In the terminology of [10], an A-semiring is an idempotent pre-semiring with 1 and δ that satisfies the basic Boolean domain axioms (BD1), (BD2) and (BD3).

Definition 24. *An* A-semiring *is a PPA-semiring S such that, for all $x, y \in S$,*

$$\overline{x \sqcap y} \sqsubseteq \overline{x \sqcap \overline{\overline{y}}} \qquad\qquad a(xy) \leq a(x \cdot d(y))$$

An A-algebra *is an A-semiring with a binary operation $-$ defined, for all $x, y \in S$, by*

$$\overline{x} - \overline{y} = \overline{\overline{\overline{x}} \sqcup \overline{y}}$$

Note that A-semirings are based on PPA-semirings. However, by the following result they form PA-semirings. Previous work has shown that (d2) follows if S is an A-semiring where $\sqcap$ distributes over $\sqcup$ and has $\bot$ as a zero (that is, a semiring not just an IL-semiring) [11]. Moreover, using results in [10] one can show that (d2) and the PA-semiring axiom follow also when only an IL-semiring is assumed. The result also locates A-algebras in our hierarchy of algebras.

Theorem 25.

1. *Every PA-semiring is a B_2-algebra.*
2. *Every A-semiring is a PA-semiring.*
3. *Every Stone algebra is an A-semiring.*
4. *S is an A-algebra if and only if S is an extended B_4-algebra.* □

Theorems 18 and 25 imply that every extended Stone algebra is an A-algebra.

9 Conclusion

We have presented a hierarchy of axiom systems as a common basis for approaches to induce a Boolean subalgebra in a larger overall algebra as the range of a complement-like operation. Except for the most basic axiomatisation, which imposes no extra structure beyond the Boolean subalgebra, the axioms assume that the overall algebra is a semilattice. The hierarchy has shed new light on the interconnections between several such approaches. The axioms are simple and perspicuous when translated into formulas of the respective theories. All of our axioms are (or can be written as) equations and hence well suited to mechanical support.

In situations which require a Boolean subalgebra our hierarchy offers a number of choices for axiom systems verified in Isabelle/HOL. Basing an axiomatisation on one of them eliminates the need to prove the intended Boolean laws for the substructure.

Working with Boolean algebras involves a choice about which operations to include in the signature and which to derive by definition. For example, [25] includes join and complement in the signature and derives meet, $\bot$ and $\top$, whereas [15] includes all of these in the signature. The standard type-class implementation of Boolean algebras in Isabelle/HOL has parameters for all of these operations, a binary difference and the orders $\sqsubseteq$ and $\sqsubset$. The separate treatment of extended structures in this paper reflects this.

Proving results such as Theorem 23 is typically highly automated in Isabelle/HOL using the built-in Sledgehammer tool [3,32]. It filters relevant lemmas, calls fully automated external theorem provers (such as E, Spass, Vampire) and SMT solvers (such as CVC4, Z3) and reconstructs proofs within Isabelle/HOL to avoid trusting external software. In several cases, Prover9 [26] was able to find a proof where the tools called by Sledgehammer failed. Since Prover9 is not integrated with Sledgehammer, we wrote a program that transforms the output generated by Prover9 to an Isabelle/HOL proof. The translation currently works for a limited range of proofs but could form the basis of an integration into Sledgehammer. Such an extension would be beneficial because Prover9 performs well for algebraic applications [6].

Acknowledgement. We thank Andreas Zelend and the anonymous referees for their helpful comments.

References

1. Balbes, R., Horn, A.: Stone lattices. Duke Math. J. **37**(3), 537–545 (1970)
2. Birkhoff, G.: Lattice Theory, Colloquium Publications, vol. XXV, 3rd edn. American Mathematical Society, Providence (1967)
3. Blanchette, J.C., Böhme, S., Paulson, L.C.: Extending Sledgehammer with SMT solvers. In: Bjørner, N., Sofronie-Stokkermans, V. (eds.) CADE 2011. LNCS (LNAI), vol. 6803, pp. 116–130. Springer, Heidelberg (2011). https://doi.org/10.1007/978-3-642-22438-6_11
4. Blanchette, J.C., Nipkow, T.: Nitpick: a counterexample generator for higher-order logic based on a relational model finder. In: Kaufmann, M., Paulson, L.C. (eds.) ITP 2010. LNCS, vol. 6172, pp. 131–146. Springer, Heidelberg (2010). https://doi.org/10.1007/978-3-642-14052-5_11
5. Byrne, L.: Two brief formulations of Boolean algebra. Bull. Am. Math. Soc. **52**(4), 269–272 (1946)
6. Dang, H.H., Höfner, P.: First-order theorem prover evaluation w.r.t. relation- and Kleene algebra. In: Berghammer, R., Möller, B., Struth, G. (eds.) PhD Programme at RelMiCS10/AKA5, pp. 48–52. Report 2008-04, Institut für Informatik, Universität Augsburg (2008)
7. Desharnais, J., Jipsen, P., Struth, G.: Domain and antidomain semigroups. In: Berghammer, R., Jaoua, A.M., Möller, B. (eds.) RelMiCS/AKA 2009. LNCS, vol. 5827, pp. 73–87. Springer, Heidelberg (2009). https://doi.org/10.1007/978-3-642-04639-1_6
8. Desharnais, J., Möller, B.: Fuzzifying modal algebra. In: Höfner, P., Jipsen, P., Kahl, W., Müller, M.E. (eds.) RAMICS 2014. LNCS, vol. 8428, pp. 395–411. Springer, Cham (2014). https://doi.org/10.1007/978-3-319-06251-8_24
9. Desharnais, J., Möller, B., Struth, G.: Kleene algebra with domain. ACM Trans. Comput. Logic **7**(4), 798–833 (2006)
10. Desharnais, J., Struth, G.: Domain axioms for a family of near-semirings. In: Meseguer, J., Roşu, G. (eds.) AMAST 2008. LNCS, vol. 5140, pp. 330–345. Springer, Heidelberg (2008). https://doi.org/10.1007/978-3-540-79980-1_25
11. Desharnais, J., Struth, G.: Modal semirings revisited. In: Audebaud, P., Paulin-Mohring, C. (eds.) MPC 2008. LNCS, vol. 5133, pp. 360–387. Springer, Heidelberg (2008). https://doi.org/10.1007/978-3-540-70594-9_19
12. Desharnais, J., Struth, G.: Internal axioms for domain semirings. Sci. Comput. Program. **76**(3), 181–203 (2011)
13. Frink, O.: Pseudo-complements in semi-lattices. Duke Math. J. **29**(4), 505–514 (1962)
14. Frink Jr., O.: Representations of Boolean algebras. Bull. Am. Math. Soc. **47**(10), 755–756 (1941)
15. Givant, S., Halmos, P.: Introduction to Boolean Algebras. Springer, New York (2009). https://doi.org/10.1007/978-0-387-68436-9
16. Grätzer, G.: Lattice Theory: First Concepts and Distributive Lattices. W. H. Freeman and Co., San Francisco (1971)
17. Guttmann, W.: Algebras for iteration and infinite computations. Acta Inf. **49**(5), 343–359 (2012)
18. Guttmann, W.: Verifying minimum spanning tree algorithms with Stone relation algebras. J. Log. Algebraic Methods Program. **101**, 132–150 (2018)
19. Guttmann, W., Struth, G., Weber, T.: Automating algebraic methods in Isabelle. In: Qin, S., Qiu, Z. (eds.) ICFEM 2011. LNCS, vol. 6991, pp. 617–632. Springer, Heidelberg (2011). https://doi.org/10.1007/978-3-642-24559-6_41

20. Haftmann, F., Wenzel, M.: Constructive type classes in Isabelle. In: Altenkirch, T., McBride, C. (eds.) TYPES 2006. LNCS, vol. 4502, pp. 160–174. Springer, Heidelberg (2007). https://doi.org/10.1007/978-3-540-74464-1_11
21. Hollenberg, M.: An equational axiomatization of dynamic negation and relational composition. J. Logic Lang. Inform. **6**(4), 381–401 (1997)
22. Huntington, E.V.: Boolean algebra. A correction. Trans. Am. Math. Soc. **35**(2), 557–558 (1933)
23. Jackson, M., Stokes, T.: Semilattice pseudo-complements on semigroups. Commun. Algebra **32**(8), 2895–2918 (2004)
24. Kozen, D.: Kleene algebra with tests. ACM Trans. Program. Lang. Syst. **19**(3), 427–443 (1997)
25. Maddux, R.D.: Relation-algebraic semantics. Theor. Comput. Sci. **160**(1–2), 1–85 (1996)
26. McCune, W.: Prover9 and Mace4 (2005–2010). https://www.cs.unm.edu/~mccune/prover9/. Accessed 16 Jan 2020
27. McCune, W., Veroff, R., Fitelson, B., Harris, K., Feist, A., Wos, L.: Short single axioms for Boolean algebra. J. Autom. Reason. **29**(1), 1–16 (2002)
28. Meredith, C.A., Prior, A.N.: Equational logic. Notre Dame J. Formal Logic **9**(3), 212–226 (1968)
29. Möller, B., Desharnais, J.: Basics of modal semirings and of Kleene/omega algebras. Report 2019-03, Institut für Informatik, Universität Augsburg (2019)
30. Möller, B., Struth, G.: Algebras of modal operators and partial correctness. Theor. Comput. Sci. **351**(2), 221–239 (2006)
31. Nipkow, T., Paulson, L.C., Wenzel, M.: Isabelle/HOL: A Proof Assistant for Higher-Order Logic. LNCS, vol. 2283. Springer, Heidelberg (2002). https://doi.org/10.1007/3-540-45949-9
32. Paulson, L.C., Blanchette, J.C.: Three years of experience with Sledgehammer, a practical link between automatic and interactive theorem provers. In: Sutcliffe, G., Ternovska, E., Schulz, S. (eds.) Proceedings of the 8th International Workshop on the Implementation of Logics, pp. 3–13 (2010)
33. Wampler-Doty, M.: A complete proof of the Robbins conjecture. Archive of Formal Proofs (2016, first version 2010)

Differential Hoare Logics and Refinement Calculi for Hybrid Systems with Isabelle/HOL

Simon Foster[1], Jonathan Julián Huerta y Munive[2(✉)], and Georg Struth[2]

[1] University of York, York, UK
[2] University of Sheffield, Sheffield, UK
jjhuertaymunive1@sheffield.ac.uk

Abstract. We present simple new Hoare logics and refinement calculi for hybrid systems in the style of differential dynamic logic. (Refinement) Kleene algebra with tests is used for reasoning about the program structure and generating verification conditions at this level. Lenses capture hybrid program stores in a generic algebraic way. The approach has been formalised with the Isabelle/HOL proof assistant. Several examples explain the workflow with the resulting verification components.

1 Introduction

Differential dynamic logic ($\mathsf{d}\mathcal{L}$) is a prominent deductive method for verifying hybrid systems [26]. It extends dynamic logic with specific inference rules for reasoning about the discrete control and continuous dynamics that characterise such systems. Continuous evolutions are modelled by $\mathsf{d}\mathcal{L}$'s evolution commands within a hybrid program syntax. These declare a vector field and a guard, which is meant to hold along the evolution. Reasoning with evolution commands in $\mathsf{d}\mathcal{L}$ requires either explicit solutions to differential equations represented by the vector field, or invariant sets [28] that describe these evolutions implicitly. Verification components inspired by $\mathsf{d}\mathcal{L}$ have already been formalised in the Isabelle proof assistant [16]. Yet the shallow embedding used in this work has shifted the focus from the original proof-theoretic approach to a semantic one, and ultimately to predicate transformer algebras supporting a different workflow.

Dynamic logics and predicate transformers are powerful tools. They support reasoning about program equivalences and transformations far beyond what standard program verification requires [4]. For the latter, much simpler Hoare logics generate precisely the verification conditions needed. Asking about the feasibility of a *differential Hoare logic* ($\mathsf{d}\mathcal{H}$) is therefore natural. As Hoare logic is strongly related to Morgan's refinement calculus [25], it is equally reasonable to ask whether and how a Morgan-style *differential refinement calculus* ($\mathsf{d}\mathcal{R}$) might allow constructing hybrid programs from specifications.

A prima facie answer to these questions seems positive: after all, the laws of Morgan's refinement calculus can be proved using the rules of Hoare logic, which

U. Fahrenberg et al. (Eds.): RAMiCS 2020, LNCS 12062, pp. 169–186, 2020.
https://doi.org/10.1007/978-3-030-43520-2_11

in turn are derivable within dynamic logic. But the formalisms envisaged might not be expressive enough for hybrid program verification or less suitable than $\mathsf{d}\mathcal{L}$ in practice. Conceptually it is also not obvious what exactly it would take to extend a standard Hoare logic or refinement calculus to hybrid programs.

Our main contribution consists in evidence that $\mathsf{d}\mathcal{H}$ and $\mathsf{d}\mathcal{R}$ are as applicable for verifying simple hybrid programs as $\mathsf{d}\mathcal{L}$, and that developing these methods requires simply adding a single Hoare-style axiom and a single refinement rule for evolution commands to the standard formalisms.

This conceptual simplicity is reflected in the Isabelle verification components for $\mathsf{d}\mathcal{H}$ and $\mathsf{d}\mathcal{R}$. These reuse components for (refinement) Kleene algebra with tests [3,13,19] ((r)KAT) for the propositional Hoare logic and refinement calculi, ignoring assignment and evolution commands. The axioms and laws for these basic commands are derived in a concrete state transformer semantics for hybrid programs [15] over a generic hybrid store model based on lenses [10], reusing other Isabelle components [8,9,15]. Data-level verification conditions are discharged using Isabelle's impressive components for ordinary differential equations [17].

This simple modular development evidences the benefits of algebraic reasoning and shallow embeddings with proof assistants. Our verification components merely require formalising a state transformer semantics for KAT and rKAT along the lines of [16] and concrete store semantics for hybrid programs. Lenses [10] give us the flexibility to switch seamlessly between stores based on real vector spaces or executable Euclidean spaces. Beyond that it suffices to derive a few algebraic laws for invariants and the Hoare-axioms and refinement laws for evolution commands in the concrete semantics. Program verification is then performed at the concrete level, but this remains hidden, as tactics generate data-level verification conditions automatically and we have programmed boiler-plate syntax for programs and correctness specifications.

Our Isabelle components support the workflows of $\mathsf{d}\mathcal{L}$ in $\mathsf{d}\mathcal{H}$ and $\mathsf{d}\mathcal{R}$. We may reason explicitly with solutions to differential equations and implicitly with invariant sets. We have formalised a third method in which solutions, that is flows, are declared ab initio in correctness specifications and need not be certified.

Our program construction and verification components have so far been evaluated on a small set of simple examples. Further work is needed to evidence scalability or compare performance with the standard $\mathsf{d}\mathcal{L}$ tool chain. We present some examples to explain the work flows supported by $\mathsf{d}\mathcal{H}$ and $\mathsf{d}\mathcal{R}$. With Isabelle tactics for automated verification condition generation in place, we notice little difference relative to our predicate transformer components [16]. The entire Isabelle formalisation is available online[1].

2 Kleene Algebra with Tests

A *Kleene algebra with tests* [19] (KAT) is a structure $(K, B, +, \cdot, 0, 1, ^*, \neg)$ where $(B, +, \cdot, 0, 1, \neg)$ is a boolean algebra with join $+$, meet $\cdot$, complementation $\neg$,

[1] https://github.com/yonoteam/HybridKATpaper.

least element 0 and greatest element 1, $B \subseteq K$, and $(K, +, \cdot, 0, 1, ^*)$ is a Kleene algebra—a semiring with idempotent addition equipped with a star operation that satisfies the axioms $1 + \alpha \cdot \alpha^* \leq \alpha^*$ and $\gamma + \alpha \cdot \beta \leq \beta \rightarrow \alpha^* \cdot \gamma \leq \beta$, as well as their opposities, with multiplication swapped. The ordering on K is defined by $\alpha \leq \beta \leftrightarrow \alpha + \beta = \beta$, as idempotent semirings are semilattices.

Elements of K represent programs; those of B tests, assertions or propositions. The operation $\cdot$ models the sequential composition of programs[2], $+$ their nondeterministic choice, $(-)^*$ their finite unbounded iteration. Program 0 aborts and 1 skips. Tests are embedded implicitly into programs. They are meant to hold in some states of a program and fail in others; $p\alpha$ (αp) restricts the execution of program α in its input (output) to those states where test p holds. The ordering $\leq$ is the opposite of the refinement ordering on programs (see Sect. 7).

Binary relations of type $\mathcal{P}(S \times S)$ form KATs [19] when $\cdot$ is interpreted as relational composition, $+$ as relational union, $(-)^*$ as reflexive-transitive closure and the elements of B as subidentities—relations below the relational unit. This grounds KAT within standard relational imperative program semantics. However, we prefer the isomorphic representation known as *state transformers* of type $S \rightarrow \mathcal{P}\, S$. Composition $\cdot$ is then interpreted as Kleisli composition

$$(f \circ_K g)\, x = \bigcup \{g\, y \mid y \in f\ x\},$$

0 as $\lambda x.\ \emptyset$ and 1 as $\eta_S = \{-\}$. Stars $f^*\, s = \bigcup_{i \in \mathbb{N}} f^i\, s$ are defined with respect to Kleisli composition using $f^0 = \eta_S$ and $f^{n+1} = f \circ_K f^n$. The boolean algebra of tests has carrier set $B_S = \{f : S \rightarrow \mathcal{P}\, S \mid f \leq \eta_S\}$, where the order on functions has been extended pointwise, and complementation is given by

$$\overline{f}\, x = \begin{cases} \eta_S\, x, & \text{if } f\, x = \emptyset, \\ \emptyset, & \text{otherwise.} \end{cases}$$

We freely identify predicates, sets and state transformers below η_S, which are isomorphic: $P \cong \{s \mid P\, s\} \cong \lambda s.\ \{x \mid x = s \wedge P\, s\}$.

Proposition 2.1. $\mathsf{Sta}\, S = ((\mathcal{P}\, S)^S, B_S, \cup, \circ_K, \lambda x.\ \emptyset, \eta_S, (-)^*, \overline{(-)})$ *forms a* KAT, *the* full state transformer KAT *over the set* S.

A *state transformer* KAT over S is any subalgebra of $\mathsf{Sta}\, S$.

KAT has been formalised via type classes in Isabelle [2]. As these allow only one type parameter, we use an alternative to the standard two-sorted approach that expands a Kleene algebra K by an *antitest* function $n : K \rightarrow K$ from which a *test* function $t : K \rightarrow K$ is defined as $t = n \circ n$. Then $K_t = \{\alpha \mid t\, \alpha = \alpha\}$ forms a boolean algebra in which n yields test complementation. It can be used in place of B. The state transformer KAT has been formalised for this article.

[2] We therefore often write ; for this operation in later sections.

3 Propositional Hoare Logic and Invariants

KAT provides a simple algebraic semantics for while programs with

$$\textbf{if } p \textbf{ then } \alpha \textbf{ else } \beta = p \cdot \alpha + \neg p \cdot \beta \qquad \text{and} \qquad \textbf{while } p \textbf{ do } \alpha = (p \cdot \alpha)^* \cdot \neg p.$$

It captures validity of Hoare triples in a partial correctness semantics as

$$\{p\}\, \alpha\, \{q\} \leftrightarrow p \cdot \alpha \cdot \neg q = 0,$$

or equivalently by $p \cdot \alpha \leq \alpha \cdot q$ or $p \cdot \alpha = p \cdot \alpha \cdot q$. It also allows deriving the rules of *propositional Hoare logic* [20]—disregarding assignments—which are useful for verification condition generation:

$$\{p\}\, skip\, \{p\}, \tag{h-skip}$$

$$p \leq p' \wedge \{p'\}\, \alpha\, \{q'\} \wedge q' \leq q \rightarrow \{p\}\, \alpha\, \{q\}, \tag{h-cons}$$

$$\{p\}\, \alpha\, \{r\} \wedge \{r\}\, \beta\, \{q\} \rightarrow \{p\}\, \alpha \cdot \beta\, \{q\}, \tag{h-seq}$$

$$\{t \cdot p\}\, \alpha\, \{q\} \wedge \{\neg t \cdot p\}\, \beta\, \{q\} \rightarrow \{p\}\, \textbf{if } t \textbf{ then } \alpha \textbf{ else } \beta\, \{q\}, \tag{h-cond}$$

$$\{t \cdot p\}\, \alpha\, \{p\} \rightarrow \{p\}\, \textbf{while } t \textbf{ do } \alpha\, \{\neg t \cdot p\}. \tag{h-while}$$

Rules for commands with invariant assertions α **inv** i are derivable in KAT, too (operationally, $\alpha\, \textbf{inv}\, i = \alpha$). An *invariant* for $\alpha \in K$ is a test $i \in B$ satisfying $\{i\}\, \alpha\, \{i\}$. Then, with **loop** α as syntactic sugar for α^*, we obtain

$$p \leq i \wedge \{i\}\, \alpha\, \{i\} \wedge i \leq q \rightarrow \{p\}\, \alpha\, \{q\}, \tag{h-inv}$$

$$\{i\}\, \alpha\, \{i\} \wedge \{j\}\, \alpha\, \{j\} \rightarrow \{i \cdot j\}\, \alpha\, \{i \cdot j\}, \tag{h-inv-mult}$$

$$\{i\}\, \alpha\, \{i\} \wedge \{j\}\, \alpha\, \{j\} \rightarrow \{i + j\}\, \alpha\, \{i + j\}, \tag{h-inv-plus}$$

$$p \leq i \wedge \{i \cdot t\}\, \alpha\, \{i\} \wedge \neg t \cdot i \leq q \rightarrow \{p\}\, \textbf{while } t \textbf{ inv } i \textbf{ do } \alpha\, \{q\}, \tag{h-while-inv}$$

$$p \leq i \wedge \{i\}\, \alpha\, \{i\} \wedge i \leq q \rightarrow \{p\}\, \textbf{loop}\, \alpha\, \textbf{inv}\, i\, \{q\}. \tag{h-loop-inv}$$

We use (h-inv) for invariants for continuous evolutions of hybrid systems in Sects. 6, 7 and 8. The rules (h-inv-mult) and (h-inv-plus) are part of a procedure, described in Sect. 6. Rule (h-while-inv) is standard for invariants for while loops; (h-loop-inv) is specific to loops of hybrid programs (see Sect. 4). The rules for propositional Hoare logic in Isabelle have already been derived for KAT [2,13], those for invariants have been formalised for this work. By Proposition 2.1, all of them hold in particular in the state transformer semantics. We have formalised this fact with Isabelle. At this stage, verification condition rules for the basic commands for assignments and evolution commands are still missing. These are formalised within the concrete state transformer semantics (see Sect. 5).

4 State Transformer Semantics for Hybrid Programs

Hybrid programs of differential dynamic logic ($\mathsf{d}\mathcal{L}$) [26] are defined by the syntax

$$\mathcal{C} ::= x := e \mid x' = f \,\&\, G \mid ?P \mid \mathcal{C};\mathcal{C} \mid \mathcal{C} + \mathcal{C} \mid \mathcal{C}^*$$

that adds *evolution commands* $x' = f \,\&\, G$ to the language of KAT—function $?(-)$ embeds tests explicitly into programs; in the tradition of KAT we leave this embedding implicit. Evolution commands introduce a time independent vector field $f : S \to S$ for an autonomous system of ordinary differential equations (ODEs) [28] together with a guard G, a predicate modelling boundary conditions or similar restrictions that hold along temporal evolutions. Guards are also known as *evolution domain restrictions* [6].

Formally, we fix a state space S of the hybrid program such as $S \subseteq \mathbb{R}^n$ for $n \in \mathbb{N}$. We model continuous variables algebraically with lenses [10] to support different state space models generically. A lens $x : A \Longrightarrow S$ is a tuple $x = (A, S, \mathit{get}, \mathit{put})$, where A is a variable type. The functions $\mathit{get}_x : S \to A$ and $\mathit{put}_x : S \to A \to S$ query and update the value of x in a particular state. They are linked by three intuitive algebraic laws [10]. For all $s \in S$ and $v, v' \in A$,

$$\mathit{get}\,(\mathit{put}\; s\; v) = v, \qquad \mathit{put}\,(\mathit{put}\; s\; v')\; v = \mathit{put}\; s\; v, \qquad \mathit{put}\; s\; (\mathit{get}\; s) = s.$$

The predicate $x \bowtie y$ checks independence of lenses x and y, which holds when x and y refer to two different regions of S. As each program variable is a lens $x : \mathbb{R} \Longrightarrow S$, state spaces $S \subseteq \mathbb{R}^n$ require n independent lenses $x_1 \cdots x_n$. Yet more general state spaces such as vector spaces are supported as well.

Systems of ODEs are modelled using vector fields: functions of type $S \to S$ on some open set S. Geometrically, vector field f assigns vectors to each point of the state space S. A solution to the *initial value problem* (IVP) for the pair (f, s) and initial value $(0, s) \in T \times S$, where T is an open interval in $\mathbb{R}$ containing 0, is then a function $X : T \to S$ that satisfies $X'\,t = f\,(X\,t)$—an autonomous system of ODEs in vector form—and $X\,0 = s$. Solution X is thus a curve in S through s, parametrised by T and tangential to f at any point in S; it is called the *trajectory* or *integral curve* of f at s whenever it is uniquely defined [28]. For IVP (f, s) with continuous vector field $f : S \to S$ and initial state $s \in S$ we define the set of solutions on T as

$$\mathsf{Sols}\, f\, T\, s = \{X \mid \forall t \in T.\; X'\,t = f\,(X\,t) \wedge X\,0 = s\}.$$

Each solution X is thus continuously differentiable and hence $f \circ X$ integrable in T. For $X \in \mathsf{Sols}\, f\, T\, s$ and guard $G : S \to \mathbb{B}$, we then define the *G-guarded orbit* of X along T in s [16] as a state transformer $\gamma_G^X : S \to \mathcal{P}\,S$ by

$$\gamma_G^X\, s = \{X\,t \mid t \in T \wedge \forall \tau \in {\downarrow}t.\; G\,(X\,\tau)\},$$

where ${\downarrow}t = \{t' \in T \mid t' \leq t\}$. Intuitively, $\gamma_G^X\, s$ is the orbit at s defined along the longest interval in T that satisfies guard G. We also define the *G-guarded orbital* of f along T in s [16] via the state transformer $\gamma_G^f : S \to \mathcal{P}\,S$ as

$$\gamma_G^f\, s = \bigcup\{\gamma_G^X\, s \mid X \in \mathsf{Sols}\ f\, T\, s\}\,.$$

In applications, $\downarrow t$ is usually an interval $[0, t] \subseteq T$. Expanding definitions,

$$\gamma_G^f\, s = \{X\, t \mid X \in \mathsf{Sols}\ f\, T\, s \wedge t \in T \wedge \forall \tau \in {\downarrow t}.\ G\,(X\,\tau)\}\,.$$

If $\top$ denotes the predicate that holds of all states in S (or the set S itself), we write γ^f instead of $\gamma_\top^f$. We define the semantics of the evolution command $x' = f\,\&\,G$ [16] for any continuous $f : S \to S$ and $G : S \to \mathbb{B}$ as

$$(x' = f\,\&\,G) = \gamma_G^f. \tag{st-evl}$$

In the evolution command $x' = f\,\&\,G$, x' is part of the traditional syntax used for specifying systems of ODEs, while de facto only a vector field f is specified. This explains why x does not appear in the right-hand side of (st-evl). Defining the state transformer semantics of assignments is standard [16], though we generalise using lenses. First, we use lenses to define state updates:

$$\sigma(x \mapsto e) = \lambda s.\ \mathit{put}_x\ (\sigma\ s)\ (e\ s)$$

for $x : A \Longrightarrow S$, $e : S \to A$, and $\sigma : S \to S$. Intuitively, this updates the value of variable x in $\sigma : S \to S$ to the value given by "expression" e in state s. For variables x and y, for example, the expression $x/(2 + y)$ is modelled by $\lambda s.\ \mathit{get}_x\ s\ /\ (2 + \mathit{get}_y\ s)$. We can also update n variables simultaneously:

$$[x_1 \mapsto e_1, x_2 \mapsto e_2, \cdots, x_n \mapsto e_n] = \mathit{id}\,(x_1 \mapsto e_1)(x_2 \mapsto e_2)\cdots(x_n \mapsto e_n),$$

where id is the identity function. State updates commute when assigning to independent lenses; they cancel one another out, when made to the same lens. We can then define a semantic analog of the substitution operator $e[f/x] = e \circ [x \mapsto f]$ that satisfies the standard laws [10]. Finally, we define the generalised simultaneous assignment operator

$$\langle\sigma\rangle = \lambda s.\ \{\sigma(s)\}. \tag{st-assgn}$$

that applies $\sigma : S \to S$ as an assignment. With our state update function, singleton assignment is a special case: $(x := e) = \langle x \mapsto e\rangle$. These concepts allow us to derive standard laws for assignments, as for instance in schematic KAT [1]:

$$\begin{aligned}
x := x \; &= \; \mathit{skip}, \\
x := e\ ;\ x := f \; &= \; x := f[e/x], \\
x := e\ ;\ y := f \; &= \; y := f\ ;\ x := e, \qquad\qquad \text{if}\, x \bowtie y, x \,\sharp\, f, y \,\sharp\, e, \\
x := e\ ;\ \textbf{if}\ t\ \textbf{then}\ \alpha\ \textbf{else}\ \beta \; &= \; \textbf{if}\ t[e/x]\ \textbf{then}\ x := e\ ;\ \alpha\ \textbf{else}\ x := e\ ;\ \beta.
\end{aligned}$$

Here, $x \,\sharp\, e$ means that the semantic expression e does not depend in its valuation on lens x [10]. An assignment of x to itself is simply skip. Two assignments to x result in a single assignment, with a semantic substitution applied.

Assignments to independent variables x and y commute provided that neither assigned expression depends on the corresponding variable. Assignment can be distributed through conditionals by a substitution to the condition. Such laws can be applied recursively for symbolic evaluation of hybrid programs.

Lenses support various store models, including records and functions [10]. We provide models for vector spaces, executable and infinite Euclidean spaces:

$$\begin{aligned} \mathit{vec\text{-}lens}_k^n &= (\mathbb{R}, \mathbb{R}^n, \lambda s.\, \mathit{vec\text{-}nth}\ s\ k, \lambda s\ v.\, \mathit{vec\text{-}upd}\ k\ v\ s), && \text{if}\, k < n, \\ \mathit{eucl\text{-}lens}_k^n &= (\mathbb{R}, V, \lambda s.\, \mathit{eucl\text{-}nth}\ s\ k, \lambda s\ v.\, \mathit{eucl\text{-}upd}\ k\ v\ s), && \text{if}\, k < n, \\ \mathit{fun\text{-}lens}_i^{(A,B)} &= (B, A \to B, (\lambda f. f i), (\lambda f\ v.\, f(i := v))). \end{aligned}$$

The vector lens selects the kth element of an n dimension vector using *vec-nth* and *vec-upd* from the HOL Analysis library [14], which provides an indexed type for the space $\mathbb{R}^n$. The Euclidean lens uses executable Euclidean spaces [18] that provide a list representation of the vectors in the n-dimensional V via an ordered basis and an inner product. The function lens selects range elements of a function associated with a domain element $i \in A$. It can be used in particular with infinite Euclidean spaces, $\mathbb{N} \to \mathbb{R}$. All three satisfy the lens axioms above.

The development in this section has been formalised with Isabelle [8,9,15], both for a state transformer and a relational semantics. An instance of the latter for particular vector fields with unique solutions forms the standard semantics of $\mathsf{d}\mathcal{L}$. By the direct connection to orbits or orbitals, the state transformer semantics is arguably conceptually simpler and more elegant.

5 Differential Hoare Logic for Flows

In the state transformer semantics of Hoare triples, the Kleisli composition in the left-hand side of $p \cdot \alpha \le \alpha \cdot q$ ensures that p holds before executing α. The right-hand side guarantees that q holds after its execution. Specifically for evolution commands, and consistently with $\mathsf{d}\mathcal{L}$, q holds along the entire orbit of a solution for f. We now complete the derivation of inference rules of $\mathsf{d}\mathcal{H}$ by adding Hoare-style rules for assignments and evolution commands in the concrete state transformer semantics.

The assignment axiom of Hoare logic needs no explanation. Our concrete state transformer semantics allows us to derive it:

$$\{P[e/x]\}\ x := e\, \{P\}. \tag{h-assgn}$$

Hence all we need to add to Hoare logic is a rule for evolution commands. We restrict our attention to Lipschitz-continuous vector fields for which unique solutions to IVPs are guaranteed by Picard-Lindelöf's theorem [28]. These are *(local) flows* $\varphi : T \to S \to S$ and $X = \varphi_s = \lambda t.\ \varphi\, t\, s$ is the trajectory at s. Guarded orbitals γ_G^f then specialise to *guarded orbits*

$$\gamma_{G,U}^f = \{\varphi_s\, t \mid t \in U \wedge \forall \tau \in {\downarrow} t.\ G\,(\varphi_s\, \tau)\},$$

where T is fixed by the Picard-Lindelöf theorem and $U \subseteq T$ is a time domain of interest, typically an interval $[0, t]$ for some $t \in T$ [16] where, by contrast to the previous section $\downarrow t = \{t' \in U \mid t' \leq t\}$ is relativised to U. This gives us the flexibility to consider dynamics over closed time intervals and it allows us to focus on time intervals and IVPs starting at $t = 0$. Accordingly, (st-evl) specialises to the following state transformer semantics for evolution commands.

$$(x' = f \,\&\, G) = \gamma^f_{G,U}. \qquad \text{(st-evl-flow)}$$

The following Hoare-style rule for evolution commands is then derivable.

Lemma 5.1. *Let $f : S \to S$ be a Lipschitz continuous vector field on $S \subseteq \mathbb{R}^n$ and $\varphi : T \to S \to S$ its local flow with $0 \in T \subseteq \mathbb{R}$. Then, for $U \subseteq T$ with $0 \in U$ and $G, Q : S \to \mathbb{B}$,*

$$\{\lambda s \in S. \forall t \in U.\ (\forall \tau \in \downarrow t.\ G\,(\varphi_s\,\tau)) \to Q\,(\varphi_s\,t)\}\ x' = f \,\&\, G\,\{Q\}. \qquad \text{(h-evl)}$$

This finishes the derivation of rules for a Hoare logic $\mathsf{d}\mathcal{H}$ for hybrid programs—to our knowledge, the first Hoare logic of this kind. As usual, there is one rule per programming construct, so that the recursive application of theh-assgnHoare logic together with (h-assgn) and (h-evl) generates proof obligations that are entirely about data-level relationships—the discrete and continuous evolution of hybrid program stores.

The rule (h-evl) supports the following procedure for reasoning with an evolution command $x' = f \,\&\, G$ and set U in $\mathsf{d}\mathcal{H}$:

1. Check that f satisfies the conditions for Picard-Lindelöf's theorem (f is Lipschitz continuous and $S \subseteq \mathbb{R}^n$ is open).
2. Supply a (local) flow φ for f with open interval of existence T around 0.
3. Check that φ_s solves the IVP (f, s) for each $s \in S$; ($\varphi_s'\,t = f\,(\varphi_s\,t)$, $\varphi_s\,0 = s$, and $U \subseteq T$).
4. If successful, apply rule (h-evl).

Example 5.2 (Thermostat verification via solutions). A thermostat regulates the temperature T of a room between bounds $T_l \leq T \leq T_h$. Variable T_0 stores an initial temperature; ϑ indicates whether the heater is switched on or off. Within time intervals of at most τ minutes, the thermostat resets time to 0, measures the temperature, and turns the heater on or off dependent on the value obtained. With $0 < T_l$, $T_h < T_u$, $0 < a$, $U = \{0..\tau\} = [0, \tau]$ we define f, for $c \in \{0, T_u\}$, as

abbreviation $f\ a\ c \equiv [T \mapsto_s -\ (a * (T\ -\ c)),\ T_0 \mapsto_s 0,\ \vartheta \mapsto_s 0,\ t \mapsto_s 1]$

The notation $x \mapsto_s f\,x$ indicates that vector field $f\,a\,c$ maps variable x to $f\,x$ for $x \in \{T, T_0, \vartheta, t\}$. Working with *vec-lens*^n_k or *eucl-lens*^n_k, we write ; instead of $\cdot$ and use guard G to restrict evolutions between T_l and T_h by setting

$$G\,T_l\,T_h\,a\,c = \left(t \leq -\frac{1}{a}\ln\left(\frac{c - \Delta_c}{c - T_0}\right)\right),$$

where $\Delta_c = T_l$ if $c = 0$, and $\Delta_c = T_h$ if $c = T_u$. The hybrid program *therm* below models the behaviour of the thermostat. To simplify notation, we separate into a loop invariant (I), discrete control (*ctrl*), and continuous dynamics (*dyn*).

abbreviation $I\ T_l\ T_h \equiv \mathbf{U}(T_l \leq T \wedge T \leq T_h \wedge (\vartheta = 0 \vee \vartheta = 1))$

abbreviation $ctrl\ T_l\ T_h \equiv$
$(t ::= 0); (T_0 ::= T);$
$(IF\ (\vartheta = 0 \wedge T_0 \leq T_l + 1)\ THEN\ (\vartheta ::= 1)\ ELSE$
$IF\ (\vartheta = 1 \wedge T_0 \geq T_h - 1)\ THEN\ (\vartheta ::= 0)\ ELSE\ skip)$

abbreviation $dyn\ T_l\ T_h\ a\ T_u\ \tau \equiv$
$IF\ (\vartheta = 0)\ THEN\ x' = f\ a\ 0\ \&\ G\ T_l\ T_h\ a\ 0\ on\ \{0..\tau\}\ UNIV\ @\ 0$
$ELSE\ x' = f\ a\ T_u\ \&\ G\ T_l\ T_h\ a\ T_u\ on\ \{0..\tau\}\ UNIV\ @\ 0$

abbreviation $therm\ T_l\ T_h\ a\ T_u\ \tau \equiv$
$LOOP\ (ctrl\ T_l\ T_h;\ dyn\ T_l\ T_h\ a\ T_u\ \tau)\ INV\ (I\ T_l\ T_h)$

The correctness specification and verification of the thermostat with $\mathsf{d}\mathcal{H}$ is then

lemma *thermostat-flow*:
assumes $0 < a$ **and** $0 \leq \tau$ **and** $0 < T_l$ **and** $T_h < T_u$
shows $\{I\ T_l\ T_h\}\ therm\ T_l\ T_h\ a\ T_u\ \tau\ \{I\ T_l\ T_h\}$
apply(*hyb-hoare* $\mathbf{U}(I\ T_l\ T_h \wedge t{=}0 \wedge T_0 = T)$)
prefer *4* **prefer** *8* **using** *local-flow-therm assms* **apply** *force+*
using *assms therm-dyn-up therm-dyn-down* **by** *rel-auto'*

The first line uses tactic *hyb-hoare* to blast away the structure of *therm* using $\mathsf{d}\mathcal{H}$. To apply *hyb-hoare*, the program must be an iteration of the composition of two programs—usually control and dynamics. The tactic requires lifting the store to an Isabelle/UTP expression [10], which is denoted by the $\mathbf{U}$ operator. Lemma *local-flow-therm*, whose proof captures the procedure described above, supplies the flow for $f\,a\,c$: $\varphi\,a\,c\,\tau = (-e^{-a\cdot\tau}(c - T) + c, \tau + t, T_0, \vartheta)^\top$, for all $\tau \in \mathbb{R}$. The remaining proof obligations are inequalities of transcendental functions. They are discharged automatically using auxiliary lemmas. □

6 Differential Hoare Logic for Invariants

Alternatively, $\mathsf{d}\mathcal{H}$ supports reasoning with invariants for evolution commands instead of supplying flows to (h-evl). The approach has been developed in [16]. Our invariants generalise the *differential invariants* of $\mathsf{d}\mathcal{L}$ [26] and the *invariant sets* of dynamical systems and (semi)group theory [28].

A predicate $I : S \to \mathbb{B}$ is an *invariant* of the continuous vector field $f : S \to S$ and guard $G : S \to \mathbb{B}$ *along* $T \subseteq \mathbb{R}$ if

$$\bigcup \mathcal{P}\, \gamma^f_G\, I \subseteq I.$$

The operation $\bigcup \circ \mathcal{P}$ is the Kleisli extension $(-)^\dagger$ in the powerset monad. Hence we could simply write $(\gamma^f_G)^\dagger\, I \subseteq I$. The definition of invariance unfolds to

$$\forall s.\ I\, s \to (\forall X \in \mathsf{Sols}\, f\, T\, s. \forall t \in T.\ (\forall \tau \in {\downarrow} t.\ G\,(X\,\tau)) \to I\,(X\,t)).$$

For $G = \top$ we call I an *invariant* of f along T. Intuitively, invariants can be seen as sets of orbits. They are compatible with the invariants from Sect. 3.

Proposition 6.1. *Let $f : S \to S$ be continuous, $G : S \to \mathbb{B}$ and $T \subseteq \mathbb{R}$. Then I is an invariant for f and G* along T *if and only if $\{I\}\, x' = f \,\&\, G\,\{I\}$.*

Hence we can use a variant of (h-inv) for verification condition generation:

$$P \leq I \wedge \{I\}\, x' = f \,\&\, G\,\{I\} \wedge (I \cdot G) \leq Q \ \to \{P\}\, x' = f \,\&\, G\,\{Q\}. \quad \text{(h-invg)}$$

The following lemma leads to a procedure.

Lemma 6.2 ([16]). *Let $f : S \to S$ be a continuous vector field, $\mu, \nu : S \to \mathbb{R}$ differentiable and $T \subseteq \mathbb{R}$ an interval such that $0 \in T$.*

1. *If $(\mu \circ X)' = (\nu \circ X)'$ for all $X \in \mathsf{Sols}\, f\, T\, s$, then $\{\mu = \nu\}\, x' = f \,\&\, G\,\{\mu = \nu\}$,*
2. *if $(\mu \circ X)'\, t \leq (\nu \circ X)'\, t$ when $t > 0$, and $(\mu \circ X)'\, t \geq (\nu \circ X)'\, t$ when $t < 0$, for all $X \in \mathsf{Sols}\, f\, T\, s$, then $\{\mu < \nu\}\, x' = f \,\&\, G\,\{\mu < \nu\}$*
3. *if $\{\mu < \nu\}\, x' = f \,\&\, G\,\{\mu < \nu\}$ and $\{\mu > \nu\}\, x' = f \,\&\, G\,\{\mu > \nu\}$, then $\{\mu \neq \nu\}\, x' = f \,\&\, G\,\{\mu \neq \nu\}$ (and conversely if 0 is the least element in T),*
4. *$\{\mu \not\leq \nu\}\, x' = f \,\&\, G\,\{\mu \not\leq \nu\}$ if and only if $\{\mu > \nu\}\, x' = f \,\&\, G\,\{\mu > \nu\}$.*

Condition (1) follows from the well known fact that two continuously differentiable functions are equal if they intersect at some point and have the same derivatives. Rules (h-invg), (h-inv-mult), (h-inv-plus), Proposition 6.1 and Lemma 6.2 yield the following procedure for verifying $\{P\}\, x' = f \,\&\, G\,\{Q\}$:

1. Check whether candidate predicate I is an invariant for f along T:
   (a) transform I into negation normal form;
   (b) reduce complex I (with (h-inv-mult), (h-inv-plus) and Lemma 6.2 (3,4);
   (c) if I is atomic, apply Lemma 6.2 (1) and (2);
   (if successful, $\{I\}\, x' = f \,\&\, G\,\{I\}$ holds by Proposition 6.1),
2. if successful, prove $P \leq I$ and $(I \cdot G) \leq Q$ to apply rule (h-invg).

Example 6.3 (Water tank verification via invariants). A controller turns a water pump on and off to keep the water level h in a tank within bounds $h_l \leq h \leq h_h$. Variable h_0 stores an initial water level; π indicates whether the pump is on or off. The rate of change of the water-level is linear with slope $k \in \{-c_o, c_i - c_o\}$ (assuming $c_i > c_o$). The vector field f for this behaviour and its invariant dI are

abbreviation $f\, k \equiv [\pi \mapsto_s 0,\ h \mapsto_s k,\ h_0 \mapsto_s 0,\ t \mapsto_s 1]$

abbreviation $dI\ h_l\ h_h\ k \equiv$
$\mathsf{U}(h = k \cdot t + h_0 \wedge 0 \leq t \wedge h_l \leq h_0 \wedge h_0 \leq h_h \wedge (\pi = 0 \vee \pi = 1))$

Program *tank-dinv* for the controller is given by guard $G\ h_x\ k$ with $h_x \in \{h_l, h_h\}$ that restricts evolutions beyond h_x, loop invariant I, control and dynamics:

abbreviation $G\ h_x\ k \equiv \mathsf{U}(t \leq (h_x - h_0)/k)$

abbreviation $I\ h_l\ h_h \equiv \mathsf{U}(h_l \leq h \wedge h \leq h_h \wedge (\pi = 0 \vee \pi = 1))$

abbreviation $dyn\ c_i\ c_o\ h_l\ h_h\ \tau \equiv IF\ (\pi = 0)\ THEN$
$x' = f\ (c_i - c_o)\ \&\ G\ h_h\ (c_i - c_o)\ on\ \{0..\tau\}\ UNIV\ @\ 0\ DINV\ (dI\ h_l\ h_h\ (c_i - c_o))$
$ELSE\ x' = f\ (-c_o)\ \&\ G\ h_l\ (-c_o)\ on\ \{0..\tau\}\ UNIV\ @\ 0\ DINV\ (dI\ h_l\ h_h\ (-c_o))$

abbreviation $ctrl\ h_l\ h_h \equiv$
$(t ::= 0);(h_0 ::= h);$
$(IF\ (\pi = 0 \wedge h_0 \leq h_l + 1)\ THEN\ (\pi ::= 1)\ ELSE$
$(IF\ (\pi = 1 \wedge h_0 \geq h_h - 1)\ THEN\ (\pi ::= 0)\ ELSE\ skip))$

abbreviation $tank\text{-}dinv\ c_i\ c_o\ h_l\ h_h\ \tau \equiv$
$LOOP\ (ctrl\ h_l\ h_h;\ dyn\ c_i\ c_o\ h_l\ h_h\ \tau)\ INV\ (I\ h_l\ h_h)$

We distinguish DINV and INV to structure specifications. The correctness specification and verification of the water tank with $\mathsf{d}\mathcal{H}$ then proceeds as follows:

lemma *tank-diff-inv*: $0 \leq \tau \Longrightarrow diff\text{-}invariant\ (dI\ h_l\ h_h\ k)\ (f\ k)\ \{0..\tau\}\ UNIV\ 0\ Guard$
⟨proof⟩

lemma *tank-inv*:
assumes $0 \leq \tau$ **and** $0 < c_o$ **and** $c_o < c_i$
shows $\{I\ h_l\ h_h\}\ tank\text{-}dinv\ c_i\ c_o\ h_l\ h_h\ \tau\ \{I\ h_l\ h_h\}$
apply(*hyb-hoare* $\mathsf{U}(I\ h_l\ h_h \wedge t = 0 \wedge h_0 = h)$)
prefer *4* **prefer** *7* **using** *tank-diff-inv assms* **apply** *force+*
using *assms tank-inv-arith1 tank-inv-arith2* **by** *rel-auto'*

Tactic *hyb-hoare* blasts away the control structure. The second proof line uses Lemma *tank-diff-inv* to check that *dI* is an invariant for any guard (*Guard* is a universally quantified variable in Lemma *tank-diff-inv*), using the procedure outlined. Auxiliary lemmas discharge the remaining proof obligations. □

7 Differential Refinement Calculi

A *refinement Kleene algebra with tests* (rKAT) [3] is a KAT (K, B) expanded by an operation $[-,-] : B \times B \to K$ that satisfies, for all $\alpha \in K$ and $p, q \in B$,

$$\{p\}\,\alpha\,\{q\} \leftrightarrow \alpha \leq [p, q].$$

The element $[p, q]$ of K corresponds to Morgan's *specification statement* [25]. It satisfies $\{p\}\,[p,q]\,\{q\}$ and $\{p\}\,\alpha\,\{q\} \to \alpha \leq [p, q]$, which makes $[p, q]$ the greatest element of K that satisfies the Hoare triple with precondition p and postcondition q. Indeed, in $\mathsf{Sta}\,S$ and for $S \subseteq \mathbb{R}^n$, $[P, Q] = \bigcup\{f : S \to \mathcal{P}\,S \mid \{P\}\,f\,\{Q\}\}$.

Variants of Morgan's laws [25] of a *propositional refinement calculus*—once more ignoring assignments—are then derivable in rKAT [3].

$$
\begin{aligned}
1 &\leq [p, p], && && \text{(r-skip)}\\
[p', q'] &\leq [p, q], && \text{if } p \leq p' \text{ and } q' \leq q, && \text{(r-cons)}\\
[p, r] \cdot [r, q] &\leq [p, q], && && \text{(r-seq)}\\
\textbf{if } t \textbf{ then } [t \cdot p, q] \textbf{ else } [\neg t \cdot p, q] &\leq [p, q], && && \text{(r-cond)}\\
\textbf{while } t \textbf{ do } [t \cdot p, p] &\leq [p, \neg t \cdot p]. && && \text{(r-while)}
\end{aligned}
$$

We have also derived $\alpha \leq [0, 1]$ and $[1, 0] \leq \alpha$, but do not use them in proofs.

For invariants and loops, we obtain the additional refinement laws

$$
\begin{aligned}
[i, i] &\leq [p, q], && \text{if } p \leq i \leq q, && \text{(r-inv)}\\
\textbf{loop } [i, i] &\leq [i, i]. && && \text{(r-loop)}
\end{aligned}
$$

In Sta S, moreover, the following assignments laws are derivable [3].

$$
\begin{aligned}
(x := e) &\leq [Q[e/x], Q], && \text{(r-assgn)}\\
(x := e) \cdot [Q, Q] &\leq [Q[e/x], Q], && \text{(r-assgnl)}\\
[Q, Q[e/x]] \cdot (x := e) &\leq [Q, Q]. && \text{(r-assgnf)}
\end{aligned}
$$

The second and third law are known as *leading* and *following* law. They introduce an assignment before and after a block of code.

Finally, we obtain the following refinement laws for evolution commands.

Lemma 7.1. *Let $f : S \to S$ be a Lipschitz continuous vector field on $S \subseteq \mathbb{R}^n$ and $\varphi : T \to S \to S$ its local flow with $0 \in T \subseteq \mathbb{R}$. Then, for $U \subseteq T$ with $0 \in U$ and $G, Q : S \to \mathbb{B}$,*

$$
\begin{aligned}
(x' = f \,\&\, G) &\leq [\lambda s. \forall t \in U.\ (\forall \tau \in {\downarrow} t.\ G\,(\varphi_s\,\tau)) \to Q\,(\varphi_s\,t), Q], && \text{(r-evl)}\\
(x' = f \,\&\, G) \cdot [Q, Q] &\leq [\lambda s. \forall t \in U.\ (\forall \tau \in {\downarrow} t.\ G\,(\varphi_s\,\tau)) \to Q\,(\varphi_s\,t), Q], && \text{(r-evll)}\\
[Q, \lambda s. \forall t \in U.\ (\forall \tau \in {\downarrow} t.\ G\,(\varphi_s\,\tau)) \to Q\,(\varphi_s\,t)] \cdot (x' = f \,\&\, G) &\leq [Q, Q]. && \text{(r-evlr)}
\end{aligned}
$$

The laws in this section form the differential refinement calculus $\mathsf{d}\mathcal{R}$. They suffice for constructing hybrid programs from initial specification statements by step-wise refinement incrementally and compositionally. A more powerful variant based on predicate transformers à la Back and von Wright [4] has been developed in [16]; yet applications remain to be explored. A previous approach to refinement in $\mathsf{d}\mathcal{L}$ [23] is quite different to the two standard calculi mentioned (see Conclusion).

Example 7.2 (Thermostat refinement via solutions). We now construct program *therm* from Example 5.2 by step-wise refinement using the rules of d$\mathcal{R}$.

lemma *R-therm-down*:
  assumes $a > 0$ **and** $0 \leq \tau$ **and** $0 < T_l$ **and** $T_h < T_u$
  shows $[\vartheta = 0 \wedge I\ T_l\ T_h \wedge t = 0 \wedge T_0 = T,\ I\ T_l\ T_h] \geq$
  $(x' = f\ a\ 0\ \&\ G\ T_l\ T_h\ a\ 0\ on\ \{0..\tau\}\ UNIV\ @\ 0)$
  apply(*rule local-flow.R-g-ode-ivl*[*OF local-flow-therm*])
  using *therm-dyn-down*[*OF assms*(*1*,*3*), *of* - T_h] *assms* **by** *rel-auto'*

lemma *R-therm-up*:
  assumes $a > 0$ **and** $0 \leq \tau$ **and** $0 < T_l$ **and** $T_h < T_u$
  shows $[\neg\ \vartheta = 0 \wedge I\ T_l\ T_h \wedge t = 0 \wedge T_0 = T,\ I\ T_l\ T_h] \geq$
  $(x' = f\ a\ T_u\ \&\ G\ T_l\ T_h\ a\ T_u\ on\ \{0..\tau\}\ UNIV\ @\ 0)$
  apply(*rule local-flow.R-g-ode-ivl*[*OF local-flow-therm*])
  using *therm-dyn-up*[*OF assms*(*1*) - - *assms*(*4*), *of* T_l] *assms* **by** *rel-auto'*

lemma *R-therm-time*: $[I\ T_l\ T_h,\ I\ T_l\ T_h \wedge t = 0] \geq (t ::= 0)$
  by (*rule R-assign-law*, *pred-simp*)

lemma *R-therm-temp*: $[I\ T_l\ T_h \wedge t = 0,\ I\ T_l\ T_h \wedge t = 0 \wedge T_0 = T] \geq (T_0 ::= T)$
  by (*rule R-assign-law*, *pred-simp*)

lemma *R-thermostat-flow*:
  assumes $a > 0$ **and** $0 \leq \tau$ **and** $0 < T_l$ **and** $T_h < T_u$
  shows $[I\ T_l\ T_h,\ I\ T_l\ T_h] \geq$ *therm* $T_l\ T_h\ a\ T_u\ \tau$
  by (*refinement*;(*rule R-therm-time*)?,(*rule R-therm-temp*)?,(*rule R-assign-law*)?,
    (*rule R-therm-up*[*OF assms*])?, (*rule R-therm-down*[*OF assms*])?) *rel-auto'*

The *refinement* tactic pushes the refinement specification through the program structure until the only remaining proof obligations are atomic refinements. We only refine the atomic programs needed to complete proofs automatically; those for the first two assignment and the evolution commands. □

Example 7.3 (Water tank refinement via invariants). Alternatively we may use differential invariants with d$\mathcal{R}$ to refine *tank-dinv* from Example 6.3. This time we supply a single structured proof to show another style of refinement. We abbreviate long expressions with schematic variables.

lemma *R-tank-inv*:
  assumes $0 \leq \tau$ **and** $0 < c_o$ **and** $c_o < c_i$
  shows $[I\ h_l\ h_h,\ I\ h_l\ h_h] \geq$ *tank-dinv* $c_i\ c_o\ h_l\ h_h\ \tau$
proof−
  have $[I\ h_l\ h_h,\ I\ h_l\ h_h] \geq$
  LOOP $((t ::= 0);[I\ h_l\ h_h \wedge t = 0,\ I\ h_l\ h_h])$ *INV* $I\ h_l\ h_h$ (**is** - $\geq$ *?R1*)
   by (*refinement*, *rel-auto'*)
  moreover have *?R1* $\geq$ *LOOP*
  $((t ::= 0);(h_0 ::= h);[I\ h_l\ h_h \wedge t = 0 \wedge h_0 = h,\ I\ h_l\ h_h])$ *INV* $I\ h_l\ h_h$ (**is** - $\geq$ *?R2*)
   by (*refinement*, *rel-auto'*)

moreover have $?R2 \geq$
$LOOP\ (ctrl\ h_l\ h_h;[I\ h_l\ h_h \wedge t = 0 \wedge h_0 = h,\ I\ h_l\ h_h])\ INV\ I\ h_l\ h_h$ (**is** - $\geq$ $?R3$)
by (*simp only*: *mult.assoc*, *refinement*; (*force*)?, (*rule R-assign-law*)?) *rel-auto'*
moreover have $?R3 \geq LOOP\ (ctrl\ h_l\ h_h;\ dyn\ c_i\ c_o\ h_l\ h_h\ \tau)\ INV\ I\ h_l\ h_h$
apply(*simp only*: *mult.assoc*, *refinement*; (*simp*)?)
prefer *4* **using** *tank-diff-inv assms* **apply** *force+*
using *tank-inv-arith1 tank-inv-arith2 assms* **by** *rel-auto'*
ultimately show $[I\ h_l\ h_h,\ I\ h_l\ h_h] \geq$ *tank-dinv* $c_i\ c_o\ h_l\ h_h\ \tau$
by *auto*
qed

The proof incrementally refines the specification of *tank-dinv* using the laws of d$\mathcal{R}$. As in Example 7.2, after refining the first two assignments, tactic *refinement* completes the construction of *ctrl*. After that, the invariant is supplied via lemma *tank-diff-inv* from Example 6.3 to construct *dyn*. The final program is then constructed by transitivity of $\leq$. A more detailed derivation is also possible. □

8 Evolution Commands for Flows

Finally, we present variants of d$\mathcal{H}$ and d$\mathcal{R}$ that start directly from flows $\varphi : T \to S \to S$ instead of vector fields. This avoids checking the conditions of the Picard-Lindelöf theorem and simplifies verification proofs considerably. Instead of $x' = f \,\&\, G$, we now use the command $\mathbf{evol}\,\varphi\, G$ in hybrid programs and define

$$(\mathbf{evol}\,\varphi\, G) = \lambda s.\ \gamma_G^{\varphi_s}\ s$$

with respect to the guarded orbit of φ_s along T in s. It then remains to derive a Hoare-style axiom and a refinement law for such evolution commands.

Lemma 8.1. *Let* $\varphi : T \to S \to S$, *where* S *is a set and* T *a preorder. Then, for* $G, P, Q : S \to \mathbb{B}$,

$$\{\lambda s \in S. \forall t \in T.\ (\forall \tau \in {\downarrow}t.\ G\,(\varphi_s\, \tau)) \to P\,(\varphi_s\, t)\}\, \mathbf{evol}\,\varphi\, G\, \{P\}, \quad \text{(h-evlfl)}$$
$$\mathbf{evol}\,\varphi\, G \leq [\lambda s. \forall t \in T.\ (\forall \tau \in {\downarrow}t.\ G\,(\varphi_s\, \tau)) \to Q\,(\varphi_s\, t), Q], \quad \text{(r-evlf)}$$
$$(\mathbf{evol}\,\varphi\, G) \cdot [Q, Q] \leq [\lambda s. \forall t \in T.\ (\forall \tau \in {\downarrow}t.\ G\,(\varphi_s\, \tau)) \to Q\,(\varphi_s\, t), Q], \quad \text{(r-evlfl)}$$
$$[Q, \lambda s. \forall t \in T.\ (\forall \tau \in {\downarrow}t.\ G\,(\varphi_s\, \tau)) \to Q\,(\varphi_s\, t)] \cdot (\mathbf{evol}\,\varphi\, G) \leq [Q, Q]. \quad \text{(r-evlfr)}$$

Example 8.2 (Bouncing ball via Hoare logic and refinement). A ball falls down from height $h \geq 0$, with x denoting its position, v its velocity and g its acceleration. Its kinematics is modelled by the flow

abbreviation $\varphi\ g\ \tau \equiv [x \mapsto_s g \cdot \tau\ \hat{}\ 2/2 + v \cdot \tau + x,\ \ v \mapsto_s g \cdot \tau + v]$

The ball bounces back elastically from the ground. This is modelled by a discrete control that checks for $x = 0$ and then flips the velocity. Guard $G = (x \geq 0)$ excludes any motion below the ground. This is modelled by the hybrid program [26]

abbreviation *bb-evol* $g\ h\ T \equiv$
  $LOOP\ (EVOL\ (\varphi\ g)\ (x \geq 0)\ T;\ (IF\ (x = 0)\ THEN\ (v ::= -v)\ ELSE\ skip))$
  $INV\ (0 \leq x \wedge 2 \cdot g \cdot x = 2 \cdot g \cdot h + v \cdot v)$

Its loop invariant conjoins the guard G with a variant of energy conservation. The correctness specification and proof with $\mathsf{d}\mathcal{H}$ and $\mathsf{d}\mathcal{R}$ are then straightforward.

lemma *bouncing-ball-dyn*:
  assumes $g < 0$ **and** $h \geq 0$
  shows $\{x = h \wedge v = 0\}$ *bb-evol* $g\ h\ T\ \{0 \leq x \wedge x \leq h\}$
  apply(*hyb-hoare* **U**$(0 \leq x \wedge 2 \cdot g \cdot x = 2 \cdot g \cdot h + v \cdot v)$)
  using *assms* **by** (*rel-auto' simp*: *bb-real-arith*)

lemma *R-bouncing-ball-dyn*:
  assumes $g < 0$ **and** $h \geq 0$
  shows $[x = h \wedge v = 0,\ 0 \leq x \wedge x \leq h] \geq$ *bb-evol* $g\ h\ T$
  apply(*refinement*; (*rule R-bb-assign*[*OF assms*])?)
  using *assms* **by** (*rel-auto' simp*: *bb-real-arith*)

In the refinement proof, the tactic leaves only the refinement for the assignment $v ::= -v$. This is supplied via lemma *R-bb-assign* and the remaining obligations are discharged with the same arithmetical facts. □

9 Conclusion

We have contributed new methods and Isabelle components to an open modular semantic framework for verifying hybrid systems that so far focussed on predicate transformer semantics [16]; more specifically a Hoare logic $\mathsf{d}\mathcal{H}$ and a Morgan-style refinement calculus $\mathsf{d}\mathcal{R}$ for hybrid programs, more generic state spaces modelled by lenses, improved Isabelle syntax for correctness specifications and hybrid programs, and increased proof automation via the tactics *hyb-hoare* and *refinement*. These components support three workflows based on certifying solutions to Lipschitz-continuous vector fields, reasoning with invariant sets for continuous vector fields, and working directly with flows without certification.

Compared to the standard $\mathsf{d}\mathcal{L}$ toolchain, $\mathsf{d}\mathcal{H}$ and $\mathsf{d}\mathcal{R}$ emphasise a natural mathematical style of semantic reasoning about dynamical systems, with minimal conceptual overhead relative to standard Hoare logics and refinement calculi. $\mathsf{d}\mathcal{H}$, in particular, is only used for automated verification condition generation. The modular approach with algebras and a shallow embedding has simplified the construction of these verification components and made it incremental relative to extant ones. Our framework is not only open to use any proof method and mathematical approach supported by Isabelle, it should also allow adding new methods, for instance based on discrete dynamical systems, hybrid automata or duration calculi [7,22], or integrate CAS's for finding solutions.

The relevance of $\mathsf{d}\mathcal{H}$ and $\mathsf{d}\mathcal{R}$ to hybrid systems verification is further evidenced by the fact that such approaches are not new: A hybrid Hoare logic has been proposed by Liu et al. [22] for a duration calculus based on hybrid CSP. It is conceptually very different from $\mathsf{d}\mathcal{H}$ and $\mathsf{d}\mathcal{L}$. A differential refinement logic based on $\mathsf{d}\mathcal{L}$ has been developed as part of Loos' PhD work [23]. It uses a proof system with inference rules for reasoning about inequalities between KAT expressions, which are interpreted as refinements between hybrid programs. According to the authors, it differs substantially from the standard approaches [4,25] in that local instead of global refinement relations can be used. Nevertheless their refinement logic has the same expressivity as $\mathsf{d}\mathcal{L}$ [23], which is essentially a predicate transformer calculus for hybrid programs [16] and thus a refinement calculus à la Back and von Wright. Ultimately, this suggests that Loos' logic is more expressive than our Morgan-style calculus, but the relative merits of the two approaches remain to be explored. The proof theory of $\mathsf{d}\mathcal{L}$ has already been deeply embedded in proof assistants [5], yet with a focus on soundness proofs for its inference rules and a mechanisation of its idiosyncratic substitution calculus, but not as prima facie verification components.

The expressivity and complexity gap between Hoare logic and predicate transformer semantics is apparent within algebra. The weakest liberal precondition operator cannot be expressed in KAT [27]. The equational theory of KAT, which captures propositional Hoare logic, is PSPACE complete [21], that of modal Kleene algebra, which yields predicate transformers, in EXPTIME [24].

Finally, while KAT and rKAT are convenient starting points for building program construction and verification components for hybrid programs, the simple and more general setting of Hoare semigroups [27] would support developing hybrid Hoare logics for total program correctness—where balls may bounce forever—or even for multirelational semantics [11,12], which are relevant needed for differential game logic [26]. This is left for future work.

Acknowledgements. The authors wish to thank the reviewers for their very thorough and insightful comments, and to Sergey Goncharov and André Platzer for discussions on a preliminary version. The second author is sponsored by CONACYT's scholarship no. 440404; the first author is supported by the EPSRC project CyPhyAssure[3] (Grant EP/S001190/1).

References

1. Angus, A., Kozen, D.: Kleene algebra with tests and program schematology. Rechnical Report 2001–1844, Computer Science Department, Cornell University (2001)
2. Armstrong, A., Gomes, V.B.F., Struth, G.: Kleene algebra with tests and demonic refinement algebras. Archive of Formal Proofs (2014)

[3] https://www.cs.york.ac.uk/circus/CyPhyAssure/.

3. Armstrong, A., Gomes, V.B.F., Struth, G.: Building program construction and verification tools from algebraic principles. Formal Aspects Comput. **28**(2), 265–293 (2016)
4. Back, R., von Wright, J.: Refinement Calculus–A Systematic Introduction. Springer, New York (1998). https://doi.org/10.1007/978-1-4612-1674-2
5. Bohrer, B., Rahli, V., Vukotic, I., Völp, M., Platzer, A.: Formally verified differential dynamic logic. In: CPP 2017, pp. 208–221. ACM (2017)
6. Doyen, L., Frehse, G., Pappas, G.J., Platzer, A.: Verification of hybrid systems. Handbook of Model Checking, pp. 1047–1110. Springer, Cham (2018). https://doi.org/10.1007/978-3-319-10575-8_30
7. Foster, S.: Hybrid relations in Isabelle/UTP. In: Ribeiro, P., Sampaio, A. (eds.) UTP 2019. LNCS, vol. 11885, pp. 130–153. Springer, Cham (2019). https://doi.org/10.1007/978-3-030-31038-7_7
8. Foster, S., Zeyda, F.: Optics in Isabelle/HOL. Archive of Formal Proofs (2018)
9. Foster, S., Zeyda, F., Nemouchi, Y., Ribeiro, P., Wolff, B.: Isabelle/UTP: Mechanised Theory Engineering for Unifying Theories of Programming. Archive of Formal Proofs (2019)
10. Foster, S., Zeyda, F., Woodcock, J.: Unifying heterogeneous state-spaces with lenses. In: Sampaio, A., Wang, F. (eds.) ICTAC 2016. LNCS, vol. 9965, pp. 295–314. Springer, Cham (2016). https://doi.org/10.1007/978-3-319-46750-4_17
11. Furusawa, H., Struth, G.: Binary multirelations. Archive of Formal Proofs (2015)
12. Furusawa, H., Struth, G.: Taming multirelations. ACM TOCL **17**(4), 28:1–28:34 (2016)
13. Gomes, V.B.F., Struth, G.: Program construction and verification components based on Kleene algebra. Archive of Formal Proofs (2016)
14. Hölzl, J., Immler, F., Huffman, B.: Type classes and filters for mathematical analysis in Isabelle/HOL. In: Blazy, S., Paulin-Mohring, C., Pichardie, D. (eds.) ITP 2013. LNCS, vol. 7998, pp. 279–294. Springer, Heidelberg (2013). https://doi.org/10.1007/978-3-642-39634-2_21
15. Huerta y Munive, J.J.:: Verification components for hybrid systems. Archive of Formal Proofs (2019)
16. Huerta y Munive, J.J., Struth, G.: Predicate transformer semantics for hybrid systems: verification components for Isabelle/HOL. arXiv:1909.05618 [cs.LO] (2019)
17. Immler, F., Hölzl, J.: Ordinary differential equations. Archive of Formal Proofs (2012)
18. Immler, F., Traut, C.: The flow of ODEs: formalization of variational equation and Poincaré map. J. Autom. Reasoning **62**(2), 215–236 (2019)
19. Kozen, D.: Kleene algebra with tests. ACM TOPLAS **19**(3), 427–443 (1997)
20. Kozen, D.: On Hoare logic and Kleene algebra with tests. ACM TOCL **1**(1), 60–76 (2000)
21. Kozen, D., Cohen, E., Smith, F.: The complexity of Kleene algebra with tests. Technical report TR96-1598, Computer Science Department, Cornell University (1996)
22. Liu, J., et al.: A calculus for hybrid CSP. In: Ueda, K. (ed.) APLAS 2010. LNCS, vol. 6461, pp. 1–15. Springer, Heidelberg (2010). https://doi.org/10.1007/978-3-642-17164-2_1
23. Loos, S.M., Platzer, A.: Differential refinement logic. In: LICS 2016, pp. 505–514. ACM (2016)
24. Möller, B., Struth, G.: Algebras of modal operators and partial correctness. Theoret. Comput. Sci. **351**(2), 221–239 (2006)

25. Morgan, C.: Programming from Specifications, 2nd edn. Prentice Hall, Upper Saddle River (1994)
26. Platzer, A.: Logical Foundations of Cyber-Physical Systems. Springer, Cham (2018). https://doi.org/10.1007/978-3-319-63588-0
27. Struth, G.: Hoare semigroups. Math. Struct. Comput. Sci. **28**(6), 775–799 (2018)
28. Teschl, G.: Ordinary Differential Equations and Dynamical Systems. AMS (2012)

Higher-Order Categorical Substructural Logic: Expanding the Horizon of Tripos Theory

Yoshihiro Maruyama(✉)

Research School of Computer Science,
The Australian National University,
Canberra, Australia
yoshihiro.maruyama@anu.edu.au

Abstract. Higher-order intuitionistic logic categorically corresponds to toposes or triposes; here we address what are toposes or triposes for higher-order substructural logics. Full Lambek calculus gives a framework to uniformly represent different logical systems as extensions of it. Here we define higher-order Full Lambek calculus, which boils down to higher-order intuitionistic logic when equipped with all the structural rules, and give categorical semantics for (any extension of) it in terms of triposes or higher-order Lawvere hyperdoctrines, which were originally conceived for intuitionistic logic, and yet are flexible enough to be adapted for substructural logics. Relativising the completeness result thus obtained to different axioms, we can obtain tripos-theoretical completeness theorems for a broad variety of higher-order logics. The framework thus developed, moreover, allows us to obtain tripos-theoretical Girard and Kolmogorov translation theorems for higher-order logics.

1 Introduction

Propositional logic corresponds to a class of algebras; for example, the algebras of classical intuitionistic logic are Heyting algebras. What are, then, the algebras of predicate logic? There is seemingly no agreed concept of algebras of predicate logic. Cylindric algebras [11] give a candidate for it. It is not very clear how far and how uniformly cylindric algebraic semantics can be extended so as to treat different sorts of logical systems, especially substructural logics (linear, relevant, fuzzy, etc.). Lawvere's hyperdoctrines [18] give another concept of algebras of predicate logic, and may be seen as a categorical extension of cylindric algebras (see, e.g., Jacobs [13], which gives a fibrational understanding of cylindric algebras; fibrations and hyperdoctrines as indexed categories are connected with each other via the Grothendieck construction). From an algebraic point of view, a hyperdoctrine is a fibred algebra, i.e., an algebra indexed by a category:

$$P : \mathbf{C}^{\mathrm{op}} \to \mathbf{Alg}.$$

Alg is a category of algebras of propositional logic (e.g., Heyting algebras or BI-algebras as in Biering et al. [2]). There are logical conditions to express quantifiers

U. Fahrenberg et al. (Eds.): RAMiCS 2020, LNCS 12062, pp. 187–203, 2020.
https://doi.org/10.1007/978-3-030-43520-2_12

and others as we shall detail below. The intuitive meaning of the base category $\mathbf{C}$ is the category of types (aka. sorts) or domains of discourse, and then $P(C)$ is the algebra of predicates on a type C. And P is called a predicate functor. Roughly, if a propositional logic L is complete with respect to a variety $\mathbf{Alg}_L$, then the corresponding fibred algebras $P : \mathbf{C}^{\mathrm{op}} \to \mathbf{Alg}_L$ yield complete semantics for the predicate logic that extends L. This may be called completeness lifting: the completeness of propositional logic with respect to **Alg** lifts to the completeness of predicate logic with respect to $P : \mathbf{C}^{\mathrm{op}} \to \mathbf{Alg}$. While this completeness lifting is demonstrated for first-order logic in [21], in the present paper, we demonstrate completeness lifting for higher-order logic of different sorts.

In order to represent different logical systems in a uniform setting, we rely upon the framework of substructural logics over Full Lambek calculus FL and their algebras (see, e.g., Galatos-Jipsen-Kowalski-Ono [8]); FL algebras (defined below) play the rôle of **Alg** above. Diverse logical systems can be represented as axiomatic extensions of FL, including classical, intuitionistic, fuzzy, relevant, paraconsistent, and linear logics. In this field, there are vital developments of the correspondence between cut elimination and algebraic completion (see Ciabattoni-Galatos-Terui [3], which focus upon the propositional case, but might possibly be extended to the first-order and higher-order cases via the framework of substructural hyperdctrines). In this paper we think of higher-order Full Lambek calculus, which boils down to higher-order intuitionistic logic (as in Lambek-Scott [17]) when equipped with all the structural rules, and give hyperdoctrine semantics for (any extension of) it. Lawvere's hyperdoctrines were originally for intuitionistic logic; yet they are flexible enough so as to be adapted for a variety of substructural logics as we shall see below. Note that, whilst toposes are impredicative, triposes can have their type theories predicative (e.g., Martin-Löf); the two-level structure of triposes allows more flexibility than toposes do.

There is a tight connection between toposes and higher-order hyperdoctrines, which are also called triposes (for triposes, see, e.g., Hyland-Johnstone-Pitts [12] and Pitts [26]; there are actually several non-equivalent definitions of triposes; we simply call higher-order hyperdoctrines triposes). Indeed, toposes and triposes correspond to each other via the two functors of taking subobject hyperdoctrines and of the tripos-to-topos construction (see, e.g., Coumans [4] and Frey [10]); note that the subobject functor Sub of a topos plays the rôle of a predicate functor P above. Both toposes and (intuitionistic) higher-order hyperdoctrines give complete semantics for higher-order intuitionistic logic; the completeness result of this paper generalises this classic result quite vastly in terms of higher-order substructural hyperdoctrines or triposes. The contributions of this paper may be summarised succinctly as follows: (i) higher-order completeness via Full Lambek triposes, which can be instantiated for a broad variety of logical systems; (ii) tripos-theoretical Girard's ! translation and Kolmogorov's $\neg\neg$ translation theorems for higher-order logic, in which the internal language of triposes is at work. As illustrated by the translation theorems, the general framework of the present paper allows us to compare different categorical logics within the one setting (many categorical logics have only been developed locally so far; there has been no global framework to compare them in the same setting).

The rest of the paper is organised as follows. We first present the syntax of Higher-order Full Lambek calculus HoFL, which obtains by adapting higher-order intuitionistic logic to Full Lambek calculus FL. And we introduce the concept of Full Lambek tripos (FL tripos for short; aka. higher-order FL hyperdoctrine; for brevity we use the former terminology), thereby obtaining the higher-order completeness theorem for HoFL. Finally, our general framework thus developed is applied, via the internal language of FL triposes, to the categorical analysis of Girard's and Kolmogorov's translation for higher-order logics.

2 Higher-Order Full Lambek Calculus

In this section we introduce Higher-order Full Lambek calculus HoFL, which extends quantified FL as in Ono [23,24] so that HoFL equipped with all the structural rules boils down to higher-order intuitionistic logic, the logic of toposes (see Lambek-Scott [17], Jacobs [13], or Johnstone [15]). Our presentation of HoFL, especially its type-theoretic part, follows the style of Pitts [25]; thus we write, e.g., "$t : \sigma\ [\Gamma]$" and "$\varphi\ [\Gamma]$", rather than "$\Gamma \vdash t : \sigma$" and "$\Gamma \vdash \varphi$", respectively, where t is a term of type σ in context Γ, and φ is a formula in context Γ.

HoFL is a so-called "logic over type theory" or "logic-enriched type theory" in Aczel's terms; there is an underlying type theory, upon which logic is built (see, e.g., Jacobs [13]). To begin with, let us give a bird's-eye view of the structure of HoFL. The type theory of HoFL is given by simply typed λ-calculus with finite product types (i.e., 1 and $\times$; these amount to the structure of CCCs, cartesian closed categories), and moreover, with the special, distinguished type

$$\mathrm{Prop}$$

which is a "proposition" type, intended to represent a truth-value object Ω on the categorical side. The logic of HoFL is given by Full Lambek calculus FL. The Prop type plays the key rôle of reflecting the logical or propositional structure into the type or term structure: every formula or proposition φ may be seen as a term of type Prop. This is essentially what the subobject classifier Ω of a topos $\mathbf{E}$ is required to satisfy, that is,

$$\mathrm{Sub}_{\mathbf{E}}(\text{-}) \simeq \mathrm{Hom}_{\mathbf{E}}(\text{-}, \Omega).$$

Spelling out the meaning of this axiom in logical terms, we have got

$$\mathrm{Pred}(\sigma) \simeq \mathrm{Term}(\sigma, \mathrm{Prop})$$

which means the structure of predicates on each type σ (or context Γ in general) is isomorphic to the structure of terms from σ to Prop. The logical meaning of Ω may thus be summarised by a sort of reflection principle, namely the reflection of the propositional structure into the type structure, which may also be called the "propositions-as-terms" or "propositions-as-functions" correspondence, arguably lying at the heart of higher-order categorical logic, for Ω would presumably be the raison d'être of higher-order categorical logic (toposes are CCCs with Ω).

The power type $P\sigma$ of a given type σ can be defined in the present framework as $\sigma \to \mathrm{Prop}$; the comprehension term $\{x : \sigma \mid \varphi\} : P\sigma$ and the membership predicate $s \in t : \mathrm{Prop}$ are definable via λ-abstraction or currying (categorically, transposing) and λ-application (categorically, evaluation), respectively. That is, $\{x : \sigma \mid \varphi\}$ may be defined as $\lambda x : \sigma.\ \varphi$ where φ is seen as a term of type Prop, and also $s \in t$ may be defined as ts where $t : \sigma \to \mathrm{Prop}$ and $s : \sigma$. These definable operations allow us to express set-theoretical reasoning in higher-order logic. There is, of course, some freedom on the choice of primitives, just as toposes can be defined in terms of either subobject classifiers or power objects. All this is to facilitate an intuitive understanding of the essential features of higher-order logic; we give a formal account below.

The syntactic details of HoFL are as follows. HoFL is equipped with the following logical connectives of Full Lambek calculus:

$$\otimes, \wedge, \vee, \backslash, /, 1, 0, \top, \bot, \forall, \exists.$$

The non-commutativity of HoFL gives rise to two kinds of implication ($\backslash$ and $/$). We have basic variables and types, denoted by letters like x and σ, respectively. And as usual $x : \sigma$ is a formal expression to say that a variable x is of type σ. Note that every variable must be typed in HoFL, unlike untyped FL. A context is a finite list of typings of variables: $x_1 : \sigma_1, ..., x_n : \sigma_n$ which is often abbreviated as Γ. Formulae and terms are then defined within specific contexts. There are relation symbols and function symbols, both in context: $R(x_1, ..., x_n)\ [x_1 : \sigma_1, ..., x_n : \sigma_n]$ is a formal expression to say that R is a relation symbol with variables $x_1, ..., x_n$ of types $\sigma_1, ..., \sigma_n$ respectively; and also $f : \tau\ [x_1 : \sigma_1, ..., x_n : \sigma_n]$ is a formal expression to say that f is a function symbol with its domain (the product of) $\sigma_1, ..., \sigma_n$ and with its codomain τ.

The type constructors of HoFL are product $\times$, function space $\to$, and the proposition type Prop, which is a nullary type constructor. The term constructors of $\times$ and $\to$ are as usual: pairing $\langle \text{-}, \text{-} \rangle$ and (first and second) projections π_1, π_2 for product $\times$, and λ-abstraction and λ-application for function space $\to$. The term constructors of Prop are all the logical connectives of Full Lambek calculus as listed above, the relation symbols taken to be of type Prop and thus working as generators of the terms of type Prop. Formulae in context, $\varphi\ [\Gamma]$, and terms in context, $t : \tau\ [\Gamma]$, are then defined in the usual, inductive manner (our terminology and notation mostly follow Pitts [25]; we are extending his framework so as to encompass higher-order substructural logics). Finally, sequents in contexts are defined as:

$$\Phi \vdash \varphi\ [\Gamma]$$

where Γ is a context, Φ is a finite list of formulae $\varphi_1, ..., \varphi_n$, and all the formulae involved are in context Γ.

So far we have not touched upon any axiom (or inference rule) involved. In the following, we first give axioms for terms, and then for sequents. The axioms for $\times$ and $\to$ are as usual (see, e.g., Pitts [25]). The axiom for Prop is as follows:

$$\frac{\varphi \vdash \psi\ [\Gamma] \quad \psi \vdash \varphi\ [\Gamma]}{\varphi = \psi : \mathrm{Prop}\ [\Gamma]}\ (prop)$$

This axiom relates the structure of propositions to that of terms, thus guaranteeing the aforementioned "propositions-as-functions" correspondence for higher-order categorical logic. There are several standard rules for contexts and substitution, which are the same as those in Pitts [25] (we do not repeat them here, referring to the Sect. 2 of Pitts [25] for the details). We now turn to inference rules for sequents. We first have the identity and cut rules as follows:

$$\frac{}{\varphi \vdash \varphi\ [\Gamma]}\ (id) \qquad \frac{\Phi_1 \vdash \varphi\ [\Gamma] \quad \Phi_2, \varphi, \Phi_3 \vdash \psi\ [\Gamma]}{\Phi_2, \Phi_1, \Phi_3 \vdash \psi\ [\Gamma]}\ (cut)$$

where ψ may be empty; this applies to the following L (Left) rules as well. Note that HoFL has no structural rule other than the cut rule. The rules governing the use of the logical connectives are as follows.

$$\frac{\Phi, \varphi, \psi, \Psi \vdash \chi\ [\Gamma]}{\Phi, \varphi \otimes \psi, \Psi \vdash \chi\ [\Gamma]}\ (\otimes L) \qquad \frac{\Phi \vdash \varphi\ [\Gamma] \quad \Psi \vdash \psi\ [\Gamma]}{\Phi, \Psi \vdash \varphi \otimes \psi\ [\Gamma]}\ (\otimes R)$$

$$\frac{\Phi, \varphi, \Psi \vdash \chi\ [\Gamma]}{\Phi, \varphi \wedge \psi, \Psi \vdash \chi\ [\Gamma]}\ (\wedge L_1) \qquad \frac{\Phi, \varphi, \Psi \vdash \chi\ [\Gamma]}{\Phi, \psi \wedge \varphi, \Psi \vdash \chi\ [\Gamma]}\ (\wedge L_2)$$

$$\frac{\Phi \vdash \varphi\ [\Gamma] \quad \Phi \vdash \psi\ [\Gamma]}{\Phi \vdash \varphi \wedge \psi\ [\Gamma]}\ (\wedge R)$$

$$\frac{\Phi, \varphi, \Psi \vdash \chi\ [\Gamma] \quad \Phi, \psi, \Psi \vdash \chi\ [\Gamma]}{\Phi, \varphi \vee \psi, \Psi \vdash \chi\ [\Gamma]}\ (\vee L)$$

$$\frac{\Phi \vdash \varphi\ [\Gamma]}{\Phi \vdash \varphi \vee \psi\ [\Gamma]}\ (\vee R_1) \qquad \frac{\Phi \vdash \varphi\ [\Gamma]}{\Phi \vdash \psi \vee \varphi\ [\Gamma]}\ (\vee R_2)$$

$$\frac{\Phi \vdash \varphi\ [\Gamma] \quad \Psi_1, \psi, \Psi_2 \vdash \chi\ [\Gamma]}{\Psi_1, \Phi, \varphi \backslash \psi, \Psi_2 \vdash \chi\ [\Gamma]}\ (\backslash L) \qquad \frac{\varphi, \Phi \vdash \psi\ [\Gamma]}{\Phi \vdash \varphi \backslash \psi\ [\Gamma]}\ (\backslash R)$$

$$\frac{\Phi \vdash \varphi\ [\Gamma] \quad \Psi_1, \psi, \Psi_2 \vdash \chi\ [\Gamma]}{\Psi_1, \psi / \varphi, \Phi, \Psi_2 \vdash \chi\ [\Gamma]}\ (/L) \qquad \frac{\Phi, \varphi \vdash \psi\ [\Gamma]}{\Phi \vdash \psi / \varphi\ [\Gamma]}\ (/R)$$

$$\frac{\Psi_1, \Psi_2 \vdash \varphi\ [\Gamma]}{\Psi_1, 1, \Psi_2 \vdash \varphi\ [\Gamma]}\ (1L) \qquad \frac{}{\vdash 1\ [\Gamma]}\ (1R)$$

$$\frac{}{0 \vdash \ [\Gamma]}\ (0L) \qquad \frac{\Phi \vdash \ [\Gamma]}{\Phi \vdash 0\ [\Gamma]}\ (0R)$$

$$\frac{}{\Phi \vdash \top\ [\Gamma]}\ (\top R) \qquad \frac{}{\Phi_1, \bot, \Phi_2 \vdash \varphi\ [\Gamma]}\ (\bot L)$$

$$\frac{\Phi_1, \varphi, \Phi_2 \vdash \psi\ [x : \sigma, \Gamma]}{\Phi_1, \forall_\sigma x \varphi, \Phi_2 \vdash \psi\ [x : \sigma, \Gamma]}\ (\forall L) \qquad \frac{\Phi \vdash \varphi\ [x : \sigma, \Gamma]}{\Phi \vdash \forall_\sigma x \varphi\ [\Gamma]}\ (\forall R)$$

$$\frac{\Phi_1, \varphi, \Phi_2 \vdash \psi\ [x:\sigma, \Gamma]}{\Phi_1, \exists_\sigma x\varphi, \Phi_2 \vdash \psi\ [\Gamma]}\ (\exists L) \qquad \frac{\Phi \vdash \varphi\ [x:\sigma, \Gamma]}{\Phi \vdash \exists_\sigma x\varphi\ [x:\sigma, \Gamma]}\ (\exists R)$$

There are eigenvariable conditions on the quantification rules: x must not appear as a free variable in the bottom sequents of the $\forall R$ and $\exists L$ rules. We write $\forall x$ and $\exists x$ when the type of x is obvious. These are all of the rules of HoFL; the provability of sequents in context is defined in the usual way. The essential difference from the first-order case is the existence of function and truth value types; they are what make the logic higher-order, enabling set-theoretical reasoning.

For a collection X of axiom schemata (which we often simply call axioms), let us denote by HoFL_X the axiomatic extension of HoFL via X. In particular, we can recover higher-order intuitionistic logic as HoFL_{ecw}, i.e., by adding to HoFL the exchange, weakening, and contraction rules (as axiom schemata).

Lemma 1. *The following sequents-in-context are deducible in* HoFL*:*

- *(i)* $\varphi \otimes (\exists x\psi) \vdash \exists x(\varphi \otimes \psi)\ [\Gamma]$ *and* $\exists x(\varphi \otimes \psi) \vdash \varphi \otimes (\exists x\psi)\ [\Gamma]$*;*
- *(ii)* $(\exists x\psi) \otimes \varphi \vdash \exists x(\psi \otimes \varphi)\ [\Gamma]$ *and* $\exists x(\psi \otimes \varphi) \vdash (\exists x\psi) \otimes \varphi\ [\Gamma]$

where it is supposed that φ *does not contain* x *as a free variable, and* Γ *contains type declarations on those free variables that appear in* φ *and* $\exists x\psi$*.*

As explained in [21], typed logic allows domains of discourse to be empty; they must be non-empty in the Tarski semantics. A type σ can be interpreted as an initial object in a category. We need no ad hoc condition on domains of discourse if we work with typed logic. This is due to Joyal as noted in Marquis and Reyes [19]. Proof-theoretically, the following is not deducible in HoFL: $\forall x\varphi \vdash \exists x\varphi\ [\,]$. Still the following is deducible: $\forall x\varphi \vdash \exists x\varphi\ [x:\sigma, \Gamma]$. That is, we can prove the sequent above when a type σ is inhabited (see [21] for more details).

3 Full Lambek Tripos

The algebras of propositional FL are FL algebras, the definition of which is reviewed below. The algebras of first-order FL are arguably FL hyperdoctrines; note that complete FL algebras only give us completeness in the presence of the *ad hoc* condition of so-called safe valuations (cf. [24]), and yet FL hyperdoctrines allow us to prove completeness without any such *ad hoc* condition, and at the same time, to recover the complete FL algebra semantics as a special, set-theoretical instance of the FL hyperdoctrine semantics (in a nutshell, the condition of safe valuations is only necessary to show completeness with respect to the restricted class of FL hyperdoctrines with the category of sets their base categories). In this section we define FL triposes, which are arguably the (fibred) algebras of higher-order FL, and prove higher-order completeness, again without any ad hoc condition such as safe valuations or Henkin-style restrictions on quantification (set-theoretical semantics is only complete under this condition).

Definition 2. *$(A, \otimes, \wedge, \vee, \backslash, /, 1, 0, \top, \bot)$ is an* FL *algebra iff the following hold:*

- *$(A, \otimes, 1)$ is a monoid; 0 is a distinguished element of A;*
- *$(A, \wedge, \vee, \top, \bot)$ is a bounded lattice;*
- *for any $a \in A$, $a\backslash(\text{-}) : A \to A$ is a right adjoint of $a \otimes (\text{-}) : A \to A$: $a \otimes b \leq c$ iff $b \leq a\backslash c$ for any $a, b, c \in A$;*
- *for any $b \in A$, $(\text{-})/b : A \to A$ is a right adjoint of $(\text{-}) \otimes b : A \to A$: $a \otimes b \leq c$ iff $a \leq c/b$. for any $a, b, c \in A$.*

A homomorphism of FL *algebras is required to preserve all the operations of* FL *algebras. Let* **FL** *denote the category of* FL *algebras and their homomorphisms.*

FL is an algebraic category (namely, a category monadic over the category of sets; see [1]), and then an axiomatic extension FL_X of FL corresponds to an algebraic subcategory of **FL**, which shall be denoted $\mathbf{FL}_X$. Note that algebraic categories are called varieties or equational classes in universal algebra.

Definition 3. *An* FL *(Full Lambek) hyperdoctrine is a contravariant functor*

$$P : \mathbf{C}^{\mathrm{op}} \to \mathbf{FL}$$

such that the base category $\mathbf{C}$ of P is a category with finite products, and that the following conditions (to express quantifiers) are satisfied:

- *For any projection $\pi : X \times Y \to Y$ in $\mathbf{C}$, $P(\pi) : P(Y) \to P(X \times Y)$ has a right adjoint, denoted $\forall_\pi : P(X \times Y) \to P(Y)$. And the corresponding Beck-Chevalley condition holds, i.e., the following diagram commutes for any arrow $f : Z \to Y$ in $\mathbf{C}$ ($\pi' : X \times Z \to Z$ below denotes a projection):*

$$\begin{array}{ccc} P(X \times Y) & \xrightarrow{\forall_\pi} & P(Y) \\ {\scriptstyle P(X\times f)}\downarrow & & \downarrow{\scriptstyle P(f)} \\ P(X \times Z) & \xrightarrow[\forall_{\pi'}]{} & P(Z) \end{array}$$

- *For any projection $\pi : X \times Y \to Y$ in $\mathbf{C}$, $P(\pi) : P(Y) \to P(X \times Y)$ has a left adjoint, denoted $\exists_\pi : P(X \times Y) \to P(Y)$. The corresponding Beck-Chevalley condition holds:*

$$\begin{array}{ccc} P(X \times Y) & \xrightarrow{\exists_\pi} & P(Y) \\ {\scriptstyle P(X\times f)}\downarrow & & \downarrow{\scriptstyle P(f)} \\ P(X \times Z) & \xrightarrow[\exists_{\pi'}]{} & P(Z) \end{array}$$

  Furthermore, the Frobenius Reciprocity conditions hold: for any projection $\pi : X \times Y \to Y$ in $\mathbf{C}$, any $a \in P(Y)$, and any $b \in P(X \times Y)$,

$$a \otimes (\exists_\pi b) = \exists_\pi (P(\pi)(a) \otimes b)$$
$$(\exists_\pi b) \otimes a = \exists_\pi (b \otimes P(\pi)(a)).$$

The logical reading of the Beck-Chevalley conditions above is that substitution commutes with quantification.

Now, FL triposes are defined as FL hyperdoctrines with their base categories CCCs, and with truth-value objects Ω (i.e., representability via $\Omega \in \mathbf{C}$):

Definition 4. *An* FL *(Full Lambek) tripos, or higher-order* FL *hyperdoctrine, is an* FL *hyperdoctrine* $P : \mathbf{C}^{\mathrm{op}} \rightarrow \mathbf{FL}$ *such that:*

- *The base category* $\mathbf{C}$ *is a CCC (Cartesian Closed Category);*
- *There is an object* $\Omega \in \mathbf{C}$ *such that*

$$P \simeq \mathrm{Hom}_{\mathbf{C}}(\text{-}, \Omega).$$

We then call Ω *the truth-value object of the* FL *tripos* P. *Given a set* X *of axioms, an* FL_X *tripos is defined by replacing* $\mathbf{FL}$ *above with* $\mathbf{FL}_X$.

For an FL tripos P, each $P(C)$ is called a fibre of the FL tripos P from a fibrational point of view; intuitively, $P(C)$ may be seen as the algebra of propositions on a type or domain of discourse C. Note that it is also possible to define FL triposes in terms of fibrations, even though the present formulation in terms of indexed categories would be categorically less demanding.

FL tripos semantics for HoFL is defined as follows.

Definition 5. *Let* $P : \mathbf{C}^{\mathrm{op}} \rightarrow \mathbf{FL}$ *be an* FL *tripos. An interpretation* $[\![\text{-}]\!]$ *of* HoFL *in the* FL *tripos* P *is defined as follows. Types and atomic symbols are interpreted in the following way:*

- *each basic type* σ *is interpreted as an object* $[\![\sigma]\!]$ *in* $\mathbf{C}$;
- *product and function types,* $\sigma \times \tau$ *and* $\sigma \rightarrow \tau$, *are interpreted, as usual, by categorical product and exponentiation;*
- *each function symbol* $f : \tau\ [\Gamma]$ *is interpreted as an arrow*

$$[\![f : \tau\ [\Gamma]]\!] : [\![\Gamma]\!] \rightarrow [\![\sigma]\!]$$

  in $\mathbf{C}$; *if the context* Γ *is* $x_1 : \sigma_1, ..., x_n : \sigma_n$, *then* $[\![\Gamma]\!]$ *denotes* $[\![\sigma_1]\!] \times ... \times [\![\sigma_n]\!]$;
- *each relation symbol* $R\ [\Gamma]$ *is interpreted as an element* $[\![R\ [\Gamma]]\!]$ *in the corresponding fibre* $P([\![\Gamma]\!])$ *of the* FL *tripos* P *at* $[\![\Gamma]\!]$.

Terms and their equality are interpreted in the following, inductive manner:

- $[\![x : \sigma\ [\Gamma_1, x : \sigma, \Gamma_2]]\!]$ *is defined as the following projection in* $\mathbf{C}$:

$$\pi : [\![\Gamma_1]\!] \times [\![\sigma]\!] \times [\![\Gamma_2]\!] \rightarrow [\![\sigma]\!].$$

- $[\![f(t_1, ..., t_n) : \tau\ [\Gamma]]\!]$ *is defined as the following arrow in* $\mathbf{C}$:

$$[\![f]\!] \circ \langle [\![t_1 : \sigma_1 \; [\Gamma]]\!], ..., [\![t_n : \sigma_n \; [\Gamma]]\!] \rangle$$

where $f : \tau \; [x_1 : \sigma_1, ..., x_n : \sigma_n]$, *and* $t_1 : \sigma_1 \; [\Gamma], ..., t_n : \sigma_n \; [\Gamma]$ *(note also that* $\langle [\![t_1 : \sigma_1 \; [\Gamma]]\!], ..., [\![t_n : \sigma_n \; [\Gamma]]\!] \rangle$ *denotes the product/pairing of arrows in* $\mathbf{C}$*).*
- λ*-abstraction,* λ*-application, projections, and pairing are interpreted, as usual, by categorical transpose, evaluation, projections, and pairing in the base CCC* $\mathbf{C}$*, respectively;*

Formulae are interpreted in the following, inductive manner:

- $[\![R(t_1, ..., t_n) \; [\Gamma]]\!]$ *is defined as*

$$P(\langle [\![t_1 : \sigma_1[\Gamma]]\!], ..., [\![t_n : \sigma_n[\Gamma]]\!] \rangle)([\![R \; [x : \sigma_1, ..., x_n : \sigma_n]]\!])$$

  where R *is a relation symbol in context* $x_1 : \sigma_1, ..., x_n : \sigma_n$.
- $[\![\varphi \otimes \psi \; [\Gamma]]\!]$ *is defined as* $[\![\varphi \; [\Gamma]]\!] \otimes [\![\psi \; [\Gamma]]\!]$*. The other binary connectives* $\wedge, \vee, \backslash, /$ *are interpreted in a similar way.* $[\![1 \; [\Gamma]]\!]$ *is defined as the monoidal unit of* $P([\![\Gamma]\!])$*. The other constants* $0, \top, \bot$ *are interpreted in a similar way.*
- $[\![\forall x \varphi \; [\Gamma]]\!]$ *is defined as* $\forall_\pi([\![\varphi \; [x : \sigma, \Gamma]]\!])$ *where* $\pi : [\![\sigma]\!] \times [\![\Gamma]\!] \to [\![\Gamma]\!]$ *is a projection in* $\mathbf{C}$*, and* φ *is a formula in context* $[x : \sigma, \Gamma]$*. Similarly,* $[\![\exists x \varphi \; [\Gamma]]\!]$ *is defined as* $\exists_\pi([\![\varphi \; [x : \sigma, \Gamma]]\!])$.

Prop *and its terms are then interpreted as follows:*

- Prop *is interpreted as the truth-value object* Ω *of the* FL *tripos* P*:*

$$[\![\mathrm{Prop}]\!] = \Omega;$$

- *each formula* φ : Prop $[\Gamma]$*, regarded as a term of type* Prop*, is interpreted as the element of* $\mathrm{Hom}_{\mathbf{C}}([\![\Gamma]\!], \Omega)$ *which corresponds to* $[\![\varphi \; [\Gamma]]\!] \in P([\![\Gamma]\!])$ *in the defining isomorphism* $P \simeq \mathrm{Hom}_{\mathbf{C}}(\text{-}, \Omega)$ *of the* FL *tripos* P*; in a nutshell,* $[\![\varphi : \mathrm{Prop} \; [\Gamma]]\!]$*'s and* $[\![\varphi \; [\Gamma]]\!]$*'s are linked via the isomorphism.*

Finally, the validity of sequents in context is defined as follows:

- $\varphi_1, ..., \varphi_n \vdash \psi \; [\Gamma]$ *is valid in an interpretation* $[\![\text{-}]\!]$ *in an* FL *tripos* P *iff the following holds in* $P([\![\Gamma]\!])$*:*

$$[\![\varphi_1 \; [\Gamma]]\!] \otimes ... \otimes [\![\varphi_n \; [\Gamma]]\!] \leq [\![\psi \; [\Gamma]]\!].$$

  In case the right-hand side of a sequent is empty, $\varphi_1, ..., \varphi_n \vdash [\Gamma]$ *is valid in* $[\![\text{-}]\!]$ *iff* $[\![\varphi_1 \; [\Gamma]]\!] \otimes ... \otimes [\![\varphi_n \; [\Gamma]]\!] \leq 0$ *in* $P([\![\Gamma]\!])$*. In case the left-hand side of a sequent is empty,* $\vdash \varphi \; [\Gamma]$ *is valid in* $[\![\text{-}]\!]$ *iff* $1 \leq [\![\varphi[\Gamma]]\!]$ *in* $P([\![\Gamma]\!])$*. When* Φ *consists of* $\varphi_1, ..., \varphi_n$*, let* $[\![\Phi \; [\Gamma]]\!]$ *denote* $[\![\varphi_1 \; [\Gamma]]\!] \otimes ... \otimes [\![\varphi_n \; [\Gamma]]\!]$.

An interpretation of HoFL_X *in an* FL_X *tripos is defined by replacing* **FL** *and* HoFL *above with* $\mathbf{FL}_X$ *and* HoFL_X*, respectively.*

The categorical conception of interpretation encompasses set-theoretical interpretations and forcing-style model constructions. First of all, interpreting logic in the **2**-valued tripos $\mathrm{Hom}_{\mathbf{Set}}(\text{-}, \mathbf{2})$ (where **2** is the two-element Boolean algebra) is precisely equivalent to the standard Tarski semantics. Yet there is a vast generalisation of this: given a quantale Ω, the representable functor

$$\mathrm{Hom}_{\mathbf{Set}}(\text{-}, \Omega) : \mathbf{Set}^{\mathrm{op}} \to \mathbf{FL}$$

forms an FL tripos, which gives rise to a universe of quantale-valued sets via the generalised tripos-to-topos construction as in [21]; if Ω is a locale in particular (i.e., complete Heyting algebra), it is known that $\mathrm{Hom}_{\mathbf{Set}}(\text{-}, \Omega)$ yields $\mathbf{Sh}(\Omega)$ (i.e., the sheaf topos on Ω). This sort of FL tripos models of set theory could hopefully be applied to solve consistency problems for substructural set theories (especially, Cantor-Łukasiewicz set theory).

Note that the base category of an FL tripos is used to interpret the type theory of HoFL, and the value category is used to interpret the logic part of HoFL. In the following, we first prove soundness and then completeness.

Proposition 6. *If $\Phi \vdash \psi\ [\Gamma]$ is provable in* HoFL *(resp.* HoFL_X*), then it is valid in any interpretation in any* FL *(resp.* FL_X*) tripos.*

Proof. Let P be an FL or FL_X tripos, and $[\![\text{-}]\!]$ an interpretation in P. Soundness for the first-order part can be proven in essentially the same way as in [21]; due to space limitations, we do not repeat it, and focus upon Prop, which is the most distinctive part of higher-order logic. So let us prove that the rule for the Prop type preserves validity. Suppose that

$$[\![\varphi\ [\Gamma]]\!] \leq [\![\psi\ [\Gamma]]\!]$$

and that

$$[\![\psi\ [\Gamma]]\!] \leq [\![\varphi\ [\Gamma]]\!].$$

It then follows that

$$[\![\varphi\ [\Gamma]]\!] = [\![\psi\ [\Gamma]]\!].$$

Note that this is a "propositional" equality, i.e., an equality in the fibre $P([\![\Gamma]\!])$ of propositions on $[\![\Gamma]\!]$. Since we have the following isomorphism

$$P([\![\Gamma]\!]) \simeq \mathrm{Hom}_{\mathbf{C}}([\![\Gamma]\!], [\![\mathrm{Prop}]\!])$$

the equality above, together with the definition of the interpretation of terms of type Prop, tells us that

$$[\![\varphi : \mathrm{Prop}\ [\Gamma]]\!] = [\![\psi : \mathrm{Prop}\ [\Gamma]]\!].$$

Note that this is a "functional" equality, i.e., an equality in $\mathrm{Hom}_{\mathbf{C}}([\![\Gamma]\!], [\![\mathrm{Prop}]\!])$. Thus, the propositional equality implies the functional equality (via the isomorphism above), and this is exactly what it is for the Prop rule to preserve validity. □

For the sake of a completeness proof, let us introduce the syntactic tripos construction (for logic over type theory), which is the combination of the syntactic category construction (for type theory) and the Lindenbaum-Tarski algebra construction (for propositional logic):

Definition 7. *The syntactic tripos of* HoFL *is defined as follows. Let us first define the syntactic base category* $\mathbf{C}$*: an object is a context* Γ *(up to* α*-equivalence); an arrow from* Γ *to* Γ' *is a list of terms (up to equality on terms)*

$$[t_1, ..., t_n]$$

where $t_1 : \sigma_1\ [\Gamma], ..., t_n : \sigma_n\ [\Gamma]$ *and* Γ' *is supposed to be* $x_1 : \sigma_1, ..., x_n : \sigma_n$*. Composition is defined via substitution. The syntactic tripos* $P_{\mathrm{HoFL}} : \mathbf{C}^{\mathrm{op}} \to \mathbf{FL}$ *is then defined as follows. Given an object* Γ *in* $\mathbf{C}$*, let* Form_Γ *denote the set of formulas in context* Γ*, and then define*

$$P_{\mathrm{HoFL}}(\Gamma) = \mathrm{Form}_\Gamma / \sim$$

where $\sim$ *is an equivalence relation on* Form_Γ *defined as follows: for* $\varphi, \psi \in \mathrm{Form}_\Gamma$*,* $\varphi \sim \psi$ *iff* $\varphi \vdash \psi\ [\Gamma]$ *and* $\psi \vdash \varphi\ [\Gamma]$ *are provable in* HoFL*. The arrow part of* P_{HoFL} *is defined as follows. Let* $[t_1, ..., t_n] : \Gamma \to \Gamma'$ *be an arrow in* $\mathbf{C}$ *where* Γ' *is* $x_1 : \sigma_1, ..., x_n : \sigma_n$*. Then we define* $P_{\mathrm{HoFL}}([t_1, ..., t_n]) : P_{\mathrm{HoFL}}(\Gamma') \to P_{\mathrm{HoFL}}(\Gamma)$ *by*

$$P_{\mathrm{HoFL}}([t_1, ..., t_n])(\varphi\ [\Gamma']) = \varphi[t_1/x_1, ..., t_n/x_n]\ [\Gamma]$$

where it is supposed that $t_1 : \sigma_1\ [\Gamma], ..., t_n : \sigma_n\ [\Gamma]$*, and that* φ *is a formula in context* $x_1 : \sigma_1, ..., x_n : \sigma_n$*. The syntactic tripos* P_{HoFL_X} *of* HoFL_X *is defined just by replacing* $\mathbf{FL}$ *and* HoFL *above with* $\mathbf{FL}_X$ *and* HoFL_X*, respectively.*

The syntactic tripos of higher-order logic is the fibrational analogue of the Lindenbaum-Tarski algebra of propositional logic; each fibre $P_{\mathrm{HoFL}}(\Gamma)$ of the syntactic tripos P_{HoFL} is the Lindenbaum-Tarski algebra of formulae in context Γ. The syntactic tripos of HoFL has the universal mapping property that inherits from the syntactic base category of the underlying type theory of HoFL, and also from the fibre-wise Lindenbaum-Tarski algebras of the logic part of HoFL. We of course have to verify that the syntactic tripos P_{HoFL} indeed carries an FL tripos structure; this is the crucial part of the completeness proof.

Lemma 8. *The syntactic tripos* $P_{\mathrm{HoFL}} : \mathbf{C}^{\mathrm{op}} \to \mathbf{FL}$ *(resp.* $\mathbf{FL}_X$*) defined above is an* FL *(resp.* FL_X*) tripos. In particular, the base category is a CCC, and there is a truth-value object* $\Omega \in \mathbf{C}$ *such that*

$$P_{\mathrm{HoFL}} \simeq \mathrm{Hom}_{\mathbf{C}}(\text{-}, \Omega).$$

Proof. The existence of products and exponentials in $\mathbf{C}$ is guaranteed by the existence of product types and function space types in the type theory of HoFL. Substitution commutes with all the logical connectives. This means that $P([t_1, ..., t_n])$ defined above is a homomorphism; so P is a contravariant functor.

P has quantifier structures as follows. Let $\pi : \Gamma \times \Gamma' \to \Gamma'$ denote the projection in $\mathbf{C}$ defined above, and consider $P(\pi)$, which has right and left adjoints in the following way. Recall Γ is $x : \sigma_1, ..., x_n : \sigma_n$. Let $\varphi \in P(\Gamma \times \Gamma')$; we identify φ with the equivalence class to which φ belongs. Define $\forall_\pi : P(\Gamma \times \Gamma') \to P(\Gamma')$ by

$$\forall_\pi(\varphi) = \forall x_1 ... \forall x_n \varphi.$$

We also define $\exists_\pi : P(\Gamma \times \Gamma') \to P(\Gamma')$ by $\exists_\pi(\varphi) = \exists x_1 ... \exists x_n \varphi$. Then, $\forall_\pi$ and $\exists_\pi$ give the right and left adjoints of $P(\pi)$, respectively.

We can verify the Beck-Chevalley condition for $\forall$ as follows. Let $\varphi \in P(\Gamma \times \Gamma')$, $\pi : \Gamma \times \Gamma' \to \Gamma'$ a projection in $\mathbf{C}$, and $\pi' : \Gamma \times \Gamma'' \to \Gamma''$ another projection in $\mathbf{C}$ for objects $\Gamma, \Gamma', \Gamma''$ in $\mathbf{C}$. Then,

$$P([t_1, ..., t_n]) \circ \forall_\pi(\varphi) = (\forall x_1 ... \forall x_n \varphi)[t_1/y_1, ..., t_n/y_m]$$

where Γ is supposed to be $x_1 : \sigma_1, ..., x_n : \sigma_n$, Γ' is $y_1 : \tau_1, ..., y_m : \tau_m$, and $t_1 : \tau_1\ [\Gamma''], ..., t_m : \tau_m\ [\Gamma'']$. Likewise we have $\forall_{\pi'} \circ P([t_1, ..., t_n])(\varphi) = \forall x_1 ... \forall x_n (\varphi[t_1/y_1, ..., t_n/y_m])$. The Beck-Chevalley condition for $\forall$ thus follows. The Beck-Chevalley condition for $\exists$ can be verified in a similar way. The two Frobenius Reciprocity conditions for $\exists$ follow immediately from Lemma 1.

In the following we prove the existence of a truth-value object Ω. Let

$$\Omega = x : \mathrm{Prop}.$$

Note that, since the objects of the base category are contexts rather than types, we cannot take Ω to be Prop *per se*; yet x : Prop practically means the same thing as Prop, thanks to α-equivalence required. We now have to show that for each context Γ,

$$P(\Gamma) \simeq \mathrm{Hom}_{\mathbf{C}}(\Gamma, x : \mathrm{Prop})$$

and this correspondence yields a natural transformation. The required isomorphism is given by mapping

$$\varphi\ [\Gamma] \in P(\Gamma)$$

to

$$\varphi : \mathrm{Prop}\ [\Gamma] \in \mathrm{Hom}_{\mathbf{C}}(\Gamma, x : \mathrm{Prop}).$$

Note that φ above is actually an equivalence class, and yet the above mapping is well defined, and also that φ : Prop $[\Gamma]$ is actually a list consisting of a single term φ : Prop $[\Gamma]$. This mapping is an isomorphism by the definition of terms of type Prop. Let us denote the above mapping by

$$\mathrm{PaF}_\Gamma : P(\Gamma) \to \mathrm{Hom}_{\mathbf{C}}(\Gamma, x : \mathrm{Prop})$$

with the idea of "Propositions-as-Functions" in mind. The naturality of this correspondence then means that the following diagram commutes for any arrow $[t_1, ..., t_n] : \Gamma' \to \Gamma$ in $\mathbf{C}$:

$$\begin{array}{ccc}
P(\Gamma) & \xrightarrow{\mathrm{PaF}_\Gamma} & \mathrm{Hom}_{\mathbf{C}}(\Gamma, x : \mathrm{Prop}) \\
{\scriptstyle P([t_1, ..., t_n])} \downarrow & & \downarrow {\scriptstyle \mathrm{Hom}_{\mathbf{C}}([t_1, ..., t_n], x:\mathrm{Prop})} \\
P(\Gamma') & \xrightarrow[\mathrm{PaF}_{\Gamma'}]{} & \mathrm{Hom}_{\mathbf{C}}(\Gamma', x : \mathrm{Prop})
\end{array}$$

By the following calculation:

$$\begin{aligned}\mathrm{Hom}_{\mathbf{C}}([t_1,...,t_n],\mathrm{Prop})\circ\mathrm{PaF}_{\Gamma}(\varphi\ [\Gamma]) &= \mathrm{Hom}_{\mathbf{C}}([t_1,...,t_n],\mathrm{Prop})(\varphi:\mathrm{Prop}\ [\Gamma])\\ &= \varphi[t_1/x_1,...,t_n/x_n]:\mathrm{Prop}\ [\Gamma']\\ &= \mathrm{PaF}_{\Gamma'}(\varphi[t_1/x_1,...,t_n/x_n]\ [\Gamma'])\\ &= \mathrm{PaF}_{\Gamma'}\circ P([t_1,...,t_n])(\varphi\ [\Gamma])\end{aligned}$$

we obtain the commutativity of the diagram and hence the naturality of the "propositions-as-functions" correspondence. □

It is straightforward to see that if $\Phi \vdash \psi\ [\Gamma]$ is valid in the canonical interpretation in the syntactic tripos P_{HoFL} (resp. P_{HoFL_X}), then it is provable in HoFL (resp. HoFL_X). And this immediately gives us completeness via the standard counter-model argument. Hence the higher-order completeness theorem:

Theorem 9. *$\Phi \vdash \psi\ [\Gamma]$ is provable in* HoFL *(resp.* HoFL_X*) iff it is valid in any interpretation in any* FL *(resp.* FL_X*) tripos.*

This higher-order completeness theorem can be applied, with a suitable choice of axioms X, for any of classical, intuitionistic, fuzzy, relevant, paraconsistent, and (both commutative and non-commutative) linear logics; higher-order completeness has not been known for these logics except the first two. The concept of (generalised) tripos, therefore, is so broadly applicable as to encompass most logical systems. Modal logics also can readily be incorporated into this framework by working with modal FL rather than plain FL. Coalgebraic dualities for modal logics (see, e.g., [14,16,20,22]) then yield models of modal triposes for them; these modal issues are to be addressed in subsequent papers.

4 Girard and Kolmogorov Translation for Triposes

We finally analyse Kolmogorov's double negation $\neg\neg$ translation (Kolmogorov found it earlier than Gödel-Gentzen; see Ferreira and Oliva [7]) and Girard's exponential ! translation from a tripos-theoretical point of view.

Propositional Kolmogorov translation algebraically means that, for any Heyting algebra A, the doubly negated algebra $\neg\neg A$, defined as $\{a \in A \mid \neg\neg a = a\}$, always forms a Boolean algebra. This $\neg\neg$ construction extends to a functor from the category **HA** of Heyting algebras to the category **BA** of Boolean algebras. And then the categorical meaning of first-order Kolmogorov translation is that, for any first-order IL hyperdoctrine $P : \mathbf{C}^{\mathrm{op}} \to \mathbf{HA}$ (where IL denotes intuitionistic logic), the following composed functor

$$\neg\neg \circ P : \mathbf{C}^{\mathrm{op}} \to \mathbf{BA}$$

forms a first-order CL hyperdoctrine (where CL denotes classical logic) as in [21]. Yet this strategy does not extend to the higher-order case: in particular, although the base category does not change in the first-order case, in which types and propositions are separated, it must nevertheless be modified in the higher-order case, in which types and propositions interact via Prop or Ω. Technicalities involved get essentially more complicated in the higher-order case. Still, we can construct from a given IL tripos $P : \mathbf{C}^{\mathrm{op}} \to \mathbf{HA}$ a CL tripos

$$P_{\neg\neg} : \mathbf{C}_{\neg\neg}^{\mathrm{op}} \to \mathbf{BA}.$$

For the sake of the description of $\mathbf{C}_{\neg\neg}$ (and $P_{\neg\neg}$), however, we work within the internal language HoFL_P of the tripos $P : \mathbf{C}^{\mathrm{op}} \to \mathbf{FL}$: in HoFL_P, we have types C and terms f corresponding to objects C and arrows f in $\mathbf{C}$, respectively, and also formulae R on a type $C \in \mathbf{C}$ corresponding to elements $R \in P(C)$.

Now we define the translation on the internal language HoFL_P of the tripos P which allows us to describe the double negation category $\mathbf{C}_{\neg\neg}$ mentioned above. The basic strategy of translation is this: we leave everything in HoFL_P as it is, unless it involves the proposition type Ω of HoFL_P; and if something involves Ω, we always put double negation on it. Formally it goes as follows:

Definition 10. *We recursively define the translation on* HoFL_P *as follows.*

- *If* $\varphi : \Omega\ [\Gamma]$ *then we put* $\neg\neg$ *on every sub-formula of* φ *(do the same for* φ *seen as formulae).*
- *If* $t : \sigma\ [\Gamma, x : \Omega, \Gamma']$ *then we replace every occurrence of* x *in* t *by* $\neg\neg x$.
- *If* $t : \Omega \times \sigma\ [\Gamma]$ *then* t *translates into* $\langle \neg\neg\pi_1 t, \pi_2 t\rangle$*; if* $t : \sigma \times \Omega\ [\Gamma]$ *then* t *translates into* $\langle \pi_1 t, \neg\neg\pi_2 t\rangle$.
- *If* $t : \sigma\ [\Gamma, x : \Omega \times \sigma, \Gamma']$ *then we replace every occurrence of* x *in* t *by* $\langle \neg\neg\pi_1 x, \pi_2 x\rangle$*; if* $t : \sigma\ [\Gamma, x : \sigma \times \Omega, \Gamma']$ *then we replace every occurrence of* x *in* t *by* $\langle \pi_1 x, \neg\neg\pi_2 x\rangle$.
- *If* $t : \sigma \to \Omega\ [\Gamma]$ *then* t *translates into* $\lambda x : \sigma.\neg\neg tx$*; if* $t : \Omega \to \sigma\ [\Gamma]$ *then* t *translates into* $\lambda x : \Omega.t\neg\neg x$.
- *If* $t : \sigma\ [\Gamma, x : \sigma \to \Omega, \Gamma']$ *then we replace every occurrence of* x *in* t *by* $\lambda y : \sigma.(\neg\neg x)y$*; if* $t : \sigma\ [\Gamma, x : \Omega \to \sigma, \Gamma']$ *then we replace every occurrence of* x *in* t *by* $\lambda y : \Omega.x\neg\neg y$.
- *Finally, if* $t : \sigma\ [\Gamma]$ *and no* Ω *appears in it, then* t *translates into itself.*

The double negation category $\mathbf{C}_{\neg\neg}$ is then defined as follows: the objects of $\mathbf{C}_{\neg\neg}$ are contexts in HoFL_P up to α-equivalence (which are essentially the same as objects in $\mathbf{C}$), and the arrows of $\mathbf{C}_{\neg\neg}$ are the translations of lists of terms in HoFL_P up to equality on terms, with their composition defined via substitution as usual. This intuitively means that those arrows in $\mathbf{C}$ that involve Ω are double negated in $\mathbf{C}_{\neg\neg}$ whilst the other part of $\mathbf{C}_{\neg\neg}$ remains the same as that of $\mathbf{C}$ (to give the rigorous definition of this, we work within the internal language). Then it is not obvious that $\mathbf{C}_{\neg\neg}$ forms a category again, let alone a CCC. Thus:

Lemma 11. $\mathbf{C}_{\neg\neg}$ *defined above forms a category, in particular a CCC.*

Proof. Since everything involving Ω is doubly negated, we have to verify that all of the relevant categorical structures, that is, composition, identity, projection, paring, evaluation, and transpose, preserve or respect double negation. Here we just give several sample proofs to show essential ideas.

Consider the case of composition. We think of single terms for simplicity. The composition of arrows $t : \sigma\ [x : \Omega]$ and $s : \sigma'\ [y : \sigma]$ in $\mathbf{C}_{\neg\neg}$ (which may be seen as $t : \Omega \to \sigma$ and $s : \sigma \to \sigma'$ in terms of the original category $\mathbf{C}$) is defined as $s[t/y] : \sigma'\ [x : \Omega]$, where every occurrence of x in $s[t/y]$ must have been replaced by $\neg\neg x$ (for $s[t/y]$ to be in $\mathbf{C}_{\neg\neg}$); this is true because every occurrence of x in t is replaced by $\neg\neg x$ by the definition of arrows in $\mathbf{C}_{\neg\neg}$. Likewise, the composition of arrows $t : \sigma'\ [x : \sigma]$ and $s : \Omega\ [y : \sigma']$ in $\mathbf{C}_{\neg\neg}$ is defined as $s[t/y] : \Omega\ [x : \sigma]$, where every sub-formula of $s[t/y]$ is doubly negated by the assumption of $s, t \in \mathbf{C}$; and hence $s[t/y] \in \mathbf{C}$. More complex cases can be proven in a similar way.

Consider the case of identity. Think of an identity on Ω, which is given by $\neg\neg x : \Omega\ [x : \Omega]$. Given $t : \Omega\ [y : \sigma']$ in $\mathbf{C}_{\neg\neg}$, $(\neg\neg x) \circ t$ is defined as $(\neg\neg x)[t/x] : \Omega\ [y : \sigma']$, which equals $\neg\neg t : \Omega\ [y : \sigma']$. By $t \in \mathbf{C}_{\neg\neg}$, t can be written as $\neg\neg t'$, and so $\neg\neg t = \neg\neg\neg\neg t' = \neg\neg t' = t$. Hence $(\neg\neg x) \circ t = t$. Likewise, given $t : \sigma'\ [y : \Omega]$ in $\mathbf{C}_{\neg\neg}$, $t \circ \neg\neg x$ is defined as $t[\neg\neg x/y] : \sigma'\ [x : \Omega]$; since every occurrence of y in t is replaced by $\neg\neg y$ because $t \in \mathbf{C}_{\neg\neg}$ and since $\neg\neg\neg\neg$ is equivalent to $\neg\neg$, we have $t[\neg\neg x/y] = t$, whence $t \circ \neg\neg x = t$. More complex cases can be shown in a similar manner.

To show the existence of finite products and exponentials involving Ω (otherwise it is trivial), it is crucial to check that doubly negated projection, pairing, evaluation, and transpose still play their own rôles, just as doubly negated identity still plays the rôle of identity as we have shown above. □

Finally we obtain the following, tripos-theoretical Kolmogorov translation theorem for higher-order logic, which may also be seen as a translation from classical set theory to intuitionistic set theory (since higher-order logic is basically set theory in logical form).

Theorem 12. *Let* $P : \mathbf{C}^{\mathrm{op}} \to \mathbf{HA}$ *be an* IL *tripos, and* $\mathbf{C}_{\neg\neg}$ *the double negation category as defined above. Then,* $P_{\neg\neg}$ *defined as*

$$\mathrm{Hom}_{\mathbf{C}_{\neg\neg}}(\text{-}, \Omega) : \mathbf{C}_{\neg\neg} \to \mathbf{BA}$$

forms a CL *tripos, called the double negation tripos of* P.

Proof. $\mathbf{C}_{\neg\neg}$ is a CCC by the lemma, and $P_{\neg\neg}$ is represented by Ω. This completes the higher-order part of the proof. Concerning the first-order part, the existence of quantifiers follows from this fact: if φ admits the double negation elimination, then $\neg\neg\forall x\varphi$ and $\neg\neg\exists x\varphi$ are equivalent to $\forall x\neg\neg\varphi$ and $\exists x\neg\neg\varphi$, respectively. □

Note that the hyperdoctrinal Kolmogorov translation does not reduce to the construction of toposes via double negation topology because there are more triposes than toposes in the adjunction between them (all toposes come from triposes, but not *vice versa*). Moreover, our hyperdoctrinal method is designed modularly enough to be applicable to Girard's translation as well as Kolmogorov's. Although Glivenko-type theorems have been shown for substructural propositional and first-order logics (see Ferreira-Ono [6] and Galatos-Ono

[9]), no such result is known for higher-order logic (as to the first-order case, [21] is typed and categorical while [6] is single-sorted and proof-theoretical).

An exponential ! on an FL algebra A is defined as a unary operation satisfying: (i) $a \leq b$ implies $!a \leq !b$; (ii) $!!a = !a \leq a$; (iii) $!\top = 1$; (iv) $!a \otimes !b = !(a \wedge b)$ (Coumans, Gehrke, and van Rooijen [5]). We denote by $\mathbf{FL}_c^!$ the category of commutative FL algebras with !, which are algebras for intuitionistic linear logic. $\mathrm{FL}_c^!$ triposes give sound and complete semantics for higher-order intuitionistic linear logic. The Girard category $\mathbf{C}_!$ of an $\mathrm{FL}_c^!$ tripos $P : \mathbf{C}^{\mathrm{op}} \to \mathbf{FL}_c^!$ is defined by replacing double negation in the above definition of $\mathbf{C}_{\neg\neg}$ with Girard's exponential !. The following is the hyperdoctrinal Girard translation theorem for higher-order logic, which can be shown in basically the same way as above; no such higher-order translation has been known so far.

Theorem 13. *Let $P : \mathbf{C}^{\mathrm{op}} \to \mathbf{FL}_c^!$ be an $\mathrm{FL}_c^!$ tripos (for intuitionistic linear logic), and $\mathbf{C}_!$ the Girard category of P. Define*

$$P_! = \mathrm{Hom}_{\mathbf{C}_!}(\text{-}, \Omega) : \mathbf{C}_! \to \mathbf{HA}.$$

Then, $P_!$ forms an IL *tripos (i.e., $\mathrm{FL}_{ecw}^!$ tripos), called the Girard tripos of P.*

Acknowledgements. The author would like to thank the reviewers of the paper for their numerous, substantial comments and suggestions for improvement. The author hereby acknowledges that this work was supported by JST PRESTO (grant code: JPMJPR17G9), JSPS Kakenhi (grant code: 17K14231), and the JSPS Core-to-Core Program "Mathematical Logic and its Applications".

References

1. Adámek, J., Herrlich, H., Strecker, G.E.: Abstract and Concrete Categories. Wiley, Hoboken (1990)
2. Biering, B., Birkedal, L., Torp-Smith, N.: BI-hyperdoctrines, higher-order separation logic, and abstraction. ACM TOPLAS **29**(5), 24 (2007)
3. Ciabattoni, A., Galatos, N., Terui, K.: Algebraic proof theory for substructural logics. Ann. Pure Appl. Logic **163**, 266–290 (2012)
4. Coumans, D.: Canonical extensions in logic - some applications and a generalisation to categories. Ph.D. thesis, Radboud Universiteit Nijmegen (2012)
5. Coumans, D., Gehrke, M., van Rooijen, L.: Relational semantics for full linear logic. J. Appl. Logic **12**, 50–66 (2014)
6. Farahani, H., Ono, H.: Glivenko theorems and negative translations in substructural predicate logics. Arch. Math. Logic **51**, 695–707 (2012)
7. Ferreira, G., Oliva, P.: On the relation between various negative translations. Logic Constr. Comput. **3**, 227–258 (2012)
8. Galatos, N., Jipsen, P., Kowalski, T., Ono, H.: Residuated Lattices: An Algebraic Glimpse at Substructural Logics. Elsevier, Amsterdam (2007)
9. Galatos, N., Ono, H.: Glivenko theorems for substructural logics over FL. J. Symb. Logic **71**, 1353–1384 (2016)
10. Frey, J.: A 2-categorical analysis of the tripos-to-topos construction arXiv:1104.2776

11. Henkin, L., Monk, J.D., Tarski, A.: Cylindric Algebras. North-Holland, Amsterdam (1971)
12. Hyland, M., Johnstone, P.T., Pitts, A.: Tripos theory. Math. Proc. Cambridge Philos. Soc. **88**, 205–232 (1980)
13. Jacobs, B.: Categorical Logic and Type Theory. Elsevier, Amsterdam (1999)
14. Johnstone, P.T.: Stone Spaces. CUP, Cambridge (1982)
15. Johnstone, P.T.: Sketches of an Elephant. OUP, Oxford (2002)
16. Kupke, C., Kurz, A., Venema, Y.: Stone coalgebras. Theoret. Comput. Sci. **327**, 109–134 (2004)
17. Lambek, J., Scott, P.J.: Introduction to Higher-Order Categorical Logic (1986)
18. Lawvere, F.W.: Adjointness in foundations. Dialectica **23**, 281–296 (1969). Reprinted with the author's retrospective commentary. In: Theory and Applications of Categories, vol. 16, pp. 1–16 (2006)
19. Marquis, J.-P., Reyes, G.: The history of categorical logic: 1963–1977. In: Handbook of the History of Logic, vol. 6, pp. 689–800. Elsevier (2011)
20. Maruyama, Y.: Natural duality, modality, and coalgebra. J. Pure Appl. Algebra **216**, 565–580 (2012)
21. Maruyama, Y.: Full lambek hyperdoctrine: categorical semantics for first-order substructural logics. In: Libkin, L., Kohlenbach, U., de Queiroz, R. (eds.) WoLLIC 2013. LNCS, vol. 8071, pp. 211–225. Springer, Heidelberg (2013). https://doi.org/10.1007/978-3-642-39992-3_19
22. Maruyama, Y.: Duality theory and categorical universal logic. EPTCS **171**, 100–112 (2014)
23. Ono, H.: Algebraic semantics for predicate logics and their completeness. RIMS Kokyuroku **927**, 88–103 (1995)
24. Ono, H.: Crawley completions of residuated lattices and algebraic completeness of substructural predicate logics. Stud. Logica **100**, 339–359 (2012)
25. Pitts, A.: Categorical logic, Chap. 2. In: Handbook of Logic in Computer Science, vol. 5. OUP (2000)
26. Pitts, A.: Tripos theory in retrospect. Math. Struct. Comput. Sci. **12**, 265–279 (2002)

Expressive Power and Succinctness of the Positive Calculus of Relations

Yoshiki Nakamura(✉)

Tokyo Institute of Technology, Tokyo, Japan
nakamura.yoshiki.ny@gmail.com

Abstract. In this paper, we study the expressive power and succinctness of *the positive calculus of relations*. We show that (1) the calculus has the same expressive power as that of three-variable existential positive (first-order) logic in terms of binary relations, and (2) the calculus is exponentially less succinct than three-variable existential positive logic, namely, there is no polynomial-size translation from three-variable existential positive logic to the calculus, whereas there is a linear-size translation in the converse direction. Additionally, we give a more fine-grained expressive power equivalence between the (full) calculus of relations and three-variable first-order logic in terms of the quantifier alternation hierarchy. It remains open whether the calculus of relations is also exponentially less succinct than three-variable first-order logic.

Keywords: Expressive power · Succinctness · The positive calculus of relations · Existential positive logic

1 Introduction

The calculus of (binary) relations (denoted by CoR, for short), which was revived by Tarski [22], is an algebraic system on binary relations. The calculus of relations and relation algebras have many applications in various areas of computer science, e.g., databases, program development and verification, and program semantics (see [8] for more details and references). Certain properties of binary relations can be simply expressed using (in)equational formulas of CoR; for example, the formula $a \cdot a \leq a$ indicates that the binary relation a is transitive, where the symbol $\cdot$ denotes the composition operator of binary relations and the symbol $\leq$ denotes the inclusion relation on sets. In fact, CoR has a high expressive power, namely, the expressive power of CoR is equivalent to *three-variable first-order logic* (denoted by FO^3) in terms of binary relations [9,15,23]. One of the downsides for this high expressive power is that the equational theory of CoR is undecidable [23], even for terms built only from one variable, union, complement, and composition [18].

In this paper, we focus on *the positive calculus of relations* (denoted by PCoR, for short) [2,19], which is a complement-free fragment of CoR. Namely, PCoR

U. Fahrenberg et al. (Eds.): RAMiCS 2020, LNCS 12062, pp. 204–220, 2020.
https://doi.org/10.1007/978-3-030-43520-2_13

terms are built from union, intersection, composition, converse, the identity relation, the empty relation, and the universal relation. PCoR is strictly less expressive than CoR, but its equational theory is decidable [2]. This decidability result also holds when adding a transitive closure operation, thus arriving at Kleene allegory terms [17].

The first contribution of this paper is to show that PCoR has the same expressive power as *three-variable existential positive (first-order) logic* (denoted by EP^3) in terms of binary relations. The standard and linear-size translation from CoR to FO^3 [22, pp. 75–76] naturally specializes into a translation from PCoR to EP^3. Conversely, translations from FO^3 to CoR also exist [23, Sect. 3.9] [9, Sect. 20] [15, Theorem 552], but they generate non-PCoR terms on the EP^3 fragment. Hence, we have to refine them. The translation we propose uses disjunctive/conjunctive normal forms where literals have at most two free variables. We define it from full FO^3 into CoR by relying on a relational sum operation ($\dagger$), which is dual to composition ($\cdot$). By specializing our translation to the various considered fragments, we obtain (1) an exponential-size translation from EP^3 to PCoR, and (2) a perfect match between the quantifier alternation hierarchy in FO^3 and *dot-dagger* ($\cdot$ - $\dagger$) *alternation hierarchy* in CoR. Roughly speaking, it shows that the two operators, $\cdot$ and $\dagger$, from the calculus of relations exactly correspond to the two quantifiers, $\exists$ and $\forall$, of first-order logic.

The second contribution of this paper is to show that PCoR is exponentially less succinct than EP^3, namely, the exponential blowup in translating from EP^3 to PCoR is unavoidable. Hence, the exponential-size translation for EP^3 given in this paper is tight.

Furthermore, we extend the two above results for both transitive closure extensions, namely, we show that PCoR with transitive closure [19] (denoted by PCoR(TC)) has the same expressive power as EP^3 with *variable-confined monadic transitive closure* (denoted by EP^3(v-MTC)) and that PCoR(TC) is exponentially less succinct than EP^3(v-MTC).

Remark 1 (On trade-off between succinctness and tractability). The combined complexity of the (binary-relation) query evaluation problem [24] is the problem to decide for a structure M, a term t, and a pair of nodes in M, whether the pair is in the binary relation denoted by t on the structure M. While PCoR is strictly less succinct than EP^3, PCoR is more tractable than EP^3 in the simple dynamic-programming algorithm for this problem (see, e.g., [14, Proposition 6.6]). While it does not imply a certain computational complexity gap between the two problems, it can be solved in $\mathcal{O}(\|t\| \times \|M\|^2)$-time for PCoR, thanks to this algorithm, if the number of occurrences of $\cdot$ is fixed, while it requires $\mathcal{O}(\|t\| \times \|M\|^3)$-time for EP^3 even if the number of occurrences of $\exists$ is fixed, where $\|t\|$ is the size of t and $\|M\|$ is the cardinality of the domain of M, respectively.

Related Work. *Expressive power* of formal systems is widely studied in mathematical logic and computer science.

An example is that the following systems have the same expressive power in terms of *recognizability over word structures*: regular expressions, deterministic finite automata, non-deterministic finite automata, and monadic second-order logic (MSO) (see, e.g., [6, Sec. 6]). However, these four systems are certainly different in terms of *succinctness*. For example, while there is an (exponential-size) translation from non-deterministic finite automata to deterministic finite automata by the powerset construction, there is no polynomial-size translation [16, Prop. 1]. This can yield significant complexity differences for various problems, e.g., membership, universality, and equivalence testing have different complexity depending on whether we start from a term, an automaton, or a formula. See [12, Thm. 16] and [7, Thm. 11] for the other succinctness gaps among regular expressions, deterministic finite automata, and non-deterministic finite automata. The succinctness gap between MSO and each of the other three systems can be shown by using the computational complexity gap that the equivalence problem is non-elementary for MSO (and even for FO [21, Thm. 5.2]), but is in PSPACE for regular expressions and non-deterministic automata, and almost linear-time for deterministic automata. Additionally, the above like expressive power equivalence is known for FO. The following have the same expressive power: star-free regular expressions, FO^3, and FO (see, e.g., [6, Sec. 6]). In [11], it is shown that FO^3 is exponentially less succinct than FO over unary alphabet words. Another example is that the following classes of formulas have the same expressive power with respect to *boolean queries*: propositional logic formulas, negation normal form formulas, and disjunctive/conjunctive normal form formulas. In [5], the succinctness among a dozen formula classes (including the above ones) is investigated.

In this paper, we compare the succinctness between PCoR and EP^3. To the best of our knowledge, it is the first comparison between the succinctness of the (positive) calculus of relations and those of other systems. Our construction in Sect. 4 is somewhat similar to the construction in [10, Sec. 4.5] in order to show that there is no polynomial-size translation from conjunctive normal form formulas to disjunctive normal form formulas, but is more complicated than the construction. This is because we should consider structures with multiple nodes, whereas it suffices to consider only singleton structures for propositional logic.

Organization. Section 2 provides the definitions of CoR and FO^3 and fragments of them (including PCoR and EP^3), the notions of the expressive power and succinctness, and the standard translation from CoR to FO^3. Section 3 gives a new translation from FO^3 to CoR. Consequently, it is shown that PCoR has the same expressive power as EP^3. Section 4 shows that PCoR is exponentially less succinct than EP^3. Section 5 extends the results in Sects. 3–4 by adding a transitive closure operator. Section 6 concludes this paper.

2 Preliminaries

$\mathbb{N}$ (resp. $\mathbb{N}_+$) denotes the set of all non-negative (resp. positive) integers. For $l, r \in \mathbb{N}$ such that $l \leq r$, $[l, r]$ denotes the set $\{l, \ldots, r\}$ and $[r]$ denotes the set $\{1, \ldots, r\}$. $\#(A)$ denotes the cardinality of a set A.

Let $\mathscr{A}$ be a countably infinite set of binary relation symbols. A *structure* M *(of binary relations)* is a tuple $\langle |M|, \{a^M\}_{a \in \mathscr{A}} \rangle$, where $|M|$ is a non-empty set, and for each $a \in \mathscr{A}$, $a^M \subseteq |M|^2$ is a binary relation on $|M|$. For two structures, M and M', we say that a function $h \colon |M| \to |M'|$ is a *homomorphism* from M to M' if for every $a \in \mathscr{A}$ and every $v, w \in |M|$, if $\langle v, w \rangle \in a^M$, then $\langle h(v), h(w) \rangle \in a^{M'}$.

The Calculus of Relations and Its Fragments. We introduce the calculus of relations (CoR) [22] and its syntactic fragments: the positive calculus of relations (PCoR) [2,19] and *the primitive positive calculus of relations* (denoted by PPCoR, a.k.a. allegory terms with top [20]). The terms of CoR consist of the following basic operations on binary relations. Let X be a set. For two binary relations R and S on the universe X, the *union* $R \cup S$, *intersection* $R \cap S$, and *complement* R^- are defined as the corresponding set-theoretic operators, respectively. The symbols $\mathbf{0}$ and $\top$ are employed to denote the *empty relation* and the *universal relation*, respectively. *Relational composition* (a.k.a. *relational multiplication*) $R \cdot S$ is defined as $\{\langle v, v' \rangle \in X^2 \mid \exists w. \langle v, w \rangle \in R \wedge \langle w, v' \rangle \in S\}$, and *relational sum* $R \dagger S$ is defined as $\{\langle v, v' \rangle \in X^2 \mid \forall w. \langle v, w \rangle \in R \vee \langle w, v' \rangle \in S\}$. In this paper, the *projection* R^π is defined as $\{\langle v_1, v_2 \rangle \in X^2 \mid \langle v_{\pi(1)}, v_{\pi(2)} \rangle \in R\}$ for each function $\pi \colon [2] \to [2]$. The symbol $\mathbf{1}$ is employed to denote the *identity relation*. We now define the syntax and semantics of CoR. The set of *terms* of CoR/PCoR/PPCoR is given by the following grammar, where $a \in \mathscr{A}$:

$$t, s \in \mathrm{Term}^{\mathrm{CoR}} ::= t^\pi \mid a \mid \mathbf{1} \mid \top \mid t \cap s \mid t \cdot s \mid \mathbf{0} \mid t \cup s \mid t \dagger s \mid t^-$$
$$t, s \in \mathrm{Term}^{\mathrm{PCoR}} ::= t^\pi \mid a \mid \mathbf{1} \mid \top \mid t \cap s \mid t \cdot s \mid \mathbf{0} \mid t \cup s$$
$$t, s \in \mathrm{Term}^{\mathrm{PPCoR}} ::= t^\pi \mid a \mid \mathbf{1} \mid \top \mid t \cap s \mid t \cdot s$$

For $k \in \mathbb{N}$, we use t^k (the *k-th iteration of t*) to denote $t^{k-1} \cdot t$ if $k \geq 1$; and $\mathbf{1}$ if $k = 0$, and use $t^{\smile}$ (the *converse* of t) to denote $t^{\{1 \mapsto 2, 2 \mapsto 1\}}$. The *semantics* $[\![t]\!]_M$ of a CoR term t on a structure M is a binary relation on $|M|$, which is defined by: $[\![a]\!]_M := a^M$; $[\![\mathbf{1}]\!]_M := \triangle(|M|)$; $[\![\top]\!]_M := |M|^2$; $[\![\mathbf{0}]\!]_M := \emptyset$; $[\![t \cup s]\!]_M := [\![t]\!]_M \cup [\![s]\!]_M$; $[\![t \cap s]\!]_M := [\![t]\!]_M \cap [\![s]\!]_M$; $[\![t^-]\!]_M := |M|^2 \setminus [\![t]\!]_M$; $[\![t^\pi]\!]_M := [\![t]\!]_M^\pi$; $[\![t \cdot s]\!]_M := [\![t]\!]_M \cdot [\![s]\!]_M$; $[\![t \dagger s]\!]_M := [\![t]\!]_M \dagger [\![s]\!]_M$, where $\triangle(X)$ denotes the *diagonal relation* (i.e., $\{\langle v, w \rangle \in X^2 \mid v = w\}$). The *size* $\|t\|$ of a CoR term t is defined by: $\|a\| := \|\mathbf{1}\| := \|\top\| := \|\mathbf{0}\| := 1$, $\|t \cup s\| := \|t \cap s\| := \|t \cdot s\| := \|t \dagger s\| := 1 + \|t\| + \|s\|$, and $\|t^-\| := \|t^\pi\| := 1 + \|t\|$.

Remark 2 (Projection and converse). As usual (e.g., [22]), t^π is defined only when t^π denotes the converse of t. This is because in the other cases, t^π can be expressed by not using the π as follows: $t^{\{1 \mapsto 1, 2 \mapsto 2\}} = t$, $t^{\{1 \mapsto 1, 2 \mapsto 1\}} = (t \cap \mathbf{1}) \cdot \top$, and $t^{\{1 \mapsto 2, 2 \mapsto 2\}} = \top \cdot (t \cap \mathbf{1})$. Nevertheless, we introduce t^π for each function $\pi \colon [2] \to [2]$ for clarifying the relationship between CoR and FO^3 in Sect. 3.1.

Since PCoR has only positive connectives, its terms define monotone operations, and we have:

Proposition 3 (e.g., [2]). *For every* PCoR *term t and every homomorphism h (from M to M'), if $\langle v, w\rangle \in [\![t]\!]_M$, then $\langle h(v), h(w)\rangle \in [\![t]\!]_{M'}$.*

Proposition 3 also implies that PCoR is strictly less expressive than CoR, because this proposition does not hold in general for CoR.

First-Order Logic and Its Fragments. Here, we introduce first-order logic (FO) and its syntactic fragments (see, e.g., [4]): *existential positive logic* (EP) and *primitive positive logic* (PP). Let $\mathcal{V}$ be a countably infinite set of *(first-order) variables.* We use x, y, z, or u to denote these variables. The set of *formulas* of FO/EP/PP is given by the following grammar, where $a \in \mathcal{A}$ and $x, y \in \mathcal{V}$:

$$\varphi, \psi \in \mathrm{Fml}^{\mathrm{FO}} ::= a(x, y) \mid x = y \mid \mathsf{tt} \mid \varphi \wedge \psi \mid \exists x.\varphi \mid \mathsf{ff} \mid \varphi \vee \psi \mid \forall x.\varphi \mid \neg\varphi$$
$$\varphi, \psi \in \mathrm{Fml}^{\mathrm{EP}} ::= a(x, y) \mid x = y \mid \mathsf{tt} \mid \varphi \wedge \psi \mid \exists x.\varphi \mid \mathsf{ff} \mid \varphi \vee \psi$$
$$\varphi, \psi \in \mathrm{Fml}^{\mathrm{PP}} ::= a(x, y) \mid x = y \mid \mathsf{tt} \mid \varphi \wedge \psi \mid \exists x.\varphi$$

$\mathbf{FV}(\varphi)$ denotes the set of *free variables* occurring in φ. For an FO formula φ and a structure M, we say that a partial function $I\colon \mathcal{V} \twoheadrightarrow |M|$ is an *interpretation* (of φ on M) if $\mathsf{dom}(I) \supseteq \mathbf{FV}(\varphi)$. Then the *semantics* ($I \models_M \varphi$) of φ on M and an interpretation I is a truth value, which is defined in a standard way as follows: $I \models_M a(x, y) :\Leftrightarrow \langle I(x), I(y)\rangle \in a^M$; $I \models_M x = y :\Leftrightarrow I(x) = I(y)$; $I \models_M \mathsf{tt} :\Leftrightarrow$ true; $I \models_M \mathsf{ff} :\Leftrightarrow$ false; $I \models_M \varphi \vee \psi :\Leftrightarrow (I \models_M \varphi)$ or $(I \models_M \psi)$; $I \models_M \varphi \wedge \psi :\Leftrightarrow (I \models_M \varphi)$ and $(I \models_M \psi)$; $I \models_M \neg\varphi :\Leftrightarrow$ (not $I \models_M \varphi$); $I \models_M \exists x.\varphi :\Leftrightarrow$ for some v, $I[v/x] \models_M \varphi$; and $I \models_M \forall x.\varphi :\Leftrightarrow$ for every v, $I[v/x] \models_M \varphi$, where $I[v/x]$ denotes the I in which the value $I(x)$ has been replaced by v. Here, an FO *(binary-relation-)term* is of the form $[\varphi]_{x,y}$, where x and y are distinct variables; and φ is an FO formula with $\mathbf{FV}(\varphi) \subseteq \{x, y\}$. (In the same manner, for a class $\mathcal{C}$ of formulas, we say that $[\varphi]_{x,y}$ is a $\mathcal{C}$ term if the formula φ is in $\mathcal{C}$.) The *semantics* $[\![[\varphi]_{x,y}]\!]_M$ of an FO term $[\varphi]_{x,y}$ is defined by the binary relation $[\![[\varphi]_{x,y}]\!]_M := \{\langle v, w\rangle \in |M|^2 \mid \{x \mapsto v, y \mapsto w\} \models \varphi\}$. The *size* $\|\varphi\|$ of an FO formula φ is defined by: $\|a(x, y)\| := \|x = y\| := \|\mathsf{tt}\| := \|\mathsf{ff}\| := 1$, $\|\varphi \vee \psi\| := \|\varphi \wedge \psi\| = 1 + \|\varphi\| + \|\psi\|$, and $\|\neg\varphi\| := \|\exists x.\varphi\| := \|\forall x.\varphi\| = 1 + \|\varphi\|$. Also, the *size* $\|[\varphi]_{x,y}\|$ of an FO term $[\varphi]_{x,y}$ is defined as $\|\varphi\|$. For the sake of brevity, we may identify formulas equivalent modulo the commutative and associative laws of $\vee$ and $\wedge$. Also, for a finite set $\Phi = \{\varphi_i \mid i \in I\}$ of formulas, we write $\bigvee \Phi$ (and similarly for $\bigwedge \Phi$) for $\varphi_{i_1} \vee \cdots \vee \varphi_{i_{\#(I)}}$ if $\#(I) > 0$, and for ff otherwise, where $I = \{i_1, \ldots, i_{\#(I)}\}$. Also, let FO^3 be the syntax fragment consisting of FO formulas such that at most three variables appear in the formula. EP^3 and PP^3 are similarly defined.

Remark 4 (Existential positive logics and conjunctive queries). The class of PP (resp. EP) formulas in prenex normal form is also known as the class of *conjunctive queries* (resp. *conjunctive queries with union*), which is a major class in database theory (see e.g., [1, Sec. 4]). However, we do not use prenex normal form because we are interested in the number of variables of formulas.

Expressive Power and Succinctness. We say that two terms t and s are *equivalent*, written $\models t \equiv s$, if for every structure M, $[\![t]\!]_M = [\![s]\!]_M$. We write $\not\models t \equiv s$ if t and s are not equivalent, and write $\models t \leq s$ if for every structure M, $[\![t]\!]_M \subseteq [\![s]\!]_M$. Also, we say that two formulas φ and ψ are *equivalent* if for every $\langle M, I \rangle$ such that $\mathsf{dom}(I) \supseteq \mathbf{FV}(\varphi) \cup \mathbf{FV}(\psi)$, $(I \models_M \varphi)$ iff $(I \models_M \psi)$.

We say that $\mathcal{C}'$ is *at least as expressive as* $\mathcal{C}$, if, for every term t in $\mathcal{C}$, there is a term t' in $\mathcal{C}'$, which is equivalent to t; $\mathcal{C}'$ has *the same expressive power as* $\mathcal{C}$, if $\mathcal{C}'$ is at least as expressive as $\mathcal{C}$ and $\mathcal{C}$ is at least as expressive as $\mathcal{C}'$; and $\mathcal{C}'$ is *strictly more expressive than* $\mathcal{C}$, if $\mathcal{C}'$ is at least as expressive as $\mathcal{C}$ and $\mathcal{C}$ is not at least as expressive as $\mathcal{C}'$.

Moreover, for a class F of functions from $\mathbb{N}$ to $\mathbb{N}$, we say that *there is an F-size translation* (preserving the semantics) from $\mathcal{C}$ to $\mathcal{C}'$ (a.k.a. $\mathcal{C}'$ is *F-succinct* than $\mathcal{C}$ [11]) if there is a function $f \in F$ such that for every term t in $\mathcal{C}$, there is a term t' in $\mathcal{C}'$ of size $\|t'\| \leq f(\|t\|)$ that is equivalent to t. In particular, we say that there is a *linear/polynomial/exponential-size* translation from $\mathcal{C}$ to $\mathcal{C}'$ if F is the set of all linear/polynomial/exponential (i.e., $\mathcal{O}(n)/n^{\mathcal{O}(1)}/2^{\mathcal{O}(n)}$) functions. We say that $\mathcal{C}'$ is *exponentially less succinct than* $\mathcal{C}$ if there is *no* $2^{o(n)}$-size translation from $\mathcal{C}$ to $\mathcal{C}'$.

The Standard Translation. We recall that from CoR to FO^3, there is an efficient translation [22]; see Fig. 1. It follows that $[\![[\mathrm{ST}_{x,y}(t)]_{x,y}]\!]_M = [\![t]\!]_M$ by simple induction on the structure of t, hence the following theorem.

$$\begin{array}{l}
\mathrm{ST}_{x,y}(\top) := \mathbf{tt} \quad \mathrm{ST}_{x,y}(\mathbf{0}) := \mathsf{ff} \quad \mathrm{ST}_{x,y}(t^{-}) := \neg\mathrm{ST}_{x,y}(t) \quad \mathrm{ST}_{x_1,x_2}(t^{\pi}) := \mathrm{ST}_{x_{\pi(1)},x_{\pi(2)}}(t) \\
\mathrm{ST}_{x,y}(a) := a(x,y) \quad \mathrm{ST}_{x,y}(t \cup s) := \mathrm{ST}_{x,y}(t) \vee \mathrm{ST}_{x,y}(s) \quad \mathrm{ST}_{x,y}(t \cdot s) := \exists z.\mathrm{ST}_{x,z}(t) \wedge \mathrm{ST}_{z,y}(s) \\
\mathrm{ST}_{x,y}(\mathbf{1}) := x = y \quad \mathrm{ST}_{x,y}(t \cap s) := \mathrm{ST}_{x,y}(t) \wedge \mathrm{ST}_{x,y}(s) \quad \mathrm{ST}_{x,y}(t \dagger s) := \forall z.\mathrm{ST}_{x,z}(t) \vee \mathrm{ST}_{z,y}(s)
\end{array}$$

Fig. 1. The standard translation, where x, y, and z are all distinct.

Theorem 5 ([22]). *There is a linear-size translation from* CoR *to* FO^3.

The following is also immediate from the standard translation (notice that $\neg$ and $\forall$ do not occur in $\mathrm{ST}_{x,y}(t)$ if $\bullet^{-}$ and $\dagger$ do not occur in t; furthermore, $\neg$, $\forall$, $\vee$, and ff do not occur in $\mathrm{ST}_{x,y}(t)$ if $\bullet^{-}$, $\dagger$, $\cup$, and $\mathbf{0}$ do not occur in t).

Corollary 6.

- *There is a linear-size translation from* PCoR *to* EP^3.
- *There is a linear-size translation from* PPCoR *to* PP^3.

3 Expressive Power Equivalence of PCoR and EP^3

In this section, we consider the converse direction of the standard translation, i.e., from FO^3 terms to CoR terms. The aim of this section is to show the following.

Theorem 7.

(1) *There is an exponential-size translation from* FO^3 *to* CoR.
(2) *There is an exponential-size translation from* EP^3 *to* PCoR.
(3) *There is a linear-size translation from* PP^3 *to* PPCoR.

This theorem (combined with Theorem 5 and Corollary 6) implies the following expressive power equivalences.

Corollary 8.

(1) CoR *has the same expressive power as* FO^3 [23].
(2) PCoR *has the same expressive power as* EP^3.
(3) PPCoR *has the same expressive power as* PP^3*; furthermore, these two have the same succinctness up to linear factors.*

From here, we prove Theorem 7 by giving a new translation from FO^3 to CoR, which is constructed in the following steps.

(‡1) Translate the given FO^3 term into a term in *negation normal form*.
(‡2) For each sub-formula of the form $\exists z.\psi$ (resp. $\forall z.\psi$), substitute ψ with an equivalent formula, which is a conjunction (resp. disjunction) of formulas having at most two free variables.
(‡3) Push the quantifiers deeper into the formula as much as possible. Then, each sub-formula $\exists z.\varphi$ (resp. $\forall z.\varphi$) is of the form $\exists z.\psi \wedge \rho$ (resp. $\forall z.\psi \vee \rho$) such that $\mathbf{FV}(\psi) \subseteq \{x, z\}$ and $\mathbf{FV}(\rho) \subseteq \{z, y\}$, where x, y, z are three distinct variables.
(‡4) Translate the FO^3 term preprocessed by the above translations to a CoR term by simple structural induction.

In the following, we describe the details of each step. We say that a formula φ is in $\text{FO}^{3(2)}$ if φ is in FO^3 and $\#(\mathbf{FV}(\varphi)) \leq 2$. Note that by the definition of FO^3 term, for every FO^3 term $[\varphi]_{x,y}$, φ is in $\text{FO}^{3(2)}$.

(‡1): We say that an FO formula is in *negation normal form* if it is in the set defined by the following grammar:

$$\varphi, \psi ::= a(x, y) \mid \neg a(x, y) \mid x = y \mid \neg x = y \mid \mathsf{tt} \mid \mathsf{ff} \mid \varphi \vee \psi \mid \varphi \wedge \psi \mid \exists x.\varphi \mid \forall x.\varphi.$$

We say that an FO formula φ is an *atomic formula* if φ is of the form $a(x, y)$, $x = y$, or tt; and is a *negated atomic formula* if φ is of the form $\neg a(x, y)$, $\neg x = y$, or ff. Every FO^3 term can be translated to an equivalent FO^3 term in negation normal form by repeatedly applying the De Morgan's law and the double negation elimination law.

Lemma 9. *There is a linear-size translation from* FO^3 *terms to* FO^3 *terms in negation normal form.*

(‡2): We say that a formula is *good* if (a) it is in negation normal form, and (b) for every sub-formula of the form $\exists z.\psi$ (resp. $\forall z.\psi$), ψ is a conjunction (resp. disjunction) of $\mathrm{FO}^{3(2)}$ formulas. According to condition (b), each sub-formula of the form $\exists z.\psi$ (resp. $\forall z.\psi$) can be written as $\exists z.\rho_1 \wedge \rho_2 \wedge \rho_3$ (resp. $\forall z.\rho_1 \vee \rho_2 \vee \rho_3$) by the associativity and commutativity of $\wedge/\vee$, where $\mathbf{FV}(\rho_1) \subseteq \{x, y\}$, $\mathbf{FV}(\rho_2) \subseteq \{y, z\}$, and $\mathbf{FV}(\rho_3) \subseteq \{z, x\}$, and x, y, z are three distinct variables. This property will be fully used in the translation (‡3). In this step, we translate negation normal form FO^3 terms into good FO^3 terms. This is the only step involving an exponential blow-up among (‡1)–(‡4).

Lemma 10. *There is an* exponential-*size translation from* FO^3 *terms in negation normal form to good* FO^3 *terms.*

Proof. We mutually define two functions, $\mathsf{T}_\exists$ and $\mathsf{T}_\forall$, from negation normal form FO^3 formulas to sets of sets of good $\mathrm{FO}^{3(2)}$ formulas and define the function T_- from negation normal form FO^3 formulas to good FO^3 formulas; see Fig. 2. Then the following are shown by simple induction on the structure of φ: φ is equivalent to the formula $\bigvee_{i \in [n]} \bigwedge_{j \in [m_i]} \psi_{i,j}$, where $\mathsf{T}_\exists(\varphi) = \{\{\psi_{i,j} \mid j \in [m_i]\} \mid i \in [n]\}$; and φ is equivalent to the formula $\bigwedge_{i \in [n]} \bigvee_{j \in [m_i]} \psi_{i,j}$, where $\mathsf{T}_\forall(\varphi) = \{\{\psi_{i,j} \mid j \in [m_i]\} \mid i \in [n]\}$. Also the translation T_- (Fig. 2) from negation normal form FO^3 formulas to good FO^3 formulas satisfies that (1) $\mathsf{T}_-(\varphi)$ is equivalent to φ; (2) $\mathbf{FV}(\mathsf{T}_-(\varphi)) \subseteq \mathbf{FV}(\varphi)$; and (3) $\|\mathsf{T}_-(\varphi)\| \leq 2^{2 \times \|\varphi\|}$ (hence, T_- is an exponential-size translation). Hence, the desired translation is obtained from T_-. □

- $\mathsf{T}_\bullet(\varphi) := \begin{cases} \varphi & (\bullet = -) \\ \{\{\varphi\}\} & (\bullet = \exists, \forall) \end{cases}$ if φ is an atomic or negated atomic formula.
- $\mathsf{T}_\bullet(\exists z.\varphi) := \begin{cases} \bigvee_{i \in [n]} \exists z. \bigwedge \Phi_i & (\bullet = -) \\ \{\{\bigvee_{i \in [n]} \exists z. \bigwedge \Phi_i\}\} & (\bullet = \exists, \forall) \end{cases}$, where $\mathsf{T}_\exists(\varphi) = \{\Phi_i \mid i \in [n]\}$.
- $\mathsf{T}_\bullet(\forall z.\varphi) := \begin{cases} \bigwedge_{i \in [n]} \forall z. \bigvee \Phi_i & (\bullet = -) \\ \{\{\bigwedge_{i \in [n]} \forall z. \bigvee \Phi_i\}\} & (\bullet = \exists, \forall) \end{cases}$, where $\mathsf{T}_\forall(\varphi) = \{\Phi_i \mid i \in [n]\}$.
- $\mathsf{T}_\bullet(\psi_1 \wedge \psi_2) := \begin{cases} \mathsf{T}_-(\psi_1) \wedge \mathsf{T}_-(\psi_2) & (\bullet = -) \\ \mathsf{T}_\forall(\psi_1) \cup \mathsf{T}_\forall(\psi_2) & (\bullet = \forall) \\ \{\Psi_1 \cup \Psi_2 \mid \Psi_1 \in \mathsf{T}_\exists(\psi_1), \Psi_2 \in \mathsf{T}_\exists(\psi_2)\} & (\bullet = \exists) \end{cases}$
- $\mathsf{T}_\bullet(\psi_1 \vee \psi_2) := \begin{cases} \mathsf{T}_-(\psi_1) \vee \mathsf{T}_-(\psi_2) & (\bullet = -) \\ \mathsf{T}_\exists(\psi_1) \cup \mathsf{T}_\exists(\psi_2) & (\bullet = \exists) \\ \{\Psi_1 \cup \Psi_2 \mid \Psi_1 \in \mathsf{T}_\forall(\psi_1), \Psi_2 \in \mathsf{T}_\forall(\psi_2)\} & (\bullet = \forall) \end{cases}$

Fig. 2. Translation to good $\mathrm{FO}^{3(2)}$ formulas.

Example 11. Let $\varphi = (a(x,z) \vee b(z,x)) \wedge c(x,y)$. Then, the formula $\exists z.\varphi$ is not a good $\mathrm{FO}^{3(2)}$ formula, but the translated formula $\mathsf{T}_{-}(\exists z.\varphi) = (\exists z.a(x,z) \wedge c(x,y)) \vee (\exists z.b(z,x) \wedge c(x,y))$ is a good $\mathrm{FO}^{3(2)}$ formula equivalent to $\exists z.\varphi$. Note that $\mathsf{T}_{-}(\exists z.\varphi)$ is calculated from $\mathsf{T}_{\exists}(\varphi) = \{\{a(x,z), c(x,y)\}, \{b(z,x), c(x,y)\}\}$.

(‡3): In this step, we translate good FO^3 terms into FO^3 terms in the following normal form.

Definition 12. *For two distinct variables x and y, we say that an $\mathrm{FO}^{3(2)}$ formula is ($\{x,y\}$-)nice if it is in the set defined by the following grammar, where $w, w' \in \{x,y\}$ and z is the variable distinct from x and y:*

$$\begin{aligned}\varphi^{\{x,y\}}, \psi^{\{x,y\}} ::= {} & a(w,w') \mid \neg a(w,w') \mid w = w' \mid \neg w = w' \mid \mathsf{tt} \mid \mathsf{ff} \mid \varphi^{\{x,y\}} \vee \psi^{\{x,y\}} \\ & \mid \varphi^{\{x,y\}} \wedge \psi^{\{x,y\}} \mid \exists z.\varphi^{\{x,z\}} \wedge \psi^{\{z,y\}} \mid \forall z.\varphi^{\{x,z\}} \vee \psi^{\{z,y\}}.\end{aligned}$$

Intuitively, if an FO^3 term is nice, then it is 'almost' a two-variable term (in that, even if a subformula of the term has three free variables, the subformula should be of the form $\varphi^{\{x,z\}} \wedge \psi^{\{z,y\}}$ or $\varphi^{\{x,z\}} \vee \psi^{\{z,y\}}$; hence its immediate subformulas have at most two free variables).

Lemma 13. *There is a linear-size translation from good* FO^3 *terms to nice* FO^3 *terms.*

Proof. Let T be the translation defined as follows: $\mathsf{T}(\varphi) := \varphi$ if φ is an atomic or negated atomic formula; $\mathsf{T}(\psi \vee \rho) := \mathsf{T}(\psi) \vee \mathsf{T}(\rho)$; $\mathsf{T}(\psi \wedge \rho) := \mathsf{T}(\psi) \wedge \mathsf{T}(\rho)$; $\mathsf{T}(\exists z.\psi_1 \wedge \psi_2 \wedge \psi_3) := \mathsf{T}(\psi_1) \wedge \exists z.\mathsf{T}(\psi_2) \wedge \mathsf{T}(\psi_3)$; and $\mathsf{T}(\forall z.\psi_1 \vee \psi_2 \vee \psi_3) := \mathsf{T}(\psi_1) \vee \forall z.\mathsf{T}(\psi_2) \vee \mathsf{T}(\psi_3)$, where $\mathbf{FV}(\psi_1) \subseteq \{x,y\}$, $\mathbf{FV}(\psi_2) \subseteq \{x,z\}$ and $\mathbf{FV}(\psi_3) \subseteq \{y,z\}$. By trivial induction on the size of φ, (1) $\mathsf{T}(\varphi)$ is equivalent to φ; and (2) $\|\mathsf{T}(\varphi)\| \leq \|\varphi\|$ (hence T is a linear-size translation). Also, for every good $\mathrm{FO}^{3(2)}$ formula φ, the formula $\mathsf{T}(\varphi)$ is exactly a nice $\mathrm{FO}^{3(2)}$ formula. Thus the desired translation is obtained from T. □

(‡4): Finally, we give a linear-time translation from nice FO^3 terms to CoR terms by simple structural induction as follows (Fig. 3).

$$\begin{gathered}\mathsf{T}([a(x_i,x_j)]_{x_1,x_2}) := a^{\pi} \quad \mathsf{T}([\neg a(x_i,x_j)]_{x_1,x_2}) := (a^{-})^{\pi} \quad \text{where } \pi = \{1 \mapsto i, 2 \mapsto j\} \\ \mathsf{T}([x_i = x_j]_{x_1,x_2}) := \mathbf{1}^{\pi} \quad \mathsf{T}([\neg x_i = x_j]_{x_1,x_2}) := (\mathbf{1}^{-})^{\pi} \\ \mathsf{T}([\mathsf{tt}]_{x,y}) := \top \quad \mathsf{T}([\varphi^{\{x,y\}} \wedge \psi^{\{x,y\}}]_{x,y}) := \mathsf{T}([\varphi^{\{x,y\}}]_{x,y}) \cap \mathsf{T}([\psi^{\{x,y\}}]_{x,y}) \\ \mathsf{T}([\mathsf{ff}]_{x,y}) := \mathbf{0} \quad \mathsf{T}([\varphi^{\{x,y\}} \vee \psi^{\{x,y\}}]_{x,y}) := \mathsf{T}([\varphi^{\{x,y\}}]_{x,y}) \cup \mathsf{T}([\psi^{\{x,y\}}]_{x,y}) \\ \mathsf{T}([\exists z.\varphi^{\{x,z\}} \wedge \psi^{\{z,y\}}]_{x,y}) := \mathsf{T}([\varphi^{\{x,z\}}]_{x,z}) \cdot \mathsf{T}([\psi^{\{z,y\}}]_{z,y}) \\ \mathsf{T}([\forall z.\varphi^{\{x,z\}} \vee \psi^{\{z,y\}}]_{x,y}) := \mathsf{T}([\varphi^{\{x,z\}}]_{x,z}) \dagger \mathsf{T}([\psi^{\{z,y\}}]_{z,y})\end{gathered}$$

Fig. 3. Translation to CoR terms.

Lemma 14. *There is a linear-size translation from nice* FO^3 *terms to* CoR *terms.*

Proof. By induction on the structure of φ, we can show that (1) for every $\{x,y\}$-nice FO^3 term $[\varphi]_{x,y}$, $\mathsf{T}([\varphi]_{x,y})$ is equivalent to $[\varphi]_{x,y}$; and (2) $\|\mathsf{T}([\varphi]_{x,y})\| \leq 2 \times \|[\varphi]_{x,y}\|$ (thus, T is a linear-size translation). Hence, the T is the desired translation. □

Proof (of Theorem 7). Theorem 7(1) has been proved by combining (‡1)–(‡4). Theorem 7(2) holds because, if a term is in EP^3 (i.e. it does not contain $\neg$ nor $\forall$), then the term translated by (‡1)–(‡3) (more precisely, (‡2)–(‡3) are sufficient) is also in EP^3, and thus the CoR term translated by (‡1)–(‡4) does not contain $\bullet^-$ nor $\dagger$, hence the translated term is a PCoR term. Also Theorem 7(3) holds because, if a term is in PP^3, then the term translated by (‡1)–(‡3) (more precisely, (‡3) is sufficient) is also in PP^3, and thus the CoR term translated by (‡1)–(‡4) is a PPCoR term.

3.1 Quantifier Alternation and Dot-Dagger Alternation Hierarchies

In this subsection, we give a more fine-grained expressive power equivalence between CoR and FO^3 in terms of *the quantifier alternation hierarchy.*

Definition 15 (quantifier alternation hierarchy, cf. [3, p. 105]). *The sets* $\{\Sigma_n, \Pi_n\}_{n\in\mathbb{N}}$ *are the minimal sets of* FO *formulas satisfying the following.*

- *If an* FO *formula* φ *contains neither* $\exists$ *nor* $\forall$*, then* $\varphi \in \Sigma_0$ *and* $\varphi \in \Pi_0$.
- *For* $n \geq 0$, $\Sigma_n \subseteq \Sigma_{n+1}$ *and* $\Pi_n \subseteq \Pi_{n+1}$.
- *For* $n \geq 1$, *if* $\varphi, \psi \in \Sigma_n$, *then* $\varphi \vee \psi, \varphi \wedge \psi, \exists x.\varphi \in \Sigma_n$ *and* $\forall x.\varphi \in \Pi_{n+1}$.
- *For* $n \geq 1$, *if* $\varphi, \psi \in \Pi_n$, *then* $\varphi \vee \psi, \varphi \wedge \psi, \forall x.\varphi \in \Pi_n$ *and* $\exists x.\varphi \in \Sigma_{n+1}$.

We also define the sets $\{\Sigma_n^3, \Pi_n^3\}_{n\in\mathbb{N}}$ *as the subclasses of* FO^3 *formulas defined by the same rules.*

We now define *dot-dagger alternation hierarchy* in CoR in the same manner as the quantifier alternation hierarchy in FO, as follows.

Definition 16 (dot-dagger alternation hierarchy). *The subclasses of* CoR *terms,* $\{\Sigma_n^{\mathrm{CoR}}, \Pi_n^{\mathrm{CoR}}\}_{n\in\mathbb{N}}$, *are the minimal sets satisfying the following.*

- *If a* CoR *term* t *contains neither* $\cdot$ *nor* $\dagger$, *then* $t \in \Sigma_0^{\mathrm{CoR}}$ *and* $t \in \Pi_0^{\mathrm{CoR}}$.
- *For* $n \geq 0$, $\Sigma_n^{\mathrm{CoR}} \subseteq \Sigma_{n+1}^{\mathrm{CoR}}$ *and* $\Pi_n^{\mathrm{CoR}} \subseteq \Pi_{n+1}^{\mathrm{CoR}}$.
- *For* $n \geq 1$, *if* $t, u \in \Sigma_n^{\mathrm{CoR}}$, *then* t^{π}, $t \cup u, t \cap u, t \cdot u \in \Sigma_n^{\mathrm{CoR}}$ *and* $t \dagger u \in \Pi_{n+1}^{\mathrm{CoR}}$.
- *For* $n \geq 1$, *if* $t, u \in \Pi_n^{\mathrm{CoR}}$, *then* t^{π}, $t \cup u, t \cap u, t \dagger u \in \Pi_n^{\mathrm{CoR}}$ *and* $t \cdot u \in \Sigma_{n+1}^{\mathrm{CoR}}$.

The following shows that the dot-dagger alternation hierarchy in CoR is expressive power equivalent to the quantifier alternation hierarchy in FO^3, uniformly.

Corollary 17. *For each* $n \geq 0$, *the class of terms in* Σ_n^{CoR} *(resp.* Π_n^{CoR}*) and the class of terms in* Σ_n^3 *(resp.* Π_n^3*) have the same expressive power.*

Proof (Sketch). Let us recall the standard translation in Sect. 2 and the translations (‡1)–(‡4) in Sect. 3. In the standard translation, if a given CoR term is in Σ_n^{CoR} (resp. Π_n^{CoR}), then the translated FO^3 term is in Σ_n^3 (resp. Π_n^3). Conversely, for each of (‡1)–(‡3), if a given $\mathrm{FO}^{3(2)}$ formula is in Σ_n^3 (resp. Π_n^3), then the translated nice $\mathrm{FO}^{3(2)}$ formula is also in Σ_n^3 (resp. Π_n^3). Also for (‡4), if a given nice FO^3 term is in Σ_n^3 (resp. Π_n^3), then the translated CoR term is in Σ_n^{CoR} (resp. Π_n^{CoR}). All the above are shown by simple induction on the size of given term/formula. □

Thus, the dot-dagger alternation hierarchy in CoR is also strict as the quantifier alternation hierarchy in FO^3.

Corollary 18 ([3, Lem. 3.9]). *$\Sigma_{n+1}^{\mathrm{CoR}}$ is strictly more expressive than Σ_n^{CoR}.*

Proof. By Corollary 17, it suffices to show that the class of Σ_{n+1}^3 formulas is strictly more expressive than the class of Σ_n^3 formulas. Let us recall the following Σ_{n+1} formula in [3, Lemma 3.9], which is not equivalent to any Σ_n formula:

$$\begin{aligned}&\exists x_0.\exists x_1.\forall x_2.\exists x_3 \cdots .Q x_{n+1}.\\ &(\mathrm{Start}(x_0, x_0) \wedge \mathrm{Move}(x_0, x_1) \wedge \mathrm{Move}(x_1, x_2) \wedge \cdots \wedge \mathrm{Move}(x_n, x_{n+1}))\\ &\to \mathrm{Win}(x_{n+1}, x_{n+1}).\end{aligned}$$

Where x_i and x_j are distinct if $i \neq j$; $Q = \exists$ if n is odd and $Q = \forall$ otherwise; the notation $\varphi \to \psi$ abbreviates $\neg\varphi \vee \psi$; and the unary relation symbols "Start" and "Win" in [3, Lemma 3.9] have been replaced with binary relation symbols, respectively. This formula is equivalent to the following formula in Σ_{n+1}^2:

$$\begin{aligned}&\exists x_0.\mathrm{Start}(x_0, x_0) \to \exists x_1.\mathrm{Move}(x_0, x_1) \to \forall x_2.\mathrm{Move}(x_1, x_2) \to \cdots \to Q x_{n+1}.\\ &\mathrm{Move}(x_n, x_{n+1}) \to \mathrm{Win}(x_{n+1}, x_{n+1}).\end{aligned}$$

Where x_i and x_j denote the same variable if $i \equiv j \pmod 2$. Therefore Σ_{n+1}^3 is strictly more expressive than Σ_n^3, because there is no Σ_n formula (hence no Σ_n^3 formula) equivalent to this Σ_{n+1}^3 formula (by [3, Lem. 3.9]). □

4 PCoR Is Exponentially Less Succinct Than EP^3

In this section, we show that the exponential blow-up of the translation from EP^3 to PCoR given in Sect. 3 is unavoidable.

Theorem 19. *There is no $2^{o(n)}$-size translation from EP^3 terms to equivalent PCoR terms. (Hence, PCoR is exponentially less succinct than EP^3.)*

For each $n \in \mathbb{N}_+$, let t_n be the following EP^3 term:

$$\left[(x = y) \vee \left(\bigvee_{i \in [n]} \mathsf{a}_i(x, y) \vee \mathsf{b}_i(x, y)\right) \vee \exists z. \left(\bigwedge_{i \in [n]} \mathsf{a}_i(x, z) \vee \mathsf{b}_i(z, y)\right)\right]_{x,y}.$$

Here, x, y, z are three distinct variables and $\mathsf{a}_1, \mathsf{b}_1, \mathsf{a}_2, \mathsf{b}_2, \ldots$ are pointwise distinct binary relation symbols in $\mathscr{A}$. To prove Theorem 19, we will actually show that there is no $2^{o(n)}$-size translation from the set $\{\mathsf{t}_n \mid n \in \mathbb{N}_+\}$ to PCoR. Note that each t_n is equivalent to the following PCoR term:

$$\mathbf{1} \cup \left(\bigcup_{i \in [n]} \mathsf{a}_i \cup \mathsf{b}_i\right) \cup \left(\bigcup_{\langle I, J\rangle \in \mathrm{Part}([n])} (\bigcap_{i \in I} \mathsf{a}_i) \cdot (\bigcap_{j \in J} \mathsf{b}_j)\right).$$

Here, $\mathrm{Part}(X)$ denotes the set of all ordered partitions of size 2 of X (i.e., the set of all pairs $\langle I, J\rangle$ s.t. $I \cup J = X$, $I \cap J = \emptyset$, $I \neq \emptyset$, and $J \neq \emptyset$). However, unfortunately, this is not a $2^{o(n)}$-size translation, because $\#(\mathrm{Part}([n])) = 2^n - 2$.

Let us consider the parameter $\mathrm{w}_n(t)$:

$$\mathrm{w}_n(t) := \#(\{\langle I, J\rangle \in \mathrm{Part}([n]) \mid \langle 1, 3\rangle \in [\![t]\!]_{M_{\langle I,J\rangle}}\}).$$

Here, $M_{\langle I,J\rangle}$ is the structure $\langle [3], \{a^{M_{\langle I,J\rangle}}\}_{a \in \mathscr{A}}\rangle$, where $a^{M_{\langle I,J\rangle}} = \{\langle 1, 2\rangle\}$ if $a \in \{\mathsf{a}_i \mid i \in I\}$, $a^{M_{\langle I,J\rangle}} = \{\langle 2, 3\rangle\}$ if $a \in \{\mathsf{b}_j \mid j \in J\}$, and $a^{M_{\langle I,J\rangle}} = \emptyset$ otherwise. Note that $\mathrm{w}_n(\mathsf{t}_n) = \#(\mathrm{Part}([n])) = 2^n - 2$ by the construction of t_n.

The following is the key lemma, which will be shown in the next subsection.

Lemma 20. *For every* PCoR *term s, if $\models s \leq \mathsf{t}_n$, then $\|s\| \geq \mathrm{w}_n(s)/8$.*

Theorem 19 can be proved by Lemma 20.

Proof (of Theorem 19 by using Lemma 20). As a consequence of Lemma 20, for every PCoR term s equivalent to t_n, $\|s\| \geq (2^n - 2)/8 \geq 2^{n-4}$, where $n \geq 2$. Note that $\mathrm{w}_n(s) = \mathrm{w}_n(\mathsf{t}_n)$ since $\models s \equiv \mathsf{t}_n$. We assume, towards contradiction, that there exists a $2^{o(n)}$-size translation f from EP^3 terms to PCoR terms. From this, there exists a monotone function $g \colon \mathbb{N} \to \mathbb{N}$ in $2^{o(n)}$ such that $\|f(\mathsf{t}_n)\| \leq g(\|\mathsf{t}_n\|)$. Also by the construction of the EP^3 term t_n, $\|\mathsf{t}_n\| \leq l(n)$ holds for some linear function $l \colon \mathbb{N} \to \mathbb{N}$. Combining the above, $2^{n-4} \leq \|f(\mathsf{t}_n)\| \leq g(\|\mathsf{t}_n\|) \leq g(l(n)) = (g \circ l)(n)$ and $g \circ l$ is a function in $2^{o(n)}$, but thus reaching a contradiction. □

We prove Lemma 20 in the rest of this section.

4.1 Proof of Lemma 20

We say that a PCoR term is in *projection normal form* if it is in the set defined by the following grammar: $t, s ::= a \mid a^{\smile} \mid \mathbf{1} \mid \top \mid \mathbf{0} \mid t \cup s \mid t \cap s \mid t \cdot s$.

Proposition 21. *There is a linear-size translation l from* PCoR *terms to* PCoR *terms in projection normal form such that $\|l(t)\| \leq 8 \times \|t\|$.*

Proof. First we replace each sub-term t^{π} with $(t \cap \mathbf{1}) \cdot \top$ if $\pi = \{1 \mapsto 1, 2 \mapsto 1\}$; t if $\pi = \{1 \mapsto 1, 2 \mapsto 2\}$; and $\top \cdot (t \cap \mathbf{1})$ if $\pi = \{1 \mapsto 2, 2 \mapsto 2\}$ (then the π of each sub-term t^{π} is converse). Secondly, we push converse operators deeper into the term by the following rewriting rules: $\mathbf{1}^{\smile} \rightsquigarrow \mathbf{1}$; $\top^{\smile} \rightsquigarrow \top$; $\mathbf{0}^{\smile} \rightsquigarrow \mathbf{0}$; $(t^{\smile})^{\smile} \rightsquigarrow t$; $(t \cup s)^{\smile} \rightsquigarrow t^{\smile} \cup s^{\smile}$; $(t \cap s)^{\smile} \rightsquigarrow t^{\smile} \cap s^{\smile}$; and $(t \cdot s)^{\smile} \rightsquigarrow s^{\smile} \cdot t^{\smile}$. Note that the first step induces a factor of 4 and that the second step induces a factor of 2. □

From this, to prove Lemma 20, it suffices to prove the following lemma.

Lemma 22. *For every s in projection normal form, if* $\models s \leq \mathsf{t}_n$, $\|s\| \geq \mathrm{w}_n(s)$.

To prove Lemma 22, we introduce a few notions; and then give a few properties (Lemmas 23 and 24) with respect to t_n.

The *disjoint union* of structures M_1 and M_2, written $M_1 \uplus M_2$, is the structure $\langle |M_1 \uplus M_2|, \{a^{M_1 \uplus M_2}\}_{a \in \mathscr{A}}\rangle$, where $|M_1 \uplus M_2| := \{\langle 1, v\rangle \mid v \in |M_1|\} \cup \{\langle 2, v\rangle \mid v \in |M_2|\}$ and $a^{M_1 \uplus M_2} := \{\langle\langle l, v\rangle, \langle l, v'\rangle\rangle \mid l \in [2], \langle v, v'\rangle \in a^{M_l}\}$. The *quotient* of a structure M w.r.t. an equivalence relation $\sim$, written $M/{\sim}$, is the structure $\langle |M/{\sim}|, \{a^{M/\sim}\}_{a \in \mathscr{A}}\rangle$, where $|M/{\sim}| := \{[v]_\sim \mid v \in |M|\}$ ($[v]_\sim$ denotes the equivalence class of v w.r.t. $\sim$) and $a^{M/\sim} := \{\langle [v]_\sim, [v']_\sim\rangle \mid \langle v, v'\rangle \in a^M\}$.

Lemma 23. *Let* $\mathrm{d}_M(v, v') := \min(\{k \in \mathbb{N} \mid \langle v, v'\rangle \in [\![(\bigcup_{i \in [n]} \mathsf{a}_i \cup \mathsf{b}_i)^k]\!]_M\} \cup \{\omega\})$.

- *If* $\mathrm{d}_M(v, v') < 2$, *then* $\langle v, v'\rangle \in [\![\mathsf{t}_n]\!]_M$.
- *If* $\mathrm{d}_M(v, v') > 2$, *then* $\langle v, v'\rangle \notin [\![\mathsf{t}_n]\!]_M$.

Proof. Immediate from the definition of t_n. □

Lemma 24. *For every two* PCoR *terms s_1 and s_2, the following hold.*

(1) *If* $\models s_1 \cup s_2 \leq \mathsf{t}_n$, *then* $\models s_1 \leq \mathsf{t}_n$ *and* $\models s_2 \leq \mathsf{t}_n$.
(2) *If* $\models s_1 \cap s_2 \leq \mathsf{t}_n$, *then* $\models s_1 \leq \mathsf{t}_n$ *or* $\models s_2 \leq \mathsf{t}_n$.
(3) *If* $\models s_1 \cdot s_2 \leq \mathsf{t}_n$ *and* $\not\models s_1 \cdot s_2 = \mathbf{0}$, *then* $\models s_1 \leq \mathsf{t}_n$ *and* $\models s_2 \leq \mathsf{t}_n$.

Proof.

(1) By $\models s_l \leq s_1 \cup s_2$ for $l \in [2]$.
(2) We show the contraposition. Let $\langle M_l, v_l, v'_l\rangle$ be such that $\langle v_l, v'_l\rangle \in [\![s_l]\!]_{M_l} \setminus [\![\mathsf{t}_n]\!]_{M_l}$ for each $l \in [2]$. Let M be the structure $(M_1 \uplus M_2)/{\sim}$, where $\sim$ is the minimal equivalence relation satisfying $\langle 1, v_1\rangle \sim \langle 2, v_2\rangle$ and $\langle 1, v'_1\rangle \sim \langle 2, v'_2\rangle$ (also we let $v = [\langle 1, v_1\rangle]_\sim$ and $v' = [\langle 1, v'_1\rangle]_\sim$). Then (2-1) $\langle v, v'\rangle \in [\![s_1 \cap s_2]\!]_M$; (2-2) $\mathrm{d}_M(v, v') \geq 2$; and (2-3) $\langle v, v'\rangle \notin [\![\mathsf{t}_n]\!]_M$ hold. For (2-1), it is because $\langle v, v'\rangle \in [\![s_1]\!]_M$ and $\langle v, v'\rangle \in [\![s_2]\!]_M$ by the construction of M and Proposition 3. For (2-2), it is because $\mathrm{d}_M(v, v') = \min(\mathrm{d}_{M_1}(v_1, v'_1), \mathrm{d}_{M_2}(v_2, v'_2))$ by the construction of M; and for $l \in [2]$, $\mathrm{d}_{M_l}(v_l, v'_l) \geq 2$ by $\langle v_l, v'_l\rangle \notin [\![\mathsf{t}_n]\!]_{M_l}$ (Lemma 23). For (2-3), by (2-2), it suffices to show that $\langle v, v'\rangle \notin [\![(\bigcap_{i \in I} \mathsf{a}_i) \cdot (\bigcap_{j \in J} \mathsf{b}_j)]\!]_M$ for every $\langle I, J\rangle$ of a partition of $[n]$. We assume, toward contradiction, that $\langle v, v'\rangle \in [\![(\bigcap_{i \in I} \mathsf{a}_i) \cdot (\bigcap_{j \in J} \mathsf{b}_j)]\!]_M$. Let w be such that $\langle v, w\rangle \in [\![\bigcap_{i \in I} \mathsf{a}_i]\!]_M$ and $\langle w, v'\rangle \in [\![\bigcap_{j \in J} \mathsf{b}_j]\!]_M$. Then w is distinct from v and v' by $\mathrm{d}_M(v, v') \geq 2$, so $w = \{\langle l, w_l\rangle\}$ for some l and some w_l. Then $\langle v_l, w_l\rangle \in [\![\bigcap_{i \in I} \mathsf{a}_i]\!]_{M_l}$ and $\langle w_l, v'_l\rangle \in [\![\bigcap_{j \in J} \mathsf{b}_j]\!]_{M_l}$ should hold, so $\langle v_l, v'_l\rangle \in [\![(\bigcap_{i \in I} \mathsf{a}_i) \cdot (\bigcap_{j \in J} \mathsf{b}_j)]\!]_{M_l}$. This contradicts to $\langle v_l, v'_l\rangle \notin [\![\mathsf{t}_n]\!]_{M_l}$. Hence $\langle v, v'\rangle \in [\![s_1 \cap s_2]\!]_M \setminus [\![\mathsf{t}_n]\!]_M$.
(3) We show the contraposition. We only write the case of $\not\models s_1 \leq \mathsf{t}_n$ (the case of $\not\models s_2 \leq \mathsf{t}_n$ is shown by same arguments). Let $\langle M_1, v_1, v'_1\rangle$ be such that $\langle v_1, v'_1\rangle \in [\![s_1]\!]_{M_1} \setminus [\![\mathsf{t}_n]\!]_{M_1}$ and let $\langle M_2, v_2, v'_2\rangle$ be such that $\langle v_2, v'_2\rangle \in [\![s_2]\!]_{M_2}$ (note that $\not\models s_2 = \mathbf{0}$ since $\not\models s_1 \cdot s_2 = \mathbf{0}$). Let M be

the structure $(M_1 \uplus M_2)/\sim$, where $\sim$ is the minimal equivalence relation satisfying $\langle 1, v_1' \rangle \sim \langle 2, v_2 \rangle$ (also we let $v = [\langle 1, v_1 \rangle]_\sim$ and $v' = [\langle 2, v_2' \rangle]_\sim$). Then (3-1) $\langle v, v' \rangle \in [\![s_1 \cdot s_2]\!]_M$; (3-2) $\mathrm{d}_M(v, v') \geq 2$; and (3-3) $\langle v, v' \rangle \notin [\![\mathsf{t}_n]\!]_M$ hold. (3-1) is shown by the construction of M and Proposition 3. (3-2) is shown by $\mathrm{d}_M(v, v') = \mathrm{d}_{M_1}(v_1, v_1') + \mathrm{d}_{M_2}(v_2, v_2')$ (by the construction of M) and $\mathrm{d}_{M_1}(v_1, v_1') \geq 2$ (by $\langle v_1, v_1' \rangle \notin [\![\mathsf{t}_n]\!]_{M_l}$ and Lemma 23). For (3-3), by (3-2), it suffices to show that $\langle v, v' \rangle \notin [\![(\bigcap_{i \in I} \mathsf{a}_i) \cdot (\bigcap_{j \in J} \mathsf{b}_j)]\!]_M$ for every $\langle I, J \rangle$ of a partition of $[n]$. We assume, toward contradiction, that $\langle v, v' \rangle \in [\![(\bigcap_{i \in I} \mathsf{a}_i) \cdot (\bigcap_{j \in J} \mathsf{b}_j)]\!]_M$. Let w be such that $\langle v, w \rangle \in [\![\bigcap_{i \in I} \mathsf{a}_i]\!]_M$ and $\langle w, v' \rangle \in [\![\bigcap_{j \in J} \mathsf{b}_j]\!]_M$. Then by $w \neq [\langle 1, v_1' \rangle]_\sim$ (notice $\mathrm{d}_{M_1}(v_1, v_1') \geq 2$), $w = \{\langle 1, w_1 \rangle\}$ for some $w_1 \in |M_1|$. From this, $\langle v_1, w_1 \rangle \in [\![\bigcap_{i \in I} \mathsf{a}_i]\!]_{M_1}$ and $\langle w_1, v_1' \rangle \in [\![\bigcap_{j \in J} \mathsf{b}_j]\!]_{M_1}$ should hold, so $\langle v_1, v_1' \rangle \in [\![(\bigcap_{i \in I} \mathsf{a}_i) \cdot (\bigcap_{j \in J} \mathsf{b}_j)]\!]_{M_1}$. This contradicts to $\langle v_1, v_1' \rangle \notin [\![\mathsf{t}_n]\!]_{M_1}$. Hence $\langle v, v' \rangle \in [\![s_1 \cdot s_2]\!]_M \setminus [\![\mathsf{t}_n]\!]_M$.

□

We are now ready to prove Lemma 22.

Proof (of Lemma 22). By induction on the structure of s.

Case $s = \mathbf{1}$, $s = \mathbf{0}$, $s = a$, or $s = a^\smile$: By $\mathrm{w}_n(s) = 0$.

Case $s = \top$: By $\not\models \top \leq \mathsf{t}_n$.

Case $s = s_1 \cup s_2$: $\mathrm{w}_n(s) \leq \mathrm{w}_n(s_1) + \mathrm{w}_n(s_2)$ holds by that, for every M, if $\langle v, w \rangle \in [\![s_1 \cup s_2]\!]_M$, then $\langle v, w \rangle \in [\![s_1]\!]_M$ or $\langle v, w \rangle \in [\![s_2]\!]_M$. Therefore by Lemma 24(1) and I.H., $\mathrm{w}_n(s) \leq \mathrm{w}_n(s_1) + \mathrm{w}_n(s_2) \leq \|s_1\| + \|s_2\| \leq \|s\|$.

Case $s = s_1 \cap s_2$: By Lemma 24(2), let l be such that $\models s_l \leq \mathsf{t}_n$. By $\models s \leq s_l$, $\mathrm{w}_n(s) \leq \mathrm{w}_n(s_l)$. Therefore by I.H., $\mathrm{w}_n(s) \leq \mathrm{w}_n(s_l) \leq \|s_l\| \leq \|s\|$.

Case $s = s_1 \cdot s_2$: If $\mathrm{w}_n(s) \leq 1$, then $\mathrm{w}_n(s) \leq \|s\|$ is trivial. Otherwise ($\mathrm{w}_n(s) \geq 2$), let $\Xi(s_1, s_2) := \{\langle \langle I, J \rangle, w \rangle \in \mathrm{Part}([n]) \times [3] \mid \langle 1, w \rangle \in [\![s_1]\!]_{M_{\langle I,J \rangle}} \wedge \langle w, 3 \rangle \in [\![s_2]\!]_{M_{\langle I,J \rangle}}\}$. Note that $\not\models s = \mathbf{0}$ and $\#(\Xi(s_1, s_2)) \geq 2$. Assume, toward contradiction, that, there are $\langle \langle I, J \rangle, v \rangle$ and $\langle \langle I', J' \rangle, v' \rangle$ such that $v \neq v'$. Without loss of generality, we can assume that $v > v'$. Let M be the structure $(M_{\langle I,J \rangle} \uplus M_{\langle I',J' \rangle})/\sim$, where $\sim$ is the minimal equivalence relation satisfying $\langle 1, v \rangle \sim \langle 2, v' \rangle$. By the construction of M and Proposition 3, $\langle [\langle 1, 1 \rangle]_\sim, [\langle 1, v \rangle]_\sim \rangle \in [\![s_1]\!]_M$ and $\langle [\langle 2, v' \rangle]_\sim, [\langle 2, 3 \rangle]_\sim \rangle \in [\![s_2]\!]_M$ hold, hence $\langle [\langle 1, 1 \rangle]_\sim, [\langle 2, 3 \rangle]_\sim \rangle \in [\![s]\!]_M$. On the other hand, by $\mathrm{d}_M([\langle 1, 1 \rangle]_\sim, [\langle 2, 3 \rangle]_\sim) = \mathrm{d}_{M_1}(1, v) + \mathrm{d}_{M_2}(v', 3) > 2$ and Lemma 23, $\langle [\langle 1, 1 \rangle]_\sim, [\langle 2, 3 \rangle]_\sim \rangle \notin [\![s]\!]_M$, thus reaching a contradiction.

Let $k \in [3]$ be the unique one such that, if $\langle \langle I, J \rangle, v \rangle \in \Xi(s_1, s_2)$, then $v = k$. We do case analysis on k.

Sub-Case $k = 1$: Then, for every $\langle I, J \rangle$, if $\langle 1, 3 \rangle \in [\![s_1 \cdot s_2]\!]_{M_{\langle I,J \rangle}}$, then $\langle 1, 3 \rangle \in [\![s_2]\!]_{M_{\langle I,J \rangle}}$. Thus $\mathrm{w}_n(s) \leq \mathrm{w}_n(s_2)$. Therefore by I.H. (notice $\models s_2 \leq \mathsf{t}_n$ by Lemma 24(3)), $\mathrm{w}_n(s) \leq \mathrm{w}_n(s_2) \leq \|s_2\| \leq \|s\|$.

Sub-Case $k = 2$: Let $\langle \langle I, J \rangle, 2 \rangle$ and $\langle \langle I', J' \rangle, 2 \rangle$ be *distinct* ones in $\Xi(s_1, s_2)$. Let M be the structure $(M_{\langle I,J \rangle} \uplus M_{\langle I',J' \rangle})/\sim$, where $\sim$ is the minimal equivalence relation satisfying $\langle 1, 2 \rangle \sim \langle 2, 2 \rangle$ (see Fig. 4). By the construction of M and Proposition 3, both $\langle [\langle 1, 1 \rangle]_\sim, [\langle 2, 3 \rangle]_\sim \rangle \in [\![s]\!]_M$ and $\langle [\langle 2, 1 \rangle]_\sim, [\langle 1, 3 \rangle]_\sim \rangle \in [\![s]\!]_M$ hold. On the other hand, $I \cup J' \not\subseteq [n]$ or $I' \cup J \not\subseteq [n]$ holds, because $\langle I, J \rangle$ and

$\langle I', J' \rangle$ are *distinct* partitions of $[n]$, and thus $\langle [\langle 1,1 \rangle]_{\sim}, [\langle 2,3 \rangle]_{\sim} \rangle \notin [\![\mathsf{t}_n]\!]_M$ or $\langle [\langle 2,1 \rangle]_{\sim}, [\langle 1,3 \rangle]_{\sim} \rangle \notin [\![\mathsf{t}_n]\!]_M$ should hold. This contradicts to $\models s \leq \mathsf{t}_n$.

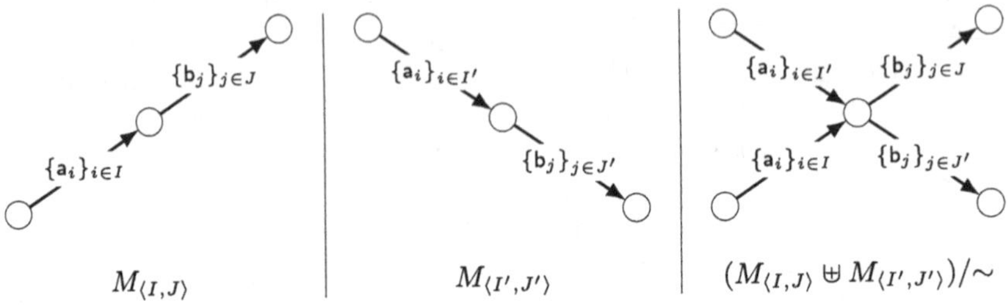

Fig. 4. Construction of $(M_{\langle I,J \rangle} \uplus M_{\langle I',J' \rangle})/{\sim}$.

Sub-Case $k = 3$: In the same way as Sub-Case $k = 1$. □

As a consequence of Lemma 22 (and Proposition 21), Lemma 20 has been proved.

5 On the Transitive Closure Extension

In this section, we remark that the results in Sects. 3 and 4 can be extended to the positive calculus of relations with transitive closure [19] (denoted by PCoR(TC), for short). We will show that the calculus has the same expressive power as three-variable existential positive (first-order) logic with (variable-confined) monadic transitive closure (denoted by EP^3(v-MTC)) (see, e.g., [6, Sec. 9] for transitive closure logic). The *syntax* of PCoR(TC) is given by: $t, s ::= a \mid \mathbf{1} \mid \top \mid \mathbf{0} \mid t \cup s \mid t \cap s \mid t^{\pi} \mid t \cdot s \mid t^{+}$. The *semantics* $[\![t]\!]_M$ and the *size* $\|t\|$ are defined in the same way as for CoR, respectively, where $[\![t^+]\!]_M := \bigcup_{k \in \mathbb{N}_+} [\![t]\!]_M^k$ and $\|t^+\| := 1 + \|t\|$. Also, the *syntax* of EP(v-MTC) is given by the following grammar, where $x, y, z, u \in \mathcal{V}$; z and u are distinct; and each $\mathrm{TC}_{z,u}(\varphi)$ is *variable-confined* (i.e., $\mathbf{FV}(\varphi) \subseteq \{z, u\}$[1]): $\varphi, \psi ::= a(x, y) \mid x = y \mid \mathsf{tt} \mid \mathsf{ff} \mid \varphi \vee \psi \mid \varphi \wedge \psi \mid \exists x.\varphi \mid [\mathrm{TC}_{z,u}(\varphi)](x, y)$. The *semantics* ($I \models_M \varphi$) and the *size* $\|\varphi\|$ are defined in the same way as for EP, where $I \models_M [\mathrm{TC}_{z,u}(\varphi)](x, y) :\Leftrightarrow \langle I(x), I(y) \rangle \in \bigcup_{k \in \mathbb{N}_+} [\![[\varphi]_{z,u}]\!]_M^k$ and $\|[\mathrm{TC}_{z,u}(\varphi)](x, y)\| := 1 + \|\varphi\|$. As in Sects. 3–4, the following are shown. The proofs are proceeded using the same strategy as that in the previous sections, extending the proofs in an appropriate way. In particular, Theorem 26 is proved by that we can extend Lemma 22 for PCoR(TC), because for every PCoR(TC) term t, if $\models t^+ \geq \mathsf{t}_n$, then $\mathrm{w}_n(t^+) = 0$.

[1] Here, z and u in $\mathrm{TC}_{z,u}(\psi)$ are viewed as bound variables (i.e., $\mathbf{FV}([\mathrm{TC}_{z,u}(\psi)](x, y))$ is defined by $\mathbf{FV}([\mathrm{TC}_{z,u}(\psi)](x, y)) := (\mathbf{FV}(\psi) \setminus \{z, u\}) \cup \{x, y\}$). See also [6, Sec. 9].

Theorem 25.

(1) *There is a linear-size translation from* PCoR(TC) *to* EP^3(v-MTC).
(2) *There is an exponential-size translation from* EP^3(v-MTC) *to* PCoR(TC).
(3) *Hence,* PCoR(TC) *has the same expressive power as* EP^3(v-MTC).

Theorem 26. *There is no* $2^{o(n)}$*-size translation from* EP^3(v-MTC) *terms to equivalent* PCoR(TC) *terms.*

6 Conclusion

We have shown that (1) the positive calculus of relations has the same expressive power as three-variable existential positive logic, and (2) the positive calculus of relations is exponentially less succinct than three-variable existential positive logic. To the best of our knowledge, it is open whether the calculus of relations is exponentially less succinct than three-variable first-order logic. It would also be interesting to construct a calculus like the (positive) calculus of relations (or cylindric algebra [13]) such that it has the same expressive power as k-variable (existential positive) first-order logic and there is a succinctness-gap between them.

References

1. Abiteboul, S., Hull, R., Vianu, V.: Foundations of Databases: The Logical Level. Addison-Wesley, Boston (1995)
2. Andréka, H., Bredikhin, D.A.: The equational theory of union-free algebras of relations. Algebra Universalis **33**(4), 516–532 (1995). https://doi.org/10.1007/BF01225472
3. Chandra, A., Harel, D.: Structure and complexity of relational queries. J. Comput. Syst. Sci. **25**(1), 99–128 (1982). https://doi.org/10.1016/0022-0000(82)90012-5
4. Chen, H.: On the complexity of existential positive queries. ACM Trans. Comput. Logic **15**(1), 1–20 (2014). https://doi.org/10.1145/2559946
5. Darwiche, A., Marquis, P.: A knowledge compilation map. J. Artif. Intell. Res. **17**, 229–264 (2002). https://doi.org/10.1613/jair.989
6. Ebbinghaus, H.-D., Flum, J.: Finite Model Theory. SMM, 2nd edn. Springer, Heidelberg (1995). https://doi.org/10.1007/3-540-28788-4
7. Ellul, K., Krawetz, B., Shallit, J., Wang, M.: Regular expressions: new results and open problems. J. Automata Lang. Comb. **9**(2–3), 233–256 (2004)
8. Givant, S.: Introduction to Relation Algebras: Relation Algebras, vol. 1. Springer, Heidelberg (2017). https://doi.org/10.1007/978-3-319-65235-1
9. Givant, S.: The calculus of relations as a foundation for mathematics. J. Autom. Reason. **37**(4), 277–322 (2007). https://doi.org/10.1007/s10817-006-9062-x
10. Gogic, G., Kautz, C., Papadimitriou, H., Selman, B.: The comparative linguistics of knowledge representation. In: Proceedings of the 14th International Joint Conference on Artificial Intelligence, (IJCAI 1995), vol. 1, pp. 862–869. Morgan Kaufmann Publishers Inc. (1995)

11. Grohe, M., Schweikardt, N.: The succinctness of first-order logic on linear orders. Logical Methods Comput. Sci. **1**(1) (2005). https://doi.org/10.2168/LMCS-1(1:6)2005
12. Gruber, H., Holzer, M.: Finite automata, digraph connectivity, and regular expression size. In: Aceto, L., Damgård, I., Goldberg, L.A., Halldórsson, M.M., Ingólfsdóttir, A., Walukiewicz, I. (eds.) ICALP 2008. LNCS, vol. 5126, pp. 39–50. Springer, Heidelberg (2008). https://doi.org/10.1007/978-3-540-70583-3_4
13. Henkin, L., Donald Monk, J., Tarski, A.: Cylindric Algebras. Part 2. North-Holland, Amsterdam (1985)
14. Libkin, L.: Elements of Finite Model Theory. Springer, Heidelberg (2012). https://doi.org/10.1007/978-3-662-07003-1
15. Maddux, R.D.: Calculus of relations, Chap. 1. In: Relation Algebras. Studies in Logic and the Foundations of Mathematics, vol. 150, pp. 1–33. Elsevier (2006). https://doi.org/10.1016/S0049-237X(06)80023-6
16. Meyer, A.R., Fischer, M.J.: Economy of description by automata, grammars, and formal systems. In: 12th Annual Symposium on Switching and Automata Theory (SWAT 1971), pp. 188–191. IEEE (1971). https://doi.org/10.1109/SWAT.1971.11
17. Nakamura, Y.: Partial derivatives on graphs for Kleene allegories. In: 32nd Annual ACM/IEEE Symposium on Logic in Computer Science (LICS 2017), pp. 1–12. IEEE (2017). https://doi.org/10.1109/LICS.2017.8005132
18. Nakamura, Y.: The undecidability of FO3 and the calculus of relations with just one binary relation. In: Khan, M.A., Manuel, A. (eds.) ICLA 2019. LNCS, vol. 11600, pp. 108–120. Springer, Heidelberg (2019). https://doi.org/10.1007/978-3-662-58771-3_11
19. Pous, D.: On the positive calculus of relations with transitive closure. In: 35th Symposium on Theoretical Aspects of Computer Science (STACS 2018), vol. 96, pp. 3:1–3:16. Schloss Dagstuhl - Leibniz-Zentrum fuer Informatik (2018). https://doi.org/10.4230/LIPICS.STACS.2018.3
20. Pous, D., Vignudelli, V.: Allegories: decidability and graph homomorphisms. In: Proceedings of the 33rd Annual ACM/IEEE Symposium on Logic in Computer Science (LICS 2018), pp. 829–838. ACM Press (2018). https://doi.org/10.1145/3209108.3209172
21. Stockmeyer, L.J.: The complexity of decision problems in automata theory and logic. Ph.D. thesis. Massachusetts Institute of Technology (1974)
22. Tarski, A.: On the calculus of relations. J. Symb. Logic **6**(3), 73–89 (1941). https://doi.org/10.2307/2268577
23. Tarski, A., Givant, S.: A Formalization of Set Theory Without Variables, vol. 41. Colloquium Publications/American Mathematical Society (1987)
24. Vardi, M.Y.: The complexity of relational query languages (extended abstract). In: Proceedings of the Fourteenth Annual ACM Symposium on Theory of Computing (STOC 1982), pp. 137–146. ACM Press (1982). https://doi.org/10.1145/800070.802186

Stone Dualities from Opfibrations

Koki Nishizawa[1(✉)], Shin-ya Katsumata[2], and Yuichi Komorida[3]

[1] Department of Information Systems Creation, Faculty of Engineering, Kanagawa University, Yokohama, Japan
nishizawa@kanagawa-u.ac.jp
[2] National Institute of Informatics, Tokyo, Japan
s-katsumata@nii.ac.jp
[3] The Graduate University for Advanced Studies, SOKENDAI, Tokyo, Japan
komorin@nii.ac.jp

Abstract. Stone dualities are dual equivalences between certain categories of algebras and those of topological spaces. A Stone duality is often derived from a dual adjunction between such categories by cutting down unnecessary objects. This dual adjunction is called the fundamental adjunction of the duality, but building it often requires concrete topological arguments. The aim of this paper is to construct fundamental adjunctions generically using (co)fibered category theory. This paper defines an abstract notion of formal spaces (including ordinary topological spaces as the leading example), and gives a construction of a fundamental adjunction between the category of algebras and the category of corresponding formal spaces.

1 Introduction

Dual equivalences between categories of spaces and those of algebras are ubiquitous in mathematics - following the famous book by Johnstone [1], they are collectively called *Stone dualities* after the Stone Representation Theorem of Boolean algebras. Technically, they often arise as the restriction of dual adjunctions called *fundamental adjunctions*. For example, the Stone duality between sober topological spaces and spatial frames is obtained by cutting down the fundamental adjunction between the category of topological spaces and that of frames, which is the heart of pointless topology [2]. Categorical settings to capture various fundamental adjunctions of Stone dualities has been studied in [3–8]. The basic idea of these settings is to formulate fundamental adjunctions as dual adjunctions that are representable through functors to **Set**. The objects representing adjoint functors are called *dualizing object* [6].

In this paper, we give a new construction of fundamental adjunctions by *(Grothendieck) opfibrations*[1]. Roughly speaking, our construction takes a category of algebras equipped with an abstract notion of subalgebra, then derives

[1] Grothendieck originally called it cofibred categories, but here we use the word opfibration to avoid confusion with cofibration in homotopy theory.

U. Fahrenberg et al. (Eds.): RAMiCS 2020, LNCS 12062, pp. 221–236, 2020.
https://doi.org/10.1007/978-3-030-43520-2_14

both the category of spaces *and* a fundamental adjunction between them. Despite its abstract nature, the constructed fundamental adjunctions reflect several properties seen in concrete ones. One such property is that, in a certain setting, our construction yields fundamental adjunctions that enjoy representability. Another is the full-faithfulness of the algebra-to-space construction. It is characterized in terms of the mono-ness of the unit arrow $X \to \Omega^{A(-,\Omega)}$, where A is a category of algebras and Ω is the dualizing object. This generalizes the full-faithfulness argument of the constructions of topological spaces from Boolean algebras [1].

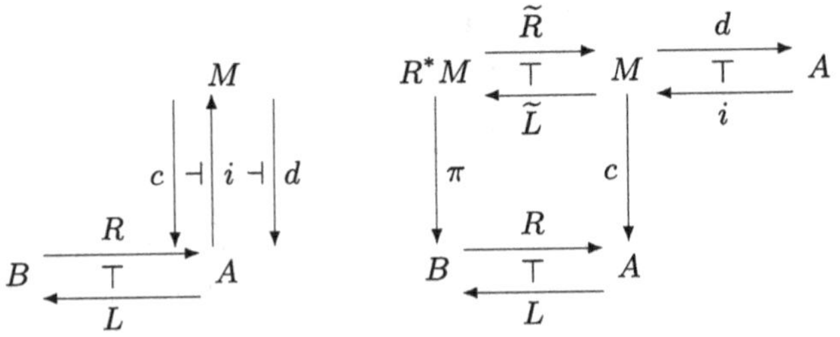

Fig. 1. Sketch of the construction of fundamental adjunction; its input data are on the left and the construction process is on the right

Above we sketch our construction of fundamental adjunctions. It takes two inputs (c, i, d) and (L, R) depicted on the left of Fig. 1. The first is a tuple (c, i, d) called *opfibered comprehension* (Definition 1). It is an opfibration $c : M \to A$ with further adjunctions $c \dashv i \dashv d$ satisfying certain properties. The category A plays the role of an algebraic category, and is equipped with a notion of subalgebra generalized by an opfibration $c : M \to A$. A typical example of c is the subobject opfibration $\mathbf{cod} : \mathbf{Sub}(A) \to A$. The second input to our construction is an adjunction $L \dashv R : B \to A$. This is often set to the *hom-power* adjunction $(A(-,\Omega))^{\mathrm{op}} \dashv \Omega^{(-)} : \mathbf{Set}^{\mathrm{op}} \to A$.

Our construction proceeds by taking the pullback of c along R (right of Fig. 1). We call $(R^*M)^{\mathrm{op}}$ the category of *formal spaces*. The horizontal leg $\tilde{R}$ of the pullback has a left adjoint by Hermida's adjoint lift theorem [9]. The fundamental adjunction then appears as the composite adjunction between A and R^*M.

We will illustrate several examples of fundamental adjunctions arising from our construction. The one between the category $\mathbf{Frm}$ of frames and that of topological spaces, which is the standard example of Stone duality, is an immediate instance: take c to be the subobject opfibration $c : \mathbf{Sub}(\mathbf{Frm}) \to \mathbf{Frm}$ and R to be the power functor $\mathbf{2}^{(-)} : \mathbf{Set}^{\mathrm{op}} \to \mathbf{Frm}$ with the Sierpinski frame $\mathbf{2}$.

Related Work. Categorical formulations of fundamental adjunctions via *dualizing objects* were studied by many authors [3–8]. For the formulation of representable dual adjunctions, see e.g. [5], where Dimov and Tholen also showed a general condition to obtain fundamental adjunctions using the *lift conditions*. In [8], Maruyama improved the lift conditions by (1) breaking the symmetry of Dimov and Tholen's framework, and (2) imposing different conditions on the algebraic and spatial categories.

Categories of *Chu spaces* [10] are self-dual, and can accommodate various dualities in them. Pratt [11] demonstrates that the self-dual category $\mathbf{Chu}(\mathbf{Set}, 2)$ can accommodate (1) sets and complete atomic Boolean algebras, (2) Stone spaces and Boolean algebras, and (3) sober spaces and spatial frames. We relate the category of Chu spaces and that of formal spaces in Sect. 4.1. However, our construction does not explain the self-duality.

The theory of *natural dualities* [12] aims to go roughly the same way as us, to make a "category of spaces" from a given category of algebras. The scope of their theory is narrower than ours, while their theory brings finer results. It is a future work to find connections between their framework and ours. It seems that neither of them can derive the other.

Organization. This paper is organized as follows. In Sect. 2, we define *opfibered comprehensions*, which is the input of our construction. In Sect. 3, we construct the category of *formal spaces* from an opfibered comprehension and derive the *fundamental adjunction.* In Sects. 4 and 5, we list various examples of our framework. In Sect. 6, we show how to relate two fundamental adjunctions. Section 7 summarizes this work and future work.

Preliminaries. The identity functor on a category A is denoted by $\mathbf{Id}_A$, and the identity natural transformation on a functor f is denoted by $\mathbf{id}_f$. We write $L \dashv R\colon B \to A$ (or simply $L \dashv R$) to mean that the functor $R\colon B \to A$ has L as a left adjoint. Its unit and counit are denoted by $\eta^{L \dashv R}, \epsilon^{L \dashv R}$, respectively.

2 Opfibered Comprehension

One of inputs to our construction of fundamental adjunction (Fig. 1) is an *opfibered comprehension.*

Definition 1 (opfibered comprehension). *An* opfibered comprehension *is defined to be a tuple* $(c\colon M \to A, i, d)$ *of functors such that*

1. $c\colon M \to A$ *is an opfibration,*
2. $i\colon A \to M$ *is the right adjoint to* c *whose counit is the identity, and*
3. $d\colon M \to A$ *is the right adjoint to* i *whose unit is the identity.*

We here recall the definition of (Grothendieck) opfibration; a good reference is [13]. A functor $c\colon M \to A$ is an opfibration if it satisfies the following *cocartesian lifting property*: for any A-arrow $f\colon a \to a'$ and $m \in M$ such that $cm = a$, there is a cocartesian lifting of f with m. A *cocartesian lifting of* f *with* m is the arrow written as $\underline{f}(m)\colon m \to f_*(m)$ satisfying $c(f_*(m)) = a'$ and the universal property:

for any A-arrow $g\colon a' \to a''$ and M-arrow $h\colon m \to m''$ satisfying $cm'' = a''$ and $ch = g \circ f$, there exists a unique arrow $k\colon f_*(m) \to m''$ satisfying $ck = g$ and $k \circ \underline{f}(m) = h$.

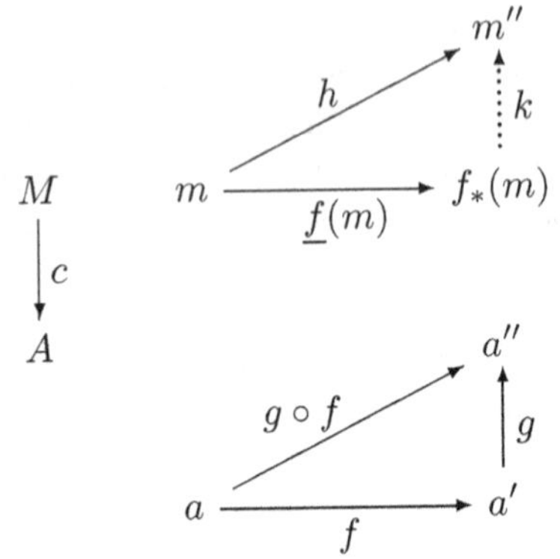

(1)

The conditions 2 and 3 imply several equalities between functors and natural transformations. The last one is proved in [14].

$$c \circ i = \mathbf{Id}_A = d \circ i$$
$$c\eta^{c \dashv i} = \mathbf{id}_c, \quad \eta_i^{c \dashv i} = \mathbf{id}_i = \epsilon_i^{i \dashv d} \quad d\epsilon^{i \dashv d} = \mathbf{id}_d \quad d\eta^{c \dashv i} = c\epsilon^{i \dashv d}$$

Perhaps the simplest example is the following:

Example 1. We write $A^{\to}$ for the arrow category of A: objects are arrows in A, and an arrow from f to g is a pair (p, q) of arrows such that $p \circ f = g \circ q$. The functors $\mathbf{cod}, \mathbf{dom}\colon A^{\to} \to A$ respectively map an object arrow $f\colon a \to a'$ to a' and a, and an arrow (p, q) to p and q. The functor $\triangle\colon A \to A^{\to}$ maps an object a to the identity arrow $\mathbf{id}_a$ on a and an arrow $f\colon a \to a'$ to $(f, f)\colon \mathbf{id}_a \to \mathbf{id}_{a'}$. It is easy to see that $(\mathbf{cod}\colon A^{\to} \to A, \triangle, \mathbf{dom})$ is an opfibered comprehension; the cocartesian lifting of an A-arrow $g\colon a' \to b'$ with an $A^{\to}$-object $f\colon a \to a'$ is given by $\underline{g}(f) = (g, \mathbf{id}_a)\colon f \to g \circ f$. We call it the *arrow opfibered comprehension.*

We can restrict the objects of $A^{\to}$ to monomorphisms if A has a (strong epi, mono)-factorization system. This is our leading example of an opfibered comprehension.

Example 2. We write $\mathbf{Sub}(A)$ for the full subcategory of $A^{\to}$ whose objects are just monomorphisms in A, since an equivalence class of monomorphisms of A is called a *subobject.* If any arrow $f\colon a \to a'$ of A can be factorized to a strong epimorphism $\mathbf{e}(f)\colon a \to \mathbf{Im}(f)$ and a monomorphism $\mathbf{m}(f)\colon \mathbf{Im}(f) \to a'$, then $(\mathbf{cod}\colon \mathbf{Sub}(A) \to A, \triangle, \mathbf{dom})$ is an opfibered comprehension; the cocartesian lifting of an A-arrow $f\colon a \to a'$ with $\mathbf{Sub}(A)$-object $m\colon x \to a$ is given by $\underline{f}(m) = (f, \mathbf{e}(f \circ m))\colon m \to \mathbf{m}(f \circ m)$, since for any $\mathbf{Sub}(A)$-object $n\colon x'' \to a''$ and any arrow $h\colon x \to x'', g\colon a' \to a''$ satisfying $g \circ f \circ m = n \circ h$, the property of the strong epimorphism $\mathbf{e}(f \circ m)$ implies the existence of the unique arrow k satisfying $k \circ \mathbf{e}(f \circ m) = h$ and $n \circ k = g \circ \mathbf{m}(f \circ m)$.

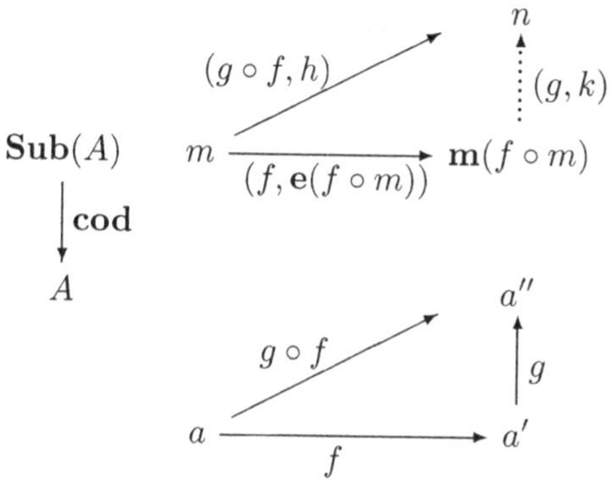

We call it the *subobject opfibered comprehension.*

These two examples above are generalized to the following:

Example 3. Let (E, M) be a *factorization system* on A in the sense of [15]. We obtain an opfibered comprehension in the same way as Example 2: the triple $(\mathbf{cod}\colon M \to A, \triangle, \mathbf{dom})$ is an opfibered comprehension, where we regard M as the full subcategory of $A^{\to}$ whose objects are arrows in M. We note that the unique diagonal fill-in property of the factorization system guarantees the uniqueness of the mediating morphism k in (1). In general, c fails to be an opfibration when (E, M) is merely a weak factorization system in the sense of [16].

On the other hand, in any opfibered comprehension arrows are factored in the following sense.

Lemma 1. *Let $(c : M \to A, i, d)$ be an opfibered comprehension. Every arrow $f : a \to a'$ in A factors as $f = d(\eta^{c \dashv i}_{f_*(ia)}) \circ d(\underline{f}(ia))$.*

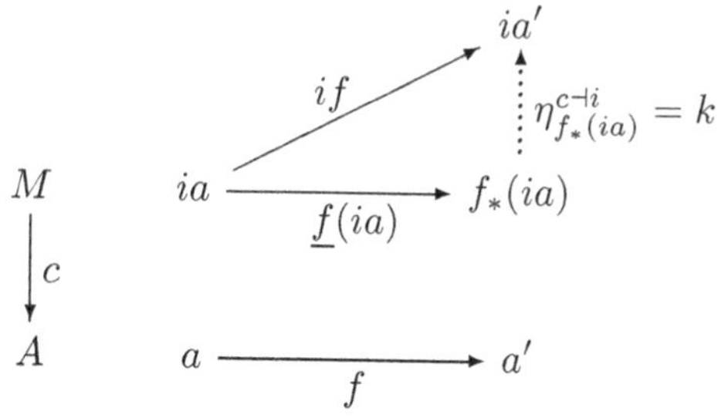

Proof. Consider the cocartesian lifting $\underline{f}(ia)$ of f with ia as above. From the universal property of the cocartesian arrow, we obtain a unique vertical arrow $k : f_*(ia) \to ia'$ such that $k \circ \underline{f}(ia) = if$. Now we have $c\eta^{c \dashv i}_{f_*(ia)} = \mathbf{id}_{a'}$ and $\eta^{c \dashv i}_{f_*(ia)} \circ \underline{f}(ia) = ic\underline{f}(ia) \circ \eta^{c \dashv i}_{ia} = if$. Therefore $k = \eta^{c \dashv i}_{f_*(ia)}$. This factorization of if in M yields the desired factorization of $dif = f$ in A.

3 Formal Space and Fundamental Adjunction

Throughout this section, we fix an opfibered comprehension $(c : M \to A, i, d)$ and an adjunction $L \dashv R : B \to A$. Our first step is to *derive* the category of spaces from the opfibered comprehension. For this, we take the pullback of c along $R : B \to A$, as done in (2). We identify the opposite of the vertex of this pullback as the category of formal spaces and formally continuous maps.

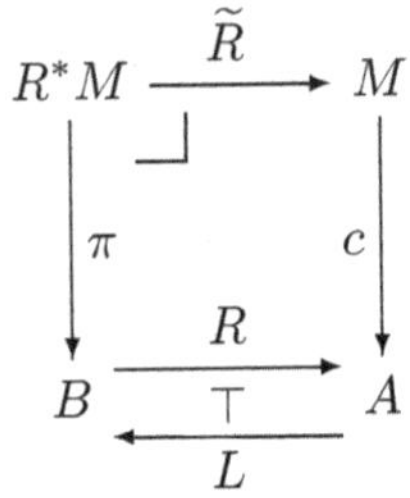

$$
\begin{array}{ccc}
R^*M & \xrightarrow{\;\widetilde{R}\;} & M \\
\big\downarrow{\scriptstyle\pi} & \lrcorner & \big\downarrow{\scriptstyle c} \\
B & \underset{L}{\overset{R}{\leftrightarrows}} & A
\end{array}
\tag{2}
$$

Definition 2. *We define the category* $\mathbf{FS}(R, c)$ *of formal spaces to be* $(R^*M)^{\mathrm{op}}$, *the opposite of the vertex category of the above pullback in the category* $\mathbf{Cat}$ *of locally small categories. We call objects and arrows in* $\mathbf{FS}(R, c)$ *formal spaces and formally continuous maps, respectively.*

We give the following concrete presentation of $\mathbf{FS}(R, c)$.

- A formal space is a tuple (b, m) of $b \in B$ and $m \in M$ satisfying $Rb = cm$.
- A formally continuous map from (b', m') to (b, m) is a tuple (f, g) of arrows $f : b \to b'$ in B and $g : m \to m'$ in M satisfying $Rf = cg$.

Example 4. The leading example of formal spaces is topological spaces. Let **Frm** be the category of frames and frame homomorphisms, and **2** be the two-point frame $\{\bot \leq \top\}$. By applying the above pullback construction to the power functor $2^{(-)} : \mathbf{Set}^{\mathrm{op}} \to \mathbf{Frm}$ and the subobject opfibered comprehension $\mathbf{cod} : \mathbf{Sub}(\mathbf{Frm}) \to \mathbf{Frm}$, we obtain the category $\mathbf{FS}(2^{(-)}, \mathbf{cod})$ of formal spaces. This is isomorphic to the category **Top** of topological spaces and continuous maps. See details in Sect. 5.

We next construct the fundamental adjunction between $\mathbf{FS}(R, c)^{\mathrm{op}}$ and A. The pullback diagram induces two extra adjunctions.

The first is a right adjoint of $\pi : R^*M \to B$. From $R \circ \mathbf{Id}_B = c \circ i \circ R$, the universal property of the pullback yields the mediating functor $\rho : B \to R^*M$ such that $\pi \circ \rho = \mathbf{Id}_B$ and $\widetilde{R} \circ \rho = i \circ R$.

Proposition 1. *We have an adjunction* $\pi \dashv \rho$ *whose counit is the identity.*

Proof. We show that $\pi \dashv \rho$ is the adjunction whose counit is $\mathrm{id}_b : b = \pi(b, iRb) = \pi\rho b \to b$. For any $b' \in B$, $m \in M$, $f : b' = \pi(b', m) \to b$ satisfying $Rb' = cm$, there exists the unique pair of $h : b' \to b$ in B and $g : m \to iRb$ in M satisfying $Rh = cg : cm \to Rb$ and $\mathrm{id}_b \circ \pi(h, g) = f$, since $c \dashv i$ and $\epsilon^{c \dashv i} = \mathbf{id}$. $\qquad\square$

The second is the left adjoint of $\tilde{R}: R^*M \to M$. This is the keystone of the fundamental adjunction. The general result of Hermida shows that the horizontal leg of the change-of-base of any opfibration along right adjoint has a left adjoint:

Theorem 1 (Corollary 3.2.5, [9]). *In* (2), $\tilde{R}: R^*M \to M$ *has a left adjoint* $\tilde{L}$ *satisfying* $\pi \circ \tilde{L} = L \circ c$.

For reference, we put his construction here. The candidate left adjoint $\tilde{L}$ maps $m \in M$ to the pair $(Lcm, (\eta_{cm})_*(m))$, where $\underline{\eta_{cm}^{L\dashv R}}(m): m \to (\eta_{cm}^{L\dashv R})_*(m)$ is the cocartesian lifting of $\eta_{cm}^{L\dashv R}$ with m. Then, $(Lcm, (\eta_{cm})_*(m))$ is an object in $\mathbf{FS}(R,c)$ and $\underline{\eta_{cm}^{L\dashv R}}(m): m \to (\eta_{cm}^{L\dashv R})_*(m) = \tilde{R}\tilde{L}m$ satisfies the universal property: for any $(b, m') \in \mathbf{FS}(R,c)$ and $f: m \to m' = \tilde{R}(b, m')$, there exists the pair of $g: Lcm \to b$ in B and $h: (\eta_{cm})_*(m) \to m'$ in M satisfying $Rg = ch$ and $h \circ \underline{\eta_{cm}^{L\dashv R}}(m) = f$. The existence and the uniqueness of g is proven by $L \dashv R$. The existence and the uniqueness of h is proven by the cocartesian lifting property of $\underline{\eta_{cm}^{L\dashv R}}(m)$.

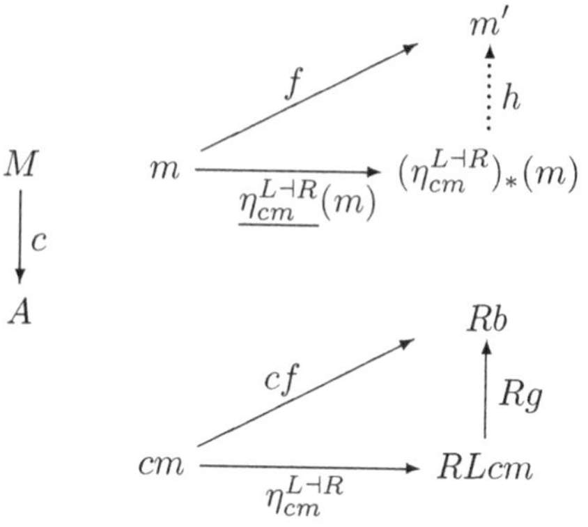

To summarize, we obtain the following sequence of adjunctions, which factors $L \dashv R$:

$$B \underset{\pi}{\overset{\rho}{\rightleftarrows}} \top \mathbf{FS}(R,c)^{\mathrm{op}} \underset{\tilde{L}}{\overset{\tilde{R}}{\rightleftarrows}} \top M \underset{i}{\overset{d}{\rightleftarrows}} \top A$$

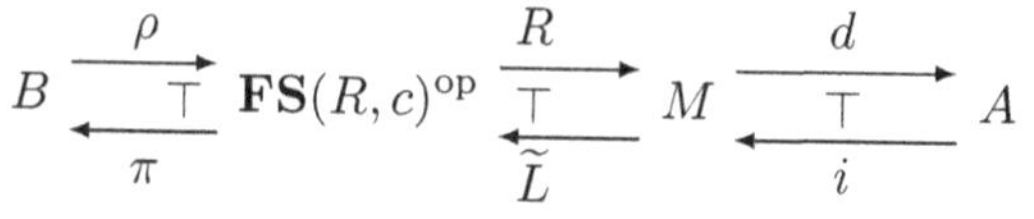

Theorem 2 (formal space factorization). *We have the factorization of* $L \dashv R$ *as* $R = d \circ \tilde{R} \circ \rho$ *and* $L = \pi \circ \tilde{L} \circ i$.

Proof. By the definition of ρ, we have $d \circ \tilde{R} \circ \rho = d \circ i \circ R = R$. By the definition of $\tilde{L}$, we have $\pi \circ \tilde{L} \circ i = L \circ c \circ i = L$.

$\square$

Starting from an opfibered comprehension $(c: M \to A, i, d)$ and an adjunction $L \dashv R: B \to A$, this factorization theorem yields an adjunction between $\mathbf{FS}(R,c)^{\mathrm{op}}$ and A. This is the main subject of this paper, the *fundamental adjunction*.

Definition 3 (fundamental adjunction). *We call* $\tilde{L} \circ i \dashv d \circ \tilde{R}$ *the* fundamental adjunction. *The left and right fundamental adjoints are denoted by*

$$\mathbf{Sp}^{R,c} \triangleq \tilde{L} \circ i \colon A \to \mathbf{FS}(R,c)^{\mathrm{op}}, \quad \mathbf{Al}^{R,c} \triangleq d \circ \tilde{R} \colon \mathbf{FS}(R,c)^{\mathrm{op}} \to A.$$

3.1 Coreflexiveness of Fundamental Adjunction

When the left fundamental adjoint $\mathbf{Sp}^{R,c}$ is full faithful, it yields an equivalence between A and its full image. We here consider when $\mathbf{Sp}^{R,c}$ is full faithful.

As shown in Lemma 1, in any opfibered comprehension $(c\colon M \to A, i, d)$ an arrow $f : a \to a'$ in A factors as $d(\eta^{c \dashv i}_{f_*(ia)}) \circ d(\underline{f}(ia))$.

Definition 4. *Let* $(c\colon M \to A, i, d)$ *be an opfibered comprehension. We say that an arrow* $f\colon a \to a'$ *in* A belongs to M *if* $d(\underline{f}(ia)) : a \to d(f_*(ia))$ *is invertible.*

Example 5. In the arrow opfibered comprehension $\mathbf{cod} : A^{\to} \to A$, any A-arrow belongs to $A^{\to}$, while for the subobject opfibered comprehension $\mathbf{cod} : \mathbf{Sub}(A) \to A$, an arrow belongs to $\mathbf{Sub}(A)$ if and only if it is mono.

Proposition 2. *The left fundamental adjoint* $\mathbf{Sp}^{R,c}$ *is full faithful if and only if* $\eta^{L \dashv R}_a$ *belongs to* M *for each* $a \in A$.

Proof. The unit of the adjunction $\mathbf{Sp}^{R,c} \dashv \mathbf{Al}^{R,c}$ is $d(\eta^{\tilde{L} \dashv \tilde{R}}_i) \circ \eta^{i \dashv d}$. Since $\eta^{i \dashv d} = \mathbf{id}$, it suffices to show that $d(\eta^{\tilde{L} \dashv \tilde{R}}_i)$ is invertible at each $a \in A$. Now recall that the m-th component $\eta^{\tilde{L} \dashv \tilde{R}}_m$ of the unit of $\tilde{L} \dashv \tilde{R}$ is the cocartesian lifting $\underline{\eta^{L \dashv R}_{cm}}(m)$ of the unit of $L \dashv R$ with m. Therefore $d(\eta^{\tilde{L} \dashv \tilde{R}}_{ia}) = d(\underline{\eta^{L \dashv R}_{cia}}(ia)) = d(\underline{\eta^{L \dashv R}_{a}}(ia))$, and, then the left hand side is invertible, if and only if $\eta^{L \dashv R}_a$ belongs to M.

Corollary 1. *Let* $\mathbf{cod} : A^{\to} \to A$ *be the arrow opfibered comprehension and* $L \dashv R$ *be an adjunction. The left fundamental adjoint* $\mathbf{Sp}^{R,c}$ *is full faithful.*

Corollary 2. *Let* $\mathbf{cod} : \mathbf{Sub}(A) \to A$ *be the subobject opfibered comprehension and* $L \dashv R$ *be an adjunction. The left fundamental adjoint* $\mathbf{Sp}^{R,c}$ *is full faithful if and only if each component of the unit of* $L \dashv R$ *is mono.*

3.2 Representability of Fundamental Adjunction

Next, we study the representability of the fundamental adjunction. We say that a dual adjunction $L \dashv R : C^{\mathrm{op}} \to D$ is *representable* through $\gamma : C \to \mathbf{Set}$ and $\delta : D \to \mathbf{Set}$ if there is a pair $\Omega_C \in C, \Omega_D \in D$ of objects such that

$$\gamma \circ L^{\mathrm{op}} \cong D(-, \Omega_D), \quad \delta \circ R \cong C(-, \Omega_C).$$

Note that it follows $\gamma\Omega_C \cong \delta\Omega_D$; see e.g. [5, Lemma 2.3].

The fundamental adjunction enjoys this representability when it arises from the following situation. Let $c : M \to A$ be an opfibered comprehension and $\Omega \in A$. We assume that (1) A comes with an adjunction $F \dashv U : A \to \mathbf{Set}$, and (2) the representable functor $(A(-,\Omega))^{\mathrm{op}} : A \to \mathbf{Set}^{\mathrm{op}}$ has a right adjoint $\Omega^{(-)} : \mathbf{Set}^{\mathrm{op}} \to A$. We call this *hom-power adjunction.* The second assumption means that A admits small powers of Ω; see [17, Section III.4] for detail. Under these assumptions, the fundamental adjunction becomes representable through $\pi : \mathbf{FS}(\Omega^{(-)}, c) \to \mathbf{Set}$ and $U : A \to \mathbf{Set}$.

Theorem 3. *We define the object $\Omega_{\mathbf{FS}} \in \mathbf{FS}(\Omega^{(-)}, c)$ to be $\mathbf{Sp}^{\Omega^{(-)},c}(F1)$, where 1 is the terminal object of* $\mathbf{Set}$. *Then*

$$\pi \circ (\mathbf{Sp}^{\Omega^{(-)},c})^{\mathrm{op}} = A(-,\Omega), \quad U \circ \mathbf{Al}^{\Omega^{(-)},c} \cong \mathbf{FS}(\Omega^{(-)}, c)(-, \Omega_{\mathbf{FS}}).$$

Proof. The first is by the factorization of $A(-,\Omega)$ (Theorem 2). We show the second.

$$\begin{aligned} U \circ \mathbf{Al}^{\Omega^{(-)},c} &\cong \mathbf{Set}(1, U \circ \mathbf{Al}^{\Omega^{(-)},c}(-)) \\ &\cong \mathbf{FS}(\Omega^{(-)}, c)^{\mathrm{op}}(\mathbf{Sp}^{\Omega^{(-)},c}(F1), -) = \mathbf{FS}(\Omega^{(-)}, c)(-, \Omega_{\mathbf{FS}}). \end{aligned}$$

4 Formal Spaces from Arrow Opfibered Comprehension

When the construction of the category of formal spaces is applied to some concrete arrow opfibered comprehension $\mathbf{cod}\colon A^{\to} \to A$, derived concepts of formal space coincide with existing structures. We illustrate two such examples: one is *Chu spaces* [10] and the other is *topological systems* introduced by Vickers [18]. These suggests that, for the arrow opfibered comprehension, our construction can be regarded as a "non-symmetric" generalization of Chu construction.

4.1 Chu Spaces

Let $(A, I, \otimes, [-,-])$ be a symmetric monoidal closed category, and $\Sigma \in A$ be an object. It plays the role of a dualizing object. In [19], Pavlovic showed that a category of Chu spaces can be obtained as the comma category of $\mathbf{Id}_A : A \to A$ and the internal hom functor $[-,\Sigma] : A^{\mathrm{op}} \to A$. In general, the comma category of the form $\mathbf{Id}_A \downarrow F$ is isomorphic to the vertex of the pullback of the codomain functor $\mathbf{cod} : A^{\to} \to A$ along F. Therefore the category $\mathrm{Chu}(A, \Sigma)$ of Chu spaces in [19] is isomorphic to the category $\mathbf{FS}([-,\Sigma], \mathbf{cod})$ of formal spaces.

4.2 Topological Systems

In [18], Vickers introduces the category of *topological system* to model state spaces paired with notions of *observations.* This category is defined by the following data, and has a similar flavor to the category of Chu spaces.

- A topological system is a tuple $(x, a, s\colon x \times a \to \mathbf{2})$ of a set, a frame, and a function such that, for each $p \in x$, the function $s(p,-)\colon a \to \mathbf{2}$ preserves finite meets and arbitrary unions.

- A map from a topological system (x', a', s') to another one (x, a, s) is a tuple (f, g) such that $f : x' \to x$ is a function, $g : a \to a'$ is a frame homomorphism and they satisfy $s \circ (f \times a) = s' \circ (x' \times g)$.

We can easily see that it is isomorphic to the category $\mathbf{FS}(\mathbf{2}^{(-)}, \mathbf{cod})$ of formal spaces, where $\mathbf{cod} : \mathbf{Frm}^{\to} \to \mathbf{Frm}$ is the arrow opfibered comprehension and $\mathbf{2}^{(-)} : \mathbf{Set}^{\mathrm{op}} \to \mathbf{Frm}$ is the power functor to $\mathbf{2}$.

The left fundamental adjoint $\mathbf{Sp}^{\mathbf{2}^{(-)},\mathbf{cod}} : \mathbf{Frm} \to \mathbf{FS}(\mathbf{2}^{(-)}, \mathbf{cod})^{\mathrm{op}}$ sends each frame to the corresponding *locale* as defined in [18]. By Proposition 2 we can see that the functor is fully faithful, which implies that the definition of the category of locales there is indeed equivalent to the more common definition: $\mathbf{Loc} = \mathbf{Frm}^{\mathrm{op}}$.

5 Formal Spaces from Subobject Opfibered Comprehension

Let $U : A \to \mathbf{Set}$ be a monadic functor and $\Omega \in A$. Then (1) U has a left adjoint, (2) A admits powers of Ω since A has small limits, and (3) A admits a strong epi-mono factorization. Therefore from such U and Ω, we obtain two ingredients needed for constructing the category of formal spaces and the fundamental adjunction: (1) the subobject opfibered comprehension $\mathbf{cod} : \mathbf{Sub}(A) \to A$, and (2) the hom-power adjunction $(A(-, \Omega))^{\mathrm{op}} \dashv \Omega^{(-)}$.

$$\mathbf{FS}(\Omega^{(-)}, \mathbf{cod})^{\mathrm{op}} \underset{\widetilde{(A(-,\Omega))^{\mathrm{op}}}}{\overset{\widetilde{\Omega^{(-)}}}{\rightleftarrows}} \mathbf{Sub}(A) \underset{\triangle}{\overset{\mathbf{dom}}{\rightleftarrows}} A \qquad (\top,\ \top)$$

The derived category $\mathbf{FS}(\Omega^{(-)}, \mathbf{cod})$ of formal spaces and the fundamental adjunction have the following concrete description.

- A formal space is a pair (I, X) of a set I and a subobject X of Ω^I.
- A formally continuous map $f : (I, X) \to (J, Y)$ is a function $f : I \to J$ such that $\Omega^f : \Omega^J \to \Omega^I$ restricts to an A-arrow of type $Y \to X$.
- The right fundamental adjoint satisfies $\mathbf{Al}^{\Omega^{(-)},\mathbf{cod}}(I, X) = X$, simply extracting the subobject part.
- The left fundamental adjoint takes $X \in A$ and computes the following pushforward in the opfibration c (recall the notation $\mathbf{e}(f), \mathbf{m}(f), \mathbf{Im}(f)$ about the factorization of f in Example 2):

$$\begin{array}{ccc} \mathbf{Sub}(A) & \triangle X \xrightarrow{(\eta_X^{A(-,\Omega)\dashv\Omega^{(-)}},\ \mathbf{e}(\eta_X^{A(-,\Omega)\dashv\Omega^{(-)}}))} & \mathbf{m}(\eta_X^{A(-,\Omega)\dashv\Omega^{(-)}}) \\ \downarrow{\scriptstyle \mathbf{cod}} & & \\ A & X \xrightarrow[\eta_X^{A(-,\Omega)\dashv\Omega^{(-)}}]{} & \Omega^{A(X,\Omega)} \end{array}$$

Then it returns the formal space $(A(X, \Omega), \mathbf{Im}(\eta_X^{A(-,\Omega)\dashv\Omega^{(-)}}))$.

From Theorem 3, the fundamental adjunction is representable, and from Corollary 2, the left fundamental adjoint $\mathbf{Sp}^{\Omega^{(-)},\mathbf{cod}}$ is fully faithful if and only if the unit $\eta : X \to \Omega^{A(X,\Omega)}$ of the hom-power adjunction is mono.

Table 1. Various algebraic categories

Category	Object	Arrow
BA	Boolean algebras	Boolean homomorphisms
SLat	Join semilattices	Join-preserving functions
Lat	Bounded lattices	Bounded lattice homomorphisms
DLat	Distributive bounded lattices	Bounded lattice homomorphisms
CSLat	Complete lattices	Join-preserving functions
Frm	Frames	Frame homomorphisms

We demonstrate that some known fundamental adjunctions between categories of algebras and those of spaces are instances of the above fundamental adjunction. To save space, we write $\mathbf{FS}(\Omega, A), \mathbf{Sp}^{\Omega,A}$ and $\mathbf{Al}^{\Omega,A}$ to mean $\mathbf{FS}(\Omega^{(-)}, \mathbf{cod}), \mathbf{Sp}^{\Omega^{(-)},\mathbf{cod}}$ and $\mathbf{Al}^{\Omega^{(-)},\mathbf{cod}}$, respectively. Let us introduce various categories by Table 1. These categories have the special object $\mathbf{2} = (\{\bot, \top\}, \bot \leq \top)$. In particular, $\mathbf{FS}(\mathbf{2}, \mathbf{BA})$ is isomorphic to the category **Fld** of fields of sets and Boolean homomorphisms [20], and $\mathbf{FS}(\mathbf{2}, \mathbf{Frm}) \cong \mathbf{Top}$ as in Example 4. Since the lattice $\mathbf{2}^X$ and its sublattices are always distributive, we have $\mathbf{FS}(\mathbf{2}, \mathbf{DLat}) \cong \mathbf{FS}(\mathbf{2}, \mathbf{Lat})$.

Example 6. When $A = \mathbf{BA}, \mathbf{SLat}, \mathbf{DLat}$, or $\mathbf{CSLat}$, $L = A(-, \mathbf{2})$ is faithful and the left fundamental adjoint $\mathbf{Sp}^{\mathbf{2},A}$ is fully faithful by Proposition 2. On the other hand, when $A = \mathbf{Lat}$, L is not faithful, since components for non distributive bounded lattices of the unit $\eta^{\mathbf{Lat}(-,\mathbf{2}) \dashv \mathbf{2}^{(-)}}$ are monomorphisms [21]. When $A = \mathbf{Frm}$, L is not faithful and $\mathbf{Sp}^{\mathbf{2},\mathbf{Frm}}$ is not fully faithful, since components for non spatial frames of the unit $\eta^{\mathbf{Sp}^{\mathbf{2},\mathbf{Frm}} \dashv \mathbf{Al}^{\mathbf{2},\mathbf{Frm}}}$ are not isomorphisms [1].

Example 7. We write **Poset** for the category of partially ordered sets and monotone maps. The representable functor $L = \mathbf{Frm}(-, \mathbf{2}) \colon \mathbf{Frm} \to \mathbf{Poset}^{\mathrm{op}}$ for **Poset**-enriched **Frm** has the right adjoint $R = \mathbf{Up}$, whose $\mathbf{Up}(X, \leq)$ is the frame of all up-closed subsets of $(X, \leq)$. Then, $\mathbf{FS}(\mathbf{Up}, \mathbf{cod})$ is the following category **PoTop**:

- its objects are $(X, \leq, \alpha)$ such that $(X, \leq) \in \mathbf{Poset}$, $(X, \alpha) \in \mathbf{Top}$, and $\alpha \subseteq \mathbf{Up}(X, \leq)$.
- its arrows $f \colon (X, \leq, \alpha) \to (Y, \leq, \beta)$ satisfy $f \colon (X, \leq) \to (Y, \leq) \in \mathbf{Poset}$ $f \colon (X, \alpha) \to (Y, \beta) \in \mathbf{Top}$.

Example 8. A fundamental adjunction $\mathbf{Sp}^{\Omega,A} \dashv \mathbf{Al}^{\Omega,A} \colon \mathbf{FS}(\Omega, A)^{\mathrm{op}} \to A$ is extendible by composing another adjunction $F \dashv U \colon A \to A'$, for example, ideal completion. A subset of a join semilattice $X = (X, \vee, \bot)$ is called an *ideal* [21] in X, if it is down-closed and finite join closed. The set of all ideals in X forms a complete join semilattice, where for a set α of ideals, its join $\bigvee \alpha$ is given not by its union $\bigcup \alpha$, but by $\{x \mid \exists \beta \subseteq \bigcup \alpha, \beta \,\text{is finite}, x \leq \bigvee \beta\}$. This construction gives adjunctions **ideals** $\dashv$ **forget**: **CSLat** $\to$ **SLat** and **ideals** $\dashv$ **forget**: **Frm** $\to$ **DLat**. Therefore, we have the following extended fundamental adjunctions.

$$\mathbf{Sp}^{\mathbf{2},\mathbf{CSLat}} \circ \mathbf{ideals} \dashv \mathbf{forget} \circ \mathbf{Al}^{\mathbf{2},\mathbf{CSLat}} \colon \mathbf{FS}(\mathbf{2},\mathbf{CSLat})^{\mathrm{op}} \to \mathbf{SLat}$$

$$\mathbf{Sp}^{\mathbf{2},\mathbf{Frm}} \circ \mathbf{ideals} \dashv \mathbf{forget} \circ \mathbf{Al}^{\mathbf{2},\mathbf{Frm}} \colon \mathbf{Top}^{\mathrm{op}} \to \mathbf{DLat}$$

$$\mathbf{Sp}^{\mathbf{Up},\mathbf{cod}} \circ \mathbf{ideals} \dashv \mathbf{forget} \circ \mathbf{Al}^{\mathbf{Up},\mathbf{cod}} \colon \mathbf{PoTop}^{\mathrm{op}} \to \mathbf{DLat}$$

6 Change of Bases

In this section, we give some construction of adjunctions among categories of different formal spaces. Below we show that an adjunction between opfibered comprehensions induces an adjunction between categories of formal spaces.

Theorem 4 (base change theorem for opfibered comprehensions). *Let*

- *$(c\colon M \to A, i, d)$ and $(c'\colon M' \to A', i', d')$ be opfibered comprehensions,*
- *$F_A \dashv U_A\colon A \to A'$ and $F_M \dashv U_M\colon M \to M'$ be adjunctions and*
- *$R\colon B \to A$ be a functor*

such that (c, c') is a map of adjunction (see [17, Section IV.7]) from $F_M \dashv U_M$ to $F_A \dashv U_A$, that is, the following equalities hold:

$$c' \circ U_M = U_A \circ c, \quad c \circ F_M = F_A \circ c', \quad c\epsilon^{F_M \dashv U_M} = \epsilon_c^{F_A \dashv U_A}.$$

Then there is an adjunction $F^* \dashv U^*\colon R^*M \to R'^*M'$ satisfying $\pi' \circ U^* = \pi$ and $\widetilde{R'} \circ U^* = U_M \circ \widetilde{R}$, where $R' = U_A \circ R$.

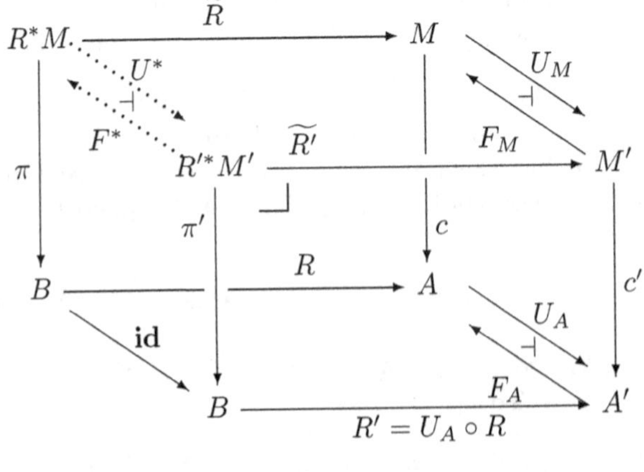

Proof. U_M satisfies $c' \circ U_M \circ \widetilde{R} = U_A \circ c \circ \widetilde{R} = U_A \circ R \circ \pi = R' \circ \pi$. Since $(R'^*M', \pi', \widetilde{R'})$ is the pullback of R' and c' in **Cat**, there exists the unique U^* satisfying $\pi' \circ U^* = \pi$ and $\widetilde{R'} \circ U^* = U_M \circ \widetilde{R}$, where U^* maps an object (b, m) to $(b, U_M m)$ and maps arrows similarly.

To save notational burden, let η^A and ϵ^A be the unit and the counit of $F_A \dashv U_A$, and η^M and ϵ^M be the unit and the counit of $F_M \dashv U_M$. That (c, c') is a map of adjunction implies that $\eta^A_{c'} \colon c' \Rightarrow U_A F_A c'$ is the same as $c'\eta^M \colon c' \Rightarrow c' U_M F_M$.

For $(b, m') \in R'^*M'$, $(b, (\epsilon^A_{Rb})_* F_M m')$ is an object of R^*M. Note that m' is above $U_A Rb$, $F_M m'$ is above $F_A U_A Rb$, and $(\epsilon^A_{Rb})_* F_M m'$ is above Rb. Let $e_{m'} \colon F_M m' \to (\epsilon^A_{Rb})_* F_M m'$ be the cocartesian arrow above ϵ^A_{Rb} defined canonically. Here,

$$\begin{aligned} c'(U_M e_{m'} \circ \eta^M_{m'}) &= c' U_M e_{m'} \circ c' \eta^M_{m'} \\ &= U_A c e_{m'} \circ \eta^A_{c'm'} \\ &= U_A \epsilon^A_{Rb} \circ \eta^A_{U_A Rb} \\ &= \mathbf{id}_{U_A Rb} \end{aligned}$$

holds. Thus, we have an arrow

$$(\mathbf{id}_b, U_M e_{m'} \circ \eta^M_{m'}) \colon (b, m') \to (b, U_M((\epsilon^A_{Rb})_* F_M m')) = U^*(b, (\epsilon^A_{Rb})_* F_M m')$$

in the category R'^*M'.

We will show that this is a universal arrow from (b, m') to U^*. Assume that we have an arrow $(f, g) \colon (b, m') \to (b_2, U_M m) = U^*(b_2, m)$. Using the adjunction $F_M \dashv U_M$, we can decompose $g \colon m' \to U_M m$ to

$$m' \xrightarrow{\eta^M_{m'}} U_M F_M m' \xrightarrow{U_M F_M g} U_M F_M U_M m \xrightarrow{U_M \epsilon^M_m} U_M m.$$

The composite

$$F_M m' \xrightarrow{F_M g} F_M U_M m \xrightarrow{\epsilon^M_m} m$$

in M is sent by c to

$$F_A U_A Rb \xrightarrow{F_A U_A Rf} F_A U_A Rb_2 \xrightarrow{\epsilon^A_{Rb_2}} Rb_2$$

in A. By naturality it is equal to

$$F_A U_A Rb \xrightarrow{\epsilon^A_{Rb}} Rb \xrightarrow{Rf} Rb_2.$$

Thus, by the universality of cocartesian lifting, $\epsilon^M_m \circ (F_M g)$ decomposes through $e_{m'}$. Combining this with the first decomposition yields the decomposition we want. Uniqueness can be shown by a similar means.

Therefore, U^* has a left adjoint F^* satisfying $F^*(b, m') = (b, (\epsilon^A_{Rb})_* F_M m')$. □

By specializing Theorem 4, we have some more usable results.

Corollary 3 (base change theorem for arrow opfibered comprehension). *Let $F_A \dashv U_A \colon A \to A'$ be an adjunction and $R \colon B \to A$ be a functor. There exist adjunctions $F_{A^{\to}} \dashv U_{A^{\to}} \colon A^{\to} \to (A')^{\to}$ and $F^* \dashv U^* \colon R^* A^{\to} \to R'^*(A')^{\to}$ satisfying the conditions in Theorem 4. Moreover, $F^*(b, m') = (b, \epsilon_{Rb} \circ F_A m')$ for each $(b, m') \in R'^*(A')^{\to}$.*

Proof. We can canonically obtain $F_{A^{\to}} \dashv U_{A^{\to}} \colon A^{\to} \to (A')^{\to}$ by defining their object parts by the arrow parts of $F_A \dashv U_A \colon A \to A'$. These satisfy the conditions in Theorem 4, so we can apply it to obtain $F^* \dashv U^* \colon R^* A^{\to} \to R'^*(A')^{\to}$.

The pushout in the construction in Theorem 4 turns out to coincide with postcomposition. This yields the equality for F^*.

□

Corollary 4 (base change theorem for subobject opfibered comprehension). *Let $F_A \dashv U_A \colon A \to A'$ be an adjunction and $R \colon B \to A$ be a functor. Assume that*

- *A and A' have (strong epi, mono)-factorization systems,*
- *F_A preserves monomorphisms, and*
- *the counit of $F_A \dashv U_A$ is componentwise monic.*

Then there exist adjunctions $F_{\mathbf{Sub}(A)} \dashv U_{\mathbf{Sub}(A)} \colon \mathbf{Sub}(A) \to \mathbf{Sub}(A')$ and $F^* \dashv U^* \colon R^*\mathbf{Sub}(A) \to R'^*\mathbf{Sub}(A')$ satisfying the conditions in Theorem 4. Moreover, $F^*(b, m') = (b, \epsilon_{Rb} \circ F_A m')$ for each $(b, m') \in R'^*\mathbf{Sub}(A')$.

Proof. Since both F_A and U_A preserve monomorphisms, we can let $F_{\mathbf{Sub}(A)}(m') = m'$ and $U_{\mathbf{Sub}(A)}(m) = m$. These satisfy the conditions in Theorem 4, so we can apply it to obtain $F^* \dashv U^* \colon R^*\mathbf{Sub}(A) \to R'^*\mathbf{Sub}(A')$.

By the assumption on the counit, the pushout in the construction in Theorem 4 turns out to coincide with postcomposition. This yields the equality for F^*. □

Example 9. The leading example of Corollary 4 is the adjunction between $\mathbf{Fld} \cong \mathbf{FS}(\mathbf{2}, \mathbf{BA})$ and $\mathbf{FS}(\mathbf{2}, \mathbf{DLat})$.

The forgetful functor $\mathbf{forget} \colon \mathbf{BA} \to \mathbf{DLat}$ has a right adjoint $\mathbf{comp}$, which maps a distributive lattice $(X, \vee, \wedge, \bot, \top)$ to the Boolean algebra of its complemented elements [20], that is to say, $\{x \in X \mid \exists x' \in X, x \vee x' = \top, x \wedge x' = \bot\}$. Since this adjunction satisfies the condition of Corollary 4, we have the adjunction $\mathbf{forget}^* \dashv \mathbf{comp}^* \colon \mathbf{FS}(\mathbf{2}, \mathbf{DLat})^{\mathrm{op}} \to \mathbf{FS}(\mathbf{2}, \mathbf{BA})^{\mathrm{op}}$ satisfying $\mathbf{forget}^* \circ \mathbf{Sp}^{\mathbf{2},\mathbf{BA}} = \mathbf{Sp}^{\mathbf{2},\mathbf{DLat}} \circ \mathbf{forget}$.

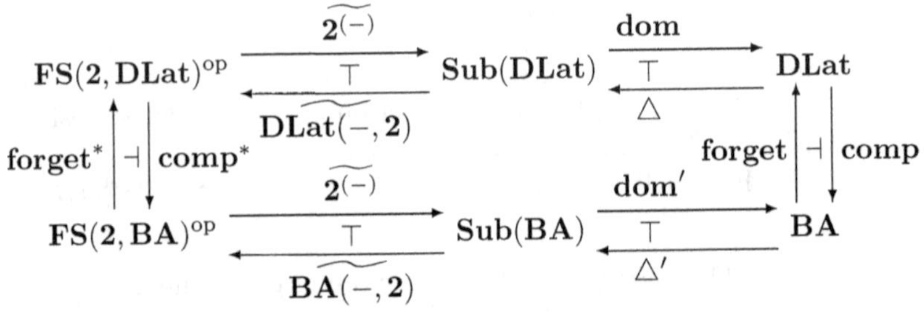

By composing **forget** $\dashv$ **comp** with the adjunction $\mathbf{Sp}^{\mathbf{2},\mathbf{Frm}} \circ \mathbf{ideals} \dashv \mathbf{forget} \circ \mathbf{Al}^{\mathbf{2},\mathbf{Frm}} \colon \mathbf{Top}^{\mathrm{op}} \to \mathbf{DLat}$ in Example 8, we also have the following adjunction.

$$\mathbf{Sp}^{\mathbf{2},\mathbf{Frm}} \circ \mathbf{ideals} \circ \mathbf{forget} \dashv \mathbf{comp} \circ \mathbf{forget} \circ \mathbf{Al}^{\mathbf{2},\mathbf{Frm}} \colon \mathbf{Top}^{\mathrm{op}} \to \mathbf{BA}$$

7 Conclusion and Future Work

This paper has defined the notion of opfibered comprehension $(c\colon M \to A, i, d)$ including the example of arrows $M = A^{\to}$ or the example of subobjects $M = \mathbf{Sub}(A)$. For any functor $R\colon B \to A$, we have constructed its formal spaces $\mathbf{FS}(R, c)$ and a fundamental adjunction $\mathbf{Sp}^{R,c} \dashv \mathbf{Al}^{R,c}\colon \mathbf{FS}(R, c)^{\mathrm{op}} \to A$. The leading example is the adjunction between $\mathbf{Top}^{\mathrm{op}}$ and $\mathbf{Frm}$.

We also have given the sufficient condition to construct different formal spaces $\mathbf{FS}(R, c)$ and $\mathbf{FS}(R', c')$. Its leading example is the adjunction between $\mathbf{FS}(\mathbf{2}, \mathbf{DLat})$ and $\mathbf{FS}(\mathbf{2}, \mathbf{BA})$.

It is future work to compare with other dualities, for example, Priestley duality [21], algebra/coalgebra duality [22], natural dualities [12], and so on.

Acknowledgments. This work was the supported by JSPS KAKENHI Grant Number JP24700017. The second and third authors carried out this research under the support of JST ERATO HASUO Metamathematics for Systems Design Project (No. JPMJER1603).

References

1. Johnstone, P.: Stone Spaces. Cambridge Studies in Advanced Mathematics. Cambridge University Press, Cambridge (1986)
2. Johnstone, P.: The point of pointless topology. Bull. Am. Math. Soc. (N.S.) **8**(1), 41–53 (1983)
3. Isbell, J.R.: General functorial semantics I. Am. J. Math. **97**, 535–590 (1972)
4. Simmons, H.: A couple of triples. Topol. Appl. **13**(2), 201–223 (1982)
5. Dimov, G., Tholen, W.: A characterization of representable dualities. In: Adamek, J., MacLane, S. (eds.) Categorical Topology and Its Relation to Analysis, pp. 336–357. World Scientific, Algebra and Combinatorics (1989)
6. Porst, H.E., Tholen, W.: Concrete dualities. In: Herrlich, H., Porst, H.E. (eds.) Category Theory at Work, pp. 111–136. Heldermann Verlag, Berlin (1991)
7. Barr, M., Kennison, J.F., Raphael, R.: Isbell duality. Theor. Appl. Categories **20**(15), 504–542 (2008)
8. Maruyama, Y.: Categorical duality theory: with applications to domains, convexity, and the distribution monad, pp. 500–520 (2013)
9. Hermida, C.: Fibrations, logical predicates and related topics. Ph.D. thesis, University of Edinburgh (1993). Technical report ECS-LFCS-93-277. Also available as Aarhus Univ. DAIMI Technical report PB-462 (1993)
10. Barr, M.: *-Autonomous Categories. Number 752 in Lecture Notes in Mathematics. Springer, Heidelberg (1979). https://doi.org/10.1007/BFb0064579
11. Pratt, V.: Chu spaces. In: Notes for the School on Category Theory and Applications, University of Coimbra, July 13–17 1999 (1999)

12. Clark, D.M., Davey, B.A.: Natural Dualities for the Working Algebraist. Number 57 in Cambridge Studies in Advanced Mathematics. Cambridge University Press, Cambridge (1998)
13. Jacobs, B.: Categorical Logic and Type Theory. Number 141 in Studies in Logic and the Foundations of Mathematics, North Holland, Amsterdam (1999)
14. Johnstone, P.: Remarks on punctual local connectedness. Theor. Appl. Categories **25**(3), 51–63 (2011)
15. Borceux, F.: Handbook of Categorical Algebra. Volume 1 of Encyclopedia of Mathematics and its Applications. Cambridge University Press, Cambridge (1994)
16. Rosický, J., Tholen, W.: Factorization, fibration and torsion. J. Homotopy Relat. Struct. **2**(2), 295–314 (2007)
17. Mac Lane, S.: Categories for the Working Mathematician. Springer, New York (1978)
18. Vickers, S.: Topology via Logic. Cambridge University Press, Cambridge (1989)
19. Pavlovic, D.: Chu I: cofree equivalences, dualities and *-autonomous categories. Math. Struct. Comput. Sci. **7**(1), 4973 (1997)
20. Blyth, T.: Lattices and Ordered Algebraic Structures. Springer, London (2005). https://doi.org/10.1007/b139095
21. Davey, B.A., Priestley, H.A.: Introduction to Lattices and Order, 2nd edn. Cambridge University Press, Cambridge (2002)
22. Kupke, C., Kurz, A., Venema, Y.: Stone coalgebras. Theor. Comput. Sci. **327**(1–2), 109–134 (2004)
23. Rocca, S.R.D. (ed.): Computer Science Logic 2013 (CSL 2013), CSL 2013, Torino, Italy, 2–5 September 2013. Volume 23 of LIPIcs, Schloss Dagstuhl - Leibniz-Zentrum fuer Informatik (2013)

Preorders, Partial Semigroups, and Quantales

Koki Nishizawa[1(✉)], Koji Yasuda[2], and Hitoshi Furusawa[3]

[1] Department of Information Systems Creation, Faculty of Engineering, Kanagawa University, Yokohama, Japan
nishizawa@kanagawa-u.ac.jp

[2] Field of Information Systems Creation, Course of Engineering, Graduate School of Engineering, Kanagawa University, Yokohama, Japan
r201970127mt@jindai.jp

[3] Department of Mathematics and Computer Science, Kagoshima University, Kagoshima, Japan
furusawa@sci.kagoshima-u.ac.jp

Abstract. It is known that each powerset quantale is embeddable into some relational unital quantale whose underlying set is the powerset of some preorder. An aim of this paper is to understand the relational embedding as a relationship between quantales and preorders. For that, this paper introduces the notion of weak preorders, a functor from the category of weak preorders to the category of partial semigroups, and a functor from the category of partial semigroups to the category of quantales and lax homomorphisms. By using these two functors, this paper shows a correspondence among four classes of weak preorders (including the class of ordinary preorders), four classes of partial semigroups, and four classes of quantales. As a corollary of the correspondence, we can understand the relational embedding map as a natural transformation between functors onto certain category of quantales.

1 Introduction

A unital quantale is defined to be a complete join semilattice together with a monoid structure satisfying the distributive laws. It was introduced by Conway under the name S-algebras [1]. A relational example of unital quantale is a powerset $\wp(A \times A)$, which is the set of all binary relations on a set A and whose monoid structure is given by relational composition and the identity relation. We call it a relational quantale. A relational quantales play an important role in computer science, for example, it is a model for the semantics of non-deterministic while-programs [2,3].

In the paper [4], the relational representation theorem for powerset quantales is shown, where a powerset quantale is defined to be a unital quantale whose complete join semilattice part is isomorphic to the powerset of some set. The relational representation is given as an embedding (i.e., injective unital homomorphism) η from a powerset quantale $\wp(A)$ to some relational

U. Fahrenberg et al. (Eds.): RAMiCS 2020, LNCS 12062, pp. 237–252, 2020.
https://doi.org/10.1007/978-3-030-43520-2_15

quantale $\wp(A \times A)$. The paper [5] includes this embedding η in the special case where A is a free monoid.

$(\wp(\mathbf{N}), \subseteq, \bigcup, [\![+]\!], \{0\})$ is the leading example of powerset quantale, where $\mathbf{N}$ is the set of all natural numbers (including 0) and $S \, [\![+]\!] \, S' \stackrel{\text{def}}{=} \{n + n' \mid n \in S, n' \in S'\}$. When the relational representation theorem [4] is applied to $\wp(\mathbf{N})$, the embedding $\eta\colon \wp(\mathbf{N}) \to \wp(\mathbf{N} \times \mathbf{N})$ embeds a subset S into the left and right reversed relation of $\{(m, n) \mid n - m \in S\}$. The subtraction $-$ for natural numbers plays an important role.

In general, the representation theorem helps to understand the algebra, not only that, but it is meaningful in the field of computer science, because it suggests the possibility of representing infinite entities with finite data. Embedding from power set quantale into binary relations also shows the possibility that infinite entities can be represented by finite data. For example, as an application example of the above $\eta\colon \wp(\mathbf{N}) \to \wp(\mathbf{N} \times \mathbf{N})$, the finite set $\{2\}$ is representing $\eta(\{2\})$, i.e., the infinite binary relation (or the infinite directed graph) $\{(m, m+2) \mid m \in \mathbf{N}\}$. In addition, the multiplication in the quantale can replace the relational composition operation. This shows the same effect as the correspondence between linear mapping and its matrix representation. Analyzing quantales' representation theorem is meaningful in the field of computer science.

The definition of relational quantale is extendable for the powerset $\wp(\leq)$ of each preorder $\leq$, where a preorder is regarded as a subset of $A \times A$ satisfying reflexivity and transitivity. An aim of this paper is to understand the relational embedding η as a relationship between quantales and preorders.

Since direct construction of a relationship between quantales and preorders is not very obvious, this paper introduces the notion of weak preorder and partial semigroup as a relaxation of preorder and semigroup, respectively. We also introduce the notion of lax homomorphism between (not unital) quantales as a relaxation of ordinary homomorphism between quantales. And, we give a functor **comp** from the category **WPreOrd** of weak preorders to the category **PSG** of partial semigroups, and a contravariant functor $\wp$ from **PSG** to the category $\mathbf{Qt_{lax}}$ of quantales and lax homomorphisms. Moreover, we also define three categories (including the category **UQt** of unital quantales and unital homomorphisms) as restrictions of $\mathbf{Qt_{lax}}$, in a step-by-step manner. By proving the following six pullbacks in **Cat**, we show the relationship among orders, semigroups, and quantales.

Recall that a pullback of $F\colon B \to A$ and $G\colon C \to A$ in **Cat** is isomorphic to the category $B \times_A C$ which consists of pairs of $b \in B$ and $c \in C$ satisfying $Fb = Gc$ with projections $\pi\colon B \times_A C \to B$ and $\pi'\colon B \times_A C \to C$, since for functors $H\colon D \to B$ and $K\colon D \to C$ satisfying $F \circ H = G \circ K$, the functor $L(d) = (Hd, Kd)$ is the unique functor $L\colon D \to B \times_A C$ satisfying $\pi \circ L = H$ and $\pi' \circ L = K$.

By using these functors, we understand the relational embedding map η as a natural transformation.

This paper is organized as follows. Section 2 defines the notions of weak preorder, partial semigroup, and lax homomorphism between quantales. In Sect. 3,

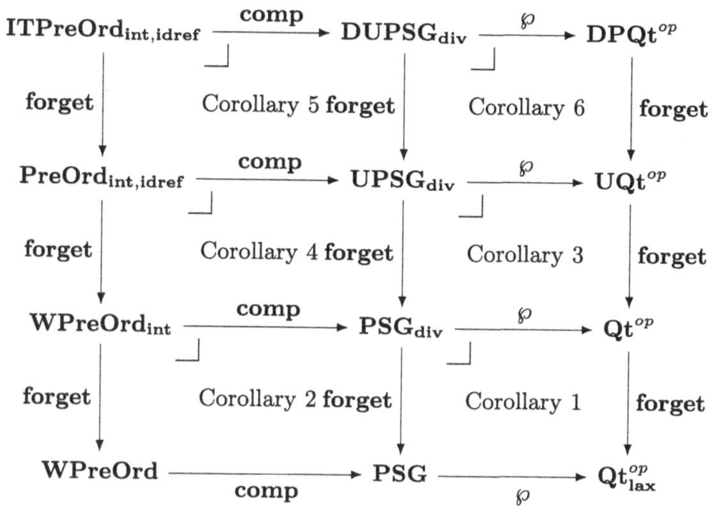

Fig. 1. The six pullbacks in **Cat** shown in this paper

we restrict arrows between quantales to ordinary homomorphisms and give the corresponding classes of arrows between weak preorders or partial semigroups. Section 4 extends quantales to unital quantales and give the corresponding extension of weak preorders or partial semigroups. In Sect. 5, we introduce sufficient classes to induce the relational embedding. Section 6 summarizes this work and discusses about future work.

2 Weak Preorders and Partial Semigroups

In this section, we define the notions of weak preorder, partial semigroup, and lax homomorphism between quantales. By using the three notions, we give three categories and two functors among them.

First, we recall the definition of quantale [6–8] and define the notion of lax homomorphism between quantales.

Definition 1 ((non unital) quantale and lax homomorphism). *A* quantale *is defined to be a tuple* $(Q, \leq, \bigvee, \odot)$ *such that*

1. $(Q, \odot)$ *is a semigroup (i.e., a binary function* $\odot$ *on* Q *that is associative),*
2. $(Q, \leq, \bigvee)$ *is a complete join semilattice (i.e., a partially ordered set* $(Q, \leq)$ *has the least upper bound* $\bigvee S$ *for arbitrary subset* S *of* Q*),*
3. $(\bigvee S) \odot q = \bigvee\{s \odot q | s \in S\}$ *for each element* q *and each subset* S *of* Q*, and*
4. $q \odot (\bigvee S) = \bigvee\{q \odot s | s \in S\}$ *for each element* q *and each subset* S *of* Q*.*

For two quantales $(Q, \leq, \bigvee, \odot)$*,* $(Q', \leq', \bigvee', \odot')$*, a* lax homomorphism *from* $(Q, \leq, \bigvee, \odot)$ *to* $(Q', \leq', \bigvee', \odot')$ *is defined to be a map* $f\colon Q \to Q'$ *such that*

1. $f(\bigvee S) = \bigvee'\{f(s) | s \in S\}$ *for each subset* S *of* Q *(join-preserving), and*
2. $f(q_1) \odot' f(q_2) \leq' f(q_1 \odot q_2)$ *for each elements* q_1, q_2 *of* Q *(closed map).*

Lax homomorphisms between quantales are closed under composition as maps. The identity map is a lax homomorphism on a quantale. Therefore, we can define the category whose objects are quantales and whose arrows are lax homomorphisms between them and we write $\mathbf{Qt}_{\mathbf{lax}}$ for it.

As the sufficient structure to construct $\odot$ in the powerset case $Q = \wp(X)$, we define the notion of partial semigroup.

Definition 2 (partial semigroup and homomorphism). *A* partial semigroup *is defined to be a tuple* $(X, \cdot)$ *such that*

1. X *is a set,*
2. $\cdot$ *is a partial binary function on* X *(i.e.,* $x \cdot y$ *may be undefined),*
3. $x \cdot y$ *and* $(x \cdot y) \cdot z$ *are defined if and only if* $y \cdot z$ *and* $x \cdot (y \cdot z)$ *are defined for* $x, y, z \in X$*, and*
4. $(x \cdot y) \cdot z = x \cdot (y \cdot z)$ *if they are defined.*

For partial semigroups $(X, \cdot)$, $(X', \cdot')$, *a* homomorphism *from* $(X, \cdot)$ *to* $(X', \cdot')$ *is defined to be a map* $f\colon X \to X'$ *such that*

1. $f(x) \cdot' f(y)$ *is defined for* $x, y \in X$ *such that* $x \cdot y$ *is defined, and*
2. $f(x) \cdot' f(y) = f(x \cdot y)$ *if they are defined.*

Homomorphisms between partial semigroups are closed under composition as maps. The identity map is a homomorphism on a partial semigroup. We write **PSG** for the category whose objects are partial semigroups and whose arrows are homomorphisms between them.

Example 1. For each set X, $(X, \cdot)$ is a partial semigroup, where $x \cdot y \stackrel{\text{def}}{=} x$ if $x = y$ and $x \cdot y$ is undefined otherwise.

Example 2. For a set A, $(A \times A, ;)$ is a partial semigroup, where $(a, b); (c, d) \stackrel{\text{def}}{=} (a, d)$ if $b = c$ and $(a, b); (c, d)$ is undefined otherwise.

Next, we present the construction of a powerset (not unital) quantale from a partial semigroup. Here, we denote by the binary map $[\![\cdot]\!]$ on $\wp(X)$ so that $S_1 \, [\![\cdot]\!] \, S_2 \stackrel{\text{def}}{=} \{s_1 \cdot s_2 \mid s_1 \in S_1, s_2 \in S_2, s_1 \cdot s_2 \text{ is defined}\}$ for a partial semigroup $(X, \cdot)$.

Theorem 1. *The following data form a functor* $\wp\colon \mathbf{PSG}^{op} \to \mathbf{Qt}_{\mathbf{lax}}$.

- *For an object* $(X, \cdot)$, $\wp(X, \cdot) \stackrel{\text{def}}{=} (\wp(X), \subseteq, \bigcup, [\![\cdot]\!])$
- *For an arrow* $f\colon (X, \cdot) \to (X', \cdot')$, $\wp(f)\colon \wp(X', \cdot') \to \wp(X, \cdot)$ *is a map* $\wp(f)(S') = \{x \in X \mid f(x) \in S'\}$.

Proof. Take an arrow $f\colon (X, \cdot) \to (X', \cdot')$ in **PSG**. $\wp(f)$ preserves $\bigcup$. Take $S'_1, S'_2 \in \wp(X')$ and $x \in \wp(f)(S'_1) \, [\![\cdot]\!] \, \wp(f)(S'_2)$. There are $s_1, s_2 \in X$ satisfying $f(s_1) \in S'_1$, $f(s_2) \in S'_2$, and $x = s_1 \cdot s_2$. They satisfy $f(x) = f(s_1 \cdot s_2) =$

$f(s_1) \cdot' f(s_2) \in S'_1 \llbracket \cdot' \rrbracket S'_2$ and then $x \in \wp(f)(S'_1 \llbracket \cdot' \rrbracket S'_2)$. Therefore, we have shown $\wp(f)(S'_1) \llbracket \cdot \rrbracket \wp(f)(S'_2) \subseteq \wp(f)(S'_1 \llbracket \cdot' \rrbracket S'_2)$. $\wp$ preserves composition and identities. □

A transitive relation R on a set X forms a partial semigroup $(R, ;)$ with the same ; as Example 2. However, we define weak preorders, as a stronger notion than transitive relations and a weaker notion than preorders.

Definition 3 (weak preorder). *A binary relation $R \subseteq X \times X$ on X is called a* weak preorder *on X, if R is transitive and R satisfies $\forall x \in X. \exists y \in X. (x, y) \in R$ or $(y, x) \in R$. A* weak preordered set *is defined to be a tuple (X, R) of a set X and a weak preorder R on X.*

Example 3. Let $\mathbf{N}$ be the set of all natural numbers (including 0). We write $m <_{\mathbf{N}} n$ when m is less than n as natural numbers and write $m \leq_{\mathbf{N}} n$ when $m <_{\mathbf{N}} n$ or $m = n$. We also regard $<_{\mathbf{N}}$ as the set of (m, n) satisfying $m <_{\mathbf{N}} n$ and regard $\leq_{\mathbf{N}}$ as the set of (m, n) satisfying $m \leq_{\mathbf{N}} n$.

$\leq_{\mathbf{N}}$ is a preorder on $\mathbf{N}$. On the other hand, $<_{\mathbf{N}}$ is not a preorder, but a weak preorder on $\mathbf{N}$.

Definition 4. *For weak preordered sets (X, R), (X', R'), a* monotone *map from (X, R) to (X', R') is defined to be a map $f \colon X \to X'$ such that $(x, y) \in R$ implies $(f(x), f(y)) \in R'$ for each $x, y \in X$.*

Monotone maps between weak preordered sets are closed under composition as maps. The identity map is a monotone map on a weak preordered set. Therefore, weak preordered sets and monotone maps between them form a category. We write **WPreOrd** for it.

Next, we present the construction of a partial semigroup from a weak preordered set.

Theorem 2. *The following data form a functor* **comp**: **WPreOrd** $\to$ **PSG**.

- *For an object (X, R), $\mathbf{comp}(X, R) \overset{\text{def}}{=} (R, ;)$ where $(w, x); (y, z)$ is defined and equal to (w, z) if $x = y$.*
- *For an arrow $f \colon (X, R) \to (X', R')$, $\mathbf{comp}(f) \colon (R, ;) \to (R', ;)$ sends (x, y) to $(f(x), f(y))$.*

Proof. Take an arrow $f \colon (X, R) \to (X', R')$ in **WPreOrd**. Take $(x, y) \in R$. By monotonicity of f, $\mathbf{comp}(f)(x, y) = (f(x), f(y)) \in R'$. Take $(w, x), (y, z) \in R$ such that $(w, x); (y, z)$ is defined. Then, $x = y$ and $(w, x); (y, z) = (w, z)$. $\mathbf{comp}(f)(w, x); \mathbf{comp}(f)(y, z) = (f(w), f(x)); (f(x), f(z))$ is defined and equal to $(f(w), f(z)) = \mathbf{comp}(f)(w, z)$. **comp** preserves composition and identities. □

Example 4. We define the arrow $\mathbf{plus1} \colon (\mathbf{N}, <_{\mathbf{N}}) \to (\mathbf{N}, \leq_{\mathbf{N}})$ in **WPreOrd** by $\mathbf{plus1}(n) \overset{\text{def}}{=} n + 1$.

$\mathbf{comp}(\mathbf{plus1})$ is the arrow from $\mathbf{comp}(\mathbf{N}, <_{\mathbf{N}}) = (<_{\mathbf{N}}, ;)$ to $\mathbf{comp}(\mathbf{N}, \leq_{\mathbf{N}}) = (\leq_{\mathbf{N}}, ;)$ in **PSG** such that $\mathbf{comp}(\mathbf{plus1})(m, n) = (m + 1, n + 1)$ for (m, n) satisfying $m <_{\mathbf{N}} n$.

$\wp(\mathbf{comp(plus1)})$ is the arrow from $\wp(\leq_{\mathbf{N}}, ;) = (\wp(\leq_{\mathbf{N}}), \subseteq, \bigcup, [\![;]\!])$ to $\wp(<_{\mathbf{N}}, ;) = (\wp(<_{\mathbf{N}}), \subseteq, \bigcup, [\![;]\!])$ in $\mathbf{Qt_{lax}}$ such that $\wp(\mathbf{comp(plus1)})(S) = \{(m,n) \mid m <_{\mathbf{N}} n, (m+1, n+1) \in S\}$ for each subset S of $\leq_{\mathbf{N}}$. Since $\wp(\mathbf{comp(plus1)})$ is a lax homomorphism, all subsets S_1, S_2 of $\leq_{\mathbf{N}}$ satisfy $\{(l,m) \mid l <_{\mathbf{N}} m, (l+1, m+1) \in S_1\} \, [\![;]\!] \, \{(m,n) \mid m <_{\mathbf{N}} n, (m+1, n+1) \in S_2\} \subseteq \{(m,n) \mid m <_{\mathbf{N}} n, (m+1, n+1) \in S_1 \, [\![;]\!] \, S_2\}$.

3 The Intermediate Value Property and Dividing Maps

In this section, we recall the category of quantales and ordinary homomorphisms of quantales. It is a subcategory of $\mathbf{Qt_{lax}}$ in Sect. 2. We study which maps between partial semigroups correspond to homomorphisms between quantales, and which maps between weak preorders correspond to those maps between partial semigroups.

Definition 5 (homomorphism between quantales). *For quantales* $(Q, \leq, \bigvee, \odot)$, $(Q', \leq', \bigvee', \odot')$, *a* homomorphism *from* $(Q, \leq, \bigvee, \odot)$ *to* $(Q', \leq', \bigvee', \odot')$ *is defined to be a map* $f\colon Q \to Q'$ *satisfying*

1. $f(\bigvee S) = \bigvee' \{f(s) | s \in S\}$ *for each subset* S *of* Q, *and*
2. $f(q_1 \odot q_2) = f(q_1) \odot' f(q_2)$ *for each elements* q_1, q_2 *of* Q.

We write $\mathbf{Qt}$ *for the subcategory of* $\mathbf{Qt_{lax}}$, *whose arrows are homomorphisms.*

Next, we introduce the additional condition for homomorphisms between partial semigroups which corresponds to homomorphisms between quantales. We call it the dividing condition.

Definition 6 (dividing map between partial semigroups). *A partial semigroup homomorphism* $f\colon (X, \cdot) \to (X', \cdot')$ *is called* dividing, *if for each* $x', y' \in X'$ *and* $z \in X$ *satisfying* $x' \cdot' y' = f(z)$, *there exist* $x, y \in X$ *such that* $f(x) = x'$, $f(y) = y'$, *and* $x \cdot y = z$.

Dividing maps between partial semigroups are closed under composition as maps. The identity map is dividing. We write $\mathbf{PSG_{div}}$ for the subcategory of $\mathbf{PSG}$, whose arrows are only dividing maps.

Next, we show that dividing maps between partial semigroups correspond to homomorphisms between quantales.

Theorem 3. *For an arrow* $f\colon (X, \cdot) \to (X', \cdot')$ *in* $\mathbf{PSG}$, *the following statements are equivalent.*

1. f *is an arrow* $f\colon (X, \cdot) \to (X', \cdot')$ *in* $\mathbf{PSG_{div}}$.
2. $\wp(f)$ *is an arrow* $\wp(f)\colon \wp(X', \cdot') \to \wp(X, \cdot)$ *in* $\mathbf{Qt}$.

Proof. ($1 \Rightarrow 2$) Assume $f\colon (X, \cdot) \to (X', \cdot')$ in $\mathbf{PSG_{div}}$. Take $S'_1, S'_2 \in \wp(X')$ and $x \in \wp(f)(S'_1 \, [\![\cdot']\!] \, S'_2)$. There exist $x'_1 \in S'_1$, $x'_2 \in S'_2$ satisfying $f(x) = x'_1 \cdot' x'_2$. Since $f \in \mathbf{PSG_{div}}$, there exist $x_1, x_2 \in X$ satisfying $f(x_1) = x'_1$, $f(x_2) = x'_2$, and $x = x_1 \cdot x_2$. Therefore, $x = x_1 \cdot x_2 \in \{x_1\} \, [\![\cdot]\!] \, \{x_2\} \subseteq \wp(f)(S'_1) \, [\![\cdot]\!] \, \wp(f)(S'_2)$.

(2 ⇒ 1) Assume $\wp(f)\colon \wp(X', \cdot') \to \wp(X, \cdot)$ in **Qt**. Take $x', y' \in X'$ and $z \in X$ satisfying $x' \cdot' y' = f(z)$. Then, $z \in \wp(f)(\{f(z)\}) = \wp(f)(\{x' \cdot' y'\}) = \wp(f)(\{x'\}[\![\cdot']\!]\{y'\}) = \wp(f)(\{x'\})[\![\cdot]\!]\wp(f)(\{y'\})$. By the definition of $[\![\cdot]\!]$, there exist x, y such that $x \in \wp(f)(\{x'\})$, $y \in \wp(f)(\{y'\})$, and $x \cdot y = z$. Then, $f(x) = x'$ and $f(y) = y'$. □

Corollary 1. *Let* $\wp\colon \mathbf{PSG_{div}} \to \mathbf{Qt}^{op}$ *be the restriction of* $\wp\colon \mathbf{PSG} \to \mathbf{Qt}^{op}_{\mathbf{lax}}$. *The forgetful functor* **forget**: $\mathbf{PSG_{div}} \to \mathbf{PSG}$ *and* $\wp\colon \mathbf{PSG_{div}} \to \mathbf{Qt}^{op}$ *is a pullback of* $\wp\colon \mathbf{PSG} \to \mathbf{Qt}^{op}_{\mathbf{lax}}$ *and the forgetful functor* **forget**: $\mathbf{Qt}^{op} \to \mathbf{Qt}^{op}_{lax}$ *in* **Cat** *(Fig. 1).*

Next, we introduce the additional condition for monotone maps between weak preorders which corresponds to dividing maps between partial semigroups. The condition is the intermediate value property.

Definition 7 (the intermediate value property). *For weak preordered sets* (X, R), (X', R'), *a monotone map* f *from* (X, R) *to* (X', R') *is said to* satisfy the intermediate value property, *if for* x, y, z' *s.t.* $(x, y) \in R$, $(f(x), z') \in R'$ *and* $(z', f(y)) \in R'$, *there exists* $z \in X$ *s.t.* $f(z) = z'$, $(x, z) \in R$, *and* $(z, y) \in R$.

Monotone maps satisfying the intermediate value property between weak preordered sets are closed under composition as maps. The identity map on a weak preordered set satisfies the intermediate value property. We write $\mathbf{WPreOrd_{int}}$ for the subcategory of **WPreOrd**, whose arrows are only arrows satisfying the intermediate value property.

Next, we show that monotone maps satisfying the intermediate value property correspond to dividing maps between partial semigroups.

Theorem 4. *For* $f\colon (X, R) \to (X', R')$ *in* **WPreOrd**, *the following statements are equivalent.*

1. f *is an arrow* $f\colon (X, R) \to (X', R')$ *in* $\mathbf{WPreOrd_{int}}$.
2. $\mathbf{comp}(f)$ *is an arrow* $\mathbf{comp}(f)\colon \mathbf{comp}(X, R) \to \mathbf{comp}(X', R')$ *in* $\mathbf{PSG_{div}}$.

Proof. (1 ⇒ 2) Assume $f\colon (X, R) \to (X', R')$ in $\mathbf{WPreOrd_{int}}$. Take (w', x'), $(y', z') \in R'$ and $(w, z) \in R$ satisfying $(w', x'); (y', z') = \mathbf{comp}(f)(w, z)$. Then, $x' = y'$, $f(w) = w'$, and $f(z) = z'$. By the intermediate value property of f, there exists $x \in X$ such that $f(x) = x'$, $(w, x) \in R$, and $(x, z) \in R$. They satisfy $\mathbf{comp}(f)(w, x) = (f(w), f(x)) = (w', x')$, $\mathbf{comp}(f)(x, z) = (f(x), f(z)) = (x', z') = (y', z')$, and $(w, x); (x, z) = (w, z)$.

(2 ⇒ 1) Assume $\mathbf{comp}(f)\colon \mathbf{comp}(X, R) \to \mathbf{comp}(X', R')$ in $\mathbf{PSG_{div}}$. Take $(x, y) \in R$ and $z' \in X'$ satisfying $(f(x), z') \in R'$ and $(z', f(y)) \in R'$. They satisfy $\mathbf{comp}(f)(x, y) = (f(x), f(y)) = (f(x), z'); (z', f(y))$. Since $\mathbf{comp}(f)$ is dividing, there exist $(x, z), (z, y) \in R$ such that $\mathbf{comp}(f)(x, z) = (f(x), z')$, $\mathbf{comp}(f)(z, y) = (z', f(y))$, and $(x, z); (z, y) = (x, y)$. Therefore, $f(z) = z'$. □

Corollary 2. *Let* $\mathbf{comp}\colon \mathbf{WPreOrd_{int}} \to \mathbf{PSG_{div}}$ *be the restriction of* $\mathbf{comp}\colon \mathbf{WPreOrd} \to \mathbf{PSG}$. *The forgetful functor* **forget**: $\mathbf{WPreOrd_{int}} \to \mathbf{WPreOrd}$ *and* $\mathbf{comp}\colon \mathbf{WPreOrd_{int}} \to \mathbf{PSG_{div}}$ *is a pullback of* $\mathbf{comp}\colon \mathbf{WPreOrd} \to \mathbf{PSG}$ *and the forgetful functor* **forget**: $\mathbf{PSG_{div}} \to \mathbf{PSG}$ *in* **Cat** *(Fig. 1).*

Example 5. In $\mathbf{WPreOrd_{int}}$, the map $\mathbf{plus1}(n) \stackrel{\text{def}}{=} n+1$ is an arrow from $(\mathbf{N}, <_{\mathbf{N}})$ to $(\mathbf{N}, <_{\mathbf{N}})$, but not to $(\mathbf{N}, \leq_{\mathbf{N}})$, since $1 \leq_{\mathbf{N}} 1 \leq_{\mathbf{N}} 2$ but not $0 <_{\mathbf{N}} 0$.

For the same reason, the map $\mathbf{comp}(\mathbf{plus1})(m,n) = (m+1, n+1)$ is a dividing map from $(<_{\mathbf{N}}, ;)$ to $(<_{\mathbf{N}}, ;)$, but not to $(\leq_{\mathbf{N}}, ;)$.

Similarly, the map $\wp(\mathbf{comp}(\mathbf{plus1}))(S) = \{(m,n) \mid m <_{\mathbf{N}} n, (m+1, n+1) \in S\}$ is a quantale homomorphism from $(\wp(<_{\mathbf{N}}), \subseteq, \bigcup, [\![;]\!])$ to $(\wp(<_{\mathbf{N}}), \subseteq, \bigcup, [\![;]\!])$, but not to $(\wp(\leq_{\mathbf{N}}), \subseteq, \bigcup, [\![;]\!])$.

4 Preorders and Unital Partial Semigroups

In this section, we recall the definition of unital quantale. We show which weak preorders and which partial semigroups correspond to unital quantales. We also study the correspondence among their maps.

Definition 8 (unital quantale). *A* unital quantale *is defined to be a tuple* $(Q, \leq, \bigvee, \odot, 1)$ *such that*

1. $(Q, \leq, \bigvee, \odot)$ *is a quantale, and*
2. $1 \in Q$ *satisfies for each* $q \in Q$, $1 \odot q = q = q \odot 1$ *(1 is called* the unit *of* $\odot$*).*

For unital quantales $(Q, \leq, \bigvee, \odot, 1)$, $(Q', \leq', \bigvee', \odot', 1')$, *a* unital *homomorphism from* $(Q, \leq, \bigvee, \odot, 1)$ *to* $(Q', \leq', \bigvee', \odot', 1')$ *is defined to be a homomorphism from* $(Q, \leq, \bigvee, \odot)$ *to* $(Q', \leq', \bigvee', \odot')$ *satisfying* $f(1) = 1'$.

Unital homomorphisms between unital quantales are closed under composition as maps. The identity map is a unital homomorphism on a unital quantale. We write **UQt** for the category whose objects are unital quantales and whose arrows are unital homomorphisms between them.

Next, we introduce the additional structure for a partial semigroup which corresponds to the unit of the multiplication of a quantale. We call the structure a unital subset.

Definition 9 (unital subset). *A* unital subset *of a partial semigroup* $(X, \cdot)$ *is defined to be a subset* $U \subseteq X$ *such that*

1. *if* $u \in U$ *and* $u \cdot x$ *is defined, then* $u \cdot x = x$,
2. *if* $u \in U$ *and* $x \cdot u$ *is defined, then* $x \cdot u = x$,
3. *for any* $x \in X$, *there exists* $u \in U$ *such that* $u \cdot x$ *is defined, and*
4. *for any* $x \in X$, *there exists* $u \in U$ *such that* $x \cdot u$ *is defined.*

Lemma 1 (uniqueness of unital subset). *For each partial semigroup* $(X, \cdot)$, *if* U *and* U' *are unital subsets of* $(X, \cdot)$, *then* $U = U'$.

Proof. Assume that $(X, \cdot)$ is a partial semigroup and that U and U' are unital subsets of $(X, \cdot)$. Take an element u of U. Since U' is a unital subset of $(X, \cdot)$, there exists $u' \in U'$ such that $u' \cdot u$ is defined and $u' \cdot u = u$. Since U is also a unital subset of $(X, \cdot)$ and $u \in U$, $u' \cdot u$ is equal to $u' \in U'$. Therefore, $U \subseteq U'$. The converse is proven, similarly. □

Definition 10 (unital partial semigroup and dividing map). *A* unital partial semigroup *is defined to be a tuple* $(X, \cdot, U)$ *such that*

1. $(X, \cdot)$ *is a partial semigroup, and*
2. U *is the unital subset of* $(X, \cdot)$.

For unital partial semigroups $(X, \cdot, U)$, $(X', \cdot', U')$, *a* dividing map *from* $(X, \cdot, U)$ *to* $(X', \cdot', U')$ *is defined to be an arrow* $f\colon (X, \cdot) \to (X', \cdot') \in \mathbf{PSG_{div}}$ *such that for any* $x \in X$, *it satisfies* $x \in U$ *if and only if* $f(x) \in U'$.

Note that a unital partial semigroup is not equal to a partial monoid, which is a base structure of an effect algebra [9], since the unital subset of a unital partial semigroup is not always singleton.

Dividing maps between unital partial semigroups are closed under composition as maps. The identity map is dividing on a unital partial semigroup. We write $\mathbf{UPSG_{div}}$ whose objects are unital partial semigroups and whose arrows are dividing maps between them.

Example 6. The set of all arrows of a small category forms the unital partial semigroup whose partial binary operator is the composition of arrows and whose unital subset is the set of identities. On the other hand, a unital partial semigroup is not always regarded as the set of arrows. For example, $(\{0, 1\}, \cdot, \{1\})$ is a unital partial semigroup, when $0 \cdot 1 = 1 \cdot 0 = 0$, $1 \cdot 1 = 1$ and $0 \cdot 0$ is undefined. To regard $\{1\}$ as the set of all identities, however, the category can have only one object, and then $0 \cdot 0$ must be defined. Therefore, a unital partial semigroup is not equal to a poloid [10]. If $x \cdot y$ and $y \cdot z$ are defined in a poloid, then $(x \cdot y) \cdot z$ must be defined. In this unital partial semigroup $(\{0, 1\}, \cdot, \{1\})$, however, $0 \cdot 1$ and $1 \cdot 0$ are defined, but $(0 \cdot 1) \cdot 0$ is undefined.

Next, we show that a unital partial semigroup corresponds to a unital quantale.

Theorem 5. *For a partial semigroup* $(X, \cdot)$ *and* $U \subseteq X$, *the following statements are equivalent.*

1. $(X, \cdot, U)$ *is a unital partial semigroup.*
2. $(\wp(X), \subseteq, \bigcup, [\![\cdot]\!], U)$ *is a unital quantale.*

Proof. $(1 \Rightarrow 2)$ Assume that $(X, \cdot, U)$ is a unital partial semigroup. Take $S \subseteq X$. The condition 1 of Definition 9 implies $U \, [\![\cdot]\!] \, S \subseteq S$, the condition 2 implies $S \, [\![\cdot]\!] \, U \subseteq S$, the conditions 1,3 imply $S \subseteq U \, [\![\cdot]\!] \, S$, and the conditions 2,4 imply $S \subseteq S \, [\![\cdot]\!] \, U$. Therefore, U is the unit of $[\![\cdot]\!]$.

$(2 \Rightarrow 1)$ Assume that $(\wp(X), \subseteq, \bigcup, [\![\cdot]\!], U)$ is a unital quantale. If $u \in U$ and $u \cdot x$ is defined, then $u \cdot x = x$, since $U \, [\![\cdot]\!] \, \{x\} \subseteq \{x\}$. If $u \in U$ and $x \cdot u$ is defined, then $x \cdot u = x$, since $\{x\} \, [\![\cdot]\!] \, U \subseteq \{x\}$. For any $x \in X$, there exists $u \in U$ such that $u \cdot x$ is defined, since $\{x\} \subseteq U \, [\![\cdot]\!] \, \{x\}$. For any $x \in X$, there exists $u \in U$ such that $x \cdot u$ is defined, since $\{x\} \subseteq \{x\} \, [\![\cdot]\!] \, U$. Therefore, U is the unital subset of $(X, \cdot)$. □

Next, we show that a dividing map between unital partial semigroups corresponds to a unital homomorphism between unital quantales.

Theorem 6. *For unital partial semigroups* $(X, \cdot, U), (X', \cdot', U')$ *and* $f\colon (X, \cdot) \to (X', \cdot')$ *in* $\mathbf{PSG_{div}}$*, the following statements are equivalent.*

1. f *is an arrow* $f\colon (X, \cdot, U) \to (X', \cdot', U')$ *in* $\mathbf{UPSG_{div}}$.
2. $\wp(f)$ *is an arrow* $\wp(f)\colon (\wp(X'), \subseteq, \bigcup, [\![\cdot']\!], U') \to (\wp(X), \subseteq, \bigcup, [\![\cdot]\!], U)$ *in* $\mathbf{UQt}$.

Proof. The condition $x \in U \Leftrightarrow f(x) \in U'$ is equivalent to $\wp(f)(U') = U$. □

Corollary 3. *Let* $\wp\colon \mathbf{UPSG_{div}} \to \mathbf{UQt}^{op}$ *be the extension of* $\wp\colon \mathbf{PSG_{div}} \to \mathbf{Qt}^{op}$ *by* $\wp(X, \cdot, U) = (\wp(X), \subseteq, \bigcup, [\![\cdot]\!], U)$. $\wp\colon \mathbf{UPSG_{div}} \to \mathbf{UQt}^{op}$ *and the forgetful functor* **forget**: $\mathbf{UPSG_{div}} \to \mathbf{PSG_{div}}$ *is a pullback of* $\wp\colon \mathbf{PSG_{div}} \to \mathbf{Qt}^{op}$ *and the forgetful functor* **forget**: $\mathbf{UQt}^{op} \to \mathbf{Qt}^{op}$ *in* **Cat** *(Fig. 1).*

Next, we show that a preordered set which corresponds to a unital partial semigroup. A *preordered set* is defined to be a tuple $(X, \leq)$ of a set X and a preorder $\leq$ on X, that is to say, $\leq$ is reflexive and transitive. We write $x \leq y$ for $(x, y) \in \leq$.

Theorem 7. *A weak preordered set* (X, R) *is a preordered set, if and only if* $\triangle_X \stackrel{\text{def}}{=} \{(x, x) \mid x \in X\}$ *is the unital subset of* $\mathbf{comp}(X, R) \stackrel{\text{def}}{=} (R, ;)$.

Proof. ($\Longrightarrow$) Assume that $(X, \leq)$ is a preordered set. Since $\leq$ is reflexive, the set $\triangle_X = \{(x, x) \mid x \in X\}$ is a subset of $\leq$. If $(x, x) \in \triangle_X$ and $(x, x); (y, z)$ is defined, then $x = y$ and $(x, x); (y, z) = (y, z)$. For any $x \leq y$, $(x, x); (x, y)$ is defined and $(x, x) \in \triangle_X$. The remaining condition is proven similarly.

($\Longleftarrow$) Take $x \in X$. There is $y \in X$ such that $(x, y) \in R$ or $(y, x) \in R$. When $(x, y) \in R$, there is $(u, u') \in \triangle_X$ such that $(u, u'); (x, y)$ is defined and equal to (x, y). Then, $u = x = u'$. When $(y, x) \in R$, there is $(u, u') \in \triangle_X$ such that $(y, x); (u, u')$ is defined and equal to (y, x). Then, $u = x = u'$. In both cases, (x, x) is equal to $(u, u') \in \triangle_X \subseteq R$. Therefore, R is reflexive. □

Next, we introduce the condition for maps between preordered sets corresponds to dividing maps between unital partial semigroups. The condition is monotone, satisfying the intermediate value property, and Id-reflecting.

Definition 11 (Id-reflecting map between preordered sets). *For preordered sets* $(X, \leq)$, $(X', \leq')$, *an* Id-reflecting *map from* $(X, \leq)$ *to* $(X', \leq')$ *is defined to be a monotone map* $f\colon X \to X'$ *such that if* $x \leq y$ *and* $f(x) = f(y)$*, then* $x = y$.

Id-reflecting maps satisfying the intermediate value property between preordered sets are closed under composition as maps. The identity map on a preordered set is an Id-reflecting map satisfying the intermediate value property. Therefore, we write $\mathbf{PreOrd_{int,idref}}$ for the category whose objects are preordered sets and whose arrows are Id-reflecting maps satisfying the intermediate value property between them.

Next, we show that Id-reflecting maps satisfying the intermediate value property correspond to dividing maps between unital partial semigroups.

Theorem 8. *For* $(X, \leq), (X', \leq')$ *in* $\mathbf{PreOrd_{int,idref}}$, $f\colon (X, \leq) \to (X', \leq')$ *in* $\mathbf{WPreOrd_{int}}$, *the following statements are equivalent.*

1. f *is an arrow* $f\colon (X, \leq) \to (X', \leq')$ *in* $\mathbf{PreOrd_{int,idref}}$.
2. $\mathbf{comp}(f)$ *is an arrow* $\mathbf{comp}(f)\colon (\leq, ;, \triangle_X) \to (\leq', ;, \triangle_{X'})$ *in* $\mathbf{UPSG_{div}}$.

Proof. ($1 \Rightarrow 2$) Assume that f is Id-reflecting and $x \leq y$. If $(x, y) \in \triangle_X$, then $x = y$ and $\mathbf{comp}(f)(x, y) = (f(x), f(y)) = (f(x), f(x)) \in \triangle_{X'}$. Conversely, if $\mathbf{comp}(f)(x, y) = (f(x), f(y)) \in \triangle_{X'}$, then $x = y$ and $(x, y) \in \triangle_X$, since $f(x) = f(y)$ and f is Id-reflecting. Therefore, we have $(x, y) \in \triangle_X$ if and only if $\mathbf{comp}(f)(x, y) \in \triangle_{X'}$.

($2 \Rightarrow 1$) Assume that f satisfies $(x, y) \in \triangle_X \Leftrightarrow \mathbf{comp}(f)(x, y) \in \triangle_{X'}$ for any $x \leq y$. Take $x \leq y$ such that $f(x) = f(y)$. They satisfy $\mathbf{comp}(f)(x, y) = (f(x), f(y)) = (f(x), f(x)) \in \triangle_{X'}$. Since f is a dividing maps between unital partial semigroups, x, y satisfy $(x, y) \in \triangle_X$ and $x = y$. Therefore, f is Id-reflecting. □

Corollary 4. *Let* $\mathbf{comp}\colon \mathbf{PreOrd_{int,idref}} \to \mathbf{UPSG_{div}}$ *be the extension of* $\mathbf{comp}\colon \mathbf{WPreOrd_{int}} \to \mathbf{PSG_{div}}$ *by* $\mathbf{comp}(X, \leq) = (\leq, ;, \triangle_X)$. *The forgetful functor* $\mathbf{forget}\colon \mathbf{PreOrd_{int,idref}} \to \mathbf{WPreOrd_{int}}$ *and* $\mathbf{comp}\colon \mathbf{PreOrd_{int,idref}} \to \mathbf{UPSG_{div}}$ *is a pullback of* $\mathbf{comp}\colon \mathbf{WPreOrd_{int}} \to \mathbf{PSG_{div}}$ *and the forgetful functor* $\mathbf{forget}\colon \mathbf{UPSG_{div}} \to \mathbf{PSG_{div}}$ *in* **Cat** *(Fig. 1).*

Example 7. $(\mathbf{N}, <_\mathbf{N})$ is an object of $\mathbf{WPreOrd_{int}}$, but not of $\mathbf{PreOrd_{int,idref}}$, since $<_\mathbf{N}$ is not reflexive. For the same reason, $\mathbf{comp}(\mathbf{N}, <_\mathbf{N}) = (<_\mathbf{N}, ;)$ has no unital subset.

The map $\mathbf{plus1}(n) \stackrel{\text{def}}{=} n + 1$ is Id-reflecting on $(\mathbf{N}, \leq_\mathbf{N})$ in $\mathbf{PreOrd_{int,idref}}$, since $m \leq n$ and $m + 1 = n + 1$ imply $m = n$. For the same reason, the map $\mathbf{comp}(\mathbf{plus1})(m, n) = (m + 1, n + 1)$ is a dividing map on $(\leq_\mathbf{N}, ;, \triangle_\mathbf{N})$ in $\mathbf{UPSG_{div}}$.

On the other hand, the map $\mathbf{zero}(n) \stackrel{\text{def}}{=} 0$ is not Id-reflecting on $(\mathbf{N}, \leq_\mathbf{N})$ in $\mathbf{PreOrd_{int,idref}}$, since $0 \leq 1$ and $\mathbf{zero}(0) = \mathbf{zero}(1)$ but not $0 = 1$. For the same reason, the map $\mathbf{comp}(\mathbf{zero})(m, n) = (0, 0)$ is not a dividing map on $(\leq_\mathbf{N}, ;, \triangle_\mathbf{N})$ in $\mathbf{UPSG_{div}}$.

5 Partial Semigroups with Partial Subtraction and Relational Embedding of Quantales

In this section, we define a subcategory of $\mathbf{UPSG_{div}}$ and the corresponding subcategory of $\mathbf{PreOrd_{int,idref}}$. Between them, we give a functor **suff** in the converse direction of **comp**. Moreover, we give the natural transformation $\dot{-}\colon \mathbf{comp} \circ \mathbf{suff} \to \mathrm{Id}$ which induces the relational embedding maps mentioned in Sect. 1.

Definition 12 (diagonal unital partial semigroup). *A partial semigroup* $(X, \cdot)$ *is called* diagonal, *if*

1. *if* $w \cdot x = y \cdot z$, *then there exists* v *such that* $w = y \cdot v$ *or* $y = w \cdot v$, *and*
2. *if* $w \cdot x = w \cdot y$, *then* $x = y$.

A unital partial semigroup $(X, \cdot, U)$ *is called* diagonal, *if* $(X, \cdot)$ *is diagonal. We write* $\mathbf{DUPSG_{div}}$ *for the subcategory of* $\mathbf{UPSG_{div}}$ *whose objects are only diagonal unital partial semigroups.*

Definition 13 (interval-total preorder). *A preorder* $\leq$ *on* X *is called* interval-total, *if* $x \leq y \leq z$ *and* $x \leq y' \leq z$ *imply* $y \leq y'$ *or* $y' \leq y$ *for each* $x, y, y', z \in X$. *A preordered set* $(X, \leq)$ *is called an* interval-totally preordered set, *if* $\leq$ *is interval-total. We write* $\mathbf{ITPreOrd_{int,idref}}$ *for the subcategory of* $\mathbf{PreOrd_{int,idref}}$ *whose objects are only interval-totally preordered sets.*

Theorem 9. *A preordered set* $(X, \leq)$ *is an interval-totally preordered set, if and only if the unital partial semigroup* $\mathbf{comp}(X, \leq) = (\leq, ;, \triangle_X)$ *is diagonal.*

Proof. $(\Longrightarrow)$

(1) Assume that $w, x, y, z \in \leq$ satisfy $w; x = y; z$. There exists $a, b, b', c \in X$ such that $w = (a, b)$, $x = (b, c)$, $y = (a, b')$, and $z = (b', c)$, that is to say, $a \leq b \leq c$ and $a \leq b' \leq c$. Since $\leq$ is interval-total, $b \leq b'$ or $b' \leq b$. When $b \leq b'$, $v = (b, b') \in \leq$ satisfies $y = (a, b') = (a, b); (b, b') = w; v$. When $b' \leq b$, $v = (b', b) \in \leq$ satisfies $w = (a, b) = (a, b'); (b', b) = y; v$.

(2) Assume that $w, x, y \in \leq$ satisfy $w; x = w; y$. There exists $a, b, c \in X$ such that $w = (a, b)$, $x = (b, c)$, and $y = (b, c)$. Therefore, $x = y$.

$(\Longleftarrow)$ Assume that $(\leq, ;, \triangle_X)$ is diagonal. Assume that $x, y, y', z \in X$ satisfy $x \leq y \leq z$ and $x \leq y' \leq z$. Then, we have that $(x, y); (y, z) = (x, z) = (x, y'); (y', z)$. Since $(\leq, ;, \triangle_X)$ is diagonal, there exists $v \leq w$ such that (x, y') $; (v, w) = (x, y)$ or $(x, y); (v, w) = (x, y')$. If $(x, y'); (v, w) = (x, y)$, then $y' = v \leq w = y$. On the other hand, if $(x, y); (v, w) = (x, y')$, then $y = v \leq w = y'$. Therefore, $\leq$ is interval-total. □

Corollary 5. *Let* $\mathbf{comp}\colon \mathbf{ITPreOrd_{int,idref}} \to \mathbf{DUPSG_{div}}$ *be the restriction of* $\mathbf{comp}\colon \mathbf{PreOrd_{int,idref}} \to \mathbf{UPSG_{div}}$.
The forgetful functor $\mathbf{forget}\colon \mathbf{ITPreOrd_{int,idref}} \to \mathbf{PreOrd_{int,idref}}$ *and* $\mathbf{comp} : \mathbf{ITPreOrd_{int,idref}} \to \mathbf{DUPSG_{div}}$ *is a pullback of* $\mathbf{comp}\colon \mathbf{PreOrd_{int,idref}} \to \mathbf{UPSG_{div}}$ *and the forgetful functor* $\mathbf{forget}\colon \mathbf{DUPSG_{div}} \to \mathbf{UPSG_{div}}$ *in* $\mathbf{Cat}$ *(Fig. 1).*

Next, we give a functor **suff** in the converse direction of **comp**.

Theorem 10. *The following data form a functor* **suff** *from* $\mathbf{DUPSG_{div}}$ *to* $\mathbf{ITPreOrd_{int,idref}}$.

- *For an object* $(X, \cdot, U)$, $\mathbf{suff}(X, \cdot, U) \stackrel{\text{def}}{=} (X, \leq_\cdot)$
  where $x \leq_\cdot y \stackrel{\text{def}}{\Longleftrightarrow} \exists z \in X. x \cdot z = y$.
- *For an arrow* $f\colon (X, \cdot, U) \to (X', \cdot', U')$, $\mathbf{suff}(f)\colon (X, \leq_\cdot) \to (X', \leq_{\cdot'})$ *is* f.

Proof. (object part) Assume that $(X, \cdot, U)$ is a diagonal unital partial semigroup. Since $(X, \cdot, U)$ is a unital partial semigroup, $\leq_\cdot$ is transitive and reflexive. If $x \leq_\cdot y \leq_\cdot z$ and $x \leq_\cdot y' \leq_\cdot z$, then there exist v, w such that $y \cdot v = z = y' \cdot w$. Since $(X, \cdot, U)$ is diagonal, there exists u such that $y = y' \cdot u$ or $y' = y \cdot u$, that is to say, $y' \leq_\cdot y$ or $y \leq_\cdot y'$. Therefore, $\leq_\cdot$ is interval-total.

(arrow part) Take $f\colon (X, \cdot, U) \to (X', \cdot', U')$ in $\mathbf{DUPSG_{div}}$. f is monotone as $f\colon (X, \leq_\cdot) \to (X', \leq_{\cdot'})$, since $x \cdot z = y$ implies $f(y) = f(x \cdot z) = f(x) \cdot' f(z)$.

We show that $\mathbf{suff}(f) = f$ is Id-reflecting. Assume that $x \leq_\cdot y$ and $f(x) = f(y)$. There exist $z \in X$ such that $x \cdot z = y$. Therefore, $f(x) = f(y) = f(x \cdot z) = f(x) \cdot' f(z)$. Since $(X', \cdot', U')$ is a unital partial semigroup, there exists $u' \in U'$ such that $f(x) \cdot' u' = f(x) = f(x) \cdot' f(z)$. Since $(X', \cdot', U')$ is diagonal, $u', f(z)$ satisfy $u' = f(z)$. Since f is a dividing map between unital partial semigroups, $f(z) = u' \in U'$ implies $z \in U$ and $x = x \cdot z = y$. Therefore, f is Id-reflecting.

We show that f satisfies the intermediate value property. Assume that $x \leq_\cdot y$, $f(x) \leq_{\cdot'} z'$, and $z' \leq_{\cdot'} f(y)$. There exist $u \in X, v', w' \in X'$ such that $x \cdot u = y$, $f(x) \cdot' v' = z'$, and $z' \cdot' w' = f(y)$. Since f is dividing, there exist $w, z \in X$ such that $f(w) = w'$, $f(z) = z'$, and $z \cdot w = y$. Therefore, $z \leq_\cdot y$. Since f is dividing and $f(z) = z' = f(x) \cdot' v'$, there exist $v, t \in X$ such that $f(t) = f(x)$, $f(v) = v'$, and $z = t \cdot v$. Since $(X, \cdot, U)$ is diagonal and $x \cdot u = y = z \cdot w = (t \cdot v) \cdot w = t \cdot (v \cdot w)$, x, t satisfy $x \leq_\cdot t$ or $t \leq_\cdot x$. Since f is Id-reflecting and $f(t) = f(x)$, x satisfies $t = x$ and $z = t \cdot v = x \cdot v$, that is to say, $x \leq_\cdot z$. Therefore, f satisfies the intermediate value property.

$\mathbf{suff}(f)$ preserves composition and identities. □

We call the following operation $\dot{-}$ the *partial subtraction* on a diagonal unital partial semigroup.

Theorem 11. *The following data form a natural transformation* $\dot{-}\colon \mathbf{comp} \circ \mathbf{suff} \to \mathrm{Id}\colon \mathbf{DUPSG_{div}} \to \mathbf{DUPSG_{div}}$.

- *For an object* $(X, \cdot, U)$, *its component* $\dot{-}\colon (\leq_\cdot, ;, \Delta_X) \to (X, \cdot, U)$ *sends* $y \leq_\cdot x$ *to* $x \dot{-} y \overset{\text{def}}{=} z$ *such that* $y \cdot z = x$.

Proof. (well-definedness) For $y \leq_\cdot x$, there exists $z \in X$ such that $y \cdot z = x$. Since $(X, \cdot, U)$ is diagonal, if $y \cdot z = y \cdot z'$ then $z = z'$.

(homomorphism of **PSG**) Assume that $(z, y); (x, w)$ is defined, $z \leq_\cdot y$, and $x \leq_\cdot w$. Then, $z \leq_\cdot y = x$ and w is representable as $w = x \cdot (w \dot{-} x) = (z \cdot (x \dot{-} z)) \cdot (w \dot{-} x)$. Therefore, $(x \dot{-} z) \cdot (w \dot{-} x)$ and $z \cdot ((x \dot{-} z) \cdot (w \dot{-} x))$ are also defined and $z \cdot ((x \dot{-} z) \cdot (w \dot{-} x)) = (z \cdot (x \dot{-} z)) \cdot (w \dot{-} x) = w$. Therefore, $(x \dot{-} z) \cdot (w \dot{-} x) = w \dot{-} z$.

(dividing) Assume that $x \leq_\cdot w$ and $w \dot{-} x = z \cdot y$. Then, $x \cdot (w \dot{-} x)$ is defined and equal to w. Therefore, $x \cdot (z \cdot y)$, $x \cdot z$, and $(x \cdot z) \cdot y$ are also defined. Since $(x \cdot z) \cdot y = x \cdot (z \cdot y) = x \cdot (w \dot{-} x) = w$, they satisfy $x \cdot z \leq_\cdot w$ and $w \dot{-} (x \cdot z) = y$. Since $x \cdot z$ is defined, obviously $x \leq_\cdot x \cdot z$ and $(x \cdot z) \dot{-} x = z$. $(x, x \cdot z); (x \cdot z, w)$ is defined and equal to (x, w).

(unital) For $y \leq_\cdot x$, we show $(y, x) \in \Delta_X \iff x \dot{-} y \in U$. Assume $(y, x) \in \Delta_X$. Then, $x = y$. There exists $u \in U$ satisfying $x \cdot u = x$. Since $(X, \cdot, U)$ is diagonal and $x \cdot (x \dot{-} x) = x = x \cdot u$, they satisfy $x \dot{-} y = x \dot{-} x = u \in U$. Conversely, assume $x \dot{-} y \in U$. $y \cdot (x \dot{-} y)$ is defined and equal to x. By the definition of unital subset, $y \cdot (x \dot{-} y) = y$. Therefore, $x = y \cdot (x \dot{-} y) = y$ and $(x, y) \in \triangle_X$.

(naturality) Assume f is an arrow $f\colon (X, \cdot, U) \to (X', \cdot', U')$ in $\mathbf{DUPSG_{div}}$. For $y \leq_{\cdot} x$, $f(y) \cdot (f(x) \dot{-} f(y)) = f(x) = f(y \cdot (x \dot{-} y)) = f(y) \cdot f(x \dot{-} y)$. Since $(X, \cdot, U)$ is diagonal, we have $f(x \dot{-} y) = f(x) \dot{-} f(y)$. □

Example 8. $(\mathbf{N}, +, 0)$ is a diagonal unital partial semigroup. Then, the interval-totally preordered set $\mathbf{suff}(\mathbf{N}, +, 0)$ is equal to $(\mathbf{N}, \leq_{\mathbf{N}})$, since $x \leq_{\mathbf{N}} y \Leftrightarrow \exists z \in \mathbf{N}. x + z = y$. Therefore, $\mathbf{comp}(\mathbf{suff}(\mathbf{N}, +, 0)) = (\leq_{\mathbf{N}}, ;)$ is also a diagonal unital partial semigroup. The partial subtraction $\dot{-}\colon (\leq_{\mathbf{N}}, ;) \to (\mathbf{N}, +, 0)$ sends $x \leq_{\mathbf{N}} y$ to $y - x$.

Definition 14 (diagonal powerset quantale). *A diagonal powerset quantale is a unital quantale $(Q, \leq, \bigvee, \odot, 1)$ such that*

1. *$(Q, \leq, \bigvee) = (\wp(X), \subseteq, \bigcup)$ for some set X,*
2. *for any $x, y \in X$, $\{x\} \odot \{y\}$ is singleton or empty,*
3. *if $\{w\} \odot \{x\} = \{y\} \odot \{z\} \neq \emptyset$, then there exists v such that $\{w\} = \{y\} \odot \{v\}$ or $\{y\} = \{w\} \odot \{v\}$, and*
4. *if $\{w\} \odot \{x\} = \{w\} \odot \{y\} \neq \emptyset$, then $x = y$.*

We define the subcategory $\mathbf{DPQt}$ *of* $\mathbf{UQt}$*, whose object is a diagonal powerset quantale.*

Theorem 12. *For a unital partial semigroup $(X, \cdot, U)$, the following statements are equivalent.*

1. *$(X, \cdot, U)$ is diagonal.*
2. *$\wp(X, \cdot, U) = (\wp(X), \subseteq, \bigcup, [\![\cdot]\!], U)$ is a diagonal powerset quantale.*

Proof. Let $(X, \cdot, U)$ be a diagonal unital partial semigroup. For any $x, y \in X$, $\{x\} [\![\cdot]\!] \{y\}$ is $\{x \cdot y\}$ if $x \cdot y$ is defined and the emptyset otherwise.

For any $w, x, y, z \in X$ satisfying $\{w\} [\![\cdot]\!] \{x\} = \{y\} [\![\cdot]\!] \{z\} \neq \emptyset$, $w \cdot x$ and $y \cdot z$ are defined and equivalent. Since $(X, \cdot, U)$ is a diagonal unital partial semigroup, there exists v such that $w = y \cdot v$ or $y = w \cdot v$. Therefore, v satisfies $\{w\} = \{y\} [\![\cdot]\!] \{v\}$ or $\{y\} = \{w\} [\![\cdot]\!] \{v\}$.

For any $w, x, y \in X$ satisfying $\{w\} [\![\cdot]\!] \{x\} = \{w\} [\![\cdot]\!] \{y\} \neq \emptyset$, $w \cdot x$ and $w \cdot y$ are defined and equivalent. Since $(X, \cdot, U)$ is a diagonal unital partial semigroup, $x = y$. □

Corollary 6. *Let* $\wp\colon \mathbf{DUPSG_{div}} \to \mathbf{DPQt}^{op}$ *be the restriction of* $\wp\colon \mathbf{UPSG_{div}} \to \mathbf{UQt}^{op}$.

The forgetful functor $\mathbf{forget}\colon \mathbf{DUPSG_{div}} \to \mathbf{UPSG_{div}}$ *and* $\wp\colon \mathbf{DUPSG_{div}} \to \mathbf{DPQt}^{op}$ *is a pullback of* $\wp\colon \mathbf{UPSG_{div}} \to \mathbf{UQt}^{op}$ *and the forgetful functor* $\mathbf{forget}\colon \mathbf{DPQt}^{op} \to \mathbf{UQt}^{op}$ *in* $\mathbf{Cat}$ *(Fig. 1).*

Theorem 13. *For any diagonal powerset quantale $(Q, \leq, \bigvee, \odot, 1)$, there exists a diagonal unital partial semigroup $(X, \cdot, U)$ satisfying $\wp(X, \cdot, U) = (Q, \leq, \bigvee, \odot, 1)$.*

Proof. Let $(Q, \leq, \bigvee, \odot, 1)$ be a diagonal powerset quantale. There exists a set X satisfying $(Q, \leq, \bigvee) = (\wp(X), \subseteq, \bigcup)$. Let $x \cdot y$ be the element of $\{x\} \odot \{y\}$ if $\{x\} \odot \{y\}$ is singleton and undefined otherwise. Assume that $x \cdot y$ and $(x \cdot y) \cdot z$ are defined. There exist $u, v \in X$ satisfying $\{x\} \odot \{y\} = \{u\}$ and $\{u\} \odot \{z\} = \{v\}$. Since $\{x\} \odot (\{y\} \odot \{z\}) = (\{x\} \odot \{y\}) \odot \{z\} = \{u\} \odot \{z\} = \{v\} \neq \emptyset$, the distributivity of $\odot$ implies $\{y\} \odot \{z\} \neq \emptyset$. Therefore, $y \cdot z$ and $x \cdot (y \cdot z)$ are defined. For any $S_1, S_2 \subseteq X$, the distributivity of $\odot$ also implies $S_1 \odot S_2 = \bigcup\{\{x\} \odot \{y\} \mid x \in S_1, y \in S_2\} = \{x \cdot y \mid x \in S_1, y \in S_2 \text{ is defined}\} = S_1 \llbracket\cdot\rrbracket S_2$. Therefore, $(X, \cdot)$ is the partial semigroup satisfying $\wp(X, \cdot) = (\wp(X), \subseteq, \bigcup, \odot)$. By Theorem 5, $(X, \cdot, 1)$ is the unital partial semigroup satisfying $\wp(X, \cdot, 1) = (Q, \leq, \bigvee, \odot, 1)$. By Theorem 12, $(X, \cdot, 1)$ is diagonal, since $(\wp(X), \subseteq, \bigcup, \llbracket\cdot\rrbracket, 1)$ is a diagonal powerset quantale. □

Theorem 14. *For any diagonal powerset quantale $(Q, \leq, \bigvee, \odot, 1)$, there exist a diagonal unital partial semigroup $(X, \cdot, U)$ and an injective arrow from $(Q, \leq, \bigvee, \odot, 1)$ to the relational quantale $(\wp(\leq.), \subseteq, \bigcup, \llbracket;\rrbracket, \triangle_X)$ in* **DPQt**.

Proof. By Theorem 11 and Corollary 6, there is a natural transformation $\wp(\dot{-})$: $\wp \circ \mathbf{comp} \circ \mathbf{suff} \to \wp$: $\mathbf{DUPSG}^{op}_{\mathbf{div}} \to \mathbf{DPQt}$. By Theorem 13, for any object $(Q, \leq, \bigvee, \odot, 1) \in \mathbf{DPQt}$, there exists an object $(X, \cdot, U) \in \mathbf{DUPSG_{div}}$ satisfying $\wp(X, \cdot, U) = (Q, \leq, \bigvee, \odot, 1)$. Therefore, the component $\wp(\dot{-})_{(X,\cdot,U)}$ is an arrow from $(Q, \leq, \bigvee, \odot, 1)$ to $\wp(\mathbf{comp}(\mathbf{suff}(X, \cdot, U)))$ in **DPQt**. By the definitions of $\wp$, **comp**, and **suff**, the powerset quantale $(Q, \leq, \bigvee, \odot, 1)$ is equal to $(\wp(X), \subseteq, \bigcup, \llbracket\cdot\rrbracket, U)$ and $\wp(\mathbf{comp}(\mathbf{suff}(X, \cdot, U)))$ is the relational quantale $(\wp(\leq.), \subseteq, \bigcup, \llbracket;\rrbracket, \triangle_X)$. The component $\wp(\dot{-})_{(X,\cdot,U)}$ is the map from $\wp(X)$ to $\wp(\leq.)$ such that $\wp(\dot{-})_{(X,\cdot,U)}(S) = \{(x, y) \mid y \in \{x\} \odot S\}$. Each subset $S \subseteq X$ is represented as follows.

$$\begin{aligned} S &= 1 \odot S = (\textstyle\bigcup\{\{x\} \mid x \in 1\}) \odot S = \bigcup\{\{x\} \odot S \mid x \in 1\} \\ &= \{y \mid \exists x \in 1, y \in \{x\} \odot S\}\} = \{y \mid \exists x \in 1, (x, y) \in \wp(\dot{-})_{(X,\cdot,U)}(S)\}\} \end{aligned}$$

$\wp(\dot{-})_{(X,\cdot,U)}$ is injective, since $\wp(\dot{-})_{(X,\cdot,U)}(S) \subseteq \wp(\dot{-})_{(X,\cdot,U)}(S')$ implies $S \subseteq S'$. □

Example 9. For a set X, $(\wp(X), \subseteq, \bigcup, \cap, X)$ is a diagonal powerset quantale.

Example 10. For a group $(G, \cdot, 1, ^{-1})$, $(\wp(G), \subseteq, \bigcup, \llbracket\cdot\rrbracket, \{1\})$ is a diagonal powerset quantale.

Example 11. For a set A, we write A^* for the set of all finite sequences on A. $(\wp(A^*), \subseteq, \bigcup, \llbracket\cdot\rrbracket, \{\epsilon\})$ is a diagonal powerset quantale, where ϵ is the empty sequence. $(\wp(\mathbf{N}), \subseteq, \bigcup, \llbracket+\rrbracket, \{0\})$ is a special case.

Example 12. For a set A, $(\wp(A \times A), \subseteq, \bigcup, ;, \triangle_A)$ is a diagonal powerset quantale, where $R; Q \stackrel{\text{def}}{=} \{(a, b) \mid \exists c \in A, (a, c) \in R, (c, b) \in Q\}$ and $\triangle_A \stackrel{\text{def}}{=} \{(a, a) \mid a \in A\}$.

6 Conclusion

This paper has introduced the notions of weak preorder, partial semigroup, lax homomorphism between quantales. And, we have shown three correspondences by proving the six pullbacks in Fig. 1.

1. The correspondence among a monotone map satisfying the intermediate value property between weak preordered sets, a dividing map between partial semigroups, and a homomorphism between quantales
2. The correspondence among a preordered set, a unital partial semigroup, and a unital quantales (including the correspondence among maps for them)
3. The correspondence among an interval-totally preordered set, a diagonal unital partial semigroup, and a diagonal powerset quantale (including the correspondence among maps for them)

We also have shown that each diagonal powerset quantale has the relational embedding which is the image of partial subtraction $\dot{-}$ by the functor $\wp$. It is a future work to generalize our result for any (possibly not diagonal) powerset quantale and to extend it to a Stone duality [11].

Acknowledgements. The authors thank Izumi Takeuti, Takeshi Tsukada, Soichiro Fujii, Mitsuhiko Fujio, and Hiroyuki Miyoshi for valuable discussion about partial semigroups.

References

1. Conway, J.H.: Regular Algebra and Finite Machines. Chapman and Hall, London (1971)
2. He, J., Hoare, C.A.R.: Weakest prespecification. Inf. Process. Lett. **24**, 71–76 (1987)
3. Vickers, S.: Topology via Logic. Cambridge University Press, Cambridge (1989)
4. Nishizawa, K., Furusawa, H.: Relational representation theorem for powerset quantales. In: Kahl, W., Griffin, T.G. (eds.) RAMICS 2012. LNCS, vol. 7560, pp. 207–218. Springer, Heidelberg (2012). https://doi.org/10.1007/978-3-642-33314-9_14
5. Pratt, V.R.: Dynamic algebras and the nature of induction. In: Proceedings of the Twelfth Annual ACM Symposium on Theory of Computing, STOC 1980, pp. 22–28. Association for Computing Machinery, New York (1980)
6. Mulvey, C.J.: In: Second Topology Conference. Rendiconti del Circolo Matematico di Palermo, vol. 2, no. 12, pp. 99–104 (1986)
7. Rosenthal, K.: Quantales and Their Applications. Pitman Research Notes in Mathematics, vol. 234. Longman Scientific & Technical (1990)
8. Eklund, P., Gutie Rrez Garcia, J., Hoehle, U., Kortelainen, J.: Semigroups in Complete Lattices: Quantales, Modules and Related Topics, vol. 54 (2018)
9. Foulis, D.J., Bennett, M.K.: Effect algebras and unsharp quantum logics. Found. Phys. **24**, 1331–1352 (1994)
10. Jonsson, D.: Poloids from the points of view of partial transformations and category theory (2017). https://arxiv.org/abs/1710.04634
11. Johnstone, P.T.: Stone Spaces. Cambridge University Press, Cambridge (1982)

Counting and Computing Join-Endomorphisms in Lattices

Santiago Quintero[2(✉)], Sergio Ramirez[1], Camilo Rueda[1], and Frank Valencia[1,3]

[1] Pontificia Universidad Javeriana Cali, Cali, Colombia
[2] LIX, École Polytechnique de Paris, Palaiseau, France
squinter@lix.polytechnique.fr
[3] CNRS-LIX, École Polytechnique de Paris, Palaiseau, France

Abstract. Structures involving a lattice and join-endomorphisms on it are ubiquitous in computer science. We study the cardinality of the set $\mathcal{E}(L)$ of all join-endomorphisms of a given finite lattice L. In particular, we show that when L is $\mathbf{M}_n$, the discrete order of n elements extended with top and bottom, $|\mathcal{E}(L)| = n!\mathcal{L}_n(-1) + (n+1)^2$ where $\mathcal{L}_n(x)$ is the Laguerre polynomial of degree n. We also study the following problem: Given a lattice L of size n and a set $S \subseteq \mathcal{E}(L)$ of size m, find the greatest lower bound $\bigsqcap_{\mathcal{E}(L)} S$. The join-endomorphism $\bigsqcap_{\mathcal{E}(L)} S$ has meaningful interpretations in epistemic logic, distributed systems, and Aumann structures. We show that this problem can be solved with worst-case time complexity in $O(n + m \log n)$ for powerset lattices, $O(mn^2)$ for lattices of sets, and $O(mn + n^3)$ for arbitrary lattices. The complexity is expressed in terms of the basic binary lattice operations performed by the algorithm.

Keywords: Join-endomorphisms · Lattice cardinality · Lattice algorithms

1 Introduction

There is a long established tradition of using lattices to model structural entities in many fields of mathematics and computer science. For example, lattices are used in concurrency theory to represent the hierarchical organization of the information resulting from agent's interactions [12]. *Mathematical morphology* (MM), a well-established theory for the analysis and processing of geometrical structures, is founded upon lattice theory [2, 14]. Lattices are also used as algebraic structures for modal and epistemic logics as well as Aumann structures (e.g., modal algebras and constraint systems [7]).

In all these and many other applications, lattice join-endomorphisms appear as fundamental. A *join-endomorphism* is a function from a lattice to itself that preserves finite joins. In MM, join-endomorphisms correspond to one of its fundamental operations; *dilations*. In modal algebra, they correspond via duality to the box modal operator. In epistemic settings, they represent belief or knowledge of agents. In fact, our own interest in lattice theory derives from using join-endomorphisms to model the perception that agents may have of a statement in a lattice of partial information [7].

This work has been partially supported by the ECOS-NORD project FACTS (C19M03).

U. Fahrenberg et al. (Eds.): RAMiCS 2020, LNCS 12062, pp. 253–269, 2020.
https://doi.org/10.1007/978-3-030-43520-2_16

For finite lattices, devising suitable algorithms to compute lattice maps with some given properties would thus be of great utility. We are interested in constructing algorithms for computing lattice morphisms. This requires, first, a careful study of the space of such maps to have a clear idea of how particular lattice structures impact on the size of the space. We are, moreover, particularly interested in computing the *maximum* join-endomorphism below a given collection of join-morphisms. This turns out to be important, among others, in spatial computation (and in epistemic logic) to model the distributed information (resp. distributed knowledge) available to a set of agents as conforming a group [8]. It could also be regarded as the maximum perception consistent with (or derivable from) a collection of perceptions of a group of agents.

Problem. Consider the set $\mathcal{E}(L)$ of all join-endomorphisms of a finite lattice L. The set $\mathcal{E}(L)$ can be made into a lattice by ordering join-endomorphisms point-wise wrt the order of L. We investigate the following maximization problem: *Given a lattice L of size n and a set $S \subseteq \mathcal{E}(L)$ of size m, find in $\mathcal{E}(L)$ the greatest lower bound of S, i.e.,* $\bigsqcap_{\mathcal{E}(L)} S$. Simply taking $\sigma : L \to L$ with $\sigma(e) \stackrel{\text{def}}{=} \bigsqcap_L \{f(e) \mid f \in S\}$ does not solve the problem as σ may not be a join-endomorphism. Furthermore, since $\mathcal{E}(L)$ can be seen as the search space, we also consider the problem of determining its cardinality. Our main results are the following.

This Paper. We present characterizations of the exact cardinality of $\mathcal{E}(L)$ for some fundamental lattices. Our contribution is to establish the cardinality of $\mathcal{E}(L)$ for the stereotypical non-distributive lattice $L = \mathbf{M}_n$. We show that $|\mathcal{E}(\mathbf{M}_n)|$ equals $r_0^n + \ldots + r_n^n + r_1^{n+1} = n!\mathcal{L}_n(-1) + (n+1)^2$ where r_k^m is the number of ways to place k non-attacking rooks on an $m \times m$ board and $\mathcal{L}_n(x)$ is the Laguerre polynomial of degree n. We also present cardinality results for powerset and linear lattices that are part of the lattice theory folklore: The number of join-endomorphisms is $n^{\log_2 n}$ for powerset lattices of size n and $\binom{2n}{n}$ for linear lattices of size $n+1$. Furthermore, we provide algorithms that, given a lattice L of size n and a set $S \subseteq \mathcal{E}(L)$ of size m, compute $\bigsqcap_{\mathcal{E}(L)} S$. Our contribution is to show that $\bigsqcap_{\mathcal{E}(L)} S$ can be computed with worst-case time complexity in $O(n + m\log n)$ for powerset lattices, $O(mn^2)$ for lattices of sets, and $O(nm + n^3)$ for arbitrary lattices.

Due to space restrictions we only include the main proofs. The missing proofs can be found in the technical report of this paper [13].

2 Background: Join-Endomorphisms and Their Space

We presuppose basic knowledge of order theory [3] and use the following notions. Let $(L, \sqsubseteq)$ be a partially ordered set (poset), and let $S \subseteq L$. We use $\bigsqcup_L S$ to denote the least upper bound (or *supremum* or *join*) of S in L, if it exists. Dually, $\bigsqcap_L S$ is the greatest lower bound (glb) (*infimum* or *meet*) of S in L, if it exists. We shall often omit the index L from $\bigsqcup_L$ and $\bigsqcap_L$ when no confusion arises. As usual, if $S = \{c, d\}$, $c \sqcup d$ and $c \sqcap d$ represent $\bigsqcup S$ and $\bigsqcap S$, respectively. If L has a greatest element (top) $\top$, and a least element (bottom) $\bot$, we have $\bigsqcup \emptyset = \bot$ and $\bigsqcap \emptyset = \top$. The poset L is *distributive* iff for every $a, b, c \in L$, $a \sqcup (b \sqcap c) = (a \sqcup b) \sqcap (a \sqcup c)$.

The poset L is a *lattice* iff each finite nonempty subset of L has a supremum and infimum in L, and it is a *complete lattice* iff each subset of L has a supremum and infimum in L. A *self-map* on L is a function $f : L \to L$. A self-map f is *monotonic* if $a \sqsubseteq b$ implies $f(a) \sqsubseteq f(b)$. We say that f *preserves* the join of $S \subseteq L$ iff $f(\bigsqcup S) = \bigsqcup\{f(c) \mid c \in S\}$. We shall use the following posets and notation. Given n, we use $\mathbf{n}$ to denote the poset $\{1, \ldots, n\}$ with the linear order $x \sqsubseteq y$ iff $x \leq y$. The poset $\bar{\mathbf{n}}$ is the set $\{1, \ldots, n\}$ with the discrete order $x \sqsubseteq y$ iff $x = y$. Given a poset L, we use $L_\perp$ for the poset that results from adding a bottom element to L. The poset $L^\top$ is similarly defined. The lattice $\mathbf{2}^n$ is the n-fold Cartesian product of $\mathbf{2}$ ordered coordinate-wise. We define $\mathbf{M}_n$ as the lattice $(\bar{\mathbf{n}}_\perp)^\top$. A *lattice of sets* is a set of sets ordered by inclusion and closed under finite unions and intersections. A *powerset lattice* is a lattice of sets that includes all the subsets of its top element.

We shall investigate the set of all join-endomorphisms of a given lattice ordered point-wise. Notice that every finite lattice is a complete lattice.

Definition 1 (Join-endomorphisms and their space). *Let L be a complete lattice. We say that a self-map is a* (lattice) join-endomorphism *iff it preserves the join of every finite subset of L. Define $\mathcal{E}(L)$ as the set of all join-endomorphisms of L. Furthermore, given $f, g \in \mathcal{E}(L)$, define $f \sqsubseteq_{\mathcal{E}} g$ iff $f(a) \sqsubseteq g(a)$ for every $a \in L$.*

The following are immediate consequences of the above definition.

Proposition 1. *Let L be a complete lattice. $f \in \mathcal{E}(L)$ iff $f(\perp) = \perp$ and $f(a \sqcup b) = f(a) \sqcup f(b)$ for all $a, b \in L$. If f is a join-endomorphism of L then f is monotonic.*

Given a set $S \subseteq \mathcal{E}(L)$, where L is a finite lattice, we are interested in finding the greatest join-endomorphism in $\mathcal{E}(L)$ below the elements of S, i.e., $\bigsqcap_{\mathcal{E}(L)} S$. Since every finite lattice is also a complete lattice, the existence of $\bigsqcap_{\mathcal{E}(L)} S$ is guaranteed by the following proposition.

Proposition 2 ([6]). *If $(L, \sqsubseteq)$ is a complete lattice, $(\mathcal{E}(L), \sqsubseteq_{\mathcal{E}})$ is a complete lattice.*

In the following sections we study the cardinality of $\mathcal{E}(L)$ for some fundamental lattices and provide efficient algorithms to compute $\bigsqcap_{\mathcal{E}(L)} S$.

3 The Size of the Function Space

The main result of this section is Theorem 1. It states the size of $\mathcal{E}(\mathbf{M}_n)$. Propositions 3 and 4 state, respectively, the size of $\mathcal{E}(L)$ for the cases when L is a powerset lattice and when L is a total order. These propositions follow from simple observations and they are part of the lattice theory folklore [1, 10, 16]. We include our original proofs of these propositions in the technical report of this paper [13].

3.1 Distributive Lattices

We begin with lattices isomorphic to $\mathbf{2}^n$. They include *finite boolean algebras* and *powerset* lattices [3]. The size of these lattices are easy to infer from the observation that the join-preserving functions on them are determined by their action on the lattices' atoms.

Proposition 3. *Suppose that $m \geq 0$. Let L be any lattice isomorphic to the product lattice $\mathbf{2}^m$. Then $|\mathcal{E}(L)| = n^{\log_2 n}$ where $n = 2^m$ is the size of L.*

Thus powerset lattices and boolean algebras have a super-polynomial, sub-exponential number of join-endomorphisms. Nevertheless, linear order lattices allow for an exponential number of join-endomorphisms given by the *central binomial coefficient*. The following proposition is also easy to prove from the observation that the join-endomorphisms over a linear order are also monotonic functions. In fact, this result appears in [1] and it is well-known among the RAMICS community [10, 16].

Proposition 4. *Suppose that $n \geq 0$. Let L be any lattice isomorphic to the linear order lattice $\mathbf{n}_\perp$. Then $|\mathcal{E}(L)| = \binom{2n}{n}$.*

It is easy to prove that $\frac{4^n}{2\sqrt{n}} \leq \binom{2n}{n} \leq 4^n$ for $n \geq 1$. Together with Proposition 4, this gives us explicit exponential lower and upper bounds for $|\mathcal{E}(L)|$ when L is a linear lattice.

3.2 Non-distributive Case

The number of join-endomorphisms for some non-distributive lattices of a given size can be much bigger than that for those distributive lattices of the same size in the previous section. We will characterize this number for an archetypal non-distributive lattice in terms of Laguerre (and rook) polynomials.

Laguerre polynomials are solutions to Laguerre's second-order linear differential equation $xy'' + (1-x)y' + ny = 0$ where y' and y'' are the first and second derivatives of an unknown function y of the variable x, and n is a non-negative integer. The Laguerre polynomial of degree n in x, $\mathcal{L}_n(x)$ is given by the summation $\sum_{k=0}^{n} \binom{n}{k} \frac{(-1)^k}{k!} x^k$.

The lattice $\mathbf{M}_n$ is non-distributive for any $n \geq 3$. The size of $\mathcal{E}(\mathbf{M}_n)$ can be succinctly expressed as follows.

Theorem 1. $|\mathcal{E}(\mathbf{M}_n)| = (n+1)^2 + n!\mathcal{L}_n(-1)$.

In combinatorics rook polynomials are generating functions of the number of ways to place non-attacking rooks on a board. A *rook polynomial* (for square boards) $\mathcal{R}_n(x)$ has the form $\sum_{k=0}^{n} x^k r(k,n)$ where the (rook) coefficient $r(k,n)$ represents the number of ways to place k non-attacking rooks on an $n \times n$ chessboard. For instance, $r(0,n) = 1$, $r(1,n) = n^2$ and $r(n,n) = n!$. In general $r(k,n) = \binom{n}{k}^2 k!$.

Rook polynomials are related to Laguerre polynomials by the equation $\mathcal{R}_n(x) = n!x^n \mathcal{L}_n(-x^{-1})$. Therefore, as a direct consequence of the above theorem, we can also characterize $|\mathcal{E}(\mathbf{M}_n)|$ in combinatorial terms as the following sum of rook coefficients.

Corollary 1. *Let* $r'(n+1,n) = r(1,n+1)$ *and* $r'(k,n) = r(k,n)$ *if* $k \leq n$. *Then* $|\mathcal{E}(\mathbf{M}_n)| = \sum_{k=0}^{n+1} r'(k,n)$.

We conclude this section with another pleasant correspondence between the endomorphisms in $\mathcal{E}(\mathbf{M}_n)$ and $\mathcal{R}_n(x)$. Let $f : L \to L$ be a function over a lattice $(L, \sqsubseteq)$. We say that f is *non-reducing* in L iff it does not map any value to a smaller one; i.e., there is no $e \in L$ such that $f(e) \sqsubset e$. The number of join-endomorphisms that are non-reducing in $\mathbf{M}_n$ is exactly the value of the rook polynomial $\mathcal{R}_n(x)$ for $x = 1$.

Corollary 2. $\mathcal{R}_n(1) = |\{ f \in \mathcal{E}(\mathbf{M}_n) \mid f$ *is non-reducing in* $\mathbf{M}_n \}|$.

Table 1 illustrates the join-endomorphisms over the lattice $\mathbf{M}_n$ as a union $\bigcup_{i=1}^{4} \mathcal{F}_i$. Corollary 2 follows from the observation that the set of non-reducing functions in $\mathbf{M}_n$ is equal to $\mathcal{F}_4$ whose size is $\mathcal{R}_n(1)$ as shown in the following proof of Theorem 1.

Proof of Theorem 1. We show that $|\mathcal{E}(\mathbf{M}_n)|$ can be expressed in terms of Laguerre polynomials: $|\mathcal{E}(\mathbf{M}_n)| = (n+1)^2 + n!\mathcal{L}_n(-1)$.

Let $\mathcal{F} = \bigcup_{i=1}^{4} \mathcal{F}_i$ where the mutually exclusive $\mathcal{F}_i$'s are defined in Table 1, and $I = \{1, \ldots, n\}$. The proof is divided in two parts: (I) $\mathcal{F} = \mathcal{E}(\mathbf{M}_n)$ and (II) $|\mathcal{F}| = (n+1)^2 + n!\mathcal{L}_n(-1)$.

Part (I). For $\mathcal{F} \subseteq \mathcal{E}(\mathbf{M}_n)$, it is easy to verify that each $f \in \mathcal{F}$ is a join-endomorphism.

For $\mathcal{E}(\mathbf{M}_n) \subseteq \mathcal{F}$ we show that for any function f from $\mathbf{M}_n$ to $\mathbf{M}_n$ if $f \notin \mathcal{F}$, then $f \notin \mathcal{E}(\mathbf{M}_n)$. Immediately, if $f(\bot) \neq \bot$ then $f \notin \mathcal{E}(\mathbf{M}_n)$.

Suppose $f(\bot) = \bot$. Let J, K, H be disjoint possibly empty sets such that $I = J \cup K \cup H$ and let $j = |J|$, $k = |K|$ and $h = |H|$. The sets J, K, H represent the elements of I mapped by f to $\top$, to elements of I, and to $\bot$, respectively. More precisely, $\mathrm{Img}(f\restriction_J) = \{\top\}$, $\mathrm{Img}(f\restriction_K) \subseteq I$ and $\mathrm{Img}(f\restriction_H) = \{\bot\}$. Furthermore, for every f either (1) $f(\top) = \bot$, (2) $f(\top) \in I$ or (3) $f(\top) = \top$. We show that $f \notin \mathcal{E}(\mathbf{M}_n)$ for case (3), proofs of cases (1) and (2) are included in [13].

Suppose $k = 0$. Notice that $f \notin \mathcal{F}_3$ and $f \notin \mathcal{F}_4$ hence $h \neq 1$ and $h \neq 0$. Thus $h > 1$ implies that there are at least two $e_1, e_2 \in H$ such that $f(e_1) = f(e_2) = \bot$. But then $f(e_1 \sqcup e_2) = f(\top) = \top \neq \bot = f(e_1) \sqcup f(e_2)$, hence $f \notin \mathcal{E}(\mathbf{M}_n)$.

Suppose $k > 0$. Assume $h = 0$. Notice that $K = I \setminus J$ and $\mathrm{Img}(f\restriction_K) \subseteq I$. Since f is a $\bot$ and $\top$ preserving function and it satisfies conditions (a) and (c) of $\mathcal{F}_4$ but $f \notin \mathcal{F}_4$, then f must violate condition (b). Thus $f\restriction_K$ is not injective. Then there are $a, b \in K$ such that $a \neq b$ but $f(a) = f(b)$. Then $f(a) \sqcup f(b) \neq \top = f(a \sqcup b)$. Consequently, $f \notin \mathcal{E}(\mathbf{M}_n)$.

Assume $h > 0$. There must be $e_1, e_2, e_3 \in I$ such that $f(e_1) = \bot$ and $f(e_2) = e_3$. Notice that $f(e_1) \sqcup f(e_2) = e_3 \neq \top = f(\top) = f(e_1 \sqcup e_2)$. Therefore, $f \notin \mathcal{E}(\mathbf{M}_n)$.

Part (II). We prove that $|\mathcal{F}| = \sum_{i=1}^{4} |\mathcal{F}_i| = (n+1)^2 + n!\mathcal{L}_n(-1)$. Recall that $n = |I|$. It is easy to prove that $|\mathcal{F}_1| = 1$, $|\mathcal{F}_2| = n^2 + n$ and $|\mathcal{F}_3| = n$. The reader is referred to [13] for details. Here we prove that $|\mathcal{F}_4| = n!\mathcal{L}_n(-1)$.

Let $f \in \mathcal{F}_4$ and let $J \subseteq I$ be a possibly empty set such that $\mathrm{Img}(f\restriction_J) = \{\top\}$ and $\mathrm{Img}(f\restriction_{I \setminus J}) \subseteq I$, where $f\restriction_{I \setminus J}$ is an injective function. We shall call $j = |J|$.

Table 1. Families $\mathcal{F}_1, \ldots, \mathcal{F}_4$ of join-endomorphisms of $\mathbf{M}_n$. $I = \{1, \ldots, n\}$. $f\restriction_A$ is the restriction of f to a subset A of its domain. $\mathrm{Img}(f)$ is the image of f. A function from each $\mathcal{F}_i$ for $\mathbf{M}_5$ is depicted with blue arrows.

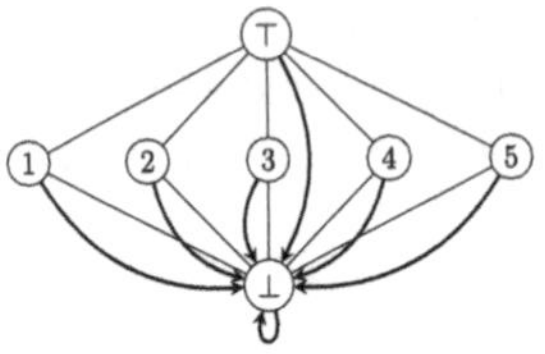	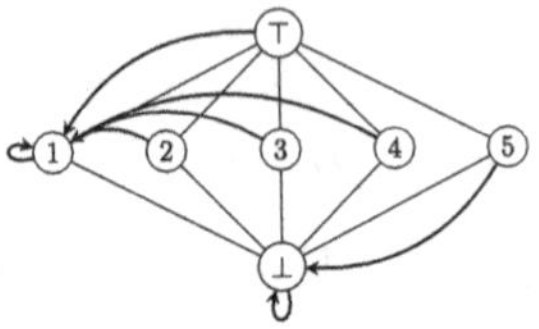
Let $\mathcal{F}_1$ be the family of functions f that for all $e \in \mathbf{M}_n$, $f(e) = \bot$.	Let $\mathcal{F}_2$ be the family of bottom preserving functions f such that for some $e, e' \in I$: (a) $f(\top) = e$, (b) $f(e') = \bot$ or $f(e') = e$, and (c) $f(e'') = e$ for all $e'' \in I \setminus \{e'\}$.
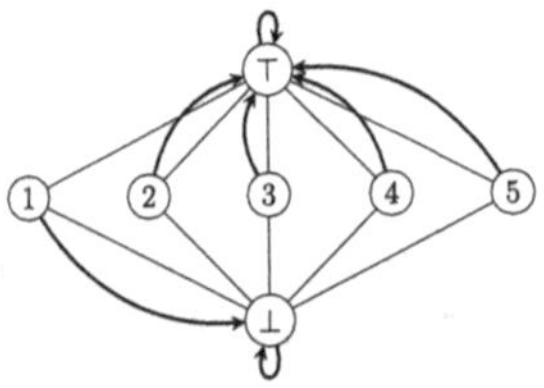	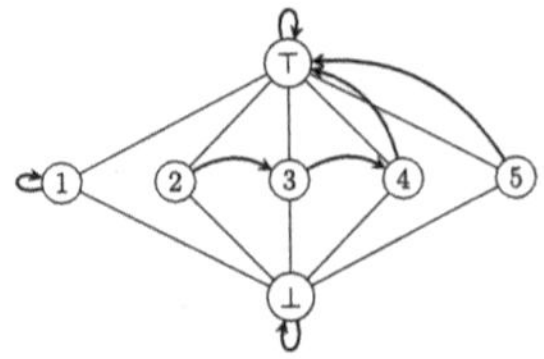
Let $\mathcal{F}_3$ be the family of top and bottom preserving functions f such that for some $e \in I$: (a) $f(e) = \bot$, and (b) $f(e') = \top$ for all $e' \in I \setminus \{e\}$.	Let $\mathcal{F}_4$ be the family of top and bottom preserving functions f that for some $J \subseteq I$: (a) $f(e) = \top$ for every $e \in J$, (b) $f\restriction_{I\setminus J}$ is injective, and (c) $\mathrm{Img}(f\restriction_{I\setminus J}) \subseteq I$.

For each of the $\binom{n}{j}$ possibilities for J, the elements of $I \setminus J$ are to be mapped to I by the injective function $f\restriction_{I\setminus J}$. The number of functions $f\restriction_{I\setminus J}$ is $\frac{n!}{j!}$. Therefore, $|\mathcal{F}_4| = \sum_{j=0}^{n} \binom{n}{j} \frac{n!}{j!}$. This sum equals $n!\mathcal{L}_n(-1)$ which in turn is equal to $\mathcal{R}_n(1)$. It follows that $|\mathcal{F}| = \sum_{i=1}^{4} |\mathcal{F}_i| = (n+1)^2 + n!\mathcal{L}_n(-1)$ as wanted. □

4 Algorithms

We shall provide efficient algorithms for the maximization problem mentioned in the introduction: Given a finite lattice L and $S \subseteq \mathcal{E}(L)$ find $\bigsqcap_{\mathcal{E}(L)} S$, i.e., the greatest join-endomorphism in the lattice $\mathcal{E}(L)$ below all the elements of S.

Finding $\bigsqcap_{\mathcal{E}(L)} S$ may not be immediate. E.g., see $\bigsqcap_{\mathcal{E}(L)} S$ in Fig. 1a for a small lattice of four elements and two join-endomorphisms. As already mentioned, a *naive approach* is to compute $\bigsqcap_{\mathcal{E}(L)} S$ by taking $\sigma_S(c) \stackrel{\text{def}}{=} \bigsqcap_L \{f(c) \mid f \in S\}$ for each $c \in L$. This does not work since σ_S is not necessarily a join-endomorphism as shown in Fig. 1b.

A *brute force* solution to compute $\bigsqcap_{\mathcal{E}(L)} S$ can be obtained by generating the set $S' = \{g \mid g \in \mathcal{E}(L) \text{ and } g \sqsubseteq f \text{ for all } f \in S\}$ and taking its join. This approach works since $\bigsqcup S' = \bigsqcap_{\mathcal{E}(L)} S$ but as shown in Sect. 3, the size of $\mathcal{E}(L)$ can be super-polynomial for distributive lattices and exponential in general.

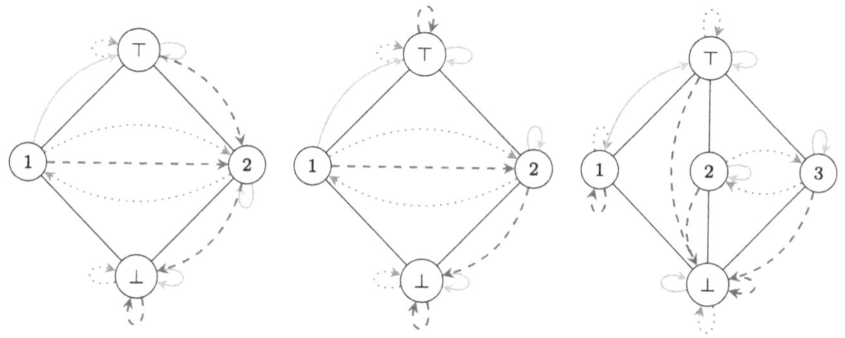

(a) f_1: ⋯→, f_2: →, f: - - → (b) f_1: ⋯→, f_2: →, σ_S: - - → (c) f_1: ⋯→, f_2: →, δ_S: - - →

Fig. 1. $S = \{f_1, f_2\} \subseteq \mathcal{E}(L)$. (a) $f = \bigsqcap_{\mathcal{E}(L)} S$. (b) $\sigma_S(c) \stackrel{\text{def}}{=} f_1(c) \sqcap f_2(c)$ is not a join-endomorphism of $\mathbf{M}_2$: $\sigma_S(1 \sqcup 2) \neq \sigma_S(1) \sqcup \sigma_S(2)$. (c) δ_S in Lemma 1 is not a join-endomorphism of the non-distributive lattice $\mathbf{M}_3$: $\delta_S(1) \sqcup \delta_S(2) = 1 \neq \bot = \delta_S(1 \sqcup 2)$.

Nevertheless, one can use lattice properties to compute $\bigsqcap_{\mathcal{E}(L)} S$ efficiently. For distributive lattices, we use the inherent compositional nature of $\bigsqcap_{\mathcal{E}(L)} S$. For arbitrary lattices, we present an algorithm that uses the function σ_S in the naive approach to compute $\bigsqcap_{\mathcal{E}(L)} S$ by approximating it from above.

We will give the time complexities in terms of the number of basic binary lattice operations (i.e., meets, joins and subtractions) performed during execution.

4.1 Meet of Join-Endomorphisms in Distributive Lattices

Here we shall illustrate some pleasant compositionality properties of the infima of join-endomorphisms that can be used for computing the join-endomorphism $\bigsqcap_{\mathcal{E}(L)} S$ in a finite distributive lattice L. In what follows we assume $n = |L|$ and $m = |S|$.

We use X^J to denote the set of tuples $(x_j)_{j \in J}$ of elements $x_j \in X$ for each $j \in J$.

Lemma 1. *Let L be a finite distributive lattice and $S = \{f_i\}_{i \in I} \subseteq \mathcal{E}(L)$. Then $\bigsqcap_{\mathcal{E}(L)} S = \delta_S$ where $\delta_S(c) \stackrel{\text{def}}{=} \bigsqcap_L \{\bigsqcup_{i \in I} f_i(a_i) \mid (a_i)_{i \in I} \in L^I$ and $\bigsqcup_{i \in I} a_i \sqsupseteq c\}$.*

The above lemma basically says that $(\bigsqcap_{\mathcal{E}(L)} S)(c)$ is the greatest element in L below all possible applications of the functions in S to elements whose join is greater or equal to c. The proof that $\delta_S \sqsupseteq_{\mathcal{E}} \bigsqcap_{\mathcal{E}(L)} S$ uses the fact that join-endomorphisms preserve joins. The proof that $\delta_S \sqsubseteq_{\mathcal{E}} \bigsqcap_{\mathcal{E}(L)} S$ proceeds by showing that δ_S is a lower bound in $\mathcal{E}(L)$ of S. Distributivity of the lattice L is crucial for this direction. In fact without it $\bigsqcap_{\mathcal{E}(L)} S = \delta_S$ does not necessarily hold as shown by the following counter-example.

Example 1. Consider the non-distributive lattice $\mathbf{M}_3$ and $S = \{f_1, f_2\}$ defined as in Fig. 1c. We obtain $\delta_S(1 \sqcup 2) = \delta_S(\top) = \bot$ and $\delta_S(1) \sqcup \delta_S(2) = 1 \sqcup \bot = 1$. Then, $\delta_S(1 \sqcup 2) \neq \delta_S(1) \sqcup \delta_S(2)$, i.e., δ_S is not a join-endomorphism.

Naive Algorithm A_1. One could use Lemma 1 directly in the obvious way to provide an algorithm for $\bigsqcap_{\mathcal{E}(L)} S$ by computing δ_S: i.e., computing the meet of elements of the form $\bigsqcup_{i \in I} f_i(a_i)$ for every tuple $(a_i)_{i \in I}$ such that $\bigsqcup_{i \in I} a_i \sqsupseteq c$. For each $c \in L$, $\delta_S(c)$ checks n^m tuples $(a_i)_{i \in I}$, each one with a cost in $O(m)$. Thus A_1 can compute $\bigsqcap_{\mathcal{E}(L)} S$ by performing $O(n \times n^m \times m) = O(mn^{m+1})$ binary lattice operations.

Nevertheless, we can use Lemma 1 to provide a recursive characterization of $\bigsqcap_{\mathcal{E}(L)} S$ that can be used in a divide-and-conquer algorithm with lower time complexity.

Proposition 5. *Let L be a finite distributive lattice and $S = S_1 \cup S_2 \subseteq \mathcal{E}(L)$. Then* $(\bigsqcap_{\mathcal{E}(L)} S)(c) = \bigsqcap_L \{(\bigsqcap_{\mathcal{E}(L)} S_1)(a) \sqcup (\bigsqcap_{\mathcal{E}(L)} S_2)(b) \mid a, b \in L$ *and* $a \sqcup b \sqsupseteq c\}$.

The above proposition bears witness to the compositional nature of $\bigsqcap_{\mathcal{E}(L)} S$. It can be proven by replacing $(\bigsqcap_{\mathcal{E}(L)} S_1)(a)$ and $(\bigsqcap_{\mathcal{E}(L)} S_2)(b)$ by $\delta_{S_1}(a)$ and $\delta_{S_2}(b)$ using Lemma 1 (see [13]).

Naive Algorithm A_2. We can use Proposition 5 to compute $\bigsqcap_{\mathcal{E}(L)} S$ with the following recursive procedure: Take any partition $\{S_1, S_2\}$ of S such that the absolute value of $|S_1| - |S_2|$ is at most 1. Then compute the meet of all $(\bigsqcap_{\mathcal{E}(L)} S_1)(a) \sqcup (\bigsqcap_{\mathcal{E}(L)} S_2)(b)$ for every a, b such that $a \sqcup b \sqsupseteq c$. Then given $c \in L$, the time complexity of a *naive* implementation of the above procedure can be obtained as the solution of the equation $T(m) = n^2(1 + 2T(m/2))$ and $T(1) = 1$ which is in $O(mn^{2\log_2 m})$. Therefore, $\bigsqcap_{\mathcal{E}(L)} S$ can be computed in $O(mn^{1+2\log_2 m})$.

The time complexity of the naive algorithm A_2 is better than that of A_1. However, by using a simple memoization technique to avoid repeating recursive calls and the following observations one can compute $\bigsqcap_{\mathcal{E}(L)} S$ in a much lower time complexity order.

4.2 Using Subtraction and Downsets to Characterize $\bigsqcap_{\mathcal{E}(L)} S$

In what follows we show that $\bigsqcap_{\mathcal{E}(L)} S$ can be computed in $O(mn^2)$ for distributive lattices and, in particular, in $O(n + m\log n)$ for powerset lattices. To achieve this we use the subtraction operator from co-Heyting algebras and the notion of down set[1].

[1] Recall that we give time complexities in terms of the number of basic binary lattice operations (i.e., meets, joins and subtractions) performed during execution.

Subtraction Operator. Notice that in Proposition 5 we are considering *all* pairs $a, b \in L$ such that $a \sqcup b \sqsupseteq c$. However, because of the monotonicity of join-endomorphisms, it suffices to take, for each $a \in L$, just *the least* b such that $a \sqcup b \sqsupseteq c$. In finite distributive lattices, and more generally in co-Heyting algebras [5], the *subtraction* operator $c \backslash a$ gives us exactly such a least element. The subtraction operator is uniquely determined by the property (*Galois connection*) $b \sqsupseteq c \backslash a$ iff $a \sqcup b \sqsupseteq c$ for all $a, b, c \in L$.

Down-Sets. Besides using just $c \backslash a$ instead of all b's such that $a \sqcup b \sqsupseteq c$, we can use a further simplification: Rather than including every $a \in L$, we only need to consider every a in the *down-set* of c. Recall that the down-set of c is defined as $\downarrow c = \{e \in L \mid e \sqsubseteq c\}$. This additional simplification is justified using properties of distributive lattices to show that for any $a' \in L$, such that $a' \not\sqsubseteq c$, there exists $a \sqsubseteq c$ such that $(\bigsqcap_{\mathcal{E}(L)} S_1)(a) \sqcup (\bigsqcap_{\mathcal{E}(L)} S_2)(c \backslash a) \sqsubseteq (\bigsqcap_{\mathcal{E}(L)} S_1)(a') \sqcup (\bigsqcap_{\mathcal{E}(L)} S_2)(c \backslash a')$.

The above observations lead us to the following theorem.

Theorem 2. *Let L be a finite distributive lattice and $S = S_1 \cup S_2 \subseteq \mathcal{E}(L)$. Then $(\bigsqcap_{\mathcal{E}(L)} S)(c) = \bigsqcap_L \{(\bigsqcap_{\mathcal{E}(L)} S_1)(a) \sqcup (\bigsqcap_{\mathcal{E}(L)} S_2)(c \backslash a) \mid a \in \downarrow c\}$.*

The above result can be used to derive a simple recursive algorithm that, given a finite distributive lattice L and $S \subseteq \mathcal{E}(L)$, computes $\bigsqcap_{\mathcal{E}(L)} S$ in worst-case time complexity $O(mn^2)$ where $m = |S|$ and $n = |L|$. We show this algorithm next.

4.3 Algorithms for Distributive Lattices

We first describe the algorithm DMEETAPP that computes the value $(\bigsqcap_{\mathcal{E}(L)} S)(c)$. We then describe the algorithm DMEET that computes the function $\bigsqcap_{\mathcal{E}(L)} S$ by calling DMEETAPP in a particular order to avoid repeating computations. We use the following definition to specify the calling order.

Definition 2. *A* binary partition tree (bpt) *of a finite set $S \neq \emptyset$ is a binary tree such that (a) its root is S, (b) if $|S| = 1$ then its root is a leaf, and (c) if $|S| > 1$ it has a left and a right subtree, themselves bpts of S_1 and S_2 resp., for a partition $\{S_1, S_2\}$ of S.*

Let Δ be a bpt of S. We use $\Delta(S')$ for the subtree of Δ rooted at $S' \subseteq S$, if it exists. We use $\langle S, \Delta_1, \Delta_2 \rangle$ for the bpt of S with Δ_1 and Δ_2 as its left and right subtrees.

The following proposition is an immediate consequence of the previous definition.

Proposition 6. *The size (number of nodes) of any bpt of S is $2m - 1$ where $m = |S|$.*

DMEETAPP(Δ, c). Let $\Delta = \langle S, \Delta_1, \Delta_2 \rangle$ be a bpt of $S \subseteq \mathcal{E}(L)$ where L is a distributive lattice. The recursive program DMEETAPP(Δ, c) defined in Algorithm 1 computes $(\bigsqcap_{\mathcal{E}(L)} S)(c)$. It uses a global lookup table T for storing the results of calls to DMEETAPP. Initially each entry of T stores a `null` value not included in L. Since S is the union of the roots of Δ_1 and Δ_2, the correctness of DMEETAPP(Δ, c) follows from Theorem 2. Termination follows from the fact that L is finite and the bpts Δ_1 and Δ_2 in the recursive calls are strictly smaller than Δ.

Algorithm 1. DMEETAPP(Δ, c) returns $\left(\bigsqcap_{\mathcal{E}(L)} S\right)(c)$ where Δ is a bpt of $S \subseteq \mathcal{E}(L)$ and L is a finite distributive lattice. The global variable T is used as a lookup table.

procedure DMEETAPP(Δ, c) ▷ $\Delta = \langle S, \Delta_1, \Delta_2 \rangle$
  if *IsNull*($T[S, c]$) **then**
    if $S = \{f\}$ **then**
      $T[S, c] \leftarrow f(c)$
    else
      $T[S, c] \leftarrow \bigsqcap_L \{\text{DMEETAPP}(\Delta_1, a) \sqcup \text{DMEETAPP}(\Delta_2, c \backslash a) \mid a \in \downarrow c\}$.

Computing $\bigsqcap_{\mathcal{E}(L)} S$ for Distributive Lattices. Let us consider an execution of DMEETAPP(Δ, c). From the definition of subtraction it follows that $c \backslash a \in \downarrow c$. Then for each recursive call DMEETAPP(Δ', a') performed by an execution of DMEETAPP(Δ, c) we have $a' \in \downarrow c$. This and the fact that T is initialized with a `null` value not in L lead us the following simple observation.

Observation 3. *Let $\Delta = \langle S, \Delta_1, \Delta_2 \rangle$ with Δ_1 and Δ_2 rooted at S_1 and S_2. Assume that $T[S_1, a'], T[S_2, a'] \in L$ for every $a' \in \downarrow c$. Then the number of binary lattice operations (meets, joins, substractions) performed by* DMEETAPP(Δ, c) *is in $O(|\downarrow c|)$.*

Algorithm 2. DMEET(L, S, P). Given a finite distributive lattice L, $P \subseteq L$ and $S \subseteq \mathcal{E}(L)$, the algorithm computes $T[S, c] = \bigsqcap_{\mathcal{E}(L)} S(c)$ for each $c \in P$. Δ is a bpt of S and T is a global lookup table.

$T[S', a] \leftarrow$ `null` ▷ for each $a \in P$ and each node S' of Δ
for each S' in a post-order traversal sequence of Δ **do** ▷ visit each S' of Δ in post-order
  for each $c \in P$ in increasing order **do** ▷ visit each $c \in P$ in increasing order w.r.t L
    DMEETAPP($\Delta(S'), c$)

DMEET(L, S, P). The values of $\left(\bigsqcap_{\mathcal{E}(L)} S\right)(c)$ for each $c \in P \subseteq L$ are computed by the program in Algorithm 2 as follows. To satisfy the assumption in Observation 3, it visits each node S' of Δ in *post-order* (i.e., before visiting a node it first visits its children). For each subtree $\Delta(S')$ of Δ, it calls DMEETAPP($\Delta(S'), c$) for every $c \in P$ in *increasing order* with respect to the order of L: I.e., before calling DMEETAPP($\Delta(S'), c$) it calls first DMEETAPP($\Delta(S'), c'$) for each $c' \in (P \cap \downarrow c) \setminus \{c\}$. The correctness of the call DMEET(L, S, P) follows from that of DMEETAPP(Δ, c).

Complexity for Distributive Lattices. Assume that L is a distributive lattice of size n and that S is a subset of $\mathcal{E}(L)$ of size m. The above-mentioned traversals of Δ and P ensure that the assumption in Observation 3 is satisfied by each call of the form DMEETAPP($\Delta(S'), c$) performed during the execution of DMEET(L, S, L). From Proposition 6 we know that the number of iterations of the outer **for** is $2m - 1$. Clearly $|\downarrow c|$ and $|P|$ are both in $O(n)$. Thus, given S' we conclude from Observation 3 that the total number of operations from all calls of the form DMEETAPP($\Delta(S'), c$), executed in the inner **for**, is in $O(n^2)$. The worst-case time complexity of DMEET(L, S, L) is then in $O(mn^2)$.

Complexity for Powerset Lattices. Assume that L is a powerset lattice. We can compute $\bigsqcap_{\mathcal{E}(L)} S$ in $O(n + m\log n)$ as follows. First call DMEET(L, S, P) where $P = J(L) \cup \{\bot\}$ and $J(L)$ is the set of *join-irreducible* elements (i.e., the singleton sets in this case) of L. Since $|J(L)| = \log_2 n$ and $|\downarrow c| = 2$ for every $c \in J(L)$, DMEET(L, S, P) can be performed in $O(m \log n)$. This produces $T[S, c] = (\bigsqcap_{\mathcal{E}(L)} S)(c)$ for each $c \in P$. To compute $T[S, e] = (\bigsqcap_{\mathcal{E}(L)} S)(e)$ for each $e \in L \setminus P$ in a total time of $O(n)$, visit each such an e in increasing order and set $T[S, e] = T[S, a] \sqcup T[S, b]$ for some $a, b \in \downarrow e \setminus \{e\}$ such that $e = a \sqcup b$. Since $e \notin P$ there must be a, b satisfying the above condition.

4.4 Algorithms for Arbitrary Lattices

The previous algorithm may fail to produce the $\bigsqcap_{\mathcal{E}(L)} S$ for non-distributive finite lattices. Nonetheless, for any arbitrary finite lattice L, $\bigsqcap_{\mathcal{E}(L)} S$ can be computed by successive approximations, starting with some self-map known to be smaller than each $f \in S$ and greater than $\bigsqcap_{\mathcal{E}(L)} S$. Assume a self-map $\sigma : L \to L$ such that $\sigma \sqsupseteq_{\mathcal{E}} \bigsqcap_{\mathcal{E}(L)} S$ and, for all $f \in S$, $\sigma \sqsubseteq_{\mathcal{E}} f$. A good starting point is $\sigma(u) = \bigsqcap\{f(u) \mid f \in S\}$, for all $u \in L$. By definition of $\sqcap$, σ is the biggest function under all functions in S, hence $\sigma \sqsupseteq_{\mathcal{E}} \bigsqcap_{\mathcal{E}(L)} S$. The program GMEET in Algorithm 3 computes decreasing upper bounds of $\bigsqcap_{\mathcal{E}(L)} S$ by correcting σ values not conforming to the following *join-endomorphism property:* $\sigma(u) \sqcup \sigma(v) = \sigma(u \sqcup v)$. The correction decreases σ and maintains the invariant $\sigma \sqsupseteq_{\mathcal{E}} \bigsqcap_{\mathcal{E}(L)} S$, as stated in Theorem 4.

Theorem 4. *Let L be a finite lattice, $u, v \in L$, $\sigma : L \to L$ and $S \subseteq \mathcal{E}(L)$. Assume $\sigma \sqsupseteq_{\mathcal{E}} \bigsqcap_{\mathcal{E}(L)} S$ holds, and consider the following updates:*

1. *when $\sigma(u) \sqcup \sigma(v) \sqsubset \sigma(u \sqcup v)$, assign $\sigma(u \sqcup v) \leftarrow \sigma(u) \sqcup \sigma(v)$*
2. *when $\sigma(u) \sqcup \sigma(v) \not\sqsubseteq \sigma(u \sqcup v)$, assign $\sigma(u) \leftarrow \sigma(u) \sqcap \sigma(u \sqcup v)$ and also $\sigma(v) \leftarrow \sigma(v) \sqcap \sigma(u \sqcup v)$*

Let σ' be the function resulting after the update. Then, (1) $\sigma' \sqsubset \sigma$ and (2) $\sigma' \sqsupseteq_{\mathcal{E}} \bigsqcap_{\mathcal{E}(L)} S$.

Algorithm 3. GMEET finds $\sigma = \bigsqcap_{\mathcal{E}(L)} S$

$\sigma(u) \leftarrow \bigsqcap\{f(u) \mid f \in S\}$ ▷ for all $u \in L$
while $u, v \in L \wedge \sigma(u) \sqcup \sigma(v) \neq \sigma(u \sqcup v)$ **do**
  if $\sigma(u) \sqcup \sigma(v) \sqsubset \sigma(u \sqcup v)$ **then** ▷ case (1)
    $\sigma(u \sqcup v) \leftarrow \sigma(u) \sqcup \sigma(v)$
  else ▷ case (2)
    $\sigma(u) \leftarrow \sigma(u) \sqcap \sigma(u \sqcup v)$
    $\sigma(v) \leftarrow \sigma(v) \sqcap \sigma(u \sqcup v)$

The procedure (see Algorithm 3) loops through pairs $u, v \in L$ while there is some pair satisfying cases (1) or (2) above for the current σ. When there is, it updates σ as mentioned in Theorem 4. At the end of the loop all pairs $u, v \in L$ satisfy the join preservation property. By the invariant mentioned in the theorem, this means $\sigma = \bigsqcap_{\mathcal{E}(L)} S$.

Algorithm 4. GMEET+ finds $\sigma = \bigsqcap_{\mathcal{E}(L)} S$

1: $\sigma(u) \leftarrow \bigsqcap\{f(u) \mid f \in S\}$ ▷ for all $u \in L$
2: Initialize $\mathtt{Sup}_w, \mathtt{Con}_w, \mathtt{Fail}_w$, for all w
3: **while** $w \in L$ such that $(u, v) \in \mathtt{Con}_w$ **do** ▷ some conflict set not empty
4: $\quad \mathtt{Con}_w \leftarrow \mathtt{Con}_w \backslash \{(u, v)\}$
5: $\quad \sigma(w) \leftarrow \sigma(u) \sqcup \sigma(v)$
6: $\quad \mathtt{Fail}_w \leftarrow \mathtt{Fail}_w \cup \mathtt{Sup}_w$ ▷ all pairs previously in $\mathtt{Sup}_w$ are now failures
7: $\quad \mathtt{Sup}_w \leftarrow \{(u, v)\}$
8: $\quad$ CHECKSUPPORTS(w) ▷ for $u \in L$, verify property $\mathtt{Sup}_{w \sqcup u}$
9: $\quad$ **while** $z \in L$ such that $(x, y) \in \mathtt{Fail}_z$ **do** ▷ some failures set not empty
10: $\quad\quad \mathtt{Fail}_z \leftarrow \mathtt{Fail}_z \backslash \{(x, y)\}$
11: $\quad\quad$ **if** $\sigma(x) \neq \sigma(x) \sqcap \sigma(z)$ **then**
12: $\quad\quad\quad \sigma(x) \leftarrow \sigma(x) \sqcap \sigma(z)$ ▷ $\sigma(x)$ decreases
13: $\quad\quad\quad \mathtt{Fail}_x \leftarrow \mathtt{Fail}_x \cup \mathtt{Sup}_x$ ▷ all pairs in $\mathtt{Sup}_x$ are now failures
14: $\quad\quad\quad \mathtt{Sup}_x \leftarrow \emptyset$
15: $\quad\quad\quad$ CHECKSUPPORTS(x) ▷ for $u \in L$, verify property $\mathtt{Sup}_{x \sqcup u}$
16: $\quad\quad$ **if** $\sigma(y) \neq \sigma(y) \sqcap \sigma(z)$ **then**
17: $\quad\quad\quad \sigma(y) \leftarrow \sigma(y) \sqcap \sigma(z)$ ▷ $\sigma(y)$ decreases
18: $\quad\quad\quad \mathtt{Fail}_y \leftarrow \mathtt{Fail}_y \cup \mathtt{Sup}_y$ ▷ all pairs in $\mathtt{Sup}_y$ are now failures
19: $\quad\quad\quad \mathtt{Sup}_y \leftarrow \emptyset$
20: $\quad\quad\quad$ CHECKSUPPORTS(y) ▷ for $u \in L$, verify property $\mathtt{Sup}_{y \sqcup u}$
21: $\quad\quad$ **if** $\sigma(x) \sqcup \sigma(y) = \sigma(z)$ **then**
22: $\quad\quad\quad \mathtt{Sup}_z \leftarrow \mathtt{Sup}_z \cup \{(x, y)\}$ ▷ (x, y) is now correct
23: $\quad\quad$ **else**
24: $\quad\quad\quad \mathtt{Con}_z \leftarrow \mathtt{Con}_z \cup \{(x, y)\}$ ▷ (x, y) is now a conflict

As for the previous algorithms in this paper the worst-time time complexity will be expressed in terms of the binary lattice operations performed during execution. Assume a fixed set S of size m. The complexity of the initialization (Line 1) of GMEET is $O(nm)$ with $n = |L|$. The value of σ for a given $w \in L$ can be updated (decreased) at most n times. Thus, there are at most n^2 updates of σ for all values of L. Finding a $w = u \sqcup v$ where $\sigma(w)$ needs an update because $\sigma(u) \sqcup \sigma(v) \neq \sigma(u \sqcup v)$ (test of the loop, Line 2) takes $O(n^2)$. Hence, the worst time complexity of the loop is in $O(n^4)$.

The program GMEET+ in Algorithm 4 uses appropriate data structures to reduce significantly the time complexity of the algorithm. Essentially, different sets are used to keep track of properties of (u, v) lattice pairs with respect to the current σ. We have a support (correct) pairs set $\mathtt{Sup}_w = \{(u, v) \mid w = u \sqcup v \wedge \sigma(u) \sqcup \sigma(v) = \sigma(w)\}$. We also have a conflicts set $\mathtt{Con}_w = \{(u, v) \mid w = u \sqcup v \wedge \sigma(u) \sqcup \sigma(v) \sqsubset \sigma(w)\}$ and failures set $\mathtt{Fail}_w = \{(u, v) \mid w = u \sqcup v \wedge \sigma(u) \sqcup \sigma(v) \not\sqsubseteq \sigma(w)\}$. Algorithm 4 updates σ as mentioned in Theorem 4 and so maintains the invariant $\sigma \sqsupseteq \bigsqcap_{\mathcal{E}(L)} S$. An additional invariant is that, for all w, sets $\mathtt{Sup}_w, \mathtt{Con}_w, \mathtt{Fail}_w$ are pairwise disjoint. When the outer loop finishes sets $\mathtt{Con}_w$ and $\mathtt{Fail}_w$ are empty (for all w) and thus every (u, v) belongs to $\mathtt{Sup}_{u \sqcup v}$, i.e. the resulting $\sigma = \bigsqcap_{\mathcal{E}(L)} S$.

Auxiliary procedure CHECKSUPPORTS(u) identifies all pairs of the form $(u, x) \in \mathtt{Sup}_{u \sqcup x}$ that may no longer satisfy the join-endomorphism property $\sigma(u) \sqcup \sigma(x) =$

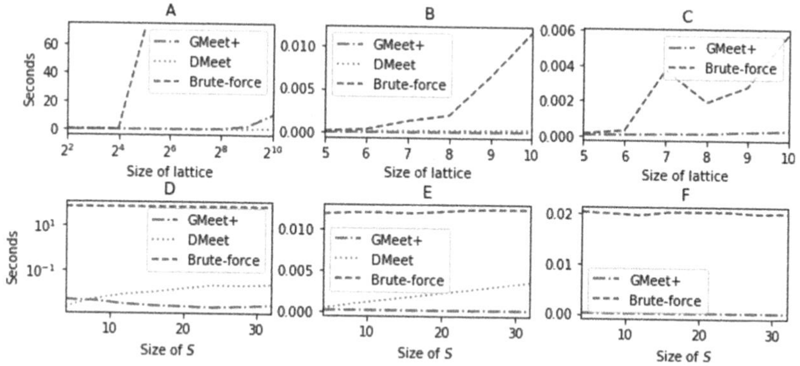

Fig. 2. Average performance time of GMEET+, DMEET and BRUTE-FORCE. Plots A and D use $\mathbf{2}^n$ lattices, B and E distributive lattices, and C and F arbitrary (possibly non-distributive) lattices. Plots A-C have a fixed number of join-endomorphisms and plots D-F have a fixed lattice size.

$\sigma(u \sqcup x)$ because of an update to $\sigma(u)$. When this happens, it adds (u, x) to the appropriate $\texttt{Con}$, or $\texttt{Fail}$ set. The time complexity of the algorithm depends on the set operations computed for each $w \in L$ chosen, either in the *conflicts* $\texttt{Con}_w$ set or in the *failures* $\texttt{Fail}_w$ set. When a w is selected (for some (u, v) such that $u \sqcup v = w$) the following holds: (1) at least one of $\sigma(w), \sigma(u), \sigma(v)$ is decreased, (2) some fix k number of elements are removed from or added to a set, (3) a union of two *disjoint* sets is computed, and (4) new support sets of w, u or v are calculated.

With an appropriate implementation, operations (1)–(2) take $O(1)$, and also operation (3), since sets are disjoint. Operation (4) clearly takes $O(n)$. In each loop of the (outer or inner) cycles of the algorithm, at least one σ reduction is computed. Furthermore, for each reduction of σ, $O(n)$ operations are performed. The maximum possible number of $\sigma(w)$ reductions, for a given w, is equal to the length d of the longest strictly decreasing chain in the lattice. The total number of possible σ reductions is thus equal to nd. The total number of operations of the algorithm is then $O(n^2 d)$. In general, d could be (at most) equal to n, therefore, after initialization, worst case complexity is $O(n^3)$. The initialization (Lines 1–2) takes $O(nm) + O(n^2)$, where $m = |S|$. Worst time complexity is thus $O(mn + n^3)$. For powerset lattices, $d = \log_2 n$, thus worst time complexity in this case is $O(mn + n^2 \log_2 n)$.

4.5 Experimental Results and Small Example

Here we present some experimental results showing the execution time of the proposed algorithms. We also discuss a small example with join-endomorphisms representing dilation operators from Mathematical Morphology [2]. We use the algorithms presented above to compute the greatest dilation below a given set of dilations and illustrate its result for a simple image.

Consider Fig. 2. In plots Fig. 2A–C, the horizontal axis is the size of the lattice. In plots Fig. 2D–F, the horizontal axis is the size of S. Curves in images Fig. 2A–C plot, for each algorithm, the average execution time of 100 runs (10 for Fig. 2A) with

Table 2. Average time in seconds over powerset lattices with $|S| = 4$

Size	A_1	A_2	GMEET	GMEET+	DMEET
16	2.01	0.958	0.00360	0.000603	0.000632
32	64.6	25.3	0.0633	0.00343	0.00181
64	1901	600	0.948	0.0154	0.00542
128	>600	>600	15.4	0.0860	0.0160
256	>600	>600	252	0.361	0.0483
512	>600	>600	>600	2.01	0.166
1024	>600	>600	>600	10.7	0.547

random sets $S \subseteq \mathcal{E}(L)$ of size 4. Images Fig. 2D–F, show the mean execution time of each algorithm for 100 runs (10 for Fig. 2D) varying the number of join-endomorphisms ($|S| = 4i$, $1 \leq i \leq 8$). The lattice size is fixed: $|L| = 10$ for Fig. 2E and F, and $|L| = 2^5$ for Fig. 2D. In all cases the lattices were randomly generated, and the parameters selected to showcase the difference between each algorithm with a sensible overall execution time. For a given lattice L and $S \subseteq \mathcal{E}(L)$, the brute-force algorithm explores the whole space $\mathcal{E}(L)$ to find all the join-endomorphism below each element of S and then computes the greatest of them. In particular, the measured spike in plot Fig. 2C corresponds to the random lattice of seven elements with the size of $\mathcal{E}(L)$ being bigger than in the other experiments in the same figure. In our experiments we observed that for a fixed S, as the size of the lattice increases, DMEET outperforms GMEET+. This is noticeable in lattices $\mathbf{2}^n$ (see Fig. 2A). Similarly, for a fixed lattice, as the size of S increases GMEET+ outperforms DMEET. GMEET+ performance can actually improve with a higher number of join-endomorphisms (see Fig. 2D) since the initial σ is usually smaller in this case.

To illustrate some performance gains, Table 2 shows the mean execution time of the algorithms discussed in this paper. We include A_1 and A_2, the algorithms outlined just after Lemma 1 and Proposition 5.

An MM Example. Mathematical morphology (MM) is a theory, based on topological, lattice-theoretical and geometric concepts, for the analysis of geometric structures. Its algebraic framework comprises [2, 14, 17], among others, complete lattices together with certain kinds of morphisms, such as *dilations*, defined as *join-endomorphisms* [14]. Our results give bounds about the number of all dilations over certain specific finite lattices and also efficient algorithms to compute their infima.

A typical application of MM is image processing. Consider the space $G = \mathbb{Z}^2$. A dilation [2] by $s_i \subseteq \mathcal{P}(G)$ is a function $\delta_{s_i} : \mathcal{P}(G) \to \mathcal{P}(G)$ such that $\delta_{s_i}(X) = \{x + e \mid x \in X \text{ and } e \in s_i\}$. The dilation $\delta_{s_i}(X)$ describes the interaction of an image X with the *structuring element* s_i. Intuitively, the dilation of X by s_i is the result of superimpose s_i on every activated pixel of X, with the center of s_i aligned with the corresponding pixel of X. Then, each pixel of every superimposed s_i is included in $\delta_{s_i}(X)$.

Let L be the powerset lattice for some finite set $D \subseteq G$. It turns out that the dilation $\bigsqcap_{\mathcal{E}(L)} S$ corresponds to the intersection of the structuring elements of the corresponding dilations in S. Figure 3 illustrates $\bigsqcap_{\mathcal{E}(L)} S$ for the two given dilations $\delta_{s_1}(I)$ and $\delta_{s_2}(I)$ with structuring elements s_1 and s_2 over the given image I.

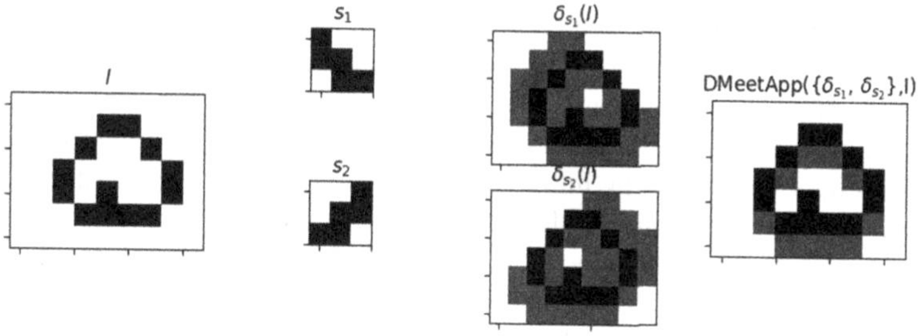

Fig. 3. Binary image I (on the left). Dilations δ_{s_1}, δ_{s_2} for structuring elements s_1, s_2. On the right $(\bigsqcap_{\mathcal{E}(L)}\{\delta_{s_1}, \delta_{s_2}\})(I)$. New elements of the image after each operation in grey and black.

5 Conclusions and Related Work

We have shown that given a lattice L of size n and a set $S \subseteq \mathcal{E}(L)$ of size m, $\bigsqcap_{\mathcal{E}(L)} S$ can be computed in the worst-case in $O(n + m\log n)$ binary lattice operations for powerset lattices, $O(mn^2)$ for lattices of sets, and $O(nm + n^3)$ for arbitrary lattices. We illustrated the experimental performance of our algorithms and a small example from mathematical morphology.

In [9] a bit-vector representation of a lattice is discussed. This work gives algorithms of logarithmic (in the size of the lattice) complexity for join and meet operations. These results count bit-vector operations. From [1] we know that $\mathcal{E}(L)$ is isomorphic to the downset of $(P \times P^{op})$, where P is the set of join-prime elements of L, and that this, in turn, is isomorphic to the set of order-preserving functions from $(P \times P^{op})$ to $\mathbf{2}$. Therefore, for the problem of computing $\bigsqcap_{\mathcal{E}(L)} S$, we get bounds $O(m \log_2(2^{(n^2)})) = O(mn^2)$ for set lattices and $O(m(\log_2 n)^2)$ for powerset lattices where $n = |L|$ and $m = |S|$. This, however, assumes a bit-vector representation of a lattice isomorphic to $\mathcal{E}(L)$. Computing this representation takes time and space proportional to the size of $\mathcal{E}(L)$ [9] which could be exponential as stated in the present paper. Notice that in our algorithms the input lattice is L instead of $\mathcal{E}(L)$.

We have stated the cardinality of the set of join-endomorphisms $\mathcal{E}(L)$ for significant families of lattices. To the best of our knowledge we are the first to establish the cardinality $(n+1)^2 + n!\mathcal{L}_n(-1)$ for the lattice $\mathbf{M}_n$. The cardinalities $n^{\log_2 n}$ for power sets (boolean algebras) and $\binom{2n}{n}$ for linear orders can also be found in the lattice literature [1,10,16]. Our original proofs for these statements can be found in the technical report of this paper [13].

The lattice $\mathcal{E}(L)$ have been studied in [6]. The authors showed that a finite lattice L is distributive iff $\mathcal{E}(L)$ is distributive. A lower bound of $2^{2n/3}$ for the number of monotonic self-maps of any finite poset L is given in [4]. Nevertheless to the best of our knowledge, no other authors have studied the problem of determining the size $\mathcal{E}(L)$ nor algorithms for computing $\bigsqcap_{\mathcal{E}(L)} S$. We believe that these problems are important, as argued in the Introduction; algebraic structures consisting of a lattice and join-endomorphisms are very common in mathematics and computer science. In fact, our interest in this subject arose in the algebraic setting of spatial and epistemic constraint systems [8] where continuous join-endomorphisms, called space functions, represent knowledge and the infima of endomorphisms correspond to distributed knowledge. We showed in [8] that distributed knowledge can be computed in $O(mn^{1+\log_2(m)})$ for distributive lattices and $O(n^4)$ in general. In this paper we have provided much lower complexity orders for computing infima of join-endomorphisms. Furthermore [8] does not provide the exact cardinality of the set of space functions of a given lattice.

As future work we plan to explore in detail the applications of our work in mathematical morphology and computer music [15]. Furthermore, in the same spirit of [11] we have developed algorithms to generate distributive and arbitrary lattices. In our experiments, we observed that for every lattice L of size n we generated, $n^{\log_2 n} \le |\mathcal{E}(L)| \le (n+1)^2 + n!\mathcal{L}_n(-1)$ and if the generated lattice was distributive, $n^{\log_2 n} \le |\mathcal{E}(L)| \le \binom{2n}{n}$. We plan to establish if these inequalities hold for every finite lattice.

Acknowledgments. We are indebted to the anonymous referees and editors of RAMICS 2020 for helping us to improve one of the complexity bounds, some proofs, and the overall quality of the paper.

References

1. Birkhoff, G.: Lattice Theory, vol. 25, p. 2. American Mathematical Society Colloquium, American Mathematical Society (1967)
2. Bloch, I., Heijmans, H., Ronse, C.: Mathematical morphology. In: Aiello, M., Pratt-Hartmann, I., Van Benthem, J. (eds.) Handbook of Spatial Logics, pp. 857–944. Springer, Dordrecht (2007). https://doi.org/10.1007/978-1-4020-5587-4_14
3. Davey, B.A., Priestley, H.A.: Introduction to Lattices and Order, 2nd edn. Cambridge University Press, Cambridge (2002)
4. Duffus, D., Rodl, V., Sands, B., Woodrow, R.: Enumeration of order preserving maps. Order **9**(1), 15–29 (1992)
5. Gierz, G., Hofmann, K.H., Keimel, K., Lawson, J.D., Mislove, M., Scott, D.S.: Continuous Lattices and Domains. Cambridge University Press, Cambridge (2003)
6. Grätzer, G., Schmidt, E.: On the lattice of all join-endomorphisms of a lattice. Proc. Am. Math. Soc. **9**, 722–722 (1958)
7. Guzmán, M., Haar, S., Perchy, S., Rueda, C., Valencia, F.D.: Belief, knowledge, lies and other utterances in an algebra for space and extrusion. J. Log. Algebr. Meth. Program. **86**(1), 107–133 (2017)
8. Guzmán, M., Knight, S., Quintero, S., Ramírez, S., Rueda, C., Valencia, F.D.: Reasoning about distributed knowledge of groups with infinitely many agents. In: 30th International Conference on Concurrency Theory, CONCUR 2019, vol. 29, pp. 1–29 (2019)
9. Habib, M., Nourine, L.: Tree structure for distributive lattices and its applications. Theor. Comput. Sci. **165**(2), 391–405 (1996)

10. Jipsen, P.: Relation algebras, idempotent semirings and generalized bunched implication algebras. In: Höfner, P., Pous, D., Struth, G. (eds.) RAMICS 2017. LNCS, vol. 10226, pp. 144–158. Springer, Cham (2017). https://doi.org/10.1007/978-3-319-57418-9_9
11. Jipsen, P., Lawless, N.: Generating all finite modular lattices of a given size. Algebra universalis **74**(3), 253–264 (2015). https://doi.org/10.1007/s00012-015-0348-x
12. Knight, S., Palamidessi, C., Panangaden, P., Valencia, F.D.: Spatial and epistemic modalities in constraint-based process calculi. In: Koutny, M., Ulidowski, I. (eds.) CONCUR 2012. LNCS, vol. 7454, pp. 317–332. Springer, Heidelberg (2012). https://doi.org/10.1007/978-3-642-32940-1_23
13. Quintero, S., Ramírez, S., Rueda, C., Valencia, F.D.: Counting and computing join-endomorphisms in lattices. Research report, LIX, Ecole polytechnique; INRIA Saclay - Ile-de-France (2019). https://hal.archives-ouvertes.fr/hal-02422624
14. Ronse, C.: Why mathematical morphology needs complete lattices. Sig. Process. **21**(2), 129–154 (1990)
15. Rueda, C., Valencia, F.: On validity in modelization of musical problems by CCP. Soft. Comput. **8**(9), 641–648 (2004)
16. Santocanale, L.: On discrete idempotent paths. In: Mercaş, R., Reidenbach, D. (eds.) WORDS 2019. LNCS, vol. 11682, pp. 312–325. Springer, Cham (2019). https://doi.org/10.1007/978-3-030-28796-2_25
17. Stell, J.: Why mathematical morphology needs quantales. In: Wilkinson, M., Roerdink, J. (eds.) International Symposium on Mathematical Morphology, ISMM09, pp. 13–16. Institute for Mathematics and Computing Science, University of Groningen (2009)

A Unary Semigroup Trace Algebra

Pedro Ribeiro(✉)

Department of Computer Science, University of York, York YO10 5GH, UK
pedro.ribeiro@york.ac.uk

Abstract. The Unifying Theories of Programming (UTP) of Hoare and He promote the unification of semantics catering for different concerns, such as, termination, data modelling, concurrency and time. Process calculi like *Circus* and CSP can be given semantics in the UTP using reactive designs whose traces can be abstractly specified using a monoid trace algebra. The prefix order over traces is defined in terms of the monoid operator. This order, however, is inadequate to characterise a broader family of timed process algebras whose traces are preordered instead. To accommodate these, we propose a unary semigroup trace algebra that is weaker than the monoid algebra. This structure satisfies some of the axioms of restriction semigroups and is a right P-Ehresmann semigroup. Reactive designs specified using it satisfy core laws that have been mechanised so far in Isabelle/UTP. More importantly, our results improve the support for unifying trace models in the UTP.

Keywords: Semantics · Process algebra · Semigroups · UTP

1 Introduction

The Unifying Theories of Programming (UTP) [1] is a relational framework for characterising different programming paradigms. It promotes the unification of semantics, while allowing different aspects, such as data, concurrency, termination, time, and so on, to be considered individually. Programs are specified via alphabetised relations in the style of Hehner's Predicative Programming [2].

Behaviour, and concurrency, in the style of CCS, ACP and CSP [3], can be defined in the UTP via the theory of reactive processes. At its core is the notion of traces, that is, sequences of events that record the history of interactions.

The time dimension has been considered in different ways [4–7]. In [4], for example, traces are sequences of pairs that encode discrete time units. The first component of a pair records the sequence of events performed during that time, and the second component the set of events refused at that point. In [7], which presents a theory that can be used to give semantics in the UTP to the hardware programming language Handel-C [8] and other synchronous languages, a more abstract view is provided by a parametric model. It requires that the operators for addition and subtraction of pairs, and traces, satisfy a set of axioms.

Semantic models employing traces typically define a prefix relation $\leq$ that specifies how a trace can be augmented, encoding some notion of causality. The

U. Fahrenberg et al. (Eds.): RAMiCS 2020, LNCS 12062, pp. 270–285, 2020.
https://doi.org/10.1007/978-3-030-43520-2_17

semantics for CSP, for example, is defined using sequences whose prefix relation is a partial order. In that setting, a trace s is a prefix of t, written $s \leq t$, exactly when $s \leq t \Leftrightarrow \exists\, u \bullet s \frown u = t$, that is, there exists a trace u, such that ($\bullet$) the concatenation ($\frown$) of s with u is t. This led Foster et al. [9] to observe that the prefix order for several trace models can be abstractly defined in terms of a left-cancellative monoid, henceforth referred to as the "monoid trace algebra".

In [7], however, a pair (s, r_0) is a prefix of (t, r_1) exactly when $s \leq t$, but the refusal sets r_0 and r_1 are not constrained. Anti-symmetry is thus not satisfied and so the prefix relation on traces is merely a pre-order. The monoid trace algebra is unsatisfactory for such a theory of synchronous languages. Solving this problem is not only of theoretical interest to establish the commonality between different trace structures, but more importantly enables key results to be reusable in synchronous process algebra, thus promoting unification of models and results, a key goal of the UTP.

Unification in the UTP can be exploited in various ways, namely via subset embeddings, weakest completion semantics [10], Galois connections and parametric theories. The approach pursued in this paper is a contribution to the latter by generalising the theory of reactive processes even further. The main contribution is the definition of a unary left-cancellative semigroup, obtained by introducing a unary function and weakening of the monoid trace algebra axioms. The pairs from [7] are shown to satisfy this structure, as are finite sequences (traces) of such pairs. There is surprisingly little impact on the proofs already established for reactive processes as demonstrated by the mechanisation of our results in Isabelle/UTP [11].

The paper is structured as follows. In Sect. 2 the theory of reactive processes is introduced, as well as the monoid trace algebra. Our unary semigroup trace algebra is defined in Sect. 3. Our theory of synchronous algebra is characterised in Sect. 4. In Sect. 5 we discuss related work. In Sect. 6 we summarize the main results and provide pointers for future work.

2 Preliminaries

In the UTP programs are specified by alphabetised relations. Variables are used to define computations, with undashed variables (x) capturing the initial value, and dashed variables (x') capturing the later, or final, value. These can be program variables, or auxiliar variables that capture information such as termination or execution time. A UTP theory is characterised by three components: an alphabet, a set of healthiness conditions, and a set of operators.

For example, in a theory of discrete time we may have variables t and t' of type $\mathbb{N}$ to record time. The relation $t' = t + 1$ describes a computation whereby time is incremented by one time unit. To define the set of valid time-monotonic computations, a function $\mathbf{HC}(P) \mathrel{\widehat=} P \wedge t \leq t'$ on predicates can be defined ($\widehat=$), so that the set of healthy predicates are the fixed points of $\mathbf{HC}$. When the healthiness conditions are idempotent and monotonic, with respect to the refinement order $\sqsubseteq$, their image forms a complete lattice, which allows reasoning about recursion.

The theory of reactive processes uses the auxiliary variables *ok* and *ok′* to capture stability, *wait* and *wait′* to record information about termination, *tr* and *tr′* to record the history of interactions with the environment, and *ref* and *ref′* to record the possibility of refusing interaction. The variable *ok* indicates whether the previous process is in a stable state, while *ok′* records this information for the current process. Similarly, *wait* records termination for the previous process and *wait′* for the current process. A process only starts executing in a state where *ok* and $\neg$ *wait* are *true*. Termination occurs when *ok′* and $\neg$ *wait′* are *true*.

The interactions with the environment are captured by sequences of events, recorded by *tr* and *tr′*. The variable *tr* records the sequence of events that took place before the current process started, while *tr′* records all the events that have been observed so far. Finally, *ref* and *ref′* record the set of events that may be refused by the process at the start, and currently.

In the theory of synchronous algebra, as already said, *tr* and *tr′* are sequences of pairs, where the first component is a sequence of events, and the second is a set of events that may be refused. The variables *ref* and *ref′* are not used.

2.1 Monoid Trace Algebra

To conciliate different trace structures for reactive processes, Foster et al. [9] propose a trace algebra, where *tr* and *tr′* are of an abstract type $\mathcal{T}$. Below we reproduce its axioms, where $^\frown$ is concatenation, and ε is the empty trace.

Definition 1 (TA). *A trace algebra* $(\mathcal{T}, {}^\frown, \varepsilon)$ *is a monoid satisfying the axioms:*

$$x \,^\frown (y \,^\frown z) = (x \,^\frown y) \,^\frown z \quad \text{(TA1)} \qquad x \,^\frown y = x \,^\frown z \Rightarrow y = z \quad \text{(TA3)}$$

$$\varepsilon \,^\frown x = x \,^\frown \varepsilon = x \quad \text{(TA2)} \qquad x \,^\frown y = \varepsilon \Rightarrow x = \varepsilon \quad \text{(TA4)}$$

Concatenation is associative (TA1), has the empty trace ε as both a left and right unit (TA2), and is left-cancellative (TA3). Axiom TA4 eliminates "negative traces" by requiring that whenever the concatenation of x and y is the empty trace then x must also be the empty trace. The dual law $x \,^\frown y = \varepsilon \Rightarrow y = \varepsilon$ can be deduced from axioms TA2 and TA4. We observe that while in [9] right-cancellation is also proposed as an axiom, the laws of the algebra, as well as the results established for the theory of reactive processes, as proved so far in Isabelle/UTP[1], do not depend on this axiom, and so we can safely omit it.

Standard finite sequences, for example, $(\operatorname{seq} A, {}^\frown, \langle\rangle)$ form a trace algebra, where $^\frown$ is sequence concatenation and $\langle\rangle$ is the empty sequence. Using the two trace algebra operators it is possible to define a trace prefix relation $(x \leq y)$ and a trace subtraction operator $(x - y)$ as reproduced below.

Definition 2 (Trace prefix). $x \leq y \mathrel{\widehat=} (\exists\, z \bullet x \,^\frown z = y)$

Definition 3 (Subtraction). $y - x \mathrel{\widehat=} \begin{cases} \iota z \bullet y = x \,^\frown z & \textit{if } \; x \leq y \\ \varepsilon & \textit{otherwise} \end{cases}$

[1] https://github.com/isabelle-utp (definitions and lemmas hyper-linked using).

A trace x is a prefix of y ($x \leq y$) whenever y can be obtained by concatenating x with some trace z. When $x \leq y$ the subtraction $y - x$ is z whose concatenation with x is y, specified using the definite description operator (ι), as z is unique by TA3, and otherwise $y - x$ is ε so that subtraction is total. In [9] it is shown that $(\mathcal{T}, \leq)$ is a partial order, and that ε is the least element. As mentioned, this is unsuitable for the synchronous algebra, so in Sect. 3 we pursue a preorder.

2.2 Generalised Reactive Processes

Using the trace algebra, it is possible to define the healthiness conditions that underpin several theories based on reactive processes, reproduced below.

Definition 4 (Generalised Reactive Processes).

$$\mathbf{R1}(P) \mathrel{\widehat{=}} tr \leq tr' \qquad \mathbf{R2}(P) \mathrel{\widehat{=}} P[\varepsilon, tr' - tr/tr, tr']$$

$$\mathbf{R2c}(P) \mathrel{\widehat{=}} \mathbf{R2}(P) \lhd tr \leq tr' \rhd P \qquad \mathbf{R2a}(P) \mathrel{\widehat{=}} \bigsqcap z \bullet P[z, z \frown (tr' - tr)/tr, tr']$$

$$\mathbf{R3}(P) \mathrel{\widehat{=}} I\!I \lhd wait \rhd P \qquad \mathbf{R}(P) \mathrel{\widehat{=}} \mathbf{R1} \circ \mathbf{R2} \circ \mathbf{R3}(P)$$

R1 requires that a trace can only be extended. **R2** requires processes to be insensitive to the initial trace and is specified by substituting tr in P with the empty trace ε and tr' with the difference $tr' - tr$. Because this difference is only well-defined when $tr \leq tr'$, the version **R2c** proposed by Foster et al. [9] applies **R2** conditionally: $P \lhd c \rhd Q$ is P if c is true, and otherwise is Q. **R2a**, defined using the greatest lower bound $\bigsqcap$, is an alternative for **R2** having the same fixed points. **R3** ensures that P may only start if the previous process has terminated ($\neg\, wait$), and otherwise behaves as the identity $I\!I$, which keeps variables unchanged. This ensures that relational composition is sequential composition. Finally, the theory is characterised by **R**, the composition of all conditions.

While these definitions are applicable to several reactive theories, **R2** and **R2a**, for example, cannot be instantiated for synchronous algebra [7], whose counterparts to **R2** and **R2a** are reproduced below with subscript $_S$. Concatenation ($\frown_S$) and subtraction ($-_S$) of their traces are also annotated with $_S$.

Definition 5.

$$\mathbf{R2}_S(P) = P[\langle(\langle\rangle, snd(last(tr)))\rangle, tr' -_S tr/tr, tr']$$

$$\mathbf{R2a}_S(P) = \bigsqcap z \bullet P[z, z \frown_S (tr' -_S tr)/tr, tr'] \land snd(last(tr)) = snd(last(z))$$

$\mathbf{R2}_S$ considers the substitution of tr with a sequence whose only element is a pair: the first component is the empty sequence, and the second component is the set of events resulting from taking the second component (snd) of the pair extracted from the $last$ element of tr, well-defined when **R1** is applied first.

Clearly the empty trace (ε) of the monoid trace algebra cannot abstractly encode an element that can take several values, such as $snd(last(tr))$. On the other hand, an examination of the algebraic laws satisfied by $\mathbf{R2}_S$, and counterparts to **R1** and **R3** in [5,7], reveals a striking similarity with the laws established for generalised reactive designs, which indicates a similar unification is feasible as we demonstrate in Sect. 4 using the algebra we define next.

3 Unary Semigroup Trace Algebra

Instead of a fixed empty trace ε, we introduce a total function $\Phi : \mathcal{T} \to \mathcal{T}$ to obtain a unary semigroup $(\mathcal{T}, ^\frown, \Phi)$. The axioms are defined next in Sect. 3.1. In Sect. 3.2 we classify it according to the literature on semigroups. In Sect. 3.3 we show that the prefix relation is a preorder, and redefine subtraction.

3.1 Axioms

The following axioms can be seen as counterparts to that of the monoid trace algebra, adapted to consider Φ and the fact that the structure is not a monoid.

Definition 6 (USTA). *A unary semigroup trace algebra* $(\mathcal{T}, ^\frown, \Phi)$ *is a left-cancellative unary semigroup satisfying the following axioms:*

$$x ^\frown (y ^\frown z) = (x ^\frown y) ^\frown z \quad \text{(USTA1)} \qquad x ^\frown y = x ^\frown z \Rightarrow y = z \quad \text{(USTA3)}$$

$$x ^\frown \Phi(x) = x \quad \text{(USTA2)} \qquad x ^\frown y = \Phi(y) \Rightarrow y = \Phi(y) \quad \text{(USTA4)}$$

Concatenation is associative (USTA1) so that we have a semigroup. Similarly to axiom TA2, we require that $\Phi(x)$ is a right identity with respect to concatenation with x (USTA2). Concatenation is also left-cancellative (USTA3). From these three axioms we can establish that $\Phi(x)$ is a left-unit for concatenation.

Lemma 1. $\Phi(x) ^\frown y = y$

Proof.

$$
\begin{aligned}
& x ^\frown y = x ^\frown y && \text{[Axiom USTA2]} \\
& \equiv (x ^\frown \Phi(x)) ^\frown y = x ^\frown y && \text{[Axioms USTA1 and USTA3]} \\
& \Rightarrow \Phi(x) ^\frown y = y && \square
\end{aligned}
$$

Similarly to axiom TA4 of the monoid trace algebra, axiom USTA4 also eliminates "negative traces", but when we draw a parallel between ε and Φ, the shape of USTA4 is different. The requirement on the second operand y of the concatenation $x ^\frown y$ (rather than the first operand x as in axiom TA4) is sufficiently weak to ensure the prefix relation $\leq$, defined in terms of $^\frown$, is not anti-symmetric.

To illustrate that axiom TA4 admits structures whose prefix relation $\leq$ is not anti-symmetric, we consider the following example.

Example 1. Consider $(\mathcal{S}, \gg, id)$, where $\mathcal{S}$ contains at least two distinct elements, $x \gg y \mathrel{\widehat{=}} y$ and id is the identity function. Such structure is a unary semigroup trace algebra. We show that $\exists\, a, b : \mathcal{S} \bullet a \leq b \land b \leq a \land a \neq b$.

Proof.

$$
\begin{aligned}
& \exists\, a, b : \mathcal{S} \bullet a \leq b \land b \leq a \land a \neq b && \text{[Definition of } \leq\text{]} \\
& = \exists\, a, b : \mathcal{S} \bullet (\exists\, z \bullet a \gg z = b) \land (\exists\, z \bullet b \gg z = a) \land a \neq b && \text{[Definition of } \gg\text{]} \\
& = \exists\, a, b : \mathcal{S} \bullet (\exists\, z \bullet z = b) \land (\exists\, z \bullet z = a) \land a \neq b && \text{[One point rule]} \\
& = \exists\, a, b : \mathcal{S} \bullet a \neq b && \text{[Assumption]} \\
& = \mathit{true} && \square
\end{aligned}
$$

An interesting generalisation of $(\mathcal{S}, \gg, id)$ is that $(\mathcal{T}, ^\frown, id)$ satisfies the axioms of a U-semigroup [12, p. 102]. In general, Φ is idempotent as we establish next.

Lemma 2. $\Phi(\Phi(x)) = \Phi(x)$

Proof. Using Axiom USTA2.

$$\begin{array}{lr} \Phi(x) ^\frown \Phi(\Phi(x)) = \Phi(x) & \text{[Lemma 1]} \\ \equiv \Phi(x) ^\frown \Phi(\Phi(x)) = \Phi(x) ^\frown \Phi(x) & \text{[Axiom USTA3]} \\ \Rightarrow \Phi(\Phi(x)) = \Phi(x) & \square \end{array}$$

Moreover, if Φ is constant we can obtain the original monoid trace algebra by having $\forall x \bullet \Phi(x) = \varepsilon$.

Theorem 1. *Provided* $\forall x \bullet \Phi(x) = \varepsilon$, *and* $(\mathcal{T}, ^\frown, \Phi)$ *is a unary semigroup trace algebra, then* $(\mathcal{T}, ^\frown, \varepsilon)$ *is a monoid trace algebra.*

Proof. Axioms TA1, TA2 and TA3 are trivially satisfied. Axiom TA4 can be satisfied by deduction using USTA2 and USTA4. □

Thus, the monoid trace algebra can be seen as a specialisation of the algebraic structure we propose. This and other results to follow have been mechanised in Isabelle[2]. Moreover, we have used Isabelle's counter-example generator **nitpick** to ascertain that axioms USTA1-USTA4 are independent. Next we discuss how the new structure can be classified according to the literature on semigroups.

3.2 Semigroup Properties

To establish key properties of the algebra, we first propose a lemma that is used in proofs to follow. The application of Φ to a trace obtained by concatenating x and y is equal to $\Phi(y)$ as stated in the lemma below.

Lemma 3. $\Phi(x ^\frown y) = \Phi(y)$

Proof. Using Axiom USTA2.

$$\begin{array}{lr} (x ^\frown y) ^\frown \Phi(x ^\frown y) = x ^\frown y & \textit{[Axioms USTA1 and USTA2]} \\ \equiv (x ^\frown y) ^\frown \Phi(x ^\frown y) = (x ^\frown y) ^\frown \Phi(y) & \textit{[Axiom USTA3]} \\ \Rightarrow \Phi(x ^\frown y) = \Phi(y) & \square \end{array}$$

From Lemmas 1 and 3 we can deduce that Φ distributes over $^\frown$, a property implicitly satisfied by left-cancellative restriction semigroups [13].

The structure is neither a left nor a right-restriction semigroup, as it satisfies only two (LR1 and LR2) out of four axioms [13] of left restriction semigroups, and three (RR1 to RR3) out of four axioms of right-restriction semigroups.

[2] https://github.com/isabelle-utp/utp-main/tree/ramics2020s.

Theorem 2 (Laws of restriction semigroups).

$$\Phi(x) \frown x = x \qquad \text{(LR1)}$$
$$\Phi(\Phi(x) \frown y) = \Phi(x) \frown \Phi(y) \qquad \text{(LR2)}$$

$$x \frown \Phi(x) = x \qquad \text{(RR1)}$$
$$\Phi(x \frown \Phi(y)) = \Phi(x) \frown \Phi(y) \qquad \text{(RR2)}$$
$$\Phi(x) \frown y = y \frown \Phi(x \frown y) \qquad \text{(RR3)}$$

Proof. (LR1) Using Lemma 1*; (RR1) using Axiom USTA2.*
(LR2) Using Lemmas 1 and 3*; (RR2) using, in addition, Lemma* 2.
(RR3)

$$\begin{aligned} & y \frown \Phi(x \frown y) && \text{[Lemma 3]} \\ &= y \frown \Phi(y) && \text{[Axiom USTA2 and Lemma 1]} \\ &= \Phi(x) \frown y \end{aligned}$$

□

A fourth axiom of restriction semigroups requires commutativity on the application of Φ with respect to $\frown$ as $\Phi(x) \frown \Phi(y) = \Phi(y) \frown \Phi(x)$. It is clear from Lemma 1 that this equality cannot hold. We have, however, that the structure satisfies the axioms of right P-Ehresmann semigroups [14], as established next.

Theorem 3. *$(\mathcal{T}, \frown, \Phi)$ is a right P-Ehresmann semigroup.*

$$x \frown \Phi(x) = x \qquad \text{(PE1)}$$
$$\Phi(x \frown y) = \Phi(\Phi(x) \frown y) \qquad \text{(PE2)}$$
$$\Phi(\Phi(x) \frown \Phi(y)) = \Phi(y) \frown \Phi(x) \frown \Phi(y) \qquad \text{(PE3)}$$
$$\Phi(x) \frown \Phi(x) = \Phi(x) \qquad \text{(PE4)}$$

Proof. (PE1) Using Axiom USTA2; (PE2) using Lemmas 1 *and* 3.
(PE3)

$$\begin{aligned} & \Phi(y) \frown \Phi(x) \frown \Phi(y) && \text{[Lemma 1]} \\ &= \Phi(x) \frown \Phi(y) && \text{[Lemma 2]} \\ &= \Phi(\Phi(x)) \frown \Phi(\Phi(y)) && \text{[Lemmas 1 and 3]} \\ &= \Phi(\Phi(x) \frown \Phi(y)) \end{aligned}$$

(PE4) Follows from Lemma 1. □

Despite proposing axioms based on a generalisation of those of the monoid trace algebra, it is pleasing to find that such a construction satisfies the axioms of a known class of semigroups. Next we study the induced prefix relation $\leq$ of the algebra, defined in terms of $\frown$, and its subtraction operator $-$.

3.3 Prefix and Subtraction

The prefix relation can be characterised exactly as in Definition 2. In what follows we study its key algebraic properties, starting by showing that it is a preorder.

Theorem 4. *Provided $(\mathcal{T}, \frown, \Phi)$ is a USTA then $(\mathcal{T}, \leq)$ is a preorder. (TP1)*

Proof. (Reflexivity) Using Definition 2 and Axiom USTA2; (Transitivity) Using Definition 2, Axiom USTA1 and predicate calculus.

Moreover, we have that $\leq$ satisfies the following laws, numbered to mirror the laws TP1-TP4 of the monoid trace algebra [9].

Theorem 5 (Trace Prefix Laws).

$$\Phi(x) \leq y \quad \text{(TP2)} \qquad x \leq x ^\frown y \quad \text{(TP3)} \qquad x ^\frown y \leq x ^\frown z \Leftrightarrow y \leq z \quad \text{(TP4)}$$

Trace $\Phi(x)$ is smaller than any other trace (TP2). Law TP3 states that concatenation constructs larger traces, and Law TP4 states that concatenation is monotonic in its right argument. Next, we introduce the subtraction operator.

Definition 7 (Subtraction). $y - x \mathrel{\widehat{=}} \begin{cases} \iota z \bullet y = x ^\frown z & \textit{if} \quad x \leq y \\ \Phi(x) & \textit{otherwise} \end{cases}$

Subtraction is defined like in Definition 3 when $x \leq y$, and otherwise is defined as $\Phi(x)$. This deliberate choice of $\Phi(x)$ is essential to ensure that the following laws TS1-TS10 (numbered after the laws TS1-TS8 in [9] as counterparts) hold. Notably absent from the following list is the counterpart to TS2 of the monoid trace algebra, which we discuss in the sequel. It does not hold in this setting, but this bears no impact on the results established for the reactive theory.

Theorem 6 (Trace Subtraction Laws).

$$x - \Phi(y) = x \tag{TS1}$$
$$x - x = \Phi(x) \tag{TS3}$$
$$(x ^\frown y) - x = y \tag{TS4}$$
$$(x - y) - z = x - (y ^\frown z) \tag{TS5}$$
$$(x ^\frown y) - (x ^\frown z) = y - z \tag{TS6}$$
$$y \leq x \land x - y = \Phi(y) \Leftrightarrow x = y \tag{TS7}$$
$$x \leq y \Rightarrow x ^\frown (y - x) = y \tag{TS8}$$
$$x \leq y \land x \leq z \Rightarrow (y - x = z - x \Leftrightarrow y = z) \tag{TS9}$$
$$x \leq y \land x \leq z \land z \leq y \Rightarrow (y - x) - (z - x) = y - z \tag{TS10}$$

Law TS1 states that the subtraction of a trace $\Phi(y)$ from another trace is ineffective. Law TS3 states that subtracting a trace from itself is equal to applying Φ. Laws TS4-TS6 and TS8 capture expected properties of concatenation and subtraction, also satisfied by the monoid trace algebra. The implication in Law TS7 states that if the subtraction $x - y$ is $\Phi(y)$, and y is a prefix of x, then x and y are the same. The reverse implication follows from Law TS3. The novel laws TS9-TS10 correspond to axioms SSub:same and SSub:subsub in [7, pp. 95–96].

The counterpart to Law TS2 ($\varepsilon - x = \varepsilon$) of the monoid trace algebra [9] could be stated in this setting as $\Phi(y) - x = \Phi(y)$. However, this equality does not hold in general. In particular, for example, if $x \leq \Phi(y)$ holds it is not necessarily the case that $(\iota z \bullet \Phi(y) = x ^\frown z) = \Phi(y)$. Existing proofs for reactive processes do not depend on Law TS2, so the fact that it does not hold in our trace algebra has no practical impact. Next, we focus on instances of the algebra.

4 Trace Models

In this section we focus on instances of our trace algebra, and show that it can be instantiated to yield the traces of [7]. To that end, we consider pairs in Sect. 4.1 whose first component is a USTA. Then in Sect. 4.2 we consider these pairs as elements of finite non-empty sequences and show the lifted structure is a USTA.

4.1 Parametric Pairs

We introduce pairs $\mathcal{P} : \mathcal{H} \times \mathcal{R}$ parametrised by types $\mathcal{H}$ and $\mathcal{R}$, whose $\mathcal{H}$ must be a USTA $(\mathcal{H}, +_{\mathcal{H}}, \Phi_{\mathcal{H}})$ where $+_{\mathcal{H}} : \mathcal{H} \times \mathcal{H} \to \mathcal{H}$ is concatenation, and $\Phi_{\mathcal{H}} : \mathcal{H} \to \mathcal{H}$ is the unary function of the USTA. To construct a USTA for parametric pairs $(\mathcal{P}, +_{\mathcal{P}}, \Phi_{\mathcal{P}})$, we define concatenation of pairs $(+_{\mathcal{P}})$ and $\Phi_{\mathcal{P}}$ as follows.

Definition 8. $(h_1, r_1) +_{\mathcal{P}} (h_2, r_2) = (h_1 +_{\mathcal{H}} h_2, r_2)$

$$\Phi_{\mathcal{P}}(h_1, r_1) = (\Phi_{\mathcal{H}}(h_1), r_1)$$

Concatenation of (h_1, r_1) and (h_2, r_2) is a pair where: the first component is the result of applying $+_{\mathcal{H}}$ to h_1 and h_2, and the second component is r_2. $\Phi_{\mathcal{P}}$ is defined as the application of $\Phi_{\mathcal{H}}$ to the first component. The definition of $+_{\mathcal{P}}$ closely follows the concatenation specified in [7]. However, unlike [7] we do not need to specify subtraction, as instead it can be derived as a lemma below.

Lemma 4. *Provided* $h_2 \leq h_1$, $(h_1, r_1) - (h_2, r_2) = (h_1 - h_2, r_1)$.

With the above construction we can establish that $(\mathcal{P}, +_{\mathcal{P}}, \Phi_{\mathcal{P}})$ is a USTA.

Theorem 7. *Provided* $(\mathcal{H}, +_{\mathcal{H}}, \Phi_{\mathcal{H}})$ *is a USTA then* $(\mathcal{P}, +_{\mathcal{P}}, \Phi_{\mathcal{P}})$ *is a USTA.*

Thus, the pairs of [7] form a USTA. Next, we consider a model for traces constructed from finite non-empty sequences whose elements are pairs of type $\mathcal{P}$.

4.2 Synchronous Traces

As already mentioned, the traces of synchronous process algebra consist of non-empty sequences of pairs [7]. In this section we construct this abstract trace structure stepwise, starting by defining a specialised model of finite non-empty sequences that is a USTA. This is then used to lift pairs of type $\mathcal{P}$ to traces.

Traces. A trace in this setting is a finite non-empty sequence defined via a recursive data type *fs* below, specified using the Z [15] notation for type constructors.

Definition 9. $fs ::= One \langle\langle \sigma \rangle\rangle \mid Cons \langle\langle \sigma \times fs \rangle\rangle$

One constructs a sequence with a single element of type σ, and *Cons* constructs a sequence where an element is followed by a sequence of type *fs*. We use angled brackets $\langle a_0, ..., a_n \rangle_{fs}$ to represent consecutive applications of *Cons*, ending in *One* a_n, and $\langle a_0 \rangle_{fs}$ for a single construction *One* a_0. The subscript $_{fs}$ distinguishes finite non-empty sequences from standard finite sequences (that may be empty).

To construct a USTA for an *fs* parametrised by a given type σ that is a USTA $(\sigma, \frown_\sigma, \Phi_\sigma)$, we need to instantiate the respective structure $(fs, \frown_{fs}, \Phi_{fs})$ in terms of $\frown_\sigma$ and Φ_σ. We define concatenation $(\frown_{fs})$ next, and Φ_{fs} in the sequel.

Definition 10 (Concatenation of non-empty sequences).

$$\_ \frown_{fs} \_ : fs \times fs \to fs$$

$$\begin{aligned} & One\,x \frown_{fs} One\,y = One\,(x +_\sigma y) \\ \forall\, x, y : \sigma;\ f, g : fs \bullet\ & One\,x \frown_{fs} Cons\,(y, f) = Cons\,(x +_\sigma y, f) \\ & Cons\,(x, f) \frown_{fs} g = Cons\,(x, f \frown_{fs} g) \end{aligned}$$

The concatenation of two sequences $\langle x \rangle_{fs}$ and $\langle y \rangle_{fs}$, with one element each, is a sequence whose only element is the result of the sum $(+_\sigma)$ of x and y. A sequence $\langle x \rangle_{fs}$ concatenated with $\langle a_0, ..., a_n \rangle_{fs}$ is defined as $\langle x +_\sigma a_0, ..., a_n \rangle_{fs}$, that is, the first element is the sum $(+_\sigma)$ of x and the first element a_0 of the second sequence. Finally, a sequence $\langle a_0, ..., a_n \rangle_{fs}$ concatenated with g has a_0 as first element followed by the concatenation of the tail of that sequence with g.

We observe that $\frown_{fs}$ is distinctive from standard sequence concatenation, so as to induce an appropriate definition for prefixing and subtraction (Definitions 2 and 7). For example, the subtraction of $\langle a \rangle_{fs}$ from itself is the sequence z whose concatenation with $\langle a \rangle_{fs}$ yields $\langle a \rangle_{fs}$ $(\iota z \bullet \langle a \rangle_{fs} = \langle a \rangle_{fs} \frown_{fs} z)$. Because *fs* sequences are non-empty, z is the sequence $\langle \Phi_\sigma(a) \rangle_{fs}$ so that $\langle a +_\sigma \Phi_\sigma(a) \rangle_{fs} = \langle a \rangle_{fs}$, as required. This in contrast to subtraction of standard sequences, where $\langle a \rangle - \langle a \rangle = \langle \rangle$. Similar reasoning applies to ensure $\leq$ is reflexive.

Indeed to show that $(fs, \frown_{fs}, \Phi_{fs})$ is a USTA given a type σ that is a USTA $(\sigma, +_\sigma, \Phi_\sigma)$, we define Φ_{fs} in terms of Φ_σ as follows.

Definition 11. $\Phi_{fs}(x) = \langle \Phi_\sigma(last(x)) \rangle_{fs}$

It is defined as the sequence whose only element is obtained by applying Φ_σ to its *last* element. By construction x is non-empty, so *last* and *head* are always well-defined. Thus, provided σ is a USTA, a sequence s of type *fs* can be split into concatenations involving its *front* and *last* element, and its *head* and *tail*.

Lemma 5. *$front(s) \frown_{fs} \langle last(s) \rangle_{fs} = s$, and $\langle head(s) \rangle_{fs} \frown_{fs} tail(s) = s$.*

The functions *front* and *tail* are tailored to non-empty sequences. For example, $front(\langle a \rangle_{fs})$ is $\langle \Phi_\sigma(a) \rangle_{fs}$, while $front(\langle a, b \rangle_{fs})$ is $\langle a, \Phi_\sigma(b) \rangle_{fs}$, and $tail(\langle a \rangle_{fs})$ is $\langle \Phi_\sigma(a) \rangle_{fs}$, while $tail(\langle a, b \rangle_{fs})$ is $\langle \Phi_\sigma(a), b \rangle_{fs}$, so that the decomposition holds.

Next we use this structure to instantiate the USTA for *fs* sequences, which corresponds to the trace structure underlying synchronous process algebra.

Theorem 8. *Provided $(\sigma, +_\sigma, \Phi_\sigma)$ is a USTA, then $(fs, \frown_{fs}, \Phi_{fs})$ is a USTA.*

As a corollary to this theorem we have that a parametric pair $\mathcal{P}$ whose type parameter $\mathcal{H}$ is a USTA $(\mathcal{H}, +_{\mathcal{H}}, \Phi_{\mathcal{H}})$ induces a $(fs, \frown_{fs}, \Phi_{fs})$ USTA.

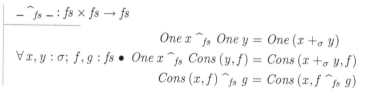

Corollary 1. *If* $(\mathcal{H}, +_{\mathcal{H}}, \Phi_{\mathcal{H}})$ *is a USTA, then* $(fs, \frown_{fs}, \Phi_{fs})$ *is a USTA.*

This demonstrates that to construct such a USTA it is sufficient to show that $\mathcal{H}$ is a USTA. This is a much more general, and concise, construction, than that proposed in [7], which instead requires satisfying nearly 26 axioms. Moreover, our results do not rely on any assumptions about the type $\mathcal{R}$, thus allowing the second component of such pairs in a trace to record arbitrary information, not only refusal sets as proposed in [7]. Next we focus on key properties of traces leading to a demonstration that we can derive core laws of [7], and the healthiness conditions of the corresponding UTP theory.

Properties. Below we establish key results on the difference of *fs* sequences.

Theorem 9. *Provided* $(\sigma, +_{\sigma}, \Phi_{\sigma})$ *is a USTA and* $s \le t$, *where* $s, t : fs$,

$$tail(t - s) = tail(t - front(s)) \tag{S1}$$
$$head(t - s) = head(t - front(s)) - last(s) \tag{S2}$$
$$last(s) \le head(t - front(s)) \tag{S3}$$

The *tail* of the difference $t - s$ is the tail of the difference between t and the *front* of s (S1). Likewise, the *head* of the difference $t - s$ is equal to the last element of s subtracted from the *head* of the difference $t - front(s)$ (S2). Related, (S3) establishes that $last(s)$ is a prefix of $head(t - front(s))$.

To illustrate the role of $S1$, we consider, as an example the subtraction of a *fs* sequence whose elements are standard sequences. The subtraction of $\langle\langle a\rangle, \langle b\rangle\rangle_{fs}$ from $\langle\langle a\rangle, \langle b, c\rangle, \langle d\rangle\rangle_{fs}$ is $\langle\langle c\rangle, \langle d\rangle\rangle_{fs}$, indicating that the first element where the sequences differ is the inner sequence $\langle b, c\rangle$. The *front* of $\langle\langle a\rangle, \langle b\rangle\rangle_{fs}$ is $\langle\langle a\rangle, \langle\rangle\rangle_{fs}$, and so the difference $\langle\langle a\rangle, \langle b, c\rangle, \langle d\rangle\rangle_{fs} - \langle\langle a\rangle, \langle\rangle\rangle_{fs}$ is $\langle\langle b, c\rangle, \langle d\rangle\rangle_{fs}$. Finally, the $tail(\langle\langle b, c\rangle, \langle d\rangle\rangle_{fs}) = \langle\langle\rangle, \langle d\rangle\rangle_{fs}$ coincides with that of $\langle\langle c\rangle, \langle d\rangle\rangle_{fs}$.

Moreover, we show below that Eq. 3 in [7] also holds in our setting of *fs* sequences of parametric pairs $\mathcal{P}$, provided $(\mathcal{H}, +_{\mathcal{H}}, \Phi_{\mathcal{H}})$ is a USTA.

Lemma 6. $s \frown_{fs} t = front(s) \frown_{fs} \langle last(s) +_{\mathcal{P}} head(t)\rangle_{fs} \frown_{fs} tail(t)$.

The concatenation of traces s and t can be decomposed into the concatenation of the *front* of s with a singleton sequence, whose only element is the result of concatenating $(+_{\mathcal{P}})$ the *last* pair of s and the *head* pair of t, and the *tail* of t.

Reactive Processes. Besides the definition of an abstract trace structure that can be instantiated to yield the trace structure in [7], we discuss next how it can be used to define a generalised theory of reactive processes. Here we focus on the instantiation of the healthiness conditions.

Healthiness Conditions. The functions **R1** and **R3** are stated like in Sect. 2, but in the context of a USTA $(\mathcal{T}, \frown, \Phi)$, with *tr* and *tr'* of type $\mathcal{T}$. **R2**, on the other hand, must be adapted to accommodate the function Φ.

Definition 12. $\mathbf{R2}(P) = P[\Phi(tr), tr' - tr/tr, tr']$

$$\mathbf{R2a}(P) = \bigsqcap z \bullet P[z, z ^\frown (tr' - tr)/tr, tr'] \wedge \Phi(tr) = \Phi(z)$$

Our definition for **R2** is stated by replacing ε with $\Phi(tr)$. Moreover, the definition for **R2a**, when compared to Definition 4, requires that, in addition z and tr agree on the application of Φ. This closely follows a solution proposed in [7, p. 83].

Despite employing a weaker trace algebra, the core properties of **R1**, **R2** and **R3**, namely idempotency and monotonicity with respect to refinement, continue to hold. Similarly, all laws of reactive processes, and those for other theories built upon reactive processes, namely CSP, continue to hold as demonstrated by the mechanisation in Isabelle/UTP, which features several hundreds of theorems.

Because the existing theories are mechanised we have been able to quickly establish that all relevant properties hold when using our algebra. Proofs of closure for sequential and parallel composition under **R2** required small adjustments to take into account Φ, but were structurally kept unchanged. Next, we illustrate a concrete instantiation of the algebra to accommodate the trace model of [4].

Concrete instantiation for Circus Time. In what follows we show how our algebra can be instantiated to yield the theory of *Circus Time*, that encompasses behaviour and data modelling in a discrete-time setting.

The parametric pair type $\mathcal{P}$ is instantiated with $\mathcal{H}$ as seq Σ, where Σ is a given type of events, which is a USTA $(\text{seq}\,\Sigma, ^\frown, \langle\rangle)$. Concatenation $(^\frown)$ is associative (USTA1), left-cancellative (USTA3) and satisfies USTA4. The empty sequence $\langle\rangle$ is a right-unit (USTA2). The parameter $\mathcal{R}$ is instantiated as $\mathbb{P}\,\Sigma$, a set of events. Thus, the first component of such a pair is a sequence and the second a set of events. For example, the pair $(\langle a, b\rangle, \{a\})$ records that having performed events a, and then b, the system can refuse to engage in event a.

Therefore, the lifted structure of finite non-empty sequences *fs* parametrised by the concrete pair structure above, gives rise to a USTA (Corollary 1). For example, in *Circus Time* the sequence $\langle(\langle a, b\rangle, \{a\}), (\langle\rangle, \Sigma)\rangle_{fs}$ encodes a situation where: during the first time unit a and b are performed, with a then being refused, and during the following time unit no events are performed $(\langle\rangle)$ with the system refusing to engage in any event (Σ).

Compared with the approach in [4], we have that both concatenation and subtraction of *fs* sequences (using the lifted structure) is total and closed under the correct type. This provides for a precise encoding of the healthiness conditions proposed in [4] using our abstract algebra. Furthermore, this makes mechanisation of the model in Isabelle/UTP an easier endeavour by eliminating the need to reprove a substantial base of existing theorems of reactive processes.

In the remainder of this section we show two key results that demonstrate **R1** and **R2** can be instantiated to yield the counterpart definitions for *Circus Time*.

Lemma 7. $s \le t \Rightarrow \mathit{front}(s) \le t \wedge \mathit{fst}(\mathit{last}(s)) \le \mathit{fst}(\mathit{head}(t - \mathit{front}(s)))$.

This corresponds to the conjunct in the definition of **R1** for *Circus Time* as defined in [16], for example, with the understanding that here *front* is total, whereas in [7,16] it is a partial function over standard sequences.

The definition of $\mathbf{R2}_S$ for *Circus Time* is derived next. First we establish a result for the subtraction of *fs* traces that depends on the following lemma.

Lemma 8. $s \le t \Rightarrow snd(head(t - front(s)) - last(s)) = snd(head(t - front(s)))$.

This lemma states that the second component of the difference, between the *head* of the difference t and the *front* of s, and $last(s)$, does not depend on $last(s)$, a result that follows from Lemma 4. For example, the subtraction of $\langle(\langle a\rangle, r_2)\rangle_{fs}$ from $\langle(\langle a, b\rangle, r_1)\rangle_{fs}$ yields the sequence $\langle(\langle b\rangle, r_1)\rangle_{fs}$ as the second component only depends on r_1, but not r_2. Next, we establish a general result for subtraction of *fs* traces.

Theorem 10. *Provided* $s \le t$, *where* $s, t : fs$,

$$t - s = \langle\begin{pmatrix} fst(head(t - front(s))) - fst(last(s)), \\ snd(head(t - front(s))) \end{pmatrix}\rangle_{fs} \frown_{fs} tail(t - front(s))$$

Proof.

$$\begin{aligned}
& t - s && \text{[Lemma 5]}\\
&= \langle head(t - s)\rangle_{fs} \frown_{fs} tail(t - s) && \text{[S2 in Theorem 9]}\\
&= \langle head(t - front(s)) - last(s)\rangle_{fs} \frown_{fs} tail(t - s) && \text{[S1 in Theorem 9]}\\
&= \langle head(t - front(s)) - last(s)\rangle_{fs} \frown_{fs} tail(t - front(s)) && \text{[Pair structure]}\\
&= \begin{pmatrix} \langle\begin{pmatrix} fst(head(t - front(s)) - last(s)), \\ snd(head(t - front(s)) - last(s)) \end{pmatrix}\rangle_{fs} \\ \frown_{fs} tail(t - front(s)) \end{pmatrix} && \text{[Lemma 4]}\\
&= \begin{pmatrix} \langle\begin{pmatrix} fst(head(t - front(s))) - fst(last(s)), \\ snd(head(t - front(s)) - last(s)) \end{pmatrix}\rangle_{fs} \\ \frown_{fs} tail(t - front(s)) \end{pmatrix} && \text{[Lemma 8]}\\
&= \begin{pmatrix} \langle\begin{pmatrix} fst(head(t - front(s))) - fst(last(s)), \\ snd(head(t - front(s))) \end{pmatrix}\rangle_{fs} \\ \frown_{fs} tail(t - front(s)) \end{pmatrix} && \square
\end{aligned}$$

The subtraction $t - s$ can be expressed in terms of the difference $t - front(s)$, and $last(s)$. The *head* of $t - front(s)$ contains the observations up until the end of the current time unit [16, p. 13]. Together with the pair instantiation as before we can derive the concrete definition of **R2** for *Circus Time*, similarly to Definition 5 where $\Phi(tr)$ becomes $\langle(\langle\rangle, snd(last(tr)))\rangle_{fs}$ following Definitions 8 and 11, and $tr' -_S tr$ is as given by Theorem 10.

5 Related Work

Traces are at the core of semantic models for reasoning about causality. Already in Hoare's CSP book [17] we can find a rich collection of operators and laws for manipulating traces. In the standard semantics [3] of CSP traces are sequences of events ordered by sequence prefixing. Richer semantic models for CSP, such as refusal testing [18,19] and the finite-linear models [3, p. 256], also record in traces the set of events refused, or accepted, in the latter, before each event.

The modelling of time in semantics for process algebra is often achieved by associating events or state observations with time. Hayes' reactive timed designs [20], comparable to action systems and TLA, define traces as mappings from time (discrete or continuous) to the values of program variables.

Sherif et al. [4] defined a semantics for *Circus Time* where traces are sequences whose elements are pairs, recording the events performed, and subsequently refused, during a time unit. Wei et al. [5] considered an equivalent model, where events and refusals are recorded separately in two distinct traces of equal length. Woodcock et al. [6] in their semantics for CML define sequences whose elements are events or refusal sets, that implicitly mark the passage of time, a structure pioneered by Lowe and Ouaknine [21] in their timed traces.

Butterfield et al. [7] proposed a parametric theory, which is the inspiration for the work presented in this paper. It generalises the model of *Circus Time* [4] to account for different observation models within a time unit. A similar approach is pursued by Zhu et al. [22], in their semantics for SystemC, who define a trace as a three dimensional sequence structure to account for macro and micro time.

Trace models for true concurrency in process algebra include the works of Barnes [23] and Smith [24]. The latter [24] defines traces whose elements are sequences, with the prefix relation allowing permutations of the inner sequences. This model can likely be instantiated as a trace algebra with elements as sets. Barnes' SCSP, on the other hand, cannot be instantiated within the setting of [7].

More recently, Foster et al. [9] proposed a left-cancellative monoid trace algebra which is at the core of the mechanisation of several reactive theories in Isabelle/UTP [11]. This enabled Foster et al. [25] to define reactive contracts, as well as a theory for hybrid relations [26]. Their prefix relation over traces, however, is an order, which is inadequate to characterise traces where the relation is not anti-symmetric. Our results are complementary and support the unification of further trace models under the Isabelle/UTP framework.

6 Conclusions

Originally motivated by the goal of mechanising *Circus Time* [4] in the theorem prover Isabelle/UTP [11], we have pursued an ambitious generalisation of the monoid trace algebra [9] to account for a broader family of timed process algebras. We have weakened the monoid axioms, inspired by the observations in [7], to construct a novel unary semigroup trace algebra that is also a right P-Ehresmann semigroup. Compared to the large set of axioms in [7], we have a much smaller set that closely mirrors the axioms of the monoid trace algebra.

Our results support the definition of a parametric UTP theory of reactive processes that abstractly characterises several trace-based semantics. Besides the trace models discussed in [9] our algebra can be instantiated to account for the models discussed in [7,8], including *Circus Time* [4]. In the future, we hope to accommodate the semantics for the system-level language SystemC [22], and perhaps even other synchronous languages such as Esterel [27].

Besides providing a foundation for the unification of further trace models in the UTP, we have also shown that our work has practical impact via its mechanisation in Isabelle/UTP [11]. It promotes the reuse of a large collection of theorems already established for the theories of reactive processes and reactive designs. It would be interesting, for example, to revisit our mechanisation of a stepwise construction for *Circus Time* [16] in this setting. Another avenue for future work is the mechanisation of the Galois connection in [4] that enables timed models to be verified using untimed tools.

The mechanisation of the timed operators of timed process calculi is likely to benefit from the definition of a timed trace algebra, consisting of an additional function from traces to time, with continuous and discrete versions. Basic processes, such as event prefixing and delay, may also be defined parametrically.

We envision it may be feasible to weaken the unary semigroup trace algebra even further to characterise additional trace structures, such as those of refusal-testing, the finite-linear model, and those of SCSP. However, it is likely that such weakenings may reveal certain laws of reactive processes no longer hold. An open question is the treatment of infinite traces, for example, which seem necessary to give a full account of the hiding operator of CSP. The Isabelle/UTP [11] mechanisation will facilitate the design space exploration of such weakenings, with immediate feedback provided to the proof engineer, a facility we used extensively during the course of developing the algebra presented in this paper.

Acknowledgements. This work is funded by the EPSRC grant EP/M025756/1. No new primary data was created as part of the study reported here. We are grateful to Ana Cavalcanti for comments on an earlier draft of this paper, and to the anonymous reviewers for their helpful and constructive feedback.

References

1. Hoare, C.A.R., Jifeng, H.: Unifying Theories of Programming. Prentice-Hall, Upper Saddle River (1998)
2. Hehner, E.C.R.: Predicative programming part I. Commun. ACM **27**(2), 134–143 (1984)
3. Roscoe, A.W.: Understanding Concurrent Systems. Springer, Heidelberg (2010). https://doi.org/10.1007/978-1-84882-258-0
4. Sherif, A., Cavalcanti, A.L.C., He, J., Sampaio, A.C.A.: A process algebraic framework for specification and validation of real-time systems. Formal Aspects Comput. **22**(2), 153–191 (2010)
5. Wei, K., Woodcock, J., Cavalcanti, A.: *Circus Time* with reactive designs. In: Wolff, B., Gaudel, M.-C., Feliachi, A. (eds.) UTP 2012. LNCS, vol. 7681, pp. 68–87. Springer, Heidelberg (2013). https://doi.org/10.1007/978-3-642-35705-3_3
6. Woodcock, J., Bryans, J., Canham, S., Foster, S.: The COMPASS modelling language: timed semantics in UTP. Open Channel Publishing, Communicating Process Architectures (2014)

7. Butterfield, A., Sherif, A., Woodcock, J.: Slotted-circus: A UTP-family of reactive theories. In: Davies, J., Gibbons, J. (eds.) IFM 2007. LNCS, vol. 4591, pp. 75–97. Springer, Heidelberg (2007). https://doi.org/10.1007/978-3-540-73210-5_5
8. Butterfield, A.: A denotational semantics for Handel-C. Formal Aspects Comput. **23**(2), 153–170 (2011)
9. Foster, S., Cavalcanti, A., Woodcock, J., Zeyda, F.: Unifying theories of time with generalised reactive processes. Inf. Process. Lett. **135**, 47–52 (2018)
10. Woodcock, J., Cavalcanti, A., Foster, S., Mota, A., Ye, K.: Probabilistic semantics for RoboChart. In: Ribeiro, P., Sampaio, A. (eds.) UTP 2019. LNCS, vol. 11885, pp. 80–105. Springer, Cham (2019). https://doi.org/10.1007/978-3-030-31038-7_5
11. Foster, S., Zeyda, F., Woodcock, J.: Isabelle/UTP: a mechanised theory engineering framework. In: Naumann, D. (ed.) UTP 2014. LNCS, vol. 8963, pp. 21–41. Springer, Cham (2015). https://doi.org/10.1007/978-3-319-14806-9_2
12. Howie, J.M.: Fundamentals of Semigroup Theory, vol. 12. Clarendon Oxford, Oxford (1995)
13. Cornock, C., Gould, V.: Proper two-sided restriction semigroups and partial actions. J. Pure Appl. Algebra **216**(4), 935–949 (2012)
14. Jones, P.R.: A common framework for restriction semigroups and regular *-semigroups. J. Pure Appl. Algebra **216**(3), 618–632 (2012)
15. Woodcock, J.C.P., Davies, J.: Using Z - Specification, Refinement, and Proof. Prentice-Hall, Upper Saddle River (1996)
16. Ribeiro, P., Cavalcanti, A., Woodcock, J.: A stepwise approach to linking theories. In: Bowen, J.P., Zhu, H. (eds.) UTP 2016. LNCS, vol. 10134, pp. 134–154. Springer, Cham (2017). https://doi.org/10.1007/978-3-319-52228-9_7
17. Hoare, C.A.R.: Communicating Sequential Processes. Prentice-Hall Inc., Upper Saddle River (1985)
18. Mukarram, A.: A refusal testing model for CSP. Ph.D. thesis, University of Oxford (1993)
19. Phillips, I.: Refusal testing. Theoret. Comput. Sci. **50**(3), 241–284 (1987)
20. Hayes, I.J., Dunne, S.E., Meinicke, L.: Unifying theories of programming that distinguish nontermination and abort. In: Bolduc, C., Desharnais, J., Ktari, B. (eds.) MPC 2010. LNCS, vol. 6120, pp. 178–194. Springer, Heidelberg (2010). https://doi.org/10.1007/978-3-642-13321-3_12
21. Lowe, G., Ouaknine, J.: On timed models and full abstraction. Electron. Notes Theor. Comput. Sci. **155**, 497–519 (2006)
22. Zhu, H., He, J., Qin, S., Brooke, P.J.: Denotational semantics and its algebraic derivation for an event-driven system-level language. Formal Aspects Comput. **27**(1), 133–166 (2015)
23. Barnes, J.E.: A mathematical theory of synchronous communication. University of Oxford (1993)
24. Smith, M.L.: A unifying theory of true concurrency based on CSP and lazy observation. In: Communicating Process Architectures 2005: WoTUG-28: Proceedings of the 28th WoTUG Technical Meeting, 18–21 September 2005, Technische Universiteit Eindhoven, the Netherlands, vol. 63, p. 177. IOS Press (2005)
25. Foster, S., Cavalcanti, A., Canham, S., Woodcock, J., Zeyda, F.: Unifying theories of reactive design contracts. Theor. Comput. Sci. **802**, 105–140 (2019)
26. Foster, S.: Hybrid relations in Isabelle/UTP. In: Ribeiro, P., Sampaio, A. (eds.) UTP 2019. LNCS, vol. 11885, pp. 130–153. Springer, Cham (2019). https://doi.org/10.1007/978-3-030-31038-7_7
27. Berry, G., Gonthier, G.: The Esterel synchronous programming language: design, semantics, implementation. Sci. Comput. Program. **19**(2), 87–152 (1992)

The Involutive Quantaloid of Completely Distributive Lattices

Luigi Santocanale(✉)

Laboratoire d'Informatique et des Systèmes, UMR 7020,
Aix-Marseille Université, CNRS, Marseille, France
luigi.santocanale@lis-lab.fr

Abstract. Let L be a complete lattice and let $\mathcal{Q}(L)$ be the unital quantale of join-continuous endo-functions of L. We prove that $\mathcal{Q}(L)$ has at most two cyclic elements, and that if it has a non-trivial cyclic element, then L is completely distributive and $\mathcal{Q}(L)$ is involutive (that is, non-commutative cyclic $\star$-autonomous). If this is the case, then the dual tensor operation corresponds, via Raney's transforms, to composition in the (dual) quantale of meet-continuous endo-functions of L.

Let $\mathsf{Latt}_{\bigvee}$ be the category of sup-lattices and join-continuous functions and let $\mathsf{Latt}_{\bigvee}^{\mathrm{cd}}$ be the full subcategory of $\mathsf{Latt}_{\bigvee}$ whose objects are the completely distributive lattices. We argue that $\mathsf{Latt}_{\bigvee}^{\mathrm{cd}}$ is itself an involutive quantaloid, thus it is the largest full-subcategory of $\mathsf{Latt}_{\bigvee}$ with this property. Since $\mathsf{Latt}_{\bigvee}^{\mathrm{cd}}$ is closed under the monoidal operations of $\mathsf{Latt}_{\bigvee}$, we also argue that if $\mathcal{Q}(L)$ is involutive, then $\mathcal{Q}(L)$ is completely distributive as well; consequently, any lattice embedding into an involutive quantale of the form $\mathcal{Q}(L)$ has, as its domain, a distributive lattice.

1 Introduction

Let C be a finite chain or the unit interval of the reals. In a series of recent works [6,21,22] we argued that the unital quantale structure of $\mathcal{Q}(C)$, the set of join-continuous functions from C to itself, plays a fundamental role to solve more complex combinatorial and geometrical problems arising in Computer Science. In [6,22] we formulated an order theoretic approach to the problem of constructing discrete approximations of curves in higher dimensional unit cubes. On the side of combinatorics, the results in [21] yield bijective proofs of counting results (that is, bijections, through which these results can easily be established) for idempotent monotone endo-functions of a finite chain [9,12] and a new algebraic interpretation of well-known combinatorial identities [3].

The quantales $\mathcal{Q}(C)$, C a finite chain or $[0,1] \subseteq \mathbb{R}$, are *involutive*—or, using another possible naming, *non-commutative cyclic $\star$-autonomous*. The involution is used in the mentioned works, even if is not clear to what extent it is necessary. It was left open in these works whether there are other complete chains C such that $\mathcal{Q}(C)$ is involutive and, at its inception, the aim of this research was to answer this question. Let us use $\mathcal{Q}(L)$ for the unital quantale of join-continuous

U. Fahrenberg et al. (Eds.): RAMiCS 2020, LNCS 12062, pp. 286–301, 2020.
https://doi.org/10.1007/978-3-030-43520-2_18

endo-functions of a complete lattice L. Recalling that involutive quantale structures on a given quantale are determined by the cyclic dualizing elements and that complete chains are completely distributive, the following statement from the monograph [4] shows that $\mathcal{Q}(C)$ is involutive for each complete chain C.

Proposition 2.6.18 in [4]**.** *Let L be a complete lattice and let o_L be the join-continuous self-mapping on L defined by $o_L(x) := \bigvee_{x \not\leq z} z$, for $x \in L$. Then the following assertions are equivalent: (i) o_L is a dualizing element of the quantale $\mathcal{Q}(L)$, (ii) L is completely distributive.*

This proposition also covers another important example studied in the literature. Let $\mathcal{D}(P)$ be the perfect completely distributive lattice of downsets of a poset P. According to the proposition, $\mathcal{Q}(\mathcal{D}(P))$ is involutive, a fact that can also be inferred via the isomorphism with the residuated lattice of weakening relations on P, known to be involutive, see [10,13,19].

We strengthen here the above statement in many ways. Firstly, we observe that the quantale $\mathcal{Q}(L)$ has at most two cyclic elements and that cyclicity of o_L is almost sufficient for $\mathcal{Q}(L)$ to be involutive:

Theorem. *If $c \in \mathcal{Q}(L)$ is cyclic, then either c is the top element of $\mathcal{Q}(L)$ or $c = o_L$. Moreover, if o_L is cyclic and not equal to the top element of $\mathcal{Q}(L)$, then L is completely distributive (and therefore $\mathcal{Q}(L)$ is involutive, as from Proposition 2.6.18 of* [4]*).*

An important consequence of the previous statement is that the quantale $\mathcal{Q}(L)$ can be made into an involutive quantale in a unique way:

Theorem. *If the quantale $\mathcal{Q}(L)$ is involutive, then its dualizing cyclic element is the join-continuous function o_L.*

In the direction from L to $\mathcal{Q}(L)$, we observe that the local involutive quantale structures on each completely distributive lattice fit together in a uniform way. A quantaloid is a category whose homsets are complete lattices and for which composition distributes on both sides with suprema. As a quantale can be considered as a one-object quantaloid, the notion of involutive quantale naturally lifts to the multi-object context—so an involutive quantale is a one-object involutive quantaloid. *Involutive quantaloids* are indeed the Girard quantaloids introduced in [19]. The following statement, proved in this paper, makes precise the intuition that the local involutive quantale structures are uniform:

Theorem. *The full subcategory of the category of complete lattices and join-continuous functions whose objects are the completely distributive lattices is an involutive quantaloid.*

The tools used in this paper rely on and emphasize Raney's characterization of completely distributive lattices [15,16]. A main remark that we develop is that if $\mathcal{Q}(L)$ is involutive, then the dual quantale structure of $\mathcal{Q}(L)$ arises from $\mathcal{Q}(L^{\partial})$, the quantale of meet-continuous endo-functions of L, via Raney's transforms (to be studied in Sect. 5).

Overall, this set of results yields an important clarification of the algebra used in our previous works [6,21,22] and, more importantly, new characterizations of completely distributive lattices adding up to the existing ones, see e.g. [7,11, 15,16,23]. These characterizations strongly rely on the algebra of quantales and residuated lattices thus on relation algebra, in a wider sense.

An ideal goal of future research is to characterize the equational theory of the involutive residuated lattices of the form $\mathcal{Q}(L)$. For the moment being, we observe that the units of the involutive quantale $\mathcal{Q}(L)$, L a completely distributive lattice, may be used to characterize properties of L:

Theorem. *A complete lattice is a chain if and only if the inclusion $0 \leq 1$ (in the language of involutive residuated lattices) holds in $\mathcal{Q}(L)$, i.e. if and only if $\mathcal{Q}(L)$ satisfies the mix law. A completely distributive lattice has no completely join-prime elements if and only if the inclusion $1 \leq 0$ holds in $\mathcal{Q}(L)$.*

It is known that the full subcategory of the category of complete lattices and join-continuous functions whose objects are the completely distributive lattices, the involutive quantaloid of completely distributive lattice, is closed under the monoidal operations inherited from the super category, see e.g. [4,7,20]. In particular, this quantaloid is itself a $\star$-autonomous category. For the sake of studying the equational theory of the $\mathcal{Q}(L)$, this fact and the previous results jointly yield the following remarkable consequence:

Corollary. *If $\mathcal{Q}(L)$ is an involutive quantale, then it is completely distributive.*

On the side of logic, it is worth observing that enforcing a linear negation (the involution, the star) on the most typical models of intuitionistic noncommutative linear logic also enforces a classical behaviour, distributivity, of the additive logical connectors. Apart from the philosophical questions about logic, the above corollary pinpoints an important obstacle in finding Cayley style representation theorems for involutive residuated lattices or a generalization of Holland's theorem [8] from lattice-ordered groups to involutive residuated lattices:

Corollary. *If a residuated lattice embedding of $\mathcal{Q}$ into some involutive residuated lattice of the form $\mathcal{Q}(L)$ exists, then $\mathcal{Q}$ is distributive.*

Finally, we observe that these mathematical results pinpoint the importance and the naturalness of considering a linear logic based on a distributive setting. This algebraic setting has already many established facets and applications. Among them, let us mention bunched implication logic, for which our last theorem provides non-standard pointless models. Let us also mention the usage of this algebra in pointfree topology: here embeddability problems for quantales dual to topological groupoids, problems analogous to the ones we are raising, have already been investigated in depth, see e.g. [14].

The paper is organised as follows. In Sect. 2 we provide definitions and elementary results. In Sect. 3 we introduce the notion of an involutive quantaloid (we shall identify an involutive quantale with a one-object involutive quantaloid).

We prove in Sect. 4 that if a quantale of the form $\mathcal{Q}(L)$ is involutive, then it has just one cyclic dualizing element. That is, there can be at most one involutive quantale structure extending the structure of $\mathcal{Q}(L)$. Moreover, we prove in this section that if $\mathcal{Q}(L)$ has a non-trivial cyclic element, then L is a completely distributive lattice. The uniqueness of the involutive structure is intimately related to the fact—analyzed at the end of Sect. 4—that the only central elements of $\mathcal{Q}(L)$ are the identity and the constant function mapping to the bottom of L. In Sect. 5 we introduce Raney's transforms and their elementary properties. Raney's transforms are the main tool used to prove, in Sect. 6, that completely distributive lattices form an involutive quantaloid. In Sect. 7 we develop some considerations on the equational theories of the lattices $\mathcal{Q}(L)$ among which, the use of the multiplicative units of $\mathcal{Q}(L)$ to characterize properties of L and the fact that $\mathcal{Q}(L)$ is completely distributive whenever it is $\mathcal{Q}(L)$ involutive.

2 Definitions and Elementary Results

Complete Lattices and the Category $\mathsf{Latt}_{\bigvee}$. A *complete lattice* is a poset L such that each $X \subseteq L$ has a supremum $\bigvee X$. A map $f : L \longrightarrow M$ is *join-continuous* if $f(\bigvee X) = \bigvee f(X)$, for each subset $X \subseteq L$. We shall denote by $\mathsf{Latt}_{\bigvee}$ the category whose objects are the complete lattices and whose morphisms are the join-continuous maps.

For a poset P, P^{∂} denotes the poset with the same elements of P but with the reverse ordering: $x \leq_{P^{\partial}} y$ iff $y \leq_P x$. In a complete lattice, the set $\bigvee\{\, y \mid y \leq x,$ for each $x \in X \,\}$ is the infimum of X. Therefore, if L is complete, then L^{∂} is also a complete lattice. Moreover, if L, M are complete lattices and $f : L \longrightarrow M$ is join-continuous, then the map $\rho(f) : M \longrightarrow L$, defined by $\rho(f)(y) := \bigvee\{\, x \in L \mid f(x) \leq y \,\}$, preserves infima and therefore it belongs to the homset $\mathsf{Latt}_{\bigvee}(M^{\partial}, L^{\partial})$. The map $\rho(f)$ is the *right adjoint* of f, meaning that, for each $x \in L$ and $y \in M$, $f(x) \leq y$ if and only if $x \leq \rho(f)(y)$. For $g : M \longrightarrow L$ meet-continuous, its *left adjoint* $\ell(g) : L \longrightarrow M$ is defined similarly, and satisfies $\ell(g)(x) \leq y$ if and only if $x \leq g(y)$, for each $x \in L$ and $y \in M$. Consequently, $\ell(\rho(f)) = f$ and $\rho(\ell(g)) = g$. Indeed, by defining with $f^{\partial} := \rho(f)$, $(\,\cdot\,)^{\partial} : \mathsf{Latt}_{\bigvee} \longrightarrow \mathsf{Latt}_{\bigvee}^{op}$ is a (contravariant) functor and a category isomorphism.

Let $\{\, f_i \mid i \in I \,\}$ be a family of join-continuous functions from L to M. The function $\bigvee_{i \in I} f_i$, defined by $(\bigvee_{i \in I} f_i)(x) := \bigvee_{i \in I} f_i(x)$, is a join-continuous map from L to M. Therefore the homset $\mathsf{Latt}_{\bigvee}(L, M)$, with the pointwise ordering, is a complete lattice, where suprema are computed by the above formula. The same formula shows that the inclusion of $\mathsf{Latt}_{\bigvee}(L, M)$ into M^L, the set of all functions form L to M, is join-continuous. It follows that, for every $f : L \longrightarrow M$, there is a (uniquely determined) greatest join-continuous function $h \in \mathsf{Latt}_{\bigvee}(L, M)$ such that $h \leq f$; in the following we shall use $\mathtt{int}(f)$ to denote such h. Observe also that, by monotonicity of composition, $\mathtt{int}(g) \circ \mathtt{int}(f) \leq g \circ f$ and therefore $\mathtt{int}(g) \circ \mathtt{int}(f) \leq \mathtt{int}(g \circ f)$.

Quantales and Involutive Quantales. A *quantale* is a complete lattice $\mathcal{Q}$ coming with a semigroup operation $\circ$ that distributes with arbitrary sups. That is, we have $(\bigvee X) \circ (\bigvee Y) = \bigvee_{x \in X, y \in Y} x \circ y$, for each $X, Y \subseteq \mathcal{Q}$. A quantale is unital if the semigroup operation has a unit. As we shall always consider unital quantales, we shall use the wording quantale as a synonym of unital quantale. In a quantale $\mathcal{Q}$, left and right residuals are defined as follows: $x \backslash y := \bigvee \{ z \in \mathcal{Q} \mid x \circ z \leq y \}$ and $y / x := \bigvee \{ z \in \mathcal{Q} \mid z \circ x \leq y \}$. Clearly, we have the following adjointness relations: $x \circ y \leq z$ iff $y \leq x \backslash z$ iff $x \leq z / y$. Let us recall that a quantale $\mathcal{Q}$ is a *residuated lattice*, that is an algebra on the signature $\wedge, \vee, 1, \circ, \backslash, /$, satisfying a finite identities, see e.g. [5, §2.2].

A standard example of quantale is $\mathcal{Q}(L)$, the set of join-continuous endofunctions of a complete lattice L. In this case, the semigroup operation is function composition; otherwise said, $\mathcal{Q}(L)$ is the homset $\mathsf{Latt}_{\bigvee}(L, L)$. We shall consider special elements of $\mathcal{Q}(L)$ and of $\mathcal{Q}(L^{\partial})$. For $x \in L$, let $c_x, a_x, \alpha_x : L \longrightarrow L$ be defined as follows:

$$c_x(t) := \begin{cases} x\,, & t \neq \bot\,, \\ \bot\,, & t = \bot\,, \end{cases} \quad a_x(t) := \begin{cases} \top\,, & t \not\leq x\,, \\ \bot\,, & t \leq x\,, \end{cases} \quad \alpha_x(t) := \begin{cases} \top\,, & x \leq t\,, \\ \bot\,, & x \not\leq t\,. \end{cases} \tag{1}$$

Clearly, $c_x, a_x \in \mathcal{Q}(L)$ while $\alpha_x \in \mathcal{Q}(L^{\partial})$. Moreover, we have $\rho(c_x) = \alpha_x$.

Completely Distributive Lattices. A complete lattice L is said to be *completely distributive* if, for each pair of families $\pi : J \longrightarrow I$ and $x : J \longrightarrow L$, the following equality holds

$$\bigwedge_{i \in I} \bigvee_{j \in J_i} x_j = \bigvee_{\psi} \bigwedge_{i \in I} x_{\psi(i)}\,,$$

where $J_i = \pi^{-1}(i)$, for each $i \in I$, and the meet on the right is over all sections ψ of π, that is, those functions such that $\pi \circ \psi = id_I$. Let us recall that the notion of a completely distributive lattice is auto-dual, meaning that a complete lattice L is completely distributive iff L^{∂} is such. For each complete lattice L, define

$$o_L(x) := \bigvee \{ t \mid x \not\leq t \}, \qquad \omega_L(y) := \bigwedge \{ t \mid t \not\leq y \}. \tag{2}$$

It is easy to see that $o_L \in \mathcal{Q}(L)$ and that $\rho(o_L) = \omega_L$. The following statement appears in [16, Theorem 4]:

Theorem 1 (Raney). *A lattice is completely distributive if and only if any of the following equivalent conditions hold:*

$$\bigvee_{x \not\leq t} \omega_L(t) = x, \qquad \bigwedge_{t \not\leq y} o_L(t) = y. \tag{3}$$

3 Involutive Quantaloids

The purpose of this section is to define *involutive quantaloids* which, not surprisingly, turn out to be the *Girard quantaloids* of [19]. Let us mention that, following [1,2] and [6,22], another possible naming for the same concept is *non-commutative, cyclic, star-autonomous quantaloid.* For the sake of conciseness, we prefer the wording involutive quantaloid.

We recall that a *quantaloid*, see e.g. [23], is a category $\mathcal{Q}$ enriched over the category of sup-lattices. This means that, for each pair of objects L, M of $\mathcal{Q}$, the homset $\mathcal{Q}(L, M)$ is a complete lattice and that composition distributes over suprema in both variables, $(\bigvee_{i \in I} g_i) \circ (\bigvee_{j \in J} f_j) = \bigvee_{i \in I, j \in J} f_i \circ g_j$. A quantale, see e.g. [18], might be seen as a one-object quantaloid. The category $\mathsf{Latt}_\bigvee$ is itself a quantaloid. The definition below mimics, in a multisorted setting, a possible definition of involutive quantale or of involutive residuated lattice. For the possible equivalent definitions of these notions, see e.g. [2] or [5, §3.3].

Definition 2. *An* involutive quantaloid *is a quantaloid $\mathcal{Q}$ coming with operations*

$$(\cdot)^{\star_{L,M}} : \mathcal{Q}(L, M) \longrightarrow \mathcal{Q}(M, L), \qquad L, M \textit{ objects of } \mathcal{Q},$$

satisfying the following conditions:

1. $(f^{\star_{L,M}})^{\star_{M,L}} = f$*, for each* $f \in \mathcal{Q}(L, M)$*,*
2. *for each* $f, g \in \mathcal{Q}(L, M)$*,*

$$f \leq g \quad \textit{iff} \quad f \circ g^{\star_{L,M}} \leq 0_M \quad \textit{iff} \quad g^{\star_{L,M}} \circ f \leq 0_L \,,$$

where $0_M := (id_M)^{\star_{M,M}}$ *and* $0_L := (id_L)^{\star_{L,L}}$.

An involutive quantale *is a one-object involutive quantaloid.*

The superscripts L and M in $(\cdot)^{\star_{L,M}}$ shall be omitted if they are clear from the context. We state next elementary facts without proofs, the reader shall have no difficulty providing them. For a category $\mathcal{C}$ enriched over posets, we use $\mathcal{C}^{co}$ for the category with same objects and homsets, but for which the order is reversed.

Lemma 3. *In an involutive quantaloid $\mathcal{Q}$, if any of the inequalities below holds, then so do the other two:*

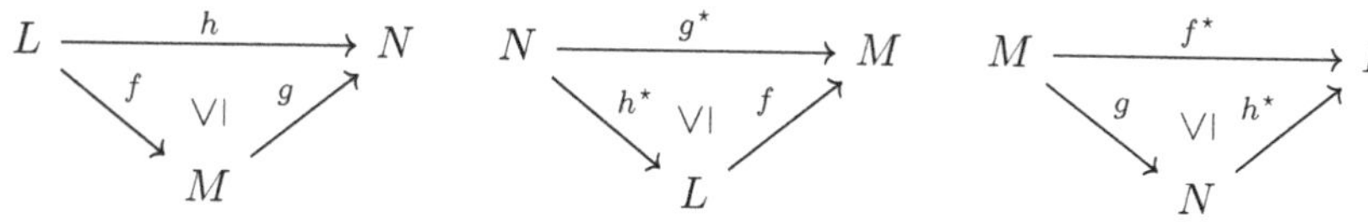

In particular, the operations $\star$ are order reversing, so $\star$ is the arrow part of a functor $\mathcal{Q} \longrightarrow (\mathcal{Q}^{op})^{co}$ which is the identity on objects.

Let us recall that in any quantaloid residuals exist being defined as follows: for $f : L \longrightarrow M$, $g : M \longrightarrow N$, and $h : L \longrightarrow N$,

$$g \backslash h : L \longrightarrow M := \bigvee \{ k \mid g \circ k \leq h \}, \quad h / f : M \longrightarrow N := \bigvee \{ k \mid k \circ f \leq h \},$$

so, the usual adjointness relations hold: $g \circ f \leq h$ iff $f \leq g \backslash h$ iff $g \leq h/f$.

Lemma 4. *In an involutive quantaloid, for $f : L \longrightarrow M$, $g : M \longrightarrow N$, and $h : L \longrightarrow N$, we have the following equalities:*

$$g\backslash h = (h^{\star_{L,N}} \circ g)^{\star_{M,L}}, \qquad h/f = (f \circ h^{\star_{L,N}})^{\star_{N,M}}.$$

In particular (for $L = N$ and $h = 0_L$) we have $g\backslash 0_L = g^{\star_{M,L}}$ and $0_L/f = f^{\star_{L,M}}$.

Let us argue that our definition coincides with the definition of a Girard quantaloid given in [19]. It is readily seen that, given an involutive quantaloid $\mathcal{Q}$, the collection $\{ 0_L = id_L^\star \mid L \text{ an object of } \mathcal{Q} \}$ is a cyclic dualizing family in the sense of [19]. Conversely, given such a family and $f : L \longrightarrow M$, we can define $f^\star := f\backslash 0_M$ and this definition yields an involutive quantaloid structure as defined here. This definition also sets a bijective correspondence between the two kind of structures.

4 Cyclic Elements of $\mathcal{Q}(L)$

We prove in this section that if a quantale of the form $\mathcal{Q}(L)$ is involutive, then $id_L^\star$ equals o_L defined in Eq. (2). From this it follows that there is at most one involutive quantale structure on $\mathcal{Q}(L)$ extending the quantale structure. Moreover, we also prove that if o_L is cyclic and distinct from $c_\top$, then L is completely distributive. To this end, let us firstly recall the following standard definitions:

Definition 5. *Let $\mathcal{Q}$ be a quantale. An element $\alpha \in \mathcal{Q}$ is said to be*

- cyclic *if $f\backslash\alpha = \alpha/f$, for each $f \in \mathcal{Q}$,*
- dualizing *if $(\alpha/f)\backslash\alpha = \alpha/(f\backslash\alpha) = f$, for each $f \in \mathcal{Q}$.*

We already mentioned that involutive quantale structures on a quantale $\mathcal{Q}$ are in bijection with cyclic dualizing elements of $\mathcal{Q}$. Let us also recall that, for an involutive quantaloid $\mathcal{Q}$ and an object L of $\mathcal{Q}$, $0_L := (id_L)^{\star_{L,L}}$ is both a cyclic and a dualizing element of the quantale $\mathcal{Q}(L, L)$.

An important first observation, stated in the next lemma, is that residuals of the form $g\backslash h$ in $\mathsf{Latt}_\vee$ can be constructed by means of the operations $\mathtt{int}(\cdot)$ (greatest join-continuous map below a given one) and $\rho(\cdot)$ (taking the right adjointof a join-continuous map).

Lemma 6. *For each $g \in \mathsf{Latt}_\vee(M, N)$, $h \in \mathsf{Latt}_\vee(L, N)$, we have*

$$g\backslash h = \mathtt{int}(\rho(g) \circ h).$$

Proof. Indeed, for each $f \in \mathsf{Latt}_\vee(L, M)$, we have $f \leq g\backslash h$ iff $g \circ f \leq h$, iff $g(f(x)) \leq h(x)$, for each $x \in L$, iff $f(x) \leq \rho(g)(h(x))$, for each $x \in L$, iff $f \leq \rho(g) \circ h$, iff $f \leq \mathtt{int}(\rho(g) \circ h)$. □

For the next lemma, recall that the join-continuous map o_L has been defined in (2) and that the maps c_t and a_t have been defined in (1).

Lemma 7. *We have $o_L = \bigvee_{t \in L} c_t \circ a_t$.*

Proof. Observe that $c_t(a_t(x)) = \bot$, if $x \leq t$, and $c_t(a_t(x)) = t$, if $x \not\leq t$. Therefore

$$(\bigvee_{t\in L} c_t \circ a_t)(x) = \bigvee_{t\in L} (c_t(a_t)(x)) = \bigvee_{\substack{t\in L,\\ c_t(a_t(x))\neq\bot}} c_t(a_t(x)) = \bigvee_{\substack{t\in L,\\ x\not\leq t}} t = o_L(x). \qquad \square$$

Lemma 8. *For each $x \in L$, $\mathtt{int}(\alpha_x) = a_{o_L(x)}$.*

Proof. Let us observe that $a_{o(x)} \leq \alpha_x$. This amounts to verifing that if $\alpha_x(t) = \bot$, then $a_{o(x)}(t) = \bot$. Now, $\alpha_x(t) = \bot$ iff $x \not\leq t$, and so $t \leq o(x)$, thus $a_{o(x)}(t) = \bot$. Next, let us suppose that $f : L \longrightarrow L$ is join-continuous and below α_x. Thus, if $\alpha_x(t) = \bot$, that is, if $x \not\leq t$, then $f(t) = \bot$. Then $f(o(x)) = f(\bigvee_{x\not\leq t} t) = \bigvee_{x\not\leq t} f(t) = \bot$. By monotonicity of f, if $t \leq o(x)$, then $f(t) = \bot$, showing that $f \leq a_{o(x)}$. $\square$

Theorem 9. *For each complete lattice L, the quantale $\mathcal{Q}(L)$ has at most two cyclic elements, among $c_\top$ and o_L.*

Proof. Now, let $h \in \mathcal{Q}(L)$ be cyclic. First we prove that $o_L \leq h$. Consider that, for each $x \in L$, $a_x \circ c_x = c_\bot \leq h$. Thus, since $g \circ f \leq h$ if and only if $f \circ g \leq h$, we also have $c_x \circ a_x \leq h$. Since this relation holds for each $x \in L$, then, using Lemma 7, the relation $o_L = \bigvee_{x\in L} c_x \circ a_x \leq h$ holds.

We argue now that if $h \neq c_\top$, then $h \leq o_L$ and therefore $h = o_L$. Let $x \in L$ and consider that $c_x \circ c_x \backslash h \leq h$. By cyclicity, we also have $c_x \backslash h \circ c_x \leq h$.

Now, $c_x \backslash h = \mathtt{int}(\rho(c_x) \circ h) = \mathtt{int}(\alpha_x \circ h)$ and therefore, using Lemma 8,

$$a_{o_L(x)} \circ h \circ c_x = \mathtt{int}(\alpha_x) \circ \mathtt{int}(h) \circ c_x \leq \mathtt{int}(\alpha_x \circ h) \circ c_x = c_x \backslash h \circ c_x \leq h\,.$$

If $t \neq \bot$, then, by evaluating the above inequality at t, we get $a_{o_L(x)}(h(x)) \leq h(t)$. Since $a_{o_L(x)}(h(x))$ takes values $\bot$ and $\top$, this means that $a_{o_L(x)}(h(x)) = \top$ implies $\top \leq h(t)$, for all $t \neq \bot$. That is, if $h(x) \not\leq o_L(x)$, then $h(t) = \top$, for all $t \neq \bot$ and $x \in L$. Otherwise stated, if $h \not\leq o_L$, then $h = c_\top$. $\square$

Let us recall that a nucleus on a quantale $\mathcal{Q}$ is a closure operator j such that $j(g) \circ j(f) \leq f(g \circ f)$. Nuclei are sort of congruences in the category of quantales while quotients into some involutive quantale bijectively correspond to nuclei j of the form $j(f) = (f\backslash 0)\backslash 0$ where 0 is cyclic [17, Theorem 1]. Thus, the above theorem exhibits the quantales $\mathcal{Q}(L)$ as sort of simple w.r.t. involutive quantales.

Lemma 10. *If L is not trivial, then $c_\top$ is not a dualizing element of $\mathcal{Q}(L)$.*

Proof. Observe that $c_\top$ is the greatest element of $\mathcal{Q}(L)$ and, for this reason, $f\backslash c_\top = c_\top / f = c_\top$, for each $f \in \mathcal{Q}(L)$. If $c_\top$ is dualizing, then $c_\bot = (c_\top / c_\bot)\backslash c_\top = c_\top$. Considering that the mapping from sending $x \in L$ to $c_x \in \mathcal{Q}(L)$ is an embedding, this shows that $\bot = \top$ in L. $\square$

Corollary 11. *If $h \in \mathcal{Q}(L)$ is a cyclic and dualizing element, then $h = o_L$. That is, if $\mathcal{Q}(L)$ is an involutive quantale, then $id_L^\star = o_L$.*

Proof. If L is trivial, then so is $\mathcal{Q}(L)$, and $h = c_\bot = o_L$. If L is not trivial, then, by Theorem 9, $h \in \{ o_L, c_\top \}$ and, by Lemma 10, $h \neq c_\top$. □

With respect to Theorem 9, we notice that $c_\top$, being the top element of $\mathcal{Q}(L)$, is always cyclic. It is therefore pertinent to ask when o_L is cyclic. Of course, this is the case if $o_L = c_\top$.

Theorem 12. *If o_L is a cyclic element of $\mathcal{Q}(L)$ and $o_L \neq c_\top$, then $x = \bigwedge_{t \not\leq x} o_L(t)$, for each $x \in L$. Consequently, L is a completely distributive lattice.*

Proof. Since o_L is cyclic, then, for each $x, y \in L$, the two conditions (a) $c_y \circ a_x \leq o_L$ and (b) $a_x \circ c_y \leq o_L$ are equivalent.

Condition (a) states that, for each $t \in L$, $t \not\leq x$ implies $y \leq o_L(t)$; that is $y \leq \bigwedge_{t \not\leq x} o_L(t)$. Condition (b) states that, for each $t \neq \bot$, if $y \not\leq x$ then $o_L(t) = \top$. This condition is equivalent to $y \not\leq x$ implies $o_L = c_\top$ or, equivalently, to $o_L \neq c_\top$ implies $y \leq x$. Thus we have that, if o_L is cyclic and $o_L \neq c_\top$, then (c) for each $x, y \in L$, $y \leq x$ iff $y \leq \bigwedge_{t \not\leq x} o_L(t)$. Now, condition (c) is easily recognized to be equivalent to the equality $x = \bigwedge_{t \not\leq x} o_L(t)$, holding for each $x \in L$. From the latter identity, complete distributivity of L follows using Raney's characterization of complete distributivity, Theorem 1. □

In this way we also obtain a refinement of one side of the equivalence stated in Proposition 2.6.18 of [4], where we do not need to refer to the cyclic dualizing element.

Corollary 13. *If $\mathcal{Q}(L)$ is an involutive quantale, then L is a completely distributive lattice.*

Proof. If $\mathcal{Q}(L)$ is an involutive quantale, then its dualizing cyclic element is, necessarily, o_L. In particular, o_L is cyclic and distinct from $c_\top$. By Theorem 12, L is completely distributive. □

We shall see that o_L is also dualizing if L is completely distributive. A remarkable fact arising from these considerations is that, on the class of pointed residuated lattices $\langle \mathcal{Q}(L), p \rangle$ (where $p \in \mathcal{Q}(L)$ is the point), the universal sentence $p \neq \top \;\&\; \forall x . x \backslash p = p / x$ implies distributivity as well as the linear double negation principle, $x = (x \backslash p) \backslash p$.

The Center of $\mathcal{Q}(L)$. Uniqueness of an involutive quantale structure extending the quantale structure of $\mathcal{Q}(L)$ can also be achieved through the observation that the unique central elements of $\mathcal{Q}(L)$ are id_L and $c_\bot$. We are thankful to Claudia Muresan for her help with investigating the center of $\mathcal{Q}(L)$.

Definition 14. *We say that an element β of a quantale $\mathcal{Q}$ is*

- central *if $\beta \circ x = x \circ \beta$, for each $x \in \mathcal{Q}$,*
- codualizing *if $x = \beta \backslash (\beta \circ x)$, for each $x \in \mathcal{Q}$.*

Lemma 15. *If $\mathcal{Q}$ is an involutive quantale, then $\alpha \in \mathcal{Q}$ is cyclic if and only if $\alpha^\star$ is central and it is dualizing if and only if $\alpha^\star$ is codualizing.*

Proof. Since $x\backslash\alpha = (\alpha^\star \circ x)^\star$, $\alpha/x = (x \circ \alpha^\star)^\star$, and $(\,\cdot\,)^\star$ is invertible, the equality $x\backslash\alpha = \alpha/x$ holds if and only if the equality $\alpha^\star \circ x = x \circ \alpha^\star$ holds.

Now α is dualizing if and only if, for each $x \in \mathcal{Q}$, $x = \alpha/(x\backslash\alpha) = \alpha^\star\backslash(x\backslash\alpha)^\star = \alpha^\star\backslash(\alpha^\star \circ x)$. □

Proposition 16. *The only central elements of* $\mathcal{Q}(L)$ *are* id_L *and* $c_\bot$.

Proof. Clearly, id_L and $c_\bot$ are central, so we shall be concerned to prove that they are the only ones with this property. To this end, for $x_0 \in L$, define

$$\nu_{x_0}(t) := \begin{cases} \bot, & t \leq x_0 \\ t, & \text{otherwise.} \end{cases}$$

Notice that if $x_0 = \bot$, then $\nu_{x_0} = id_L$, while if $x_0 = \top$, then $\nu_{x_0} = c_\bot$. We firstly claim that if β is central in $\mathcal{Q}(L)$, then $\beta = \nu_{x_0}$, for some $x_0 \in L$. Suppose β is central. For each $x \in L$, we have $c_x(x) = x$ and therefore

$$\beta(x) = (\beta \circ c_x)(x) = c_x(\beta(x))\,.$$

If $\beta(x) \neq \bot$, then, evaluating the rightmost expression, we obtain $\beta(x) = x$. Let $x_0 := \bigvee\{\, y \mid \beta(y) = \bot \,\}$, so $\beta(x_0) = \bot$. If $t \leq x_0$, then $\beta(t) \leq \beta(x_0) = \bot$ and, otherwise, $\beta(t) \neq \bot$ and so $\beta(t) = c_t(\beta(t)) = t$. Therefore, $\beta = \nu_{x_0}$.

Next, we claim that if $x_0 \notin \{\, \bot, \top \,\}$, then ν_{x_0} is not central. Observe that

$$\nu_{x_0}(f(x)) = \begin{cases} \bot, & f(x) \leq x_0\,, \\ f(x), & \text{otherwise}\,, \end{cases} \qquad f(\nu_{x_0}(x)) = \begin{cases} \bot, & x \leq x_0\,, \\ f(x), & \text{otherwise}\,. \end{cases}$$

It follows that if $\nu_{x_0} \circ f = f \circ \nu_{x_0}$, then $f(x_0) \leq x_0$. Indeed, if $f(x_0) \not\leq x_0$, then $f(x_0) \neq \bot$, $\nu_{x_0}(f(x_0)) = f(x_0) \neq \bot$, and $f(\nu_{x_0}(x_0)) = \bot$. Now, if $x_0 \notin \{\, \bot, \top \,\}$, then $c_\top$ is such that $x_0 < \top = c_\top(x_0)$, and therefore $\nu_{x_0} \circ c_\top \neq c_\top \circ \nu_{x_0}$. □

It is now possible to argue that, for a complete lattice L, there exists at most one extension of $\mathcal{Q}(L)$ to an involutive quantale as follows. Suppose that $\mathcal{Q}(L)$ is involutive, so let $(\,\cdot\,)^\star$ be a fixed involutive quantale structure. We shall argue that $id_L^\star$ is the unique cyclic and dualizing element of $\mathcal{Q}(L)$. If α is an arbitrary cyclic and dualizing element of $\mathcal{Q}(L)$, then $\beta := \alpha^\star$ is central and codualizing and $\beta \in \{\, c_\bot, id_L \,\}$ using Proposition 16. Since β is codualizing, then it is an injective function: if $\beta(x) = \beta(y)$, then $\beta \circ c_x = \beta \circ c_y$ and $c_x = \beta\backslash(\beta \circ c_x) = \beta\backslash(\beta \circ c_y) = c_y$; since the mapping sending t to c_t is an embedding, we obtain $x = y$. Thus, if L is not trivial, $\beta \neq c_\bot$ (since $c_\bot$ is constant). Whether or not L is trivial, we derive $\beta = id_L$. It follows that $\alpha = \alpha^{\star\star} = \beta^\star = id_L^\star$.

5 Raney's Transforms

Let L, M be two complete lattices. For $f : L \longrightarrow M$, define

$$f^\vee(x) := \bigvee_{x \not\leq t} f(t)\,, \qquad f^\wedge(x) := \bigwedge_{t \not\leq x} f(t)\,, \quad \text{for each } x \in L.$$

We call $f^{\vee}$ and $f^{\wedge}$ the *Raney's transforms* of f. Notice that f is not required to be monotone in order to define $f^{\vee}$ or $f^{\wedge}$ which, on the other hand, are easily seen to be monotone; these functions are even join and meet-continuous, respectively, as argued in the next lemma.

Lemma 17. *For any $f : L \longrightarrow M$, define*

$$g_f(y) := \bigwedge \{ z \mid f(z) \not\leq y \} . \tag{4}$$

Then g_f is right adjointto $f^{\vee}$ and therefore $f^{\vee}$ is join-continuous. Dually, $f^{\wedge}$ is meet-continuous.

We call the operation $(\,\cdot\,)^{\vee}$ Raney's transform for the following reason. For $\theta \subseteq L \times M$ an arbitrary relation, Raney [16] defined (up to some dualities)

$$r_\theta(x) := \bigwedge \{ y \in M \mid \forall (t,v).\, (t,v) \in \theta \text{ implies } x \leq t \text{ or } v \leq y \} . \tag{5}$$

Recall that a left adjoint $\ell : L \longrightarrow M$ can be expressed from its right adjoint $\rho : M \longrightarrow L$ by the formula $\ell(x) = \bigwedge \{ y \mid x \leq \rho(y) \}$. Using this expression with $\ell = f^{\vee}$ and $\rho = g_f$ defined in (4), we obtain

$$f^{\vee}(x) = \bigwedge \{ y \in M \mid \forall t.\, f(t) \not\leq y \text{ implies } x \leq t \} . \tag{6}$$

Clearly, if in (5) we let θ be the graph of f, defined by $(t,v) \in \theta$ if and only if $f(t) = v$, then we obtain equality between the right-hand sides of (5) and (6), and so $f^{\vee} = r_\theta$.

We list next the few properties we need to know about these transforms.

Lemma 18. *The transform $(\,\cdot\,)^{\vee}$ has the following properties:*

1. *if $f \leq g : L \longrightarrow M$, then $f^{\vee} \leq g^{\vee}$,*
2. *if $g : L \longrightarrow M$ and $f : M \longrightarrow N$ is monotone, then $(f \circ g)^{\vee} \leq f \circ (g^{\vee})$,*
3. *if $g : L \longrightarrow M$ and $f : M \longrightarrow N$ is join-continuous, then $(f \circ g)^{\vee} = f \circ (g^{\vee})$,*
4. *if $f : L \longrightarrow M$ is join-continuous (with L and M complete), then*

$$\ell(f^{\wedge}) = \rho(f)^{\vee} : M \longrightarrow L . \tag{7}$$

The proof of these properties does not present difficulties, possibly apart for the last item, for which we refer the reader to [7, Proposition 4.6 (b.iii)].

6 $\mathsf{Latt}_{\bigvee}^{\mathsf{cd}}$ Is an Involutive Quantaloid

We prove now that $\mathsf{Latt}_{\bigvee}^{\mathsf{cd}}$, the full subcategory of $\mathsf{Latt}_{\bigvee}$ whose objects are the completely distributive lattices, is an involutive quantaloid. By the results of Sect. 4, this is also the largest full subcategory of $\mathsf{Latt}_{\bigvee}$ with this property.

Recall from Theorem 1 that a complete lattice is completely distributive if and only if ${\omega_L}^{\vee} = id_L$ (or, equivalently, ${o_L}^{\wedge} = id_L$).

Lemma 19. *If L is a completely distributive lattice and $f : L \longrightarrow M$ is monotone, then* $\mathtt{int}(f) = (f \circ \omega_L)^\vee$ *and* $f^\vee = \mathtt{int}(f \circ o_L)$.

Proof. By monotonicity of f, we have $(f \circ \omega_L)^\vee \leq f \circ ({\omega_L}^\vee) = f$. Suppose that g is join-continuous and $g \leq f$. Then $g = g \circ ({\omega_L}^\vee) = (g \circ \omega_L)^\vee \leq (f \circ \omega_L)^\vee$. To see that $f^\vee = \mathtt{int}(f \circ o_L)$, observe that $f^\vee = (f \circ id)^\vee \leq f \circ ({id_L}^\vee) = f \circ o_L$, and therefore $f^\vee \leq \mathtt{int}(f \circ o_L)$. On the other hand, $\mathtt{int}(f \circ o_L) = (f \circ o_L \circ \omega_L)^\vee \leq f^\vee$, using the conunit of the adjunction, $o_L \circ \omega_L \leq id_L$. □

The interior operator so defined is quite peculiar, since for $g : L \longrightarrow M$ monotone and $f : M \longrightarrow N$ join-continuous, we have

$$\mathtt{int}(f \circ g) = (f \circ g \circ \omega_L)^\vee = f \circ (g \circ \omega_L)^\vee = f \circ \mathtt{int}(g)\,.$$

In general, if L is not a completely distributive lattice, then we would have, above, only an inequality, since $\mathtt{int}(f \circ g) \geq \mathtt{int}(f) \circ \mathtt{int}(g) = f \circ \mathtt{int}(g)$.

Lemma 20. *If L is a completely distributive lattice and $f : L \longrightarrow M$ is join-continuous, then $f = f^{\wedge\vee}$.*

Proof. We firstly show that $f^{\wedge\vee} \leq f$. If $x \not\leq t$, then $f^\wedge(t) = \bigwedge_{u \not\leq t} f(u) \leq f(x)$ and therefore $f^{\wedge\vee}(x) = \bigvee_{x \not\leq t} f^\wedge(t) \leq f(x)$, for all $x \in L$. Let us argue that $f \leq f^{\wedge\vee}$:

$$f = f \circ id_L = f \circ ({\omega_L}^\vee) = (f \circ \omega_L)^\vee \leq f^{\wedge\vee}\,,$$

where we have used the fact, dual to the relation $f^\vee = \mathtt{int}(f \circ o_L)$ established in Lemma 19, that $f^\wedge$ is the least meet-continuous function above $f \circ \omega_L$, so in particular $f \circ \omega_L \leq f^\wedge$. □

For $f : L \longrightarrow M$ join-continuous, define $f^{\star_{L,M}} : M \longrightarrow L$ as follows:

$$f^{\star_{L,M}} := \rho(f)^\vee = \ell(f^\wedge)\,.$$

Let us remark that the mappings $(\,\cdot\,)^\star$ so defined are the maps witnessing that completely distributive lattices are nuclear, see [7, Theorem 4.7]. We leave for future research to establish an exact connection between the notions of involutive quantaloid and of nuclear object in an autonomous category.

Theorem 21. *The operations $(\,\cdot\,)^{\star_{L,M}}$ so defined yield an involutive quantaloid structure on $\mathsf{Latt}^{\mathsf{cd}}_{\bigvee}$.*

Proof. Firstly, we verify that $f^{\star\star} = f$ using Lemmas 18 and 20, and the fact that the join-continuous functions are in bijection with meet-continuous functions via taking adjoints: $f^{\star\star} = \rho(\rho(f)^\vee)^\vee = \rho(\ell(f^\wedge))^\vee = f^{\wedge\vee} = f$.

We now verify that $(\,\cdot\,)^\star$ satisfies the constraints needed to have an involutive quantaloid. Let us remark that $id_L^\star = \rho(id_L)^\vee = {id_L}^\vee = o_L$.

Observe that since $(\,\cdot\,)^\star$ is defined by composing an order reversing and an order preserving function, it is order reversing. Since it is an involution, then $f \leq g$ if and only if $g^\star \leq f^\star$.

Now we assume that $f : L \longrightarrow M$ and $h : M \longrightarrow L$ and recall (see Lemma 19) that $h^{\star_{M,L}} = \rho(h)^\vee = \mathtt{int}(\rho(h) \circ o_L) : L \longrightarrow M$. Therefore, $h \circ f \leq o_L$ if and only if $f \leq \rho(h) \circ o_L$, if and only if $f \leq \mathtt{int}(\rho(h) \circ o_L) = h^\star$. Therefore, if $g : L \longrightarrow M$, then (letting $h = g^\star$) $f \leq g$ if and only if $g^\star \circ f \leq o_L$. Then, also, $f \leq g$ if and only if $g^\star \leq f^\star$ if and only if $f \circ g^\star \leq o_M$. □

Putting together Theorems 12 and 21, we obtain the following generalization of Proposition 2.6.18 in [4], where no mention of the choice of the cyclic dualizing element is required.

Corollary 22. *The quantale $\mathcal{Q}(L)$ is involutive if and only if L is a completely distributive lattice.*

For $f : L \longrightarrow M$ and $g : M \longrightarrow N$ (with L, M, N completely distributive lattices), let us define

$$g \oplus f := (f^\star \circ g^\star)^\star : L \longrightarrow M\,,$$

and observe that

$$(g \oplus f) = (f^\star \circ g^\star)^\star = \rho((\ell(f^\wedge) \circ \ell(g^\wedge)))^\vee = (g^\wedge \circ f^\wedge)^\vee\,.$$

That is:

Proposition 23. *The dual quantaloid structure arises via Raney's transforms from the composition* $\mathsf{Latt}_\vee(L^\partial, M^\partial) \times \mathsf{Latt}_\vee(M^\partial, N^\partial) \longrightarrow \mathsf{Latt}_\vee(L^\partial, N^\partial)$.

7 Remarks on the Equational Theory of the $\mathcal{Q}(L)$

We develop in this section few considerations concerning the equational theory of the involutive residuated lattices $\mathcal{Q}(L)$.

Theorem 24. *A complete lattice L is a chain if and only if $\mathcal{Q}(L)$ is an involutive quantale satisfying the mix rule, i.e. the inclusion $x \circ y \leq x \oplus y$.*

Proof. It is well known that the mix rule is equivalent to the inclusion $0 \leq 1$—where 1 is the unit for $\circ$ and 0 is the unit for $\oplus$. Therefore, an involutive quantale of the form $\mathcal{Q}(L)$ satisfies the mix rule if and only if $o_L \leq id_L$. This relation is easily seen to be equivalent to the statement that if $x \not\leq t$, then $t \leq x$, so L is a chain. For the converse, we just need to recall that every chain is a completely distributive lattice. □

Let us recall that an element x of a complete lattice L is *completely join-prime* if, for every $Y \subseteq L$, the relation $x \leq \bigvee Y$ implies $x \leq y$ for some $y \in Y$. It is not difficult to see that x is completely join-prime if and only if $x \not\leq o_L(x)$. Thus we say that a complete lattice is *smooth* if it has no completely join-prime element. For example, the interval $[0, 1]$ of the reals is a smooth completely distributive lattice. The following statement is an immediate consequence of these considerations.

Theorem 25. *A complete lattice L is smooth if and only if $id_L \leq o_L$. Thus, a completely distributive lattice L is smooth if and only if $\mathcal{Q}(L)$ satisfies the inclusion $1 \leq 0$ in the language of involutive residuated lattices.*

These statements generalize the remarks by Galatos and Jipsen, this collection, on the involutive residuated lattice of weakening relations on P. Let us recall that this involutive residuated lattice is isomorphic to $\mathcal{Q}(\mathcal{D}(P))$ where $\mathcal{D}(P)$ is the collection of downsets of P. Thus, they observe that $\mathcal{Q}(\mathcal{D}(P))$ satisfies the mix rule if and only if P is a chain, and that there are no non-trivial posets P such that $\mathcal{Q}(\mathcal{D}(P))$ satisfies the inclusion $1 \leq 0$. These facts might be seen as consequences Theorems 24 and 25, considering that $\mathcal{D}(P)$ is a chain if and only if P is a chain, and that $\mathcal{D}(P)$ is spatial, meaning that every element of $\mathcal{D}(P)$ is the join of the completely join-prime elements below it (so, $\mathcal{D}(P)$ has plenty of completely join-prime elements).

For a family $\{ f_i \in \mathsf{Latt}_{\bigvee}(L, M) \mid i \in I \}$, let us define $\bigwedge_{i \in I} f_i$ and $\bigwedge\!\!\!\!\bigwedge_{i \in I} f_i$ by

$$(\bigwedge_{i \in I} f_i)(x) := \bigwedge_{i \in I}(f_i(x))\,, \qquad (\bigwedge\!\!\!\!\bigwedge_{i \in I} f_i)(x) := \bigvee_{x \not\leq t} \bigwedge_{i \in I} f_i(\omega_L(t))\,. \tag{8}$$

Notice that $\bigwedge_{i \in i} f_i$ need not be join-continuous while $\bigwedge\!\!\!\!\bigwedge_{i \in I} f_i$ is join-continuous and $\bigwedge\!\!\!\!\bigwedge_{i \in I} f_i = \mathtt{int}(\bigwedge_{i \in i} f_i)$ if L is completely distributive, see Lemma 19. Under the latter condition, $\bigwedge\!\!\!\!\bigwedge_{i \in I} f_i$ is the infimum of $\{ f_i \mid i \in I \}$ within the complete lattice $\mathsf{Latt}_{\bigvee}(L, M)$. The explicit description of the infimum given in (8) can be exploited to prove that $\mathsf{Latt}_{\bigvee}^{\mathsf{cd}}$ is closed under the monoidal operations inherited from $\mathsf{Latt}_{\bigvee}$, see e.g. [4, 7, 20], thus it is $\star$-autonomous [1]. We expect the formula in (8) also to be useful for computational issues, see Ramirez et al., this collection.

Coming back to the equational theory of the $\mathcal{Q}(L)$, an important consequence of $\mathsf{Latt}_{\bigvee}^{\mathsf{cd}}$ being $\star$-autonomous is that $\mathcal{Q}(L)$ is completely distributive if L is completely distributive (the converse holds as well). Then, the following obstacle arises towards finding representation theorems for involutive residuated lattices via the $\mathcal{Q}(L)$:

Corollary 26. *If an involutive residuated lattice $\mathcal{Q}$ has an embedding into an involutive quantale of the form $\mathcal{Q}(L)$, then $\mathcal{Q}$ is distributive.*

Indeed, if $\mathcal{Q}(L)$ is an involutive quantale, then L is a completely distributive lattice and $\mathcal{Q}(L)$ as well. Thus, if $\mathcal{Q}$ has a lattice embedding into $\mathcal{Q}(L)$, then L is distributive.

8 Conclusions and Future Steps

The research exposed in this paper tackles and solves a natural problem encountered during our investigations of certain quantales built from complete chains [6, 21, 22]. The problem asks to characterize the complete chains whose quantale of join-continuous endomaps is involutive. Every complete chain is a completely distributive lattice and by now we know that every complete chain has this property; in particular, other properties of chains and posets, such as self-duality, are not relevant.

The solution provided, building on [4, Proposition 2.6.18], is as general as possible, in two respects. On the one hand, an exact characterization of all the complete lattices—not just the chains—L for which $\mathcal{Q}(L)$ is involutive becomes available: these are the completely distributive lattices; improving on [4, Proposition 2.6.18], we argue that the choice of a cyclic dualizing element does not matter. In particular, the characterization covers different kind of involutive quantales known in the literature, those discovered in our investigation of complete chains and those known as the residuated lattices of weakening relations—arising from the relational semantics of distributive linear logic. On the other hand, we show that the involutive quantale structures on completely distributive lattices are uniform, yielding and involutive quantaloid structure on the category of completely distributive lattices and join-continuous functions.

We have drawn several consequences from the observations developed, among them, the fact that if an involutive quantale $\mathcal{Q}$ can be embedded into an quantale of the form $\mathcal{Q}(L)$, then it is distributive. This fact calls for a characterization of the involutive residuated lattices embeddable into some $\mathcal{Q}(L)$, a research track that might require to or end up with determining the variety of involutive residuated lattices generated by the $\mathcal{Q}(L)$. A second research goal, that we might tackle in a close future, demands to investigate the algebra developed in connection with the continuous weak order [22] in the wider and abstract setting of completely distributive lattices. Let us recall that in [22] a surprising bijection was established between two kind of objects, the maximal chains in the cube lattice $[0,1]^d$ and the families $\{ f_{i,j} \in \mathcal{Q}([0,1]) \mid 1 \leq i < j \leq d \}$ such that, for $i < j < k$, $f_{j,k} \circ f_{i,j} \leq f_{i,k} \leq f_{j,k} \oplus f_{i,j}$. So, are there other surprising bijections if the interval $[0,1]$ is replaced by an arbitrary completely distributive lattice, and if we move from the involutive quantale setting to the multisorted setting of involutive quantaloids?

Acknowledgment. The author is thankful to Srecko Brlek, Claudia Muresan, and André Joyal for the fruitful discussions these scientists shared with him on this topic during winter 2018.

References

1. Barr, M.: ∗-Autonomous Categories. Lecture Notes in Mathematics, vol. 752. Springer, Berlin (1979). https://doi.org/10.1007/BFb0064579
2. Barr, M.: Nonsymmetric *-autonomous categories. Theor. Comput. Sci. **139**(1–2), 115–130 (1995)
3. Dzhumadil'daev, A.S.: Worpitzky identity for multipermutations. Math. Notes **90**(3), 448–450 (2011)
4. Eklund, P., Gutiérrez García, J., Höhle, U., Kortelainen, J.: Semigroups in Complete Lattices, Developments in Mathematics: Quantales, Modules and Related Topics, vol. 54. Springer, Cham (2018). https://doi.org/10.1007/978-3-319-78948-4. With a foreword by Jimmie Lawson
5. Galatos, N., Jipsen, P., Kowalski, T., Ono, H.: Residuated Lattices: An Algebraic Glimpse at Substructural Logics. Studies in Logic and the Foundations of Mathematics, vol. 151. Elsevier, Amsterdam (2007)

6. Gouveia, M.J., Santocanale, L.: MIX ⋆-autonomous quantales and the continuous weak order. In: Desharnais, J., Guttmann, W., Joosten, S. (eds.) RAMiCS 2018. LNCS, vol. 11194, pp. 184–201. Springer, Cham (2018). https://doi.org/10.1007/978-3-030-02149-8_12
7. Higgs, D.A., Rowe, K.A.: Nuclearity in the category of complete semilattices. J. Pure Appl. Algebra **57**(1), 67–78 (1989)
8. Holland, C.: The lattice-ordered group of automorphisms of an ordered set. Michigan Math. J. **10**, 399–408 (1963)
9. Howie, J.M.: Products of idempotents in certain semigroups of transformations. Proc. Edinburgh Math. Soc. **17**(2), 223–236 (1971)
10. Jipsen, P.: Relation algebras, idempotent semirings and generalized bunched implication algebras. In: Höfner, P., Pous, D., Struth, G. (eds.) RAMICS 2017. LNCS, vol. 10226, pp. 144–158. Springer, Cham (2017). https://doi.org/10.1007/978-3-319-57418-9_9
11. Lambrou, M.S.: Completely distributive lattices. Fund. Math. **119**(3), 227–240 (1983)
12. Laradji, A., Umar, A.: Combinatorial results for semigroups of order-preserving full transformations. Semigroup Forum **72**(1), 51–62 (2006)
13. Meloni, G., Santocanale, L.: Relational semantics for distributive linear logic, August 1995, preprint. https://hal.archives-ouvertes.fr/hal-01851509
14. Protin, M.C., Resende, P.: Quantales of open groupoids. J. Noncommut. Geom. **6**(2), 199–247 (2012)
15. Raney, G.N.: A subdirect-union representation for completely distributive complete lattices. Proc. Am. Math. Soc. **4**, 518–522 (1953)
16. Raney, G.N.: Tight Galois connections and complete distributivity. Trans. Am. Math. Soc. **97**, 418–426 (1960)
17. Rosenthal, K.I.: A note on Girard quantalesa. Cahiers Topologie Géom. Différentielle Catég. **31**(1), 3–11 (1990)
18. Rosenthal, K.I.: Quantales and Their Applications. Pitman Research Notes in Mathematics Series, vol. 234. Longman Scientific & Technical, Harlow (1990)
19. Rosenthal, K.I.: Girard quantaloids. Math. Struct. Comput. Sci. **2**(1), 93–108 (1992)
20. Rowe, K.A.: Nuclearity. Canad. Math. Bull. **31**(2), 227–235 (1988)
21. Santocanale, L.: On discrete idempotent paths. In: Mercaş, R., Reidenbach, D. (eds.) WORDS 2019. LNCS, vol. 11682, pp. 312–325. Springer, Cham (2019). https://doi.org/10.1007/978-3-030-28796-2_25
22. Santocanale, L., Gouveia, M.J.: The continuous weak order, December 2018, preprint. https://hal.archives-ouvertes.fr/hal-01944759
23. Stubbe, I.: Towards "dynamic domains": totally continuous cocomplete Q-categories. Theor. Comput. Sci. **373**(1–2), 142–160 (2007)

Computer-Supported Exploration of a Categorical Axiomatization of Modeloids

Lucca Tiemens[1,3(✉)], Dana S. Scott[2], Christoph Benzmüller[3,4], and Miroslav Benda[5]

[1] Technische Universität Berlin, Berlin, Germany
tiemens@campus.tu-berlin.de
[2] University of California, Berkeley, CA, USA
scott@andrew.cmu.edu
[3] Freie Universität Berlin, Berlin, Germany
[4] University of Luxembourg, Esch-sur-Alzette, Luxembourg
c.benzmueller@fu-berlin.de
[5] Formerly at University of Washington, Seattle, USA
miro@oasa.com

Abstract. A modeloid, a certain set of partial bijections, emerges from the idea to abstract from a structure to the set of its partial automorphisms. It comes with an operation, called the derivative, which is inspired by Ehrenfeucht-Fraïssé games. In this paper we develop a generalization of a modeloid first to an inverse semigroup and then to an inverse category using an axiomatic approach to category theory. We then show that this formulation enables a purely algebraic view on Ehrenfeucht-Fraïssé games.

1 Introduction

Modeloids have been introduced by Benda [1]. They can be seen as an abstraction from a structure to a partial automorphism semigroup created in the attempt to study properties of structures from a different, more general angle which is independent of the language that is defining the structure. We do not follow Benda's original formulation in terms of an equivalence relation but treat modeloids as a certain set of partial bijections. Our recent interest in them was triggered by Scott's suggestion to look at the modeloidal concept from a categorical perspective. The new approach aims at establishing a framework in which the relationship between different structures of the same vocabulary can be studied by means of their partial isomorphisms. The overall project is work in progress, but as a first result we obtained a purely algebraic formulation of Ehrenfeucht-Fraïssé games.

Throughout the project, computer-based theorem proving is employed in order to demonstrate and explore the virtues of automated and interactive theorem proving in context. The software used is Isabelle/HOL [13] in the 2019

U. Fahrenberg et al. (Eds.): RAMiCS 2020, LNCS 12062, pp. 302–317, 2020.
https://doi.org/10.1007/978-3-030-43520-2_19

Edition. We are generally interested in conducting as many proofs of lemmas and theorems as possible by using only the *sledgehammer*[1] command, and to study how far full proof automation scales in this area. Reporting on these practically motivated studies, however, will not be the focus of this paper. We only briefly mention here how we encoded, in Isabelle/HOL, an inverse semigroup and an inverse category, and we present a summary of our practical experience.

Inverse semigroups (see e.g. [10] for more information) play a major role in this paper. They serve as a bridge between modeloids and category theory. The justification for this is given by the fact that an inverse semigroup can be faithfully embedded into a set of partial bijections by the Wagner-Preston representation theorem. This opens up the possibility of generalizing modeloids, which are sets of partial bijections, to the language of inverse semigroup theory.

Once there, we have a natural transition from an inverse semigroup to an inverse category (for further reference see [12]). We introduce the theory of inverse categories by an equational axiomatization that enables computer-supported reasoning. This serves as the basis for our formulation of a categorical modeloid.

In each stage of generalization the derivative, a central operation in the theory of modeloids, can be adapted and reformulated. This operation is about extending the elements of a modeloid. Suppose that τ is a finite relational vocabulary meaning that τ consists only of finitely many relation and/or constant symbols. As it turns out, the derivative on a categorical modeloid on the category of finite τ-structures is equivalent to playing an Ehrenfeucht-Fraïssé game.

This paper is organized in the following way. In Sect. 2 we define both modeloids and the derivative operation. We then turn to inverse semigroups in Sect. 3 and develop the axiomatization of a modeloid in inverse semigroup language. Section 4 shows how to represent a category in Isabelle/HOL and defines the categorical modeloid. After the derivative operation is established in this context, we give an introduction to Ehrenfeucht-Fraïssé games in Sect. 5 and present the close connection between the categorical derivative and Ehrenfeucht-Fraïssé games. Proofs for the stated theorems, propositions and lemmas are presented in the extended preprint [16] of this paper (cf. also [15]); the Isabelle/HOL source files are available online.[2]

2 Modeloids

Let us first recall the definitions of a partial bijection and of partial composition.

Definition 1 (Partial bijection and partial composition). *A partial bijection $f : X \to Y$ is a partial injective function. The inverse of f, also a partial bijection and denoted by f^{-1}, is given by the preimage of the elements in the codomain of f: $f^{-1}(y) = f^{-1}(\{y\})$, $\forall y \in cod(f)$.*

[1] *Sledgehammer* [3] is linking interactive proof development in Isabelle/HOL with anonymous calls to various integrated automated theorem proving systems. Among others, the tool converts the higher-order problems given to it into first-order representations for the integrated provers, it calls them and analyses their responses, and it tries to identify minimal sets of dependencies for the theorems it proves this way.

[2] See http://christoph-benzmueller.de/papers/RAMICSadditionalMaterial.zip.

The composition between two partial functions $f : X \to Y$ and $g : Y \to Z$ is defined only on $f^{-1}(dom(g) \cap cod(f))$. Then the partial composition

$$(g \circ f)(x) = g(f(x)), \quad \forall x \in f^{-1}(dom(g) \cap cod(f))$$

is well-defined.

Furthermore, let Σ be a finite non-empty set. We then define

$$F(\Sigma) := \{f : \Sigma \to \Sigma \mid f \text{ is a partial bijection}\} \tag{1}$$

as the set of all partial bijections on Σ.

Definition 2 (Modeloid[1]). *Let $M \subseteq F(\Sigma)$. M is called a modeloid on Σ if, and only if, it satisfies the following axioms:*

1. *Closure of composition: $f, g \in M \Rightarrow f \circ g \in M$*
2. *Closure of taking inverses: $f \in M \Rightarrow f^{-1} \in M$*
3. *Inclusion property: $f \in M$ and $A \subset dom(f)$ implies $f|_A \in M$*
4. *Identity: $id_\Sigma \in M$*

As such, a modeloid is a set of partial bijections which is closed under composition and taking inverses, which has the identity on Σ as a member, and which satisfies the inclusion property. The inclusion property can be seen as a downward closure in regards of function restriction.

In order to further illustrate the definition, we present a motivating example from model theory.

Example 1. Let $S = (A, R_1, ...)$ be a finite relational structure. The set M of all partial isomorphisms on S forms a *modeloid.*

The name modeloid originates from the above example since S is also called a model. For further motivation, background information and details on modeloids, we refer to Benda's paper [1]; a nice example in there is the construction of a Scott Sentence presented through modeloidal glasses [1, p. 82]. We, on the other hand, turn to the core concept of the derivative which is defined in the following way. For convenience we represent a partial bijection as a set of tuples.

Definition 3 (Derivative). *Let M be a modeloid on Σ. Then the derivative $D(M) \subseteq F(\Sigma)$ is defined by*

$$\begin{aligned}\{(x_1, y_1), .., (x_n, y_n)\} \in D(M) :\Leftrightarrow \\ \forall a \in \Sigma\, \exists b \in \Sigma : \{(x_1, y_1), .., (x_n, y_n), (a, b)\} \in M \wedge \\ \forall a \in \Sigma\, \exists b \in \Sigma : \{(x_1, y_1), .., (x_n, y_n), (b, a)\} \in M\end{aligned}$$

A derivative $D(M)$ is thus a set which only contains partial bijections that can be extended by an arbitrary element from Σ and which then still belong to M. This extension can take place either in the domain or in the range of the function. The next two results [1, Prop 2.3] provide some insight into why modeloids and the derivative operation are in harmony.

Lemma 1. *Let M be a modeloid on Σ and $D(M)$ the derivative. Then we have that $D(M) \subseteq M$.*

Proposition 1. *If M is a modeloid then so is $D(M)$.*

The importance of these results is essentially due to the fact that they enable us to apply the derivative repeatedly.

3 Inverse Semigroups and Modeloids

In this section we show how the Wagner-Preston representation theorem justifies our generalization of a modeloid to inverse semigroup language. We also discuss how well proof automation performs in the context of inverse semigroups. Some familiarity with the Isabelle/HOL proof assistant [3,13] is assumed.

3.1 Inverse Semigroups in Isabelle/HOL

We start with the equational definition of an inverse semigroup.

Definition 4 (Inverse semigroup [6]). *Let S be a set equipped with the binary operation $* : S \times S \to S$ and the unary operation $a \mapsto a^{-1}$. $(S,^{-1},*)$ is called an inverse semigroup if, and only if, it satisfies the axioms*

1. $(x * y) * z = x * (y * z)$ *for all* $x, y, z \in S$,
2. $x * x^{-1} * x = x$ *for all* $x \in S$,
3. $(x^{-1})^{-1} = x$ *for all* $x \in S$ *and*
4. $x * x^{-1} * y * y^{-1} = y * y^{-1} * x * x^{-1}$ *for all* $x, y \in S$

An inverse in semigroup theory is a generalization of the known group theoretical definition. This generalized definition does not depend on a specified unique neutral element. Intuitively, it can be thought of as the inverse map of a partial bijection.

Definition 5 (Inverse). *Let $(S, *)$ be a semigroup and $x \in S$. Then $y \in S$ is called an inverse of x if, and only if, $x * y * x = x$ and $y * x * y = y$.*

We encode an inverse semigroup as follows in Isabelle/HOL.

```
locale inverseSemigroup =

—‹The two function symbols, the inverse and composition,
    appearing in the vocabulary of an inverse semigroup›
  fixes
    inv:: "'a⇒'a"     ("inv _" 109) and
    mult::"'a⇒'a⇒'a" (infixl "⊗" 110)

—‹Here we set the axioms that all individuals of type @{text "'a"} have to obey›
  assumes
    Associativity: "(x ⊗ y) ⊗ z = x ⊗ (y ⊗ z)" and
    Regularity:    "x ⊗ (inv x) ⊗ x = x" and
    InvInv:        "((inv (inv x)) = x)" and
    IdemComm:      "x ⊗ (inv x) ⊗ y ⊗ (inv y) = y ⊗ (inv y) ⊗ x ⊗ (inv x)"
```

The domain for individuals is chosen to be $'a$, which is a type variable. This means we have encoded a polymorphic version of inverse semigroups.

Using this implementation almost all results needed for proving the Wagner-Preston representation theorem, which we will discuss shortly, can be found by automated theorem proving. Occasionally, however, some additional lemmas to the ones usually presented in a textbook (e.g. [10]) are needed. By automated theorem proving we here mean the use of *sledgehammer* [3] for finding the proofs of the given statements without any further interaction. Regarding equivalent definitions of an inverse semigroup, we were able to automate the proofs of the following theorem (except for 2. ⇒ 1., which is due to a Skolemization issue).

Theorem 1 ([10]). *Let $(S, *)$ be a semigroup. Then the following are equivalent:*

1. *There is $^{-1} : S \to S$ such that $(S, ^{-1}, *)$ is an inverse semigroup.*
2. *Every element of S has a unique inverse.*
3. *Every element of S is regular, meaning $\forall x \in S\, \exists y \in S : x * y * x = x$, and idempotents in S commute.*

Our experiments confirm that automated theorem proving (and also model finding) can well support the exploration of an axiomatic theory as presented. However, the intellectual effort needed to model and formulate the presented mathematics in the first place is of course still crucial, and a great deal of work has gone into this intuitive aspect of the development process. A more technical challenge also is to find suitable intermediate steps that can be proven automatically by *sledgehammer*.

3.2 Modeloid as Inverse Semigroup

We now show that every modeloid $M \subseteq F(\Sigma)$ under partial composition is an inverse semigroup. We make use of Theorem 1 by using the third characterization. For this task regard $(M, \circ)$ as a semigroup. This is clear since composition of partial functions is associative. Since the partial identities of M are exactly the idempotent elements in $(M, \circ)$, commutativity is ensured by referring to the next proposition. Furthermore, also by using the next proposition, the closure of taking inverses required by a modeloid implies regularity for all elements in M. Hence, $(M, ^{-1}, \circ)$ is an inverse semigroup.

Proposition 2 ([10]). *Let $f : X \to Y$ be a partial bijection.*

1. *For a partial bijection $g : Y \to X$, the equations $f = fgf$ and $g = gfg$ hold if, and only if, $g = f^{-1}$*
2. *$1_A 1_B = 1_{A \cap B} = 1_B 1_A$ for all partial identities 1_A and 1_B where $A, B \subseteq X$*

Not only is every modeloid an inverse semigroup, but by the Wagner-Preston representation theorem also every inverse semigroup can be faithfully embedded into $F(\Sigma)$, which is itself a modeloid. This motivates the idea of formulating the axioms for a modeloid in inverse semigroup language. Our aim is to restate the derivative operation in this context. In order to achieve this, we shall translate the axioms from Definition 2, examining them one by one.

1. Closure of Composition: Because of the embedding, the composition of partial functions will simply be the $*$-operation in an inverse semigroup.
2. Closure of taking inverses: By Theorem 1 an inverse semigroup is such that the inverse exists for every element and is unique, hence resembling the inverses of partial functions and in particular the closure property.
3. The inclusion property: Here it is not immediately apparent how this can be expressed within an inverse semigroup. We shall see that the *natural partial order* is capable of doing that.
4. The identity on Σ: The identity is a certain idempotent element in an inverse semigroup. It will lead us to the notion of an *inverse monoid.*

It is Axiom 3 that we focus our attention on next. We define the *natural partial order* and present the Wagner-Preston representation theorem, which establishes a connection to function restriction in $F(\Sigma)$. We introduce notation for such a restriction. For two partial functions f, g we write $g \subseteq f$ to say that $dom(g) \subseteq dom(f)$ and $\forall x \in dom(g) : g(x) = f(x)$.

Definition 6 (Natural partial order). *Let $\Sigma = (\Sigma, ^{-1}, *)$ be an inverse semigroup and $s, t \in \Sigma$. Then we define for some idempotent $e \in \Sigma$*

$$s \leq t :\Leftrightarrow s = t * e.$$

Theorem 2 (Wagner-Preston representation theorem[10]). *Let $\Sigma = (\Sigma, ^{-1}, *)$ be an inverse semigroup. Then there is an injective homomorphism $\Omega : \Sigma \to F(\Sigma)$, such that for $a, b \in \Sigma$ we have $a \leq b \Longleftrightarrow \Omega(a) \subseteq \Omega(b)$.*

From this theorem it is clear what we mean by a faithful embedding of an inverse semigroup into the set of partial bijections $F(\Sigma)$. Faithfulness corresponds to the fact that the *natural partial order* in light of the representation theorem is equivalent to the partial order which function restriction defines. This nicely opens up the possibility to capture the essence of the inclusion property from Definition 2 by the *natural partial order.* Let $M \subset S$, where S is an inverse semigroup. Then the inclusion property can be stated as

$$\forall f \in M \, \forall g \in S : g \leq f \Longrightarrow g \in M. \tag{2}$$

By setting $S = F(\Sigma)$, the dependency of M on $F(\Sigma)$ can be seen explicitly. In the abstract formulation of a modeloid we will keep this subset property. It is immediate that a modeloid, seen as an inverse semigroup, fulfills (2) by the following proposition.

Proposition 3. *Let M be a modeloid on Σ. Then, for $f, g \in M$, we have*

$$g \leq f \Longleftrightarrow g \subseteq f$$

In a modeloid M the inclusion property implies that the empty partial bijection, which we denote by 0, is also included in M. As a result we want to establish a similar behavior in the generalized modeloid. The deeper reason for this is found in the definition of the derivative operation, because it requires the notion of an atom, which can only be defined if a zero element is present. Seeing M as an inverse semigroup, 0 is an idempotent element for which the following property holds: $\forall x \in M : 0 * x = 0$. Hence, we will call the idempotent with this property the *zero element*. When defining a modeloid in semigroup language we require the zero element to be part of it.

Turning to Axiom 4, which is $id_\Sigma \in M$, we examine which element of an inverse semigroup S is most suitable for this task. To evaluate, we again look at the modeloid M as an inverse semigroup. In this semigroup, id_Σ will be an idempotent e satisfying $\forall x \in M : e{*}x = x$. Such an element is known as a neutral element in the context of group theory. We require for the inverse semigroup, which we eventually call a modeloid, that e is part of it. What we get is known as an inverse monoid in the literature.

Remark 1. Given an inverse monoid, denoted by S^1, and the element e with $e{*}x = x$, $\forall x \in S^1$. Consider the representation theorem again: this theorem is not guaranteeing uniqueness of the embedding, and in fact there can be several ones. Hence, we cannot assume that e will be mapped to the identity id_Σ. However, for all idempotent $f \in S^1$ we have that $f \leq e$ because $f = e * f$ by the assumption about e. Hence, e is always the upper bound of all idempotents in S^1.

We have prepared everything needed for defining a modeloid again. We shall call it a semimodeloid. Note, as mentioned before, that a modeloid is a subset of $F(\Sigma)$ for some non-empty set Σ and, as discussed, we keep this subset property to state the inclusion axiom.

Definition 7 (Semimodeloid). *Let $S^1 = (\Sigma, {}^{-1}, *, e, 0)$ be an inverse monoid. Then $M \subseteq \Sigma$ is called a semimodeloid if, and only if,*

1. $\forall x, y \in M : (x * y) \in M$
2. $\forall x \in M : x^{-1} \in M$
3. $\forall x \in M\, \forall y \in S^1 : y \leq x \Rightarrow y \in M$
4. $e \in M$

Remark 2. A semimodeloid is again an inverse monoid with the zero element.

Proposition 4. *Every semimodeloid can be faithfully embedded into a modeloid. Furthermore, by the considerations above, every modeloid is a semimodeloid.*

Now we develop the derivative operation in the setting of a semimodeloid. Consider again Definition 3 in which we have introduced the derivative operation. It is evident that the elements of Σ are of crucial importance. Furthermore, we are required to be able to extend the domain of a function by one element at a time. This poses a challenge because in an inverse monoid this information is not directly accessible. But as we shall see, it is possible to obtain.

First we characterize the elements of Σ. Therefore, consider $F(\Sigma)$ and realize that all the singleton-identities $id_{\{a\}}$, for $a \in \Sigma$, are in natural bijection to the elements of Σ. The special property of such a singleton-identity is

$$\forall f \in F(\Sigma) : f \subseteq id_{\{a\}} \Rightarrow (f = id_{\{a\}} \vee f = 0) \tag{3}$$

since $dom(id_{\{a\}}) = \{a\}$. Seeing $F(\Sigma)$ as an inverse monoid with zero element leads to the following definition.

Definition 8 (Atom). *Let S^1 be an inverse monoid with zero element 0. Then a non-zero element $x \in S^1$ is an atom if, and only if,*

$$\forall f \in S^1 : f \leq x \Rightarrow (f = x \vee f = 0)$$

Our plan is to use the notion of an atom to define the derivative. The next lemma justifies this usage.

Lemma 2. *The idempotent atoms in $F(\Sigma)$ are exactly the singleton-identities.*

This suffices to define the derivative for semimodeloids. We then ensure that the definition matches Definition 3 if the semimodeloid is on $F(\Sigma)$.

Definition 9 (Derivative—semimodeloid). *Let M be a semimodeloid on the inverse monoid S^1 with zero element 0. We define the derivative $D(M)$ of M as*

$$D(M) := \{f \in M \mid \forall \text{ idempotent atoms } a \in S^1 \, \exists x \in M : (f \leq x \wedge a \leq x^{-1}x) \wedge \\ \forall \text{ idempotent atoms } b \in S^1 \, \exists y \in M : (f \leq y \wedge b \leq yy^{-1})\}$$

If we think about x in the above definition as a partial bijection, then $x^{-1}x$ is the identity on the domain of x and, hence, the condition $a \leq x^{-1}x$ expresses that a is in the domain of x. Similarly $b \leq yy^{-1}$ states that b is in the range of y.

Proposition 5. *The derivative on a modeloid M produces the same result as the semimodeloidal derivative on M.*

4 Categorical Axiomatization of a Modeloid

We use an axiomatic approach to category theory based on *free logic* [8,9,14]. As demonstrated by Benzmüller and Scott [2], this approach enables the encoding of category theory in Isabelle/HOL. Their encoding work is extended below to cover also inverse categories. Subsequently we formalize modeloids and derivatives in this setting.

4.1 Category Theory in Isabelle/HOL

When looking at the definition of a category $\mathbf{C}$, one can realize that the objects $A, B, C, ..$ are in natural bijection with the identity morphisms $\mathbf{1}_A, \mathbf{1}_B, \mathbf{1}_C, ...$ because those are unique. This enables a characterization of a category just by its morphisms and their compositions, which is used to establish a formal axiomatization. However, in this axiomatic approach we are faced with the challenge of partiality, because the composition between two morphisms $f, g \in \mathbf{C}$ is defined if, and only if,

$$\text{dom}(g) = \text{cod}(f). \tag{4}$$

As a result composition is a partial operation.

An elegant way to deal with this issue is by changing the underlying logic to *free logic*. In free logic an explicit notion of existence is introduced for the objects in the domain that we quantify over. In our case the domain consists of the morphisms of a category. The idea now is to define the composition total, that is, any two morphisms can always be composed, but only those compositions "exist" that satisfy (4). Because we can distinguish between existing and non-existing morphisms, we are able to formulate statements that take only existing morphisms into account. In this paper we want to work with a unique non-existing morphism which will be denoted by $\star$. Hence a composition of morphisms, that does not satisfy (4), will result in $\star$. We refer to Benzmüller and Scott [2] for more information on the encoding of free logic in Isabelle/HOL.

Based upon this groundwork, a category in Isabelle/HOL is defined as follows.

```
typedecl α —‹This type can be thought of representing the morphisms of a category.›

locale category =
          —‹We need three functions and constant to define a category.›
  fixes domain:: "α ⇒ α" ("dom _" [108] 109) and
        codomain:: "α ⇒ α" ("cod _" [110] 111) and
        composition:: "α ⇒ α ⇒ α" (infix "·" 110) and
        star::α ("*") —‹Symbol for non-existing elements in terms of free logic›

  assumes
—‹Here we define the axioms that the morphisms
   in interaction with the functions have to obey.›

—‹The existence of the domain of a morphism implies the existence of the morphism.›
        S1: "E(dom x) → E x" and
        S2: "E(cod y) → E y" and          —‹The same goes for the codomain.›


—‹As we have seen, the composition only exists if the two morphisms are composable.›
—‹We use ≃ to denote the existing equality which requires that both sides
   of the equation exist.›
        S3: "E(x·y) ↔ dom x ≃ cod y" and


        S4: "x·(y·z) ≅ (x·y)·z" and       —‹Composition of morphisms is associative.›
—‹We use ≅ to denote the Kleene equality which only implies equality
   if at least one side of the equation exists›


—‹The domain of a morphisms serves as the right identity for composition.›
        S5: "x·(dom x) ≅ x" and
        S6: "(cod y)·y ≅ y" and      —‹So does the codomain as a left identity.›


—‹Finally we make sure that there is only one non-existing morphism.›
        L1: "¬(E m) → (m = *)"
```

For convenience, we will assume a category to be small for the rest of this paper. As a result, a category for us has only a set of morphisms which satisfies the above axiom schema. This allows us to use notation from set theory. We write $(m : X \to Y) \in \mathbf{C}$ to mean that m is a morphism from the category $\mathbf{C}$. In addition, it says that $\mathrm{dom}(m) \cong X$ and $\mathrm{cod}(m) \cong Y$, so X is the domain of m and Y the codomain. The identity morphisms X and Y, which are representing objects in the usual sense, are characterized by the property that $X \cong \mathrm{dom}(X) \cong \mathrm{cod}(X)$, respectively for Y. Hence every $c \in \mathbf{C}$ satisfying $c \cong \mathrm{dom}(c)$ or $c \cong \mathrm{cod}(c)$ is representing an object, and we refer to such a morphism as an *object*.

We want a categorical generalization of an inverse semigroup, so let's turn to the question of how to introduce generalized inverses to a category. In the above setting we found that by adding the axioms of an inverse semigroup, which are responsible for shaping these inverses (Definition 4, Axioms 2–4), we arrive at a notion that is equivalent to the usual definition of an inverse category. Note that this definition is adopted to our free logic foundation by using *Kleene equality*, which is denoted by $\cong$. We emphasize again that this equality between terms states that, if either term exists, so does the other one and they are equal.

Definition 10 (Inverse category [7]). *A small category* $\mathbf{C}$ *is called an inverse category if for any morphism* $s : X \to Y \in \mathbf{C}$ *there exists a unique morphisms* $\hat{s} : Y \to X$ *such that* $s \cong s \cdot \hat{s} \cdot s$ *and* $\hat{s} \cong \hat{s} \cdot s \cdot \hat{s}$.

For the representation in Isabelle/HOL we skolemized the definition.

```
abbreviation (in category) inverseDef::"α ⇒ α ⇒ bool" ("_ inverseOf _")
  where "t inverseOf s ≡ s ≅ s·(t·s) ∧ t ≅ t·(s·t)"
—‹On the right hand side we find the definition for the inverse.›

locale inverseCategory = category +
—‹We introduce the new function symbol representing the inverse in the signature.›
  fixes inverse::"α ⇒ α" ("inv _")
  assumes uniqueInv:
—‹We axiomatize that every morphism has a unique inverse.›
     "∀s. (((inv s) inverseOf s) ∧ (∀w. (w inverseOf s) ⟶ w ≅ (inv s)))"
```

Next, we see the quantifier free definition.

```
locale inverseCategoryQantFree = category +
—‹We introduce the new function symbol representing the inverse in the signature.›
  fixes inverse::"α ⇒ α" ("inv _")
—‹We add the axioms that shape the inverse in an inverse semigroup.›
  assumes C1: "x·((inv x)·x) ≅ x" and
          C2: "((inv (inv x)) ≅ x)" and
          C3: "((x·inv x) · (y·(inv y))) ≅ ((y·(inv y)) · (x·(inv x)))"
```

The equivalence between the two formulations has been shown by interactive theorem proving. Again, a significant number of the required subproofs could be automated by *sledgehammer*. In addition, the minimality of the axioms for the quantifier free version above was checked effectively using *sledgehammer* and *nitpick*[3]. Inverse categories are interesting to us because of the following proposition.

Proposition 6. *Let* $\mathbf{C}$ *be an inverse category with exactly one object. Then* $\mathbf{C}$ *is an inverse semigroup.*

This allows us to generalize a semimodeloid to an inverse category by formulating the new axioms in such a way that this categorical construction will collapse to a semimodeloid under the condition of having just one object.

[3] *nitpick* [4] is a counterexample generator for higher-order logic integrated with Isabelle/HOL.

4.2 Categorical Axiomatization of a Modeloid

The notion of the *natural partial order* is also definable in an inverse category. To state it, we first introduce a definition for *idempotence*.

Definition 11 (Idempotence). *Let* $\mathbf{C}$ *be a small category. Then a morphism* $e \in \mathbf{C}$ *is called idempotent if, and only if,*

$$e \cdot e \cong e.$$

Whenever we do not assume that both sides of the equation exist, we use *Kleene equality*.

Definition 12 (Natural partial order [12]**).** *Let* $\mathbf{C}$ *be an inverse category and let* $s, t : X \to Y$ *be morphisms in* $\mathbf{C}$*. We define*

$$s \leq t :\Leftrightarrow \exists \text{ idempotent } e \in End_C(X) : s \cong t \cdot e$$

where $End_C(X) := \{m \in \mathbf{C} \,|\, m : X \to X\}$ is called an endoset.

When defining a categorical modeloid M on an inverse category $\mathbf{C}$, we will see that for each object X in $\mathbf{C}$, $End_C(X)$ is a semimodeloid. We require the category to have a zero element in each of its endosets in order to define an atom. For this we simply write that $\mathbf{C}$ has all zero elements.

Definition 13 (Categorical modeloid). *Let* $\mathbf{C}$ *be an inverse category with all zero elements. Then a* categorical modeloid M *on* $\mathbf{C}$ *is such that* $M \subseteq \mathbf{C}$ *satisfies the following axioms:*

1. $a, b \in M \Rightarrow a \cdot b \in M$
2. $a \in M \Rightarrow a^{-1} \in M$
3. $\forall a \in \mathbf{C} \forall b \in M : a \leq b \Rightarrow a \in M$
4. $\forall$ *objects* $X \in \mathbf{C} : X \in M$

It is evident that this definition is close by its appearance to a semimodeloid. However, we are now dealing with a network of semimodeloids and have thus reached a much more expressive definition.

Proposition 7. *Let* $\mathbf{C}$ *be an inverse category with all zero elements and let* M *be a categorical modeloid on* $\mathbf{C}$*. Then for each object* X *in* M *we get that* $End_M(X)$ *is a semimodeloid (on itself).*

Remark 3. Every semimodeloid can easily be seen as a categorical modeloid by the fact that an inverse monoid with zero element is an one-object inverse category.

We have formulated a generalization of a modeloid in category theory. What is left now is to define the derivative in this context. We will need the notion of a homset and of an atom, which we already introduced for semigroups.

Definition 14 (Homset). *Let* $\mathbf{C}$ *be a small category. Then the homset between two elements* $X, Y \in \mathbf{C}$*, satisfying* $X \cong dom(X)$ *and* $Y \cong dom(Y)$*, is defined as*

$$Hom_C(X,Y) := \{m \in \mathbf{C} \,|\, m : X \to Y\}$$

Hence an endoset is a special case of a homset. We only assume zero elements to be present in endosets and as a result an atom needs to be part of an endoset.

Definition 15 (Atom). *Let* $\mathbf{C}$ *be an inverse category with all zero elements. Then an element* $a \in End_C(X)$*, for some object* $X \in \mathbf{C}$*, is an* atom *if, and only if, the existence of* a *implies that* a *is not the zero element and*

$$\forall e \in End_C(X) : e \leq a \text{ implies that } e \cong a \vee e \cong 0_{End_C(X)}.$$

This concludes the preliminaries for defining the derivative on a homset.

Definition 16 (Derivative—homset). *Let* $\mathbf{C}$ *be an inverse category with all zero elements and let* M *be a categorical modeloid on* $\mathbf{C}$*. We define the derivative on* $Hom_M(X,Y)$ *for* $X, Y \in M$ *as* $D(Hom_M(X,Y)) := \{f \in Hom_M(X,Y) \,|$
$\forall$ *idempotent atoms* $a \in End_M(X)\, \exists h \in Hom_M(X,Y) : (f \leq h \wedge a \leq h^{-1}h) \wedge$
$\forall$ *idempotent atoms* $b \in End_M(Y)\, \exists g \in Hom_M(X,Y) : (f \leq g \wedge b \leq gg^{-1})\}$

Remark 4. Let $\mathbf{C}$ be an inverse category with just one object X and a zero element. Then $\mathbf{C}$ is an inverse semigroup by Proposition 6 and the derivative on the homset $D(Hom_C(X,X))$ is equal to the semimodeloidal derivative $D(\mathbf{C})$.

Now the key property of this operation is that it produces a categorical modeloid again if we apply it to *all* homsets simultaneously.

Theorem 3. *Let* $\mathbf{C}$ *be an inverse category with all zero elements and let* M *be a categorical modeloid on* $\mathbf{C}$*. Then*

$$\bigcup_{X,Y \in M} D(Hom_M(X,Y))$$

is a categorical modeloid on $\mathbf{C}$*.*

As a result we define this to be the derivative operation on categorical modeloids.

Definition 17 (Derivative—categorical modeloid). *Let* $\mathbf{C}$ *be an inverse category with all zero elements and let* M *be a categorical modeloid on* $\mathbf{C}$*. Then we set the derivative as*

$$D(M) := \bigcup_{X,Y \in M} D(Hom_M(X,Y)).$$

Let M be a categorical modeloid. We define

$$D^0(M) := M \text{ and } D^{n+1}(M) := D(D^n(M)) \tag{5}$$

for $n \in \mathbb{N}$. $D^m(M)$ thus takes the derivative m-times. This notion is used in the next section.

5 Algebraic Ehrenfeucht-Fraïssé Games

When moving from classical model theory to the finite case, some machinery for proving inexpressibility results in first-order logic, such as the *compactness theorem*, fails. However, Ehrenfeucht-Fraïssé (EF) games are still applicable and, therefore, play a central notion in finite model theory due to the possibility to show that a property is first-order axiomatizable. For more information see [11].

In this section we explicitly show that derivatives on categorical modeloids generalize EF games.

5.1 Rules of EF Game

To play an EF game, two finite τ-structures $\mathcal{A}$ and $\mathcal{B}$, where τ is a finite relational vocabulary, are needed. In general EF games are not restricted to finite structures, but for our purpose we shall only deal with this case. In order to give an intuitive understanding we imagine two players, which we call the spoiler and the duplicator, playing the game. The rules are quite simple. In $n \in \mathbb{N}$ rounds the spoiler tries to show that the two given structures are not equal, while the duplicator tries to disprove the spoiler every time. A round consists of the following:

- The spoiler picks either $\mathcal{A}$ or $\mathcal{B}$ and then makes a move by choosing an element from that structure, so $a \in \mathcal{A}$ or $b \in \mathcal{B}$.
- After the spoiler is done, the duplicator picks an element of the other structure and the round ends.

Next we define what the winning condition for each round will be. For convenience let $Part(\mathcal{A}, \mathcal{B})$ be the set of all partial isomorphisms from $\mathcal{A}$ to $\mathcal{B}$. Furthermore, given a constant symbole c from τ, we denote by $c^{\mathcal{A}}$ the interpretation of c in the structure $\mathcal{A}$.

Definition 18 (Winning position [11]**).** *Suppose the EF game was played for n rounds. Then there are moves $(a_1, .., a_n)$ picked from $\mathcal{A}$ and moves $(b_1, .., b_n)$ picked from $\mathcal{B}$. For this to be a winning position we require that for some $r \in \mathbb{N}$ the map*

$$\{(a_1, b_1), .., (a_n, b_n), (c_1^{\mathcal{A}}, c_1^{\mathcal{B}}), .., (c_r^{\mathcal{A}}, c_r^{\mathcal{B}})\} \in Part(\mathcal{A}, \mathcal{B})$$

where the c_i are all constant symbols of τ.

In order to win, the duplicator needs to defeat the spoiler in every possible course of the game. We say the duplicator has an *n-round winning strategy in the Ehrenfeucht-Fraïssé game on $\mathcal{A}$ and $\mathcal{B}$* [11], if the duplicator is in a winning position after n moves regardless of what the spoiler does. This is made precise by the back-and-forth method due to Fraïssé.

Definition 19 (Back-and-forth relation [5]**).** *We define a binary relation $\equiv_m$, $m \in \mathbb{N}$, on all τ-structures by $\mathcal{A} \equiv_m \mathcal{B}$ iff there is a sequence (I_j) for $0 \leq j \leq m$ such that*

- *Every I_j is a non-empty set of partial isomorphisms from $\mathcal{A}$ to $\mathcal{B}$*
- *(Forth property) $\forall j < m$ we have $\forall a \in \mathcal{A} \forall f \in I_{j+1} \exists g \in I_j : f \subseteq g \wedge a \in dom(g)$*
- *(Back property) $\forall j < m$ we have $\forall b \in \mathcal{B} \forall f \in I_{j+1} \exists g \in I_j : f \subseteq g \wedge b \in cod(g)$*

Hence $\mathcal{A} \equiv_n \mathcal{B}$ means that the duplicator has a n-round winning strategy.

5.2 The Derivative and Fraïssé's Method

We relate the categorical derivative to Fraïssé's method which we have just seen. In order to do this, we define a categorical modeloid on the category of finite τ-structures, where τ is a finite relational vocabulary. For that let $\mathcal{A}$ and $\mathcal{B}$ be two finite τ-structures. Denote by $F(\mathcal{A}, \mathcal{B})$ the set

$$\bigcup_{(X,Y) \in \{\mathcal{A},\mathcal{B}\}^2} Part(X, Y)$$

and let $\star \notin F(\mathcal{A}, \mathcal{B})$ be an arbitrary element. Then define $C := F(\mathcal{A}, \mathcal{B}) \cup \{\star\}$.

We construct two functions $dom : C \to C$ and $cod : C \to C$ such that for a partial isomorphism $f : X \to Y \in F(\mathcal{A}, \mathcal{B})$ we set $dom(f) = id_X$ and $cod(f) = id_Y$, and for the element $\star$ we define $dom(\star) = \star$ and $cod(\star) = \star$.

Next we define a binary operation $\cdot : C \to C$ by

$$f \cdot g = \begin{cases} f \circ g, \text{ if } dom(f) = cod(g) \text{ and } f, g \neq \star \\ \star, \quad \text{else} \end{cases}$$

where $\circ$ denotes the composition of partial functions.

Proposition 8. $\mathbf{D} := (C, dom, cod, \cdot, \star, ^{-1})$ *is an inverse category where f^{-1} denotes the inverse of each partial isomorphism f and $\star^{-1} = \star$. The existing elements are exactly all elements in $F(\mathcal{A}, \mathcal{B})$ and the compositions $f \circ g$ in case $dom(f) = cod(g)$, for $f, g \in F(\mathcal{A}, \mathcal{B})$.*

What we have just seen provides a general procedure for creating categories in our setting, which is founded on a free logic that is itself encoded in Isabelle/HOL.

Corollary 1. $\mathbf{D} := (C, dom, cod, \cdot, \star, ^{-1})$ *is also a categorical modeloid on itself.*

Remark 5. Hence we have that every inverse category having a zero element for each of its endosets is also a categorical modeloid and thus admits a derivative.

At this point we are able to use the derivative on $\mathbf{D}$. The final theorem draws the concluding connection between modeloids and Fraïssé's method. We show that in the established setting, an m-round winning strategy between $\mathcal{A}$ and $\mathcal{B}$ is given by the sets which the derivative produces if applied m times. Note the abuse of notation in the way we are using $\equiv_m$ here.

Theorem 4. *Let M be the categorical modeloid* $\mathbf{D}$. *Then*

$$\exists h : X \to Y \in D^m(M) \text{ with } h \neq \star \iff X \equiv_m Y, \quad m \in \mathbb{N}$$

As a result the derivative on this modeloid is equivalent to playing an EF game between the two structures. Hence on an arbitrary categorical modeloid the derivative can be seen as a generalization of EF games.

6 Conclusion

In this paper we have shown how to arrive at the notion of a categorical modeloid using axiomatic category theory. We started out with a set of partial bijections abstracting from a structure, then we interpreted this set as an inverse semigroup by the embedding due to the Wagner-Preston representation theorem, and, finally, we were able to axiomatize a modeloid in an inverse category. The key feature we employed is the natural partial order which also enabled us to present the derivative operation in each step of abstraction. The categorical derivative on the category of finite structures of a finite vocabulary can then be used to play Ehrenfeucht-Fraïssé games between two structures. As a result a more abstract representation of these games is possible.

Using our encoding of inverse categories in Isabelle/HOL, we are currently extending this encoding work to cover also categorical modeloids and their derivatives. This naturally extends the framework established by Benzmüller and Scott so far [2]. Furthermore, an investigation of the generalized Ehrenfeucht-Fraïssé games in terms of applicability has to be conducted. We believe that the notion of a categorical modeloid will continue to play a role when connecting model theoretical and categorical concepts.

Acknowledgment. We wish to thank the anonymous referees for their helpful comments and suggestions.

References

1. Benda, M.: Modeloids. I. Trans. Am. Math. Soc. **250**, 47–90 (1979). https://doi.org/10.1090/s0002-9947-1979-0530044-4
2. Benzmüller, C., Scott, D.S.: Automating free logic in HOL, with an experimental application in category theory. J. Autom. Reason. **64**(1), 53–72 (2020). https://doi.org/10.1007/s10817-018-09507-7
3. Blanchette, J.C., Böhme, S., Paulson, L.C.: Extending sledgehammer with SMT solvers. J. Autom. Reasoning **51**(1), 109–128 (2013). https://doi.org/10.1007/s10817-013-9278-5
4. Blanchette, J.C., Nipkow, T.: Nitpick: a counterexample generator for higher-order logic based on a relational model finder. In: Kaufmann, M., Paulson, L.C. (eds.) ITP 2010. LNCS, vol. 6172, pp. 131–146. Springer, Heidelberg (2010). https://doi.org/10.1007/978-3-642-14052-5_11
5. Ebbinghaus, H.D., Flum, J.: Finite Model Theory. Springer, Heidelberg (1995). https://doi.org/10.1007/3-540-28788-4

6. Howie, J.M.: Fundamentals of Semigroup Theory. Oxford University Press, Oxford (1995)
7. Kastl, J.: Inverse categories. Algebraische Modelle, Kategorien und Gruppoide **7**, 51–60 (1979)
8. Lambert, K.: The definition of e! in free logic. In: Abstracts: The International Congress for Logic, Methodology and Philosophy of Science. Stanford University Press, Stanford (1960)
9. Lambert, K.: Free Logic: Selected Essays. Cambridge University Press, Cambridge (2002)
10. Lawson, M.V.: Inverse Semigroups. World Scientific (1998). https://doi.org/10.1142/3645
11. Libkin, L.: Elements of Finite Model Theory. Springer, Heidelberg (2013)
12. Linckelmann, M.: On inverse categories and transfer in cohomology. Proc. Edinburgh Math. Soc. **56**(1), 187–210 (2012). https://doi.org/10.1017/s0013091512000211
13. Nipkow, T., Paulson, L.C., Wenzel, M.: Isabelle/HOL: A Proof Assistant for Higher-Order Logic. LNCS, vol. 2283. Springer, Heidelberg (2002). https://doi.org/10.1007/3-540-45949-9
14. Scott, D.: Existence and description in formal logic. In: Schoenman, R. (ed.) Bertrand Russell: Philosopher of the Century, pp. 181–200. George Allen & Unwin, London (1967). Reprinted with additions in: Lambert, K. (ed.) Philosophical Application of Free Logic, pp. 28–48. Oxford Universitry Press (1991)
15. Tiemens, L.: Computer-supported Exploration of a Categorical Axiomatization of Miroslav Benda's Modeloids. Bachelor's thesis, Freie Universität Berlin (2019). https://www.mi.fu-berlin.de/inf/groups/ag-ki/Theses/Completed-theses/Bachelor-theses/2019/Tiemens/BA-Tiemens2.pdf
16. Tiemens, L., Scott, D.S., Benzmüller, C., Benda, M.: Computer-supported exploration of a categorical axiomatization of modeloids. ArXiv abs/1910.12863 (2019). https://arxiv.org/abs/1910.12863

Sharpness in the Fuzzy World

Michael Winter(✉)

Department of Computer Science, Brock University, St. Catharines, ON L2S 3A1, Canada
mwinter@brocku.ca

Abstract. In this paper we focus on a fuzzy version of the so-called (un)sharpness property of relational products in arrow/fuzzy categories. It is shown that the fuzzy version can be reduced to a regular (un)sharpness problem. As a consequence we obtain that relational products are also sharp in the fuzzy sense if all relational products and powers exist. This result is important in applications of arrow/fuzzy categories since relational products, and, hence, the fuzzy version of the (un)sharpness problem, are integral components of these applications.

1 Introduction

In this paper we are interested in working with multi-valued relations in an abstract setting. In particular, we are interested in a framework for working with so-called L-relations algebraically. Given a complete Heyting algebra L, an L-relation Q between two sets A and B is just a function $Q : A \times B \to L$, i.e., the relation provides a membership value $Q(a, b)$ from the Heyting algebra L indicating to what degree the pair (a, b) is in relation Q. L-relations generalize (regular) relations in the following sense. As usual a relation R between two sets A and B is simply a set of pairs, i.e., $R \subseteq A \times B$. Alternatively, R can be represented by its characteristic function $R : A \times B \to \mathbb{B}$ where $\mathbb{B}$ is the set of (Boolean) truth values. If we identify *true* with the greatest element and *false* with the smallest element of the Heyting algebra L we immediately obtain L-relations as a generalization of regular relations. In particular, we call an L-relation returning 0 or 1 for every pair a crisp relation. Obviously, a crisp L-relation corresponds to a regular relation.

Allegories [3] establish a suitable framework for working with any kind of relations. In particular, they provide an axiomatic system that characterizes the usual operations on relations such as the set-theoretic operation $\sqcap$ (meet) and $\sqcup$ (join) as well as the additional operations ; (composition) and $.^{\smile}$ (converse) of relations. Arrow categories [12–14] add two operations $.^{\uparrow}$ (support) and $.^{\downarrow}$ (kernel) that allow to define crispness, and, hence, to formulate properties of L-relations. In particular, $R^{\uparrow}$ represents the smallest crisp relation containing R, and $R^{\downarrow}$ represents the greatest crisp relation contained in R. In the work mentioned above it has been shown that arrow categories provide a suitable framework for working with L-relations for a fixed Heyting algebra L.

The author gratefully acknowledges support from the Natural Sciences and Engineering Research Council of Canada (283267).

U. Fahrenberg et al. (Eds.): RAMiCS 2020, LNCS 12062, pp. 318–334, 2020.
https://doi.org/10.1007/978-3-030-43520-2_20

In the theory of fuzzy (or L-fuzzy relations) one often uses additional operations. The standard operation $\sqcap$ computes the degree of membership of a pair (a, b) in the intersection of two L-relations Q and R as the meet of the two corresponding membership degrees, i.e., we have

$$(Q \sqcap R)(a, b) = Q(a, b) \wedge R(a, b),$$

where $\wedge$ on the right-hand side denotes the meet in the Heyting algebra L. Often we are interested in an operation where the degree is computed by applying a commutative and integral quantale (or t-norm like for short) operation $*$ to the two membership degrees instead of the meet operation of the lattice. Hence, we define a new operation on relations

$$(Q * R)(a, b) = Q(a, b) * R(a, b),$$

where $(L, *)$ is a complete Heyting algebra with a t-norm like operation. Similarly, we can also define a composition operation based on $*$ by

$$(Q \overset{*}{;} S)(a, c) = \bigvee_b Q(a, b) * S(b, c).$$

In [17] a set of axioms for these operations was introduced, and we will call an arrow category together with these new operations and a suitable set of axioms a fuzzy category (cf. Definition 11). Hence, fuzzy categories are very suitable as an algebraic/categorical framework for L-fuzzy relations.

In the following we will use the notation $Q : C \to A$ to denote the fact that Q is a L-relation (or regular relation) between the sets C and A. If $A \times B$ is the cartesian product of the sets A and B, then we can define the so-called fork (or right tupling) operation $Q \otimes R : C \to A \times B$ of two relations $Q : C \to A$ and $R : C \to B$ by

$$(Q \otimes R)(c, (a, b)) = Q(c, a) \wedge R(c, b).$$

In other words, $Q \otimes R$ relates c with the pair (a, b) if c is related to a in Q and c is related to b in R. The join (or left tupling) operation $S \oslash T : A \times B \to D$ for two relations $S : A \to D$ and $T : B \to D$ is defined similarly. In the abstract theory of allegories the cartesian product can be defined abstractly by so-called relational products. This definition is based on the projection functions as relations. Using the projections it is easily possible to derive the fork and join operation algebraically. The so-called sharpness problem focuses on the following simple equation

$$(Q \otimes R); (S \oslash T) = Q; S \sqcap R; T.$$

It is easy to verify this equation for concrete relations. In the abstract setting, however, this equation needs not to be valid. A counter example was provided in [5]. Therefore, we call a product sharp (resp. unsharp) if the previous equation is valid (resp. not valid) in the abstract setting. It is worth mentioning that the sharpness property is closely related to the representation problem of allegories. The existence of all products implies representability as well as sharpness. Conversely, a representable allegory is obviously embedded in a category satisfying sharpness, i.e., satisfies sharpness itself since products are unique up to isomorphism. Anyhow, sharpness is often necessary to prove properties about certain constructions algebraically. For example, if $\mathcal{P}(A)$ denotes the relational power of A, an abstract version of the power set, then one can easily define a relation $\mathcal{M} : \mathcal{P}(A) \times \mathcal{P}(A) \to \mathcal{P}(A)$ (in fact a map) that maps a pair of subsets of A to their intersection. In order to verify typical properties of this relation algebraically sharpness of the products involved is usually needed. To our knowledge there are three different theorems showing sharpness under certain specific conditions. The first theorem is due to Zierer [18]. It requires two suitable additional products to exist in order to prove the desired property. In particular, this theorem shows that sharpness holds if the underlying allegory has all relational products. This result was later improved by Desharnais [1]. A consequence of the latter theorem is that sharpness is already valid if only one additional relational product exists. Last but not least, in [11] it was shown that sharpness always holds if one of the relations involved is between an object B and its relational power $\mathcal{P}(B)$. More precisely, if, for example, R of the sharpness equation has type $B \to \mathcal{P}(B)$, then the singleton relation $\mathrm{syQ}(\mathbb{I}_B, \varepsilon) : B \to \mathcal{P}(B)$ mapping an element of b of B to the singleton set $\{b\}$ is a map parallel to R which can be used to show sharpness (cf. Theorem 16 in the fuzzy case).

In an L-fuzzy setting we are often interested in a modified version of the relation $\mathcal{M}$ introduced above. As already mentioned above we are often interested in computing the intersection of two L-fuzzy sets by using a t-norm like operation $*$, i.e., the degree of membership of an element in the intersection of two sets A and B is computed by $*$ instead of the lattice meet. This leads to a relation $\mathcal{M}_*$ obtained from $\mathcal{M}$ by replacing the regular fork operation in the algebraic definition of $\mathcal{M}$ by a fork based on $*$, i.e., by the operation

$$(Q \otimes_* R)(c, (a, b)) = Q(c, a) * R(c, b).$$

At this point we want to mention at least one important application of $\mathcal{M}_*$. A relation algebraic approach to L-fuzzy topological spaces [6], similar to the approach taken in [11], requires the usage of $\mathcal{M}_*$ already in the definition of those spaces. One of the axioms requires that the set of open sets is closed under this modified intersection. Consequently, we are interested in showing basic properties of this ($*$-based) meet operation. Several of these properties require a fuzzy version of the sharpness problem, i.e., the equation

$$(Q \otimes_* R) \mathbin{;_*} (S \otimes_* T) = Q \mathbin{;_*} S * R \mathbin{;_*} T.$$

We will sketch one of these properties and its proof briefly in the conclusion.

As in the regular case it is easy to verify sharpness for concrete L-relations (Theorem 14) but an algebraic proof is not possible without further assumptions. In fact, since it is possible to choose the lattice meet as the t-norm like operation $*$, the regular sharpness condition is a special case of the fuzzy version above. Unfortunately, none of the three theorems mentioned above can be applied immediately since the t-norm based operations $*$ and $;_*$ do not satisfy all properties of their counterparts $\sqcap$ and ;. In fact, the first theorem mentioned above [18] is based on the fact that composition with a map from the left distributes over meet. The additional two products are used to modify the fork on the left of the composition into a suitable map. In the fuzzy setting $;_*$ composition with a map from the left only distributes over $*$ if the map is, in addition, crisp. Consequently, a straightforward generalization to the fuzzy case does not seem possible. The second theorem relies heavily on the modular inclusion and the fact that $\sqcap$ is idempotent. Only weaker versions of the modular inclusion are valid for $\mathbin{\stackrel{*}{;}}$ and $*$. Furthermore, $*$ is not necessarily idempotent so that this theorem cannot be generalized to the fuzzy case either. In this paper we will show that the last of the three theorems can be generalized to the fuzzy case (Theorem 16), which then will be used to reduce the fuzzy sharpness problem to a regular one (Theorem 18).

2 Mathematical Preliminaries

Suppose L is a complete Heyting algebra which smallest element 0, greatest element 1, meet $\wedge$ and join $\vee$. Then an L-relation R between two sets A and B is a function $R : A \times B \to L$. The relational operations on L-relations are based on the operations of L in the usual way, e.g., meet $\sqcap$ and composition ; are defined as

$$(Q \sqcap R)(x, y) = Q(x, y) \wedge R(x, y), \text{ and } (Q; S)(x, z) = \bigvee_y Q(x, y) \wedge S(y, z).$$

Please note that we use the convention that composition is from left to right, i.e., $Q; S$ means Q first, and then S. As already mentioned in the introduction the special case of $\mathbb{B}$-relations corresponds to regular (set-theoretic) relations represented by their characteristic function.

In this section we want to introduce several categories as an abstract framework to work with L-relations.

2.1 Dedekind, Arrow and Fuzzy Categories

In this section we want to recall some basic notions from categories, allegories and arrow categories [3,13,14]. Furthermore, we will introduce fuzzy categories as an extension of arrow categories adding t-norm like operations on relations.

We will write $R : A \to B$ to indicate that a morphism R of a category C has source A and target B. Composition and the identity morphism are denoted by ; and $\mathbb{I}_A$, respectively.

First, we are going to introduce Dedekind categories [7,8]. These categories are called locally complete division allegories in [3].

Definition 1. *A Dedekind category $\mathcal{R}$ is a category satisfying the following:*

1. *For all objects A and B the collection $\mathcal{R}[A, B]$ is a complete Heyting algebra. Meet, join, the induced ordering, the least and the greatest element are denoted by $\sqcap, \sqcup, \sqsubseteq, \perp\!\!\!\perp_{AB}, \top\!\!\!\top_{AB}$, respectively.*
2. *There is a monotone operation $\breve{}$ (called converse) mapping a relation $Q : A \to B$ to $Q^{\smile} : B \to A$ such that for all relations $Q : A \to B$ and $R : B \to C$ the following holds: $(Q; R)^{\smile} = R^{\smile}; Q^{\smile}$ and $(Q^{\smile})^{\smile} = Q$.*
3. *For all relations $Q : A \to B, R : B \to C$ and $S : A \to C$ the modular law $(Q; R) \sqcap S \sqsubseteq Q; (R \sqcap (Q^{\smile}; S))$ holds.*

4. *For all relations $R : B \to C$ and $S : A \to C$ there is a relation $S/R : A \to B$ (called the left residual of S and R) such that for all $X : A \to B$ the following holds: $X;R \sqsubseteq S \Leftrightarrow X \sqsubseteq S/R$.*

It is easy to verify that the collection of all L-relations between sets for a given complete Heyting algebra L forms a Dedekind category.

Throughout this paper we will use some basic properties of relations in Dedekind categories such as $\mathbb{⊥}_{AB}^{\smile} = \mathbb{⊥}_{BA}$, $\mathbb{T}_{AB}^{\smile} = \mathbb{T}_{BA}$, $\mathbb{I}_A^{\smile} = \mathbb{I}_A$, the monotonicity resp. antitonicity of the operations, and the fact that composition distributes over join from both sides without mentioning.

An important class of relations is given by maps.

Definition 2. *Let $\mathcal{R}$ be a Dedekind category. Then a relation $Q : A \to B$ is called*

1. *univalent (or partial function) iff $Q^{\smile};Q \sqsubseteq \mathbb{I}_B$,*
2. *total iff $\mathbb{I}_A \sqsubseteq Q;Q^{\smile}$,*
3. *injective iff $Q^{\smile}$ is univalent,*
4. *surjective iff $Q^{\smile}$ is total,*
5. *a map iff Q is total and univalent.*

It is well known that Q is total iff $Q;\mathbb{T}_{BC} = \mathbb{T}_{AC}$. We will use this equivalence in the remainder of the paper without mentioning.

The following property about maps is used later. A proof can be found in [13].

Lemma 3. *Let $Q : A \to B$, $R : A \to C$ be relations, and $f : B \to C$ be a map. Then we have*

$$Q;f \sqsubseteq R \textit{ iff } Q \sqsubseteq R;f^{\smile}.$$

Using converse a second residual can be defined by $Q\backslash R := (R^{\smile}/Q^{\smile})^{\smile}$. This operation is characterized by the equivalence $Q;X \sqsubseteq R$ iff $X \sqsubseteq Q\backslash R$. Together the two residuals allow the definition of the so-called symmetric quotient. This construction is defined by $\mathrm{syQ}(Q,R) := Q\backslash R \sqcap Q^{\smile}/R^{\smile}$. Consequently the symmetric quotient is characterized by $Q;X \sqsubseteq R$ and $R;X^{\smile} \sqsubseteq Q$ iff $X \sqsubseteq \mathrm{syQ}(Q,R)$.

In the following lemma we have collected some basic properties of the symmetric quotient that are needed in the remainder of the paper. A proof can be found in [2, 13].

Lemma 4. *Let $\mathcal{D}$ be a Dedekind category, $Q : A \to B, R : A \to C$, and $S : A \to D$. Then we have:*

1. $\mathrm{syQ}(Q,R)^{\smile} = \mathrm{syQ}(R,Q)$.
2. $\mathrm{syQ}(Q,R);\mathrm{syQ}(R,S) \sqsubseteq \mathrm{syQ}(Q,S)$.
3. *if Q is univalent and surjective, then* $\mathrm{syQ}(Q,Q) = \mathbb{I}_B$.

The equation $\mathbb{T}_{AB};\mathbb{T}_{BC} = \mathbb{T}_{AC}$ for all objects A, B and C can easily be shown for L-relations but does not follow from the axioms of a Dedekind category. However, two special cases of this equation where one of the two universal relations on the left-hand side is homogeneous can be verified in any Dedekind category, i.e., we have $\mathbb{T}_{AA};\mathbb{T}_{AB} = \mathbb{T}_{AB};\mathbb{T}_{BB} = \mathbb{T}_{AB}$ for all objects A and B. If the general equation holds, then we call the Dedekind category uniform.

Any Dedekind category allows to identify the membership values used in an abstract manner by using so-called scalar relations.

Definition 5. *A relation $\alpha : A \to A$ is called a scalar on A iff $\alpha \sqsubseteq \mathbb{I}_A$ and $\mathbb{T}_{AA};\alpha = \alpha;\mathbb{T}_{AA}$.*

The notion of scalars was introduced by Furusawa and Kawahara [4]. An L-relation is a scalar iff every element of A is related to itself with a fixed degree a from L. Therefore, there is a one-one correspondence, i.e., a isomorphism of complete Heyting algebras, between the scalar relations on A and the lattice L. Because of this isomorphism we will occasionally identify scalars and the elements from L by using α for the scalar as well as the corresponding element (via the isomorphism) from L.

A crisp L-relation R satisfies $R(x,y) = 0$ or $R(x,y) = 1$ for all pairs (x,y). These relations can be identified with Boolean valued relations, i.e., regular set-theoretic relations. The notion of crispness cannot be defined abstractly in the theory of Dedekind categories [12,13]. Because of this arrow categories were introduced [13,14]. These categories add two operations $(.)^\downarrow$ and $(.)^\uparrow$ to Dedekind categories. Intuitively, the down-arrow (or kernel) operation maps an L-relation R to the greatest crisp relation included in R and the up-arrow (or support) operation maps R to the least crisp relation that includes R.

Definition 6. *An arrow category $\mathcal{A}$ is a Dedekind category with $\mathbb{T}_{AB} \neq \perp\!\!\!\perp_{AB}$ for all objects A and B together with two operations $^\uparrow$ and $^\downarrow$ satisfying the following:*

1. *$R^\uparrow, R^\downarrow : A \to B$ for all $R : A \to B$.*
2. *$(^\uparrow, ^\downarrow)$ is a Galois correspondence, i.e., $Q^\uparrow \sqsubseteq R$ iff $Q \sqsubseteq R^\downarrow$ for all $Q, R : A \to B$.*
3. *$(R^{\smile};S^\downarrow)^\uparrow = R^{\uparrow\smile};S^\downarrow$ for all $R : B \to A$ and $S : B \to C$.*
4. *If $\alpha \neq \perp\!\!\!\perp_{AA}$ is a non-zero scalar then $\alpha^\uparrow = \mathbb{I}_A$.*
5. *$(Q \sqcap R^\downarrow)^\uparrow = Q^\uparrow \sqcap R^\downarrow$ for all $Q, R : A \to B$.*

A relation $R : A \to B$ of an arrow category $\mathcal{A}$ is called crisp iff $R^\uparrow = R$ (or equivalently $R^\downarrow = R$). The collection of crisp relations is closed under all operations of a Dedekind category, and, hence, forms a sub-Dedekind category of $\mathcal{A}$ [13,14]. In particular, we will need the following lemma. A proof can be found in [13,15].

Lemma 7. *Let $\mathcal{A}$ be an arrow category, $Q : A \to B$, and $R : B \to C$. Then we have:*

1. *$Q^\downarrow \sqsubseteq Q \sqsubseteq Q^\uparrow$.*
2. *$\perp\!\!\!\perp_{AB}, \mathbb{T}_{AB}, \mathbb{I}_A, Q^\downarrow$ and $Q^\uparrow$ are crisp.*
3. *$Q^{\smile\downarrow} = Q^{\downarrow\smile}$ and $Q^{\smile\uparrow} = Q^{\uparrow\smile}$.*
4. *$Q^\downarrow;R^\downarrow \sqsubseteq (Q;R)^\downarrow$.*

It was mentioned above that the collection of all L-relations forms a uniform Dedekind category. Unlike Dedekind categories in general, arrow categories are always uniform [13]. It is worth mentioning that the proof of this fact requires scalars and Axiom 4 of arrow categories, i.e., two concepts that are otherwise not used in this paper.

The abstract version of a cartesian product is given by a relational (or direct) product [9,10]. In the context of arrow categories we are usually interested in relational products for which the projections are crisp.

Definition 8. *The relational product of two objects A and B is an object $A \times B$ together with two relations $\pi : A \times B \to A$ and $\rho : A \times B \to B$ so that the following equations hold*

$$\pi^{\downarrow} = \pi, \quad \rho^{\downarrow} = \rho, \quad \pi^{\smile};\pi \sqsubseteq \mathbb{I}_A, \quad \rho^{\smile};\rho \sqsubseteq \mathbb{I}_B, \quad \pi^{\smile};\rho = \mathbb{T}_{AB}, \quad \pi;\pi^{\smile} \sqcap \rho;\rho^{\smile} = \mathbb{I}_{A\times B}.$$

An arrow category has products if the relational product for each pair of objects exists.

It follows immediately from the definition above that the two projections are crisp maps.

Using the projections of a relational product we now introduce the fork (or right tupling) operation $Q \otimes R : C \to A \times B$ of two relations $Q : C \to A$ and $R : C \to B$ by $Q \otimes R := Q;\pi^{\smile} \sqcap R;\rho^{\smile}$. Analogously, the left tupling operation $S \otimes T : A \times B \to D$ for two relations $S : A \to D$ and $T : B \to B$ is defined as $S \otimes T := \pi;S \sqcap \rho;T$. We will adopt the convention that composition binds tighter than the operations defined above. The (un)sharpness problem of relational products is the validity of the equation

$$(Q \otimes R);(S \otimes T) = Q;S \sqcap R;T.$$

The equation above is easy to verify for concrete relation but does not follow from the axioms of a relational product. As mentioned in the introduction sharpness can be shown by requiring additional structure (cf. [1,11,18]). In particular, the relational product is sharp iff all relational products exists.

An abstract version of power sets is given by the notion of a relational (or direct) power [10]. Please note that in the context of arrow categories we are interested in the set of all L-fuzzy subsets of A, i.e., in the set of all functions $f : A \to L$. This leads to the following definition of a relational power in arrow categories that differs from previous definitions [16].

Definition 9. *An object $\mathcal{P}(A)$ together with a relation $\varepsilon : A \to \mathcal{P}(A)$ is called a relational power iff*

$$\mathrm{syQ}(\varepsilon,\varepsilon)^{\downarrow} = \mathbb{I}_{\mathcal{P}(A)} \text{ and } \mathrm{syQ}(R,\varepsilon)^{\downarrow} \text{ is total for every } R : A \to B.$$

Compared to the previous definitions of the relational power the definition above adds the kernel operation to the symmetric quotient in both axioms. Without this modification the relational power in the case of L-relations would not contain all fuzzy subsets of A. On the other hand, the *is element* relation ε is no longer extensional, and, hence, the inclusion (or subset) relation $\varepsilon\backslash\varepsilon$ is just a pre-order, i.e., not necessarily antisymmetric. For a concrete example explaining this situation in detail we refer to [16].

Lemma 10. *Let $\mathcal{A}$ be an arrow category, $\mathcal{P}(A)$ the relational power of A, and $Q : A \to B$. Then we have:*

1. $\mathrm{syQ}(Q,\varepsilon)^{\downarrow}$ *is a crisp map.*
2. $\mathrm{syQ}(\mathbb{I}_A,\varepsilon)^{\downarrow}$ *is a crisp and injective map.*
3. $\varepsilon;\mathrm{syQ}(\varepsilon,Q)^{\downarrow} = Q$ *and* $\mathrm{syQ}(Q,\varepsilon)^{\downarrow};\varepsilon^{\smile} = Q^{\smile}$.

Proof. 1. By Lemma 7(2) the relation $\mathrm{syQ}(Q,\varepsilon)^{\downarrow}$ is crisp, and, by the second axiom of a relational power, total. The remaining property follows from

$$\begin{aligned}
\mathrm{syQ}(Q,\varepsilon)^{\downarrow\smile};\mathrm{syQ}(Q,\varepsilon)^{\downarrow} &= \mathrm{syQ}(\varepsilon,Q)^{\downarrow};\mathrm{syQ}(Q,\varepsilon)^{\downarrow} && \text{Lemma 7(3) and 4(1)}\\
&\sqsubseteq (\mathrm{syQ}(\varepsilon,Q);\mathrm{syQ}(Q,\varepsilon))^{\downarrow} && \text{Lemma 7(4)}\\
&\sqsubseteq \mathrm{syQ}(\varepsilon,\varepsilon)^{\downarrow} && \text{Lemma 4(2)}\\
&= \mathbb{I}_{\mathcal{P}(A)}.
\end{aligned}$$

2. Due to (1) it remains to show that $\mathrm{syQ}(\mathbb{I}_A,\varepsilon)^{\downarrow}$ is injective.

$$\begin{aligned}
\mathrm{syQ}(\mathbb{I}_A,\varepsilon)^{\downarrow};\mathrm{syQ}(\mathbb{I}_A,\varepsilon)^{\downarrow\smile} &= \mathrm{syQ}(\mathbb{I}_A,\varepsilon)^{\downarrow};\mathrm{syQ}(\varepsilon,\mathbb{I}_A)^{\downarrow} && \text{Lemma 7(3) and 4(1)}\\
&\sqsubseteq (\mathrm{syQ}(\mathbb{I}_A,\varepsilon);\mathrm{syQ}(\varepsilon,\mathbb{I}_A))^{\downarrow} && \text{Lemma 7(4)}\\
&\sqsubseteq \mathrm{syQ}(\mathbb{I}_A,\mathbb{I}_A)^{\downarrow} && \text{Lemma 4(2)}\\
&= \mathbb{I}_A^{\downarrow} && \text{Lemma 4(3)}\\
&= \mathbb{I}_A. && \text{Lemma 7(2)}
\end{aligned}$$

3. We only show the first equation since the second follows from the first by using converse. First, we have

$$\begin{aligned}
\varepsilon;\mathrm{syQ}(\varepsilon,Q)^{\downarrow} &\sqsubseteq \varepsilon;\mathrm{syQ}(\varepsilon,Q) && \text{Lemma 7(1)}\\
&\sqsubseteq Q.
\end{aligned}$$

The inclusion $Q;\mathrm{syQ}(Q,\varepsilon)^{\downarrow} \sqsubseteq \varepsilon$ follows analogously. This immediately implies $Q \sqsubseteq \varepsilon;\mathrm{syQ}(Q,\varepsilon)^{\downarrow\smile} = \varepsilon;\mathrm{syQ}(\varepsilon,Q)^{\downarrow}$ using (1) and Lemma 3. □

As already mentioned in the introduction in applications of L-fuzzy theory so-called t-norm like operations are important. If L is a complete lattice, then $\langle L,*\rangle$ with a binary operation $*$ on L is called a partially ordered (Abelian) monoid iff

1. $\langle L,*,1\rangle$ is a bounded Abelian monoid, i.e., $*$ is associative and commutative with the greatest element 1 of L as neutral element,
2. $*$ is monotonic in both parameters.

If $*$ distributes over arbitrary unions in both parameters, then we call $*$ continuous. A continuous partially ordered Abelian monoid is also known as commutative integral quantale. For simplicity and its analogy to t-norm operations in fuzzy sets we call the operation $*$ of a continuous partially ordered Abelian monoid a t-norm like operation.

Given a complete Heyting algebra L with a t-norm like operation $*$ we can define two new operations on L-relations based on $*$ by

$$(Q*R)(a,b) = Q(a,b)*R(a,b), \text{ and } (Q\overset{*}{;}S)(a,c) = \bigvee_b Q(a,b)*S(b,c).$$

In [17] the basic properties of these operations, in particular, the validity of versions of the modular inclusion for $*$ and $\overset{*}{;}$ instead of $\sqcap$ and ; was investigated. This study lead to the following abstract definition of a fuzzy category.

Definition 11. *A fuzzy category $\mathcal{F}$ is an arrow category together with two operations $*$ and $\overset{*}{;}$ so that the following holds:*

1. $*$ *maps two relations $Q : A \to B$ and $R : A \to B$ to a relation $Q * R : A \to B$.*
2. $*$ *is associative, commutative and continuous.*
3. $Q * R^{\downarrow} = Q \sqcap R^{\downarrow}$ *for all $Q, R : A \to B$.*
4. $(Q * R)^{\smile} = Q^{\smile} * R^{\smile}$ *for all $Q, R : A \to B$.*
5. $\overset{*}{;}$ *maps two relations $Q : A \to B$ and $R : B \to C$ to a relation $Q \overset{*}{;} R : A \to C$.*
6. $\overset{*}{;}$ *is associative and continuous.*
7. $Q \overset{*}{;} R^{\downarrow} = Q; R^{\downarrow}$ *for all $Q : A \to B$ and $R : B \to C$.*
8. $(Q \overset{*}{;} R)^{\smile} = R^{\smile} \overset{*}{;} Q^{\smile}$ *for all $Q : A \to B$ and $R : B \to C$.*
9. *The exchange inclusion $(Q * R) \overset{*}{;} (S * T) \sqsubseteq Q \overset{*}{;} S * R \overset{*}{;} T$ is valid for all $Q, R : A \to B$ and $S, T : B \to C$.*
10. *The following versions of the modular inclusion are valid:*
    (a) $Q; R * S \sqsubseteq Q; (R \sqcap Q^{\smile} \overset{*}{;} S)$,
    (b) $Q \overset{*}{;} R * S \sqsubseteq Q; (R * Q^{\smile} \overset{*}{;} S)$,
    (c) $(P * Q) \overset{*}{;} R * S \sqsubseteq P \overset{*}{;} (R * Q^{\smile} \overset{*}{;} S)$,
    for all $P, Q : A \to B$, $R : B \to C$, and $S : A \to C$.

The following lemma collects some immediate consequences of the definition above.

Lemma 12. *Let $\mathcal{F}$ be a fuzzy category, and $Q, R : A \to B$ and $S : B \to C$ be relations. Then we have:*

1. $Q * R \sqsubseteq Q$ *and* $Q * R \sqsubseteq R$.
2. $Q * \top\!\!\top_{AB} = Q$.
3. $Q \overset{*}{;} S \sqsubseteq Q; S$.
4. $Q^{\downarrow} \overset{*}{;} S = Q^{\downarrow}; S$.
5. $Q \overset{*}{;} \mathbb{I}_B = \mathbb{I}_A \overset{*}{;} Q = Q$.

Proof. 1. We have $Q * R \sqsubseteq Q * R^{\uparrow} = Q \sqcap R^{\uparrow} \sqsubseteq Q$ by the monotonicity of $*$ and Axiom 3. The second property follows analogously.
2. This is an immediate consequence of Axiom 3 since $\top\!\!\top_{AB}$ is crisp.
3. We compute

$$
\begin{aligned}
Q \overset{*}{;} S &= Q \overset{*}{;} S * \top\!\!\top_{AC} && \text{by (2)} \\
&\sqsubseteq Q; (S * Q^{\smile} \overset{*}{;} \top\!\!\top_{AC}) && \text{modular inclusion (b)} \\
&\sqsubseteq Q; S. && \text{by (1)}
\end{aligned}
$$

4. This is just the dual (via $.^{\smile}$) of Axiom 7.
5. This is an immediate consequence of Axiom 7 and (4) since $\mathbb{I}_A$ and $\mathbb{I}_B$ are crisp. □

In the remainder of the paper we will use the Axioms (1)–(8) and the lemma above without mentioning. In the following lemma we show some basic properties of relations in fuzzy categories that are needed in the remainder of the paper.

Lemma 13. *Let $\mathcal{F}$ be a fuzzy category, $Q : A \to B$ be crisp, $f : B \to C$ crisp and univalent, $P : A \to B$, $R, S : B \to D$, $T : A \to C$, and $U : A \to D$. Then we have:*

1. $Q;(R * S) \sqsubseteq Q;R * Q;S$.
2. $(P * T;f^{\smile});f = P;f * T$.
3. $(P * Q)\overset{*}{;}R * U \sqsubseteq (P * Q)\overset{*}{;}(R * Q^{\smile}\overset{*}{;}U)$.

Proof. 1. This follows immediately from

$$\begin{aligned} Q;(R * S) &= Q\overset{*}{;}(R * S) && Q \text{ crisp} \\ &= (Q * Q)\overset{*}{;}(R * S) && Q \text{ crisp} \\ &\sqsubseteq Q\overset{*}{;}R * Q\overset{*}{;}S && \text{exchange inclusion} \\ &= Q;R * Q;S. && Q \text{ crisp} \end{aligned}$$

2. Using (1) we get the inclusion $\sqsubseteq$ from

$$\begin{aligned} (P * T;f^{\smile});f &\sqsubseteq P;f * T;f^{\smile};f && \text{dual of (1) and } f \text{ crisp} \\ &\sqsubseteq P;f * T, && f \text{ univalent} \end{aligned}$$

and the converse inclusion is shown by

$$\begin{aligned} P;f * T &= P\overset{*}{;}f * T && f \text{ crisp} \\ &\sqsubseteq (P * T\overset{*}{;}f^{\smile});f && \text{dual of modular inclusion (b)} \\ &= (P * T;f^{\smile});f. && f \text{ crisp} \end{aligned}$$

3. This follows immediately from

$$\begin{aligned} (P * Q)\overset{*}{;}R * U &= (P * (Q \sqcap Q))\overset{*}{;}R * U \\ &= (P * Q * Q)\overset{*}{;}R * U && Q \text{ crisp} \\ &\sqsubseteq (P * Q)\overset{*}{;}(R * Q^{\smile}\overset{*}{;}U). && \text{modular inclusion (c)} \end{aligned}$$

This completes the proof. □

3 *L*-Fuzzy Sharpness Problem

In a fuzzy category we can define a fuzzy right tupling $Q \olessthan_* R : C \to A \times B$ of two relations $Q : C \to A$ and $R : C \to B$ by $Q \olessthan_* R := Q;\pi^{\smile} * R;\rho^{\smile}$. Analogously, the fuzzy left tupling operation $S \ogreaterthan_* T : A \times B \to D$ for two relations $S : A \to D$ and $T : B \to D$ is defined as $S \ogreaterthan_* T := \pi;S * \rho;T$. Please note that replacing the regular composition ; in both definitions by the t-norm based composition $\overset{*}{;}$ does not lead to a different construction due to Axiom 7 (Definition 11) of fuzzy categories and the fact that the projections are crisp.

These new definitions allow us to state a fuzzy version of the (un)sharpness problem, i.e., whether the following equation is valid

$$\text{(FS)} \qquad (Q \olessthan_* R)\overset{*}{;}(S \ogreaterthan_* T) = Q\overset{*}{;}S * R\overset{*}{;}T.$$

First, we want to verify that the equation is indeed valid for concrete *L*-relations.

Theorem 14. *If $Q : C \to A$, $R : C \to B$, $S : A \to D$ and $T : B \to D$ are L-relations and $\pi : A \times B \to A$, $\rho : A \times B \to B$ the standard (concrete) projection relations, then we have*

$$(Q \otimes_* R) \mathbin{;_*} (S \oslash_* T) = Q \mathbin{;_*} S * R \mathbin{;_*} T.$$

Proof. Suppose $c \in C, d \in D$ and compute

$$\begin{aligned}
&((Q \otimes_* R) \mathbin{;_*} (S \oslash_* T))(c,d) \\
&= \bigvee_{(a,b)\in A\times B} (Q \otimes_* R)(c,(a,b)) * (S \oslash_* T)((a,b),d) \\
&= \bigvee_{(a,b)\in A\times B} Q(c,a) * R(c,b) * S(a,d) * T(b,d) && \pi,\rho \text{ crisp maps} \\
&= \bigvee_{a\in A, b\in B} Q(c,a) * S(a,d) * R(c,b) * T(b,d) \\
&= \left(\bigvee_{a\in A} Q(c,a) * S(a,d)\right) * \left(\bigvee_{b\in B} R(c,b) * T(b,d)\right) && * \text{ continuous} \\
&= (Q \mathbin{;_*} S)(c,d) * (R \mathbin{;_*} T)(c,d) \\
&= (Q \mathbin{;_*} S * R \mathbin{;_*} T)(c,d),
\end{aligned}$$

i.e., we have $(Q \otimes_* R) \mathbin{;_*} (S \oslash_* T) = Q \mathbin{;_*} S * R \mathbin{;_*} T$. □

Similar to the regular sharpness problem, the inclusion $\sqsubseteq$ of (FS) follows without any further assumptions.

Lemma 15. *Let $\mathcal{F}$ be a fuzzy category, $A \times B$ the relational product of A and B, and $Q : C \to A$, $R : C \to B$, $S : A \to D$, and $T : B \to D$ be relations. Then we have*

$$(Q \otimes_* R) \mathbin{;_*} (S \oslash_* T) \sqsubseteq Q \mathbin{;_*} S * R \mathbin{;_*} T.$$

Proof. The following computation

$$\begin{aligned}
(Q \otimes_* R) \mathbin{;_*} (S \oslash_* T) &= (Q;\pi^\smile * R;\rho^\smile) \mathbin{;_*} (\pi;S * \rho;T) \\
&\sqsubseteq (Q;\pi^\smile) \mathbin{;_*} (\pi;S) * (R;\rho^\smile) \mathbin{;_*} (\rho;T) && \text{exchange inclusion} \\
&= Q \mathbin{;_*} \pi^\smile \mathbin{;_*} \pi \mathbin{;_*} S * R \mathbin{;_*} \rho^\smile \mathbin{;_*} \rho \mathbin{;_*} T && \pi,\rho \text{ crisp} \\
&= Q \mathbin{;_*} (\pi^\smile;\pi) \mathbin{;_*} S * R \mathbin{;_*} (\rho^\smile;\rho) \mathbin{;_*} T && \pi,\rho \text{ crisp} \\
&\sqsubseteq Q \mathbin{;_*} S * R \mathbin{;_*} T && \pi,\rho \text{ univalent}
\end{aligned}$$

shows the assertion. □

As mentioned in the introduction there are several theorems showing regular sharpness under certain additional assumptions. Only one of these results can be generalized to the fuzzy version of the unsharpness problem. Proposition 3.2.1 of [11] basically requires that an injective map exists in parallel to the relation R of the right-tupling in the sharpness equation. The precise situation in the fuzzy case is visualized in Fig. 1. In order to prove the fuzzy version (Theorem 16) of Proposition 3.2.1 we need to require that f is, in addition, crisp.

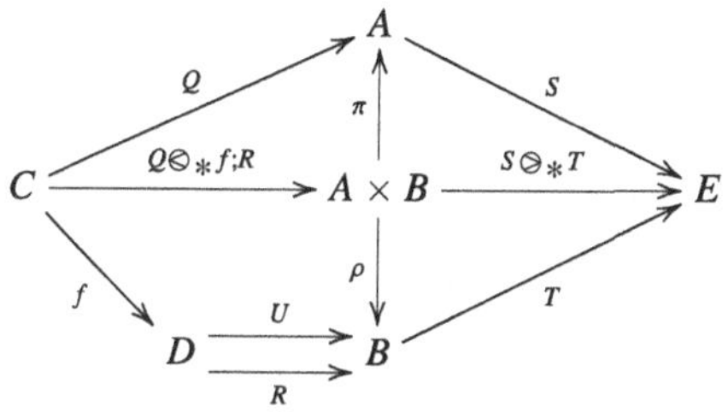

Fig. 1. Situation of Theorem 16

Theorem 16. *Let $\mathcal{F}$ be a fuzzy category, $A \times B$ the relational product of A and B, $Q : C \to A$, $R : D \to B$, $S : A \to E$, $T : B \to E$ be relations, and $f : C \to D$ be crisp and univalent. If there is an injective relation $U : D \to B$ such that $f;U$ is total, then we have*

$$(Q \otimes_* f;R) \mathbin{\overset{*}{;}} (S \otimes_* T) = Q \mathbin{\overset{*}{;}} S * f;(R \mathbin{\overset{*}{;}} T).$$

Proof. In order to show the inclusion $\sqsupseteq$, consider the following computation

$$
\begin{array}{ll}
Q \mathbin{\overset{*}{;}} S * f;(R \mathbin{\overset{*}{;}} T) & \\
= (Q \sqcap \mathbb{T}_{CA}) \mathbin{\overset{*}{;}} S * f;(R \mathbin{\overset{*}{;}} T) & \\
= (Q * \mathbb{T}_{CA}) \mathbin{\overset{*}{;}} S * f;(R \mathbin{\overset{*}{;}} T) & \mathbb{T}_{CA} \text{ crisp} \\
= (Q * f;U;\mathbb{T}_{BA}) \mathbin{\overset{*}{;}} S * f;(R \mathbin{\overset{*}{;}} T) & f;U \text{ total} \\
= (Q * f;U;\rho^{\smile};\pi) \mathbin{\overset{*}{;}} S * f;(R \mathbin{\overset{*}{;}} T) & \\
= ((Q;\pi^{\smile} * f;U;\rho^{\smile});\pi) \mathbin{\overset{*}{;}} S * f;(R \mathbin{\overset{*}{;}} T) & \text{Lemma 13(2)} \\
= (Q;\pi^{\smile} * f;U;\rho^{\smile}) \mathbin{\overset{*}{;}} \pi \mathbin{\overset{*}{;}} S * f;(R \mathbin{\overset{*}{;}} T) & \pi \text{ crisp} \\
\sqsubseteq (Q;\pi^{\smile} * f;U;\rho^{\smile}) \mathbin{\overset{*}{;}} (\pi \mathbin{\overset{*}{;}} S * (f;U;\rho^{\smile})^{\smile} \mathbin{\overset{*}{;}} (f;(R \mathbin{\overset{*}{;}} T))) & \text{Lemma 13(3)} \\
= (Q \otimes_* f;U) \mathbin{\overset{*}{;}} (\pi \mathbin{\overset{*}{;}} S * (f;U;\rho^{\smile})^{\smile} \mathbin{\overset{*}{;}} (f;(R \mathbin{\overset{*}{;}} T))) & \\
= (Q \otimes_* f;U) \mathbin{\overset{*}{;}} (\pi \mathbin{\overset{*}{;}} S * (f;U;\rho^{\smile})^{\smile} \mathbin{\overset{*}{;}} f \mathbin{\overset{*}{;}} R \mathbin{\overset{*}{;}} T) & f \text{ crisp} \\
= (Q \otimes_* f;U) \mathbin{\overset{*}{;}} (\pi \mathbin{\overset{*}{;}} S * (\rho;U^{\smile};f^{\smile}) \mathbin{\overset{*}{;}} f \mathbin{\overset{*}{;}} R \mathbin{\overset{*}{;}} T) & \\
= (Q \otimes_* f;U) \mathbin{\overset{*}{;}} (\pi \mathbin{\overset{*}{;}} S * \rho \mathbin{\overset{*}{;}} U^{\smile} \mathbin{\overset{*}{;}} f^{\smile} \mathbin{\overset{*}{;}} f \mathbin{\overset{*}{;}} R \mathbin{\overset{*}{;}} T) & f,\rho \text{ crisp} \\
= (Q \otimes_* f;U) \mathbin{\overset{*}{;}} (\pi \mathbin{\overset{*}{;}} S * \rho \mathbin{\overset{*}{;}} U^{\smile} \mathbin{\overset{*}{;}} (f^{\smile};f) \mathbin{\overset{*}{;}} R \mathbin{\overset{*}{;}} T) & f \text{ crisp} \\
\sqsubseteq (Q \otimes_* f;U) \mathbin{\overset{*}{;}} (\pi \mathbin{\overset{*}{;}} S * \rho \mathbin{\overset{*}{;}} U^{\smile} \mathbin{\overset{*}{;}} R \mathbin{\overset{*}{;}} T) & f \text{ univalent} \\
= (Q \otimes_* f;U) \mathbin{\overset{*}{;}} (\pi;S * (\rho;(U^{\smile} \mathbin{\overset{*}{;}} R)) \mathbin{\overset{*}{;}} T) & \pi,\rho \text{ crisp} \\
= (Q \otimes_* f;U) \mathbin{\overset{*}{;}} (\pi;S * (\mathbb{T}_{A\times B,B} \sqcap \rho;(U^{\smile} \mathbin{\overset{*}{;}} R)) \mathbin{\overset{*}{;}} T) & \\
= (Q \otimes_* f;U) \mathbin{\overset{*}{;}} (\pi;S * (\mathbb{T}_{A\times B,B} * \rho;(U^{\smile} \mathbin{\overset{*}{;}} R)) \mathbin{\overset{*}{;}} T) & \mathbb{T}_{A\times B,B} \text{ crisp} \\
= (Q \otimes_* f;U) \mathbin{\overset{*}{;}} (\pi;S * (\pi;\pi^{\smile};\rho * \rho;(U^{\smile} \mathbin{\overset{*}{;}} R)) \mathbin{\overset{*}{;}} T) & \pi;\pi^{\smile};\rho = \mathbb{T}_{A\times B,B}
\end{array}
$$

$$
\begin{aligned}
&= (Q \obslash_* f; U) \mathbin{;_*} (\pi; S * ((\pi;\pi^\smile * \rho; (U^\smile \mathbin{;_*} R); \rho^\smile); \rho) \mathbin{;_*} T) && \text{Lemma 13(2)}\\
&= (Q \obslash_* f; U) \mathbin{;_*} (\pi; S * (\pi;\pi^\smile * \rho; (U^\smile \mathbin{;_*} R); \rho^\smile) \mathbin{;_*} \rho \mathbin{;_*} T) && \rho \text{ crisp}\\
&= (Q \obslash_* f; U) \mathbin{;_*} (\pi; S * (\pi;\pi^\smile * \rho; (U^\smile \mathbin{;_*} R); \rho^\smile) \mathbin{;_*} (\rho; T)) && \rho \text{ crisp}\\
&\sqsubseteq (Q \obslash_* f; U) \mathbin{;_*} (\pi;\pi^\smile * \rho; (U^\smile \mathbin{;_*} R); \rho^\smile) \mathbin{;_*} ((\pi;\pi^\smile) \mathbin{;_*} (\pi; S) * \rho; T) && \text{Lemma 13(3)}\\
&= (Q \obslash_* f; U) \mathbin{;_*} (\pi^\smile \oslash_* (U^\smile \mathbin{;_*} R); \rho^\smile) \mathbin{;_*} ((\pi;\pi^\smile) \mathbin{;_*} (\pi; S) * \rho; T) &&\\
&= (Q \obslash_* f; U) \mathbin{;_*} (\pi^\smile \oslash_* (U^\smile \mathbin{;_*} R); \rho^\smile) \mathbin{;_*} (\pi;\pi^\smile;\pi; S * \rho; T) && \pi \text{ crisp}\\
&= (Q \obslash_* f; U) \mathbin{;_*} (\pi^\smile \oslash_* (U^\smile \mathbin{;_*} R); \rho^\smile) \mathbin{;_*} (\pi; S * \rho; T) &&\\
&= (Q \obslash_* f; U) \mathbin{;_*} (\pi^\smile \oslash_* (U^\smile \mathbin{;_*} R); \rho^\smile) \mathbin{;_*} (S \oslash_* T) &&\\
&\sqsubseteq (Q \mathbin{;_*} \pi^\smile * (f; U) \mathbin{;_*} ((U^\smile \mathbin{;_*} R); \rho^\smile)) \mathbin{;_*} (S \oslash_* T) && \text{Lemma 15}\\
&= (Q \mathbin{;_*} \pi^\smile * f \mathbin{;_*} U \mathbin{;_*} U^\smile \mathbin{;_*} R \mathbin{;_*} \rho^\smile) \mathbin{;_*} (S \oslash_* T) && f, \rho \text{ crisp}\\
&\sqsubseteq (Q \mathbin{;_*} \pi^\smile * f \mathbin{;_*} (U; U^\smile) \mathbin{;_*} R \mathbin{;_*} \rho^\smile) \mathbin{;_*} (S \oslash_* T) && \text{Lemma 12(3)}\\
&\sqsubseteq (Q \mathbin{;_*} \pi^\smile * f \mathbin{;_*} R \mathbin{;_*} \rho^\smile) \mathbin{;_*} (S \oslash_* T) && U \text{ injective}\\
&= (Q; \pi^\smile * f; R; \rho^\smile) \mathbin{;_*} (S \oslash_* T) && f, \pi, \rho \text{ crisp}\\
&= (Q \obslash_* f; R) \mathbin{;_*} (S \oslash_* T).
\end{aligned}
$$

The converse inclusion $\sqsubseteq$ was already shown in Lemma 15. □

As an immediate consequence of the previous theorem we obtain sharpness if one of the relations is between an object B and its relational power $\mathcal{P}(B)$. The situation of the following corollary is visualized in Fig. 2.

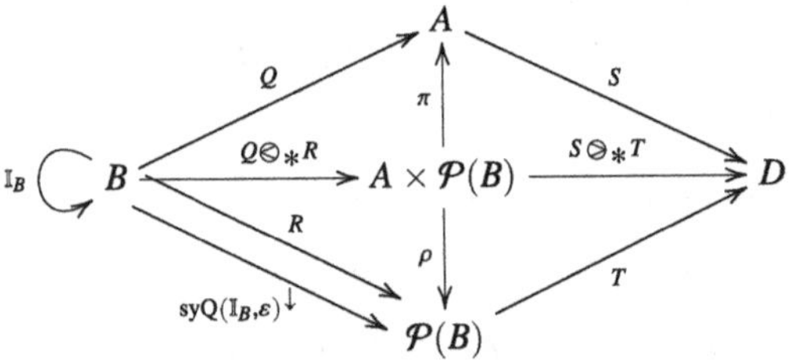

Fig. 2. Situation of Corollary 17

Corollary 17. *Let $\mathcal{F}$ be a fuzzy category, $\mathcal{P}(B)$ the relational power of B, $A \times \mathcal{P}(B)$ the relational product of A and $\mathcal{P}(B)$, $Q : B \to A$, $R : B \to \mathcal{P}(B)$, $S : A \to D$, and $T : \mathcal{P}(B) \to D$ be relations. Then we have*

$$(Q \obslash_* R) \mathbin{;_*} (S \oslash_* T) = Q \mathbin{;_*} S * R \mathbin{;_*} T.$$

Proof. This follows immediately from Theorem 16 with $f = \mathbb{I}_B$ and $U = \mathrm{syQ}(\mathbb{I}_B, \varepsilon)^{\downarrow}$ where the relation U is a crisp and injective map by Lemma 10(2). □

Please note that the previous corollary immediately implies

$$(\varepsilon \otimes_* \varepsilon) \overset{*}{;} (S \otimes_* T) = \varepsilon \overset{*}{;} S * \varepsilon \overset{*}{;} T$$

for appropriate relations S and T.

We are now able to prove the main theorem of this paper. It shows the fuzzy version of sharpness if a suitable crisp version of sharpness holds. As above the situation of the theorem and its proof is visualized in Fig. 3.

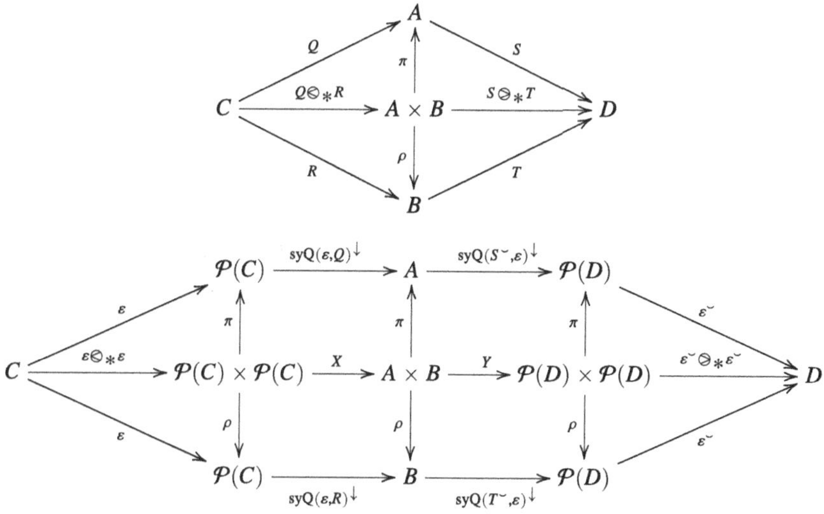

Fig. 3. Situation of Theorem 18 and its proof where $X = \pi; \mathrm{syQ}(\varepsilon, Q)^{\downarrow} \otimes \rho; \mathrm{syQ}(\varepsilon, R)^{\downarrow}$ and $Y = \mathrm{syQ}(S^{\smile}, \varepsilon)^{\downarrow}; \pi^{\smile} \otimes \mathrm{syQ}(T^{\smile}, \varepsilon)^{\downarrow}; \rho^{\smile}$.

Theorem 18. *Let $\mathcal{F}$ be a fuzzy category and $A \times B$ the relational product of A and B. If the objects $\mathcal{P}(C), \mathcal{P}(D), \mathcal{P}(C) \times \mathcal{P}(C)$, $\mathcal{P}(D) \times \mathcal{P}(D)$ exist, and if the product $A \times B$ is sharp for all crisp relations $Q' : \mathcal{P}(C) \times \mathcal{P}(C) \to A$, $R' : \mathcal{P}(C) \times \mathcal{P}(C) \to B$, $S' : A \to \mathcal{P}(D) \times \mathcal{P}(D)$, and $T' : B \to \mathcal{P}(D) \times \mathcal{P}(D)$, i.e., we have*

$$(Q' \otimes R'); (S' \otimes T') = Q'; S' \sqcap R'; T',$$

then also the fuzzy version of sharpness holds, i.e., we have

$$(Q \otimes_* R) \overset{*}{;} (S \otimes_* T) = Q \overset{*}{;} S * R \overset{*}{;} T$$

for all $Q : C \to A$, $R : C \to B$, $S : A \to D$, and $T : B \to D$.

Proof. First, consider the following computation

$$
\begin{aligned}
Q \otimes_* R &= \varepsilon; \mathrm{syQ}(\varepsilon, Q)^{\downarrow} \otimes_* \varepsilon; \mathrm{syQ}(\varepsilon, R)^{\downarrow} && \text{Lemma 10(3)}\\
&= \varepsilon; \mathrm{syQ}(\varepsilon, Q)^{\downarrow}; \pi^{\smile} * \varepsilon; \mathrm{syQ}(\varepsilon, R)^{\downarrow}; \rho^{\smile} \\
&= \varepsilon \mathbin{\overset{*}{;}} (\mathrm{syQ}(\varepsilon, Q)^{\downarrow}; \pi^{\smile}) * \varepsilon \mathbin{\overset{*}{;}} (\mathrm{syQ}(\varepsilon, R)^{\downarrow}; \rho^{\smile}) && \text{right relation crisp}\\
&= (\varepsilon \otimes_* \varepsilon) \mathbin{\overset{*}{;}} (\mathrm{syQ}(\varepsilon, Q)^{\downarrow}; \pi^{\smile} \otimes_* \mathrm{syQ}(\varepsilon, R)^{\downarrow}; \rho^{\smile}) && \text{Corollary 17}\\
&= (\varepsilon \otimes_* \varepsilon) \mathbin{\overset{*}{;}} (\pi; \mathrm{syQ}(\varepsilon, Q)^{\downarrow}; \pi^{\smile} * \rho; \mathrm{syQ}(\varepsilon, R)^{\downarrow}; \rho^{\smile}) \\
&= (\varepsilon \otimes_* \varepsilon) \mathbin{\overset{*}{;}} (\pi; \mathrm{syQ}(\varepsilon, Q)^{\downarrow}; \pi^{\smile} \sqcap \rho; \mathrm{syQ}(\varepsilon, R)^{\downarrow}; \rho^{\smile}) && \text{both relations crisp}\\
&= (\varepsilon \otimes_* \varepsilon) \mathbin{\overset{*}{;}} (\pi; \mathrm{syQ}(\varepsilon, Q)^{\downarrow} \otimes \rho; \mathrm{syQ}(\varepsilon, R)^{\downarrow}).
\end{aligned}
$$

The equation $S \otimes_* T = (\mathrm{syQ}(S^{\smile}, \varepsilon)^{\downarrow}; \pi^{\smile} \otimes \mathrm{syQ}(T^{\smile}, \varepsilon)^{\downarrow}; \rho^{\smile}) \mathbin{\overset{*}{;}} (\varepsilon^{\smile} \otimes_* \varepsilon^{\smile})$ is shown analogously. Furthermore, we have

$$
\begin{aligned}
&(\pi; \mathrm{syQ}(\varepsilon, Q)^{\downarrow} \otimes \rho; \mathrm{syQ}(\varepsilon, R)^{\downarrow}) \mathbin{\overset{*}{;}} (\mathrm{syQ}(S^{\smile}, \varepsilon)^{\downarrow}; \pi^{\smile} \otimes \mathrm{syQ}(T^{\smile}, \varepsilon)^{\downarrow}; \rho^{\smile})\\
&= (\pi; \mathrm{syQ}(\varepsilon, Q)^{\downarrow} \otimes \rho; \mathrm{syQ}(\varepsilon, R)^{\downarrow}); (\mathrm{syQ}(S^{\smile}, \varepsilon)^{\downarrow}; \pi^{\smile} \otimes \mathrm{syQ}(T^{\smile}, \varepsilon)^{\downarrow}; \rho^{\smile}) && \text{crisp rel.}\\
&= \pi; \mathrm{syQ}(\varepsilon, Q)^{\downarrow}; \mathrm{syQ}(S^{\smile}, \varepsilon)^{\downarrow}; \pi^{\smile} \sqcap \rho; \mathrm{syQ}(\varepsilon, R)^{\downarrow}; \mathrm{syQ}(T^{\smile}, \varepsilon)^{\downarrow}; \rho^{\smile} && \text{assump.}\\
&= \pi; \mathrm{syQ}(\varepsilon, Q)^{\downarrow}; \mathrm{syQ}(S^{\smile}, \varepsilon)^{\downarrow}; \pi^{\smile} * \rho; \mathrm{syQ}(\varepsilon, R)^{\downarrow}; \mathrm{syQ}(T^{\smile}, \varepsilon)^{\downarrow}; \rho^{\smile} && \text{crisp rel.}\\
&= \mathrm{syQ}(\varepsilon, Q)^{\downarrow}; \mathrm{syQ}(S^{\smile}, \varepsilon)^{\downarrow}; \pi^{\smile} \otimes_* \mathrm{syQ}(\varepsilon, R)^{\downarrow}; \mathrm{syQ}(T^{\smile}, \varepsilon)^{\downarrow}; \rho^{\smile}
\end{aligned}
$$

as well as

$$
\begin{aligned}
&(\varepsilon \otimes_* \varepsilon) \mathbin{\overset{*}{;}} (\mathrm{syQ}(\varepsilon, Q)^{\downarrow}; \mathrm{syQ}(S^{\smile}, \varepsilon)^{\downarrow}; \pi^{\smile} \otimes_* \mathrm{syQ}(\varepsilon, R)^{\downarrow}; \mathrm{syQ}(T^{\smile}, \varepsilon)^{\downarrow}; \rho^{\smile})\\
&= \varepsilon \mathbin{\overset{*}{;}} (\mathrm{syQ}(\varepsilon, Q)^{\downarrow}; \mathrm{syQ}(S^{\smile}, \varepsilon)^{\downarrow}; \pi^{\smile}) * \varepsilon \mathbin{\overset{*}{;}} (\mathrm{syQ}(\varepsilon, R)^{\downarrow}; \mathrm{syQ}(T^{\smile}, \varepsilon)^{\downarrow}; \rho^{\smile}) && \text{Corollary 17}\\
&= \varepsilon; \mathrm{syQ}(\varepsilon, Q)^{\downarrow}; \mathrm{syQ}(S^{\smile}, \varepsilon)^{\downarrow}; \pi^{\smile} * \varepsilon; \mathrm{syQ}(\varepsilon, R)^{\downarrow}; \mathrm{syQ}(T^{\smile}, \varepsilon)^{\downarrow}; \rho^{\smile} && \text{right crisp}\\
&= Q; \mathrm{syQ}(S^{\smile}, \varepsilon)^{\downarrow}; \pi^{\smile} * R; \mathrm{syQ}(T^{\smile}, \varepsilon)^{\downarrow}; \rho^{\smile} && \text{Lemma 10(3)}\\
&= Q; \mathrm{syQ}(S^{\smile}, \varepsilon)^{\downarrow} \otimes_* R; \mathrm{syQ}(T^{\smile}, \varepsilon)^{\downarrow}
\end{aligned}
$$

and finally

$$
\begin{aligned}
&(Q; \mathrm{syQ}(S^{\smile}, \varepsilon)^{\downarrow} \otimes_* R; \mathrm{syQ}(T^{\smile}, \varepsilon)^{\downarrow}) \mathbin{\overset{*}{;}} (\varepsilon^{\smile} \otimes_* \varepsilon^{\smile})\\
&= (Q; \mathrm{syQ}(S^{\smile}, \varepsilon)^{\downarrow}) \mathbin{\overset{*}{;}} \varepsilon^{\smile} * (R; \mathrm{syQ}(T^{\smile}, \varepsilon)^{\downarrow}) \mathbin{\overset{*}{;}} \varepsilon^{\smile} && \text{Corollary 17}\\
&= Q \mathbin{\overset{*}{;}} \mathrm{syQ}(S^{\smile}, \varepsilon)^{\downarrow} \mathbin{\overset{*}{;}} \varepsilon^{\smile} * R \mathbin{\overset{*}{;}} \mathrm{syQ}(T^{\smile}, \varepsilon)^{\downarrow} \mathbin{\overset{*}{;}} \varepsilon^{\smile} && \text{right relation crisp}\\
&= Q \mathbin{\overset{*}{;}} (\mathrm{syQ}(S^{\smile}, \varepsilon)^{\downarrow}; \varepsilon^{\smile}) * R \mathbin{\overset{*}{;}} (\mathrm{syQ}(T^{\smile}, \varepsilon)^{\downarrow}; \varepsilon^{\smile}) && \text{left relation crisp}\\
&= Q \mathbin{\overset{*}{;}} S * R \mathbin{\overset{*}{;}} T && \text{Lemma 10(3).}
\end{aligned}
$$

Combining the computation above immediately leads to the fuzzy version of sharpness of the product $A \times B$. □

In [1] it was shown that the regular sharpness property from C via $A \times B$ to D holds, i.e., for all relations $Q : C \to A$, $R : C \to B$, $S : A \to D$, and $T : B \to D$, if one of the additional products $A \times C$, $B \times C$, $A \times D$ or $B \times D$ exists. Combining this with the result of the main theorem we obtain that fuzzy sharpness from C via $A \times B$ to D holds if the objects $\mathcal{P}(C), \mathcal{P}(D), \mathcal{P}(C) \times \mathcal{P}(C)$, $\mathcal{P}(D) \times \mathcal{P}(D)$ and either one of the products $A \times \mathcal{P}(C) \times \mathcal{P}(C)$, $B \times \mathcal{P}(C) \times \mathcal{P}(C)$, $A \times \mathcal{P}(D) \times \mathcal{P}(D)$ or $B \times \mathcal{P}(D) \times \mathcal{P}(D)$ exists.

From this discussion we immediately obtain the following corollary.

Corollary 19. *Let $\mathcal{F}$ be a fuzzy category with relational products and powers. Then fuzzy sharpness holds.*

4 Conclusion

In this paper we have introduced the fuzzy unsharpness problem. This problem is obtained by replacing $\sqcap$ and ; in the regular unsharpness problem by the t-norm based operations $*$ and $\overset{*}{;}$. We have shown that the fuzzy unsharpness problem can be reduced to the regular unsharpness problem if some additional relational products and powers exist. As a consequence we obtained Corollary 19 stating fuzzy sharpness if the fuzzy category has all relational products and powers.

In the opinion of the author the results of this paper show once again that (un)sharpness is more a structural problem related to relational products and representability than a problem related to the specific operations used in the equation.

Last but not least, we want to state a property that requires fuzzy sharpness in its proof. The actual proof is omitted due to lack of space. Suppose $\varepsilon_1 : A \to \mathcal{P}(A)$, $\varepsilon_2 : \mathcal{P}(A) \to \mathcal{P}(\mathcal{P}(A))$, $\pi_1, \rho_1 : \mathcal{P}(A) \times \mathcal{P}(A) \to \mathcal{P}(A)$, and $\pi_2, \rho_2 : \mathcal{P}(\mathcal{P}(A)) \times \mathcal{P}(\mathcal{P}(A)) \to \mathcal{P}(\mathcal{P}(A))$ are given. Then the binary ($*$-based) meet relation $\mathcal{M}_* : \mathcal{P}(A) \times \mathcal{P}(A) \to \mathcal{P}(A)$ defined by $\mathcal{M}_* = \mathrm{syQ}(\varepsilon_1 \otimes_* \varepsilon_1, \varepsilon_1)^{\downarrow}$ maps two fuzzy subsets of A to their t-norm based intersection, i.e., $\mathcal{M}_*((B, C), D)$ iff $D(a) = B(a) * C(a)$ for all $a \in A$. Similarly, the join relation $\mathcal{J} : \mathcal{P}(\mathcal{P}(A)) \to \mathcal{P}(A)$ defined by $\mathcal{J} = \mathrm{syQ}(\varepsilon_1; \varepsilon_2, \varepsilon_1)^{\downarrow}$ maps a set of fuzzy subsets of A to their union. The property below now states that the ($*$-based) meet of the two unions of two sets of fuzzy sets M and N is equal to the union of the set obtained by taking the ($*$-based) meet of all pairs of sets from M and N:

$$(\pi_2; \mathcal{J} \otimes \rho_2; \mathcal{J}); \mathcal{M}_* = \mathrm{syQ}(\varepsilon_1; \mathcal{M}_*^{\smile}; (\pi_1; \varepsilon_2 \otimes_* \rho_1; \varepsilon_2), \varepsilon_1)^{\downarrow}$$

In the proof of this property fuzzy sharpness is needed several times. The most general case takes the form:

$$(\varepsilon_1 \otimes_* \varepsilon_1); (\varepsilon_2; \pi_2^{\smile} \otimes_* \varepsilon_2; \rho_2^{\smile}) = \varepsilon_1; \varepsilon_2; \pi_2^{\smile} * \varepsilon_1; \varepsilon_2; \rho_2^{\smile}.$$

References

1. Desharnais, J.: Monomorphic characterization of n-ary direct products. Inf. Sci. **119**(3–4), 275–288 (1999)
2. Furusawa, H., Kahl, W.: A Study on Symmetric Quotients. University of the Federal Armed Forces Munich, Bericht Nr. 1998–06 (1998)
3. Freyd, P., Scedrov, A.: Categories, Allegories. North-Holland Mathematical Library, vol. 39. North-Holland, Amsterdam (1990)

4. Kawahara, Y., Furusawa, H.: Crispness and representation theorems in Dedekind categories. DOI-TR 143, Kyushu University (1997)
5. Maddux, R.: On the derivation of identities involving projection functions. In: Csirmaz, L., Gabbay, D., de Rijke, M. (eds.) Logic Colloquium '92. Studies in Logic, Languages, and Information, pp. 143–163. CSLI Publications (1995)
6. Imangazin, N., Winter, M.: A Relation-Algebraic Approach to *L*-Fuzzy Topology. Submitted to Relational and Algebraic Methods in Computer Science (RAMiCS 2019) (2019)
7. Olivier, J.P., Serrato, D.: Catégories de Dedekind. Morphismes dans les Catégories de Schröder. C.R. Acad. Sci. Paris **290**, 939–941 (1980)
8. Olivier, J.P., Serrato, D.: Squares and rectangles in relational categories - three cases: semi-lattice, distributive lattice and Boolean non-unitary. Fuzzy Sets Syst. **72**, 167–178 (1995)
9. Schmidt, G., Ströhlein, T.: Relations and Graphs. Springer, Berlin (1993). https://doi.org/10.1007/978-3-642-77968-8
10. Schmidt, G.: Relational Mathematics. Cambridge University Press, Cambridge (2011)
11. Schmidt, G., Winter, M.: Relational Topology. LNM, vol. 2208. Springer, Heidelberg (2018). https://doi.org/10.1007/978-3-319-74451-3
12. Winter, M.: A new algebraic approach to *L*-fuzzy relations convenient to study crispness. INS Inf. Sci. **139**, 233–252 (2001)
13. Winter, M.: Goguen Categories - A Categorical Approach to *L*-fuzzy Relations. Springer, Berlin (2007). https://doi.org/10.1007/978-1-4020-6164-6
14. Winter, M.: Arrow categories. Fuzzy Sets Syst. **160**, 2893–2909 (2009)
15. Winter, M.: Membership values in arrow categories. Fuzzy Sets Syst. **267**, 41–61 (2015)
16. Winter, M.: Type-n arrow categories. In: Höfner, P., Pous, D., Struth, G. (eds.) RAMICS 2017. LNCS, vol. 10226, pp. 307–322. Springer, Cham (2017). https://doi.org/10.1007/978-3-319-57418-9_19
17. Winter, M.: T-norm based operations in arrow categories. In: Desharnais, J., Guttmann, W., Joosten, S. (eds.) RAMiCS 2018. LNCS, vol. 11194, pp. 70–86. Springer, Cham (2018). https://doi.org/10.1007/978-3-030-02149-8_5
18. Zierer, H.: Programmierung mit Funktionsobjekten: Konstruktive Erzeugung semantischer Bereiche und Anwendung auf die partielle Auswertung. Dissertation, TU München, TUM-I8803 (1988)

Author Index

Printed in the United States
By Bookmasters